Coral reef aquatic ecosystem

Ron Nolan

W. MacManiman

Tropical forest biome

Deciduous forest biome

ENVIRONMENTAL SCIENCE

An Introduction

SECOND EDITION

■

G. Tyler Miller, Jr.
St. Andrews Presbyterian College

WADSWORTH PUBLISHING COMPANY

Belmont, California
A Division of Wadsworth, Inc.

Science Editor: Jack Carey

Editorial Assistant: Judi Walcom

Production Editor: Leland Moss

Managing Designer: James Chadwick

Print Buyer: Barbara Britton

Art Editor: Marta Kongsle

Designer: Julia Scannell

Copy Editor: Anne Montague

Associate Copy Editor: Mark Nichol

Illustrators: Joan Carol, Raychel Ciemma, Florence Fujimoto, Darwen and Vally Hennings, Victor Royer, Linda Harris-Sweezey, John and Judith Waller

Photo Researcher: Stuart Kenter

Compositor: Graphic Typesetting Service

Cover: Julia Scannell

Cover Photograph: Galen Rowell/Mountain Light

Printed in the United States of America **48**

2 3 4 5 6 7 8 9 10—92 91 90 89 88

Library of Congress Cataloging-in-Publication Data

Miller, G. Tyler (George Tyler), 1931–
    Environmental science, an introduction / G. Tyler Miller, Jr.—2nd ed.
        p. cm.
    Bibliography: p.
    Includes index.
    ISBN 0-534-09066-4
    1. Ecology. 2. Human ecology. 3. Natural resources.
4. Pollution. I. Title.
QH541.M515 1988
304.2—dc19   87-22452
                                                  CIP

Books in the Wadsworth Biology Series

Biology: The Unity and Diversity of Life, 4th, Starr and Taggart
Energy and Environment: The Four Energy Crises, 2nd, Miller
Replenish the Earth: A Primer in Human Ecology, Miller
Oceanography: An Introduction, 3rd, Ingmanson and Wallace
Living in the Environment, 5th, Miller

Biology books under the editorship of William A. Jensen, University of California, Berkeley

Biology: The Foundations, 2nd, Wolfe
Biology of the Cell, 2nd, Wolfe
Botany, 2nd, Jensen and Salisbury
Plant Physiology, 2nd, Salisbury and Ross
Plant Physiology Laboratory Manual, Ross
Plants: An Evolutionary Survey, 2nd, Scagel et al.
Nonvascular Plants: An Evolutionary Survey, Scagel et al.
Introduction to Cell Biology, Wolfe

Brief Contents

Detailed Contents

Preface: To the Instructor

Goals The purposes of this book are to (1) cover environmental concepts and information in an accurate, balanced, and interesting way without the use of mathematics; (2) enable the teacher to use the material in a flexible manner; and (3) use basic ecological concepts that govern how nature works (see Chapters 3 through 6) to evaluate environmental problems and options available in dealing with them; and (4) show how most environmental problems are interrelated.

This second edition of *Environmental Science* is a briefer version (basic text of 406 pages) of the fifth edition of *Living in the Environment,* a more comprehensive textbook (basic text of 603 pages) with greater emphasis on environmental economics, politics, and ethics.

A Well-Tested Product The material in this textbook and *Living in the Environment* has been used and class-tested by over 1 million students (including used copies) at over two-thirds of the country's colleges and universities. By a very large margin, these two related books have been the most widely used environmental science textbooks in the United States since 1975, when the first edition of *Living in the Environment* was published.

Flexibility To provide teaching flexibility, this book is divided into five major parts:

- Humans and Nature: An Overview (2 chapters)
- Basic Concepts (4 chapters)
- Population (2 chapters)
- Resources (9 chapters)
- Pollution (4 chapters)

Once Parts One and Two have been covered, the remainder of the book may be used in any order that meets the needs of the instructor. Major parts, chapters within these parts, and many sections within chapters can be moved around or omitted to accommodate courses with varying lengths and emphases.

Other Major Features This textbook (1) *consistently uses fundamental concepts* (Chapters 3–6) to illustrate the interrelationships of environmental and resource problems and their possible solutions; (2) *provides balanced discussions of the opposing views* on major environmental and resource issues; (3) *is based on an extensive review of the professional literature* (from more than 10,000 research sources; key readings for each chapter are listed at the end of the text); and (4) *has received extensive manuscript review* by 177 experts and instructors (see List of Reviewers on pp. xiii–xiv)—plus unsolicited suggestions from hundreds of students and teachers who have used this material.

Major Changes in the Second Edition Despite the overwhelming success of this textbook, the publisher and I feel obligated to improve each edition to meet changing needs indicated by the extensive reviews and surveys of users. The major changes in this edition include:

1. Material throughout the book has been updated, rewritten, and in many cases condensed to make this an even better textbook.
2. Material has been rearranged for pedagogical purposes and to reduce the total number of chapters from 24 to 21.
3. Throughout the text, important and controversial issues and further insights into environmental problems are highlighted in *Spotlights.*
4. In addition to general updating, an entirely new chapter on risk-benefit and cost-benefit analysis (Chapter 6) has been added. New material has also been added on desertification (Chapter 9), the Chernobyl accident (Chapter 16), solar-assisted water stoves (Chapter 17), indoor air pollution from radioactive radon (Chapter 18), the Rhine River and Chesapeake Bay (Chapter 19), and the Bhopal tragedy (Chapter 21).
5. The number of illustrations and photographs has been increased from 160 in the first edition to 327

in this edition. Special emphasis has been placed on using illustrations that simplify key ideas.

6. Two new study aids have been added: **(a)** *general objectives*, a list of major questions at the beginning of each chapter, and **(b)** a *student preface* to acquaint students with the book's overall goals and numerous study aids.

As you and your students deal with the crucial and exciting issues discussed in this book, I hope you will take the time to point out and suggest improvements for future editions. Please send such information to me, care of Jack Carey, Science Editor, Wadsworth Publishing Company, 10 Davis Drive, Belmont, CA 94002.

Supplementary Materials   Dr. Robert Janiskee at the University of South Carolina has written an excellent instructor's manual for use with this text. It contains sample multiple-choice test questions with answers; suggested projects, field trips, and experiments; and a list of suitable topics for term papers and reports for each chapter. In addition, a series of master sheets for making overhead transparencies of many key diagrams is available from the publisher.

Acknowledgments   I wish to thank the many students and teachers who responded so favorably to the first edition of *Environmental Science* and offered suggestions for improvement. I am also deeply indebted to the numerous reviewers, who pointed out errors and suggested many important improvements. Any errors and deficiencies remaining are mine, not theirs.

It has also been a pleasure to work with a team of talented people who have helped improve this book. I am particularly indebted to production editor Leland Moss, art editor Marta Kongsle, photo researcher Stuart Kenter, designers Julia Scannell and James Chadwick, copy editor Anne Montague, and to artists Darwin and Vally Hennings, John and Judy Waller, Linda Harris Sweezey, Rachel Ciemma, and Florence Fujimoto for their outstanding artwork. Above all I wish to thank Jack Carey, science editor at Wadsworth, for his encouragement, help, friendship, and superb reviewing system.

G. Tyler Miller, Jr.

Preface: To the Student

Major Goals I have written this book to show you that learning about environmental concepts and issues is (1) fun and interesting; (2) need not be difficult; (3) does not require use of mathematics or complex chemical and biological information; (4) is important and relevant to every aspect of your life; and (5) can help you make wiser decisions in vital matters that affect you, your loved ones, and society as a whole. This is not just another college course to be passed for credit. It is an introduction to how nature works, how it has been and is being abused, and what we can do to protect and improve it for ourselves, future generations, and other living things. I am convinced that nothing else deserves more of our energy, time, care, and personal involvement.

I have gone to considerable effort to present opposing views on these complex and highly controversial life-and-death issues in a balanced way. My goal is not to tell you what to think but to provide you with ecological concepts and information you can use to reach your own conclusions.

Emphasis on Concepts The purpose of useful learning is not to stuff ourselves full of information but to learn and understand a small number of basic concepts or principles with which we can integrate numerous facts into meaningful patterns. In this book a small number of key scientific concepts (presented in Chapters 3 through 6 and summarized in the Epilogue) are used throughout to evaluate environmental and resource problems and options for dealing with them.

A Realistic but Hopeful National and Global Outlook Our actions are based primarily on our worldview—our beliefs about how the world works. In this book I offer a realistic but hopeful view of the future based primarily on how much has been done since 1965 (frankly, much more than I expected at that time), when the American public first became aware of many environmental problems. Much more needs

to be done, but there is hope, if enough of us care. The key is to *think globally and act locally*. Most environmental and resource problems and their possible solutions are interrelated and must be considered on a local, national, and global scale—as this book does. Pollution, for example, does not respect national boundaries, as clearly illustrated by the radiation that drifted over much of the world after being released in the 1986 accident at the Chernobyl nuclear power plant in the Soviet Union.

How This Book Is Organized To get a better idea of what you will be learning, I suggest that you take a few minutes to look at the brief table of contents and the detailed table of contents. This book has been designed to be flexible enough for use in courses with different lengths and emphases. After the six chapters in Parts One and Two have been studied, the remainder of the book can be covered in essentially any order, so do not be concerned if your instructor skips around and omits material (I hope you will go ahead and read it on your own).

This book is written for you. To help you learn more efficiently and effectively, I have provided a number of learning aids, described in the remainder of this preface.

General Objectives and Vocabulary Each chapter begins with a few general questions or learning objectives written in nontechnical language and designed to give you an idea of what you will be learning. After you finish a chapter, you can go back and try to answer these questions to review what you have learned.

In each chapter you will be introduced to a number of new terms whose meanings you need to know and understand. To help you identify these key terms, each is printed in boldface when it is introduced. The pages where key terms are defined are also shown in boldface in the index, and a glossary of all key terms appears at the end of the book.

Visual Aids and Discussion Topics   Great emphasis has been placed on developing a variety of diagrams that illustrate complex ideas in a simple manner. Many carefully selected photos have also been used to give you a better picture of how topics discussed in this book relate to the real world.

The questions at the end of each chapter are not designed to test your recall of facts. That is left to your instructor. Instead, these questions are designed to make you think, to apply what you have learned to your personal lifestyle, to take sides on controversial issues, and to back up your conclusions and beliefs.

Further Readings   At the end of this book you will find a list of readings, grouped by chapter, which you can use to increase your knowledge of a particular topic and to prepare reports or term papers. Most of these works cite other works that can enhance your knowledge.

Save This Book   This book will be a useful reference long after you have completed this course, because you will have to deal with the vital issues discussed in it throughout your life. So instead of throwing it away or reselling it to the bookstore for about half what you paid for it, I suggest you keep it in your personal library. Learning is a lifelong process and you should be building a collection of books that will be useful to you now and in the future. You may also purchase the latest updated version of this text (pub- lished every three years) directly from Wadsworth Publishing Company, 10 Davis Drive, Belmont, CA 94002.

Help Me Improve This Book   I need your help in improving this book in future editions. Writing and publishing a book is such a complex process that some typographical and other errors are almost certain to be present. If you come across what you believe to be an error, write it down, send it to me, and consider turning in a copy to your instructor. Hundreds of other students have helped me improve this book since it was first published. I hope you will continue this tradition by letting me know what you like and dislike most about the book. Send any errors you find and your suggestions for improvement to Jack Carey, Science Editor, Wadsworth Publishing Company, 10 Davis Drive, Belmont, CA 94002. He will send them on to me. Unfortunately, time does not permit me to answer your letters, but be aware of how much I appreciate learning from you.

And Now   Relax and enjoy yourself as you learn more about the exciting and challenging issues we all face in preserving the earth's life-support system for ourselves, future generations, and the millions of different types of plants and animals we share the planet with and depend on for our survival.

G. Tyler Miller, Jr.

Reviewers

Barbara J. Abraham *Hampton College*
Larry G. Allen *California State University, Northridge*
James R. Anderson *U.S. Geological Survey*
Kenneth B. Armitage *University of Kansas*
Virgil R. Baker *Arizona State University*
Ian G. Barbour *Carleton College*
Albert J. Beck *California State University, Chico*
R. W. Behan *Northern Arizona University*
Keith L. Bildstein *Winthrop College*
Jeff Bland *University of Puget Sound*
Roger G. Bland *Central Michigan University*
Georg Borgstrom *Michigan State University*
Arthur C. Borror *University of New Hampshire*
John H. Bounds *Sam Houston State University*
Leon F. Bouvier *Population Reference Bureau*
Michael F. Brewer *Resources for the Future, Inc.*
Patrick E. Brunelle *Contra Costa College*
Terrence J. Burgess *Saddleback College North*
Lynton K. Caldwell *Indiana University*
Faith Thompson Campbell *Natural Resources Defense Council, Inc.*
E. Ray Canterbery *Florida State University*
Ted J. Case *University of San Diego*
Ann Causey *Auburn University*
Richard A. Cellarius *Evergreen State University*
William U. Chandler *Worldwatch Institute*
R. F. Christman *University of North Carolina, Chapel Hill*
Preston Cloud *University of California, Santa Barbara*
Bernard C. Cohen *University of Pittsburgh*
Richard A. Cooley *University of California, Santa Cruz*
Dennis J. Corrigan
John D. Cunningham *Keene State College*
Herman E. Daly *Louisiana State University*
Raymond F. Dasmann *University of California, Santa Cruz*
Kingsley Davis *University of California, Berkeley*
Edward E. DeMartini *University of California, Santa Barbara*
Thomas R. Detwyler *University of Michigan*
Lon D. Drake *University of Iowa*
W. T. Edmonson *University of Washington*
Thomas Eisner *Cornell University*
David E. Fairbrothers *Rutgers University*
Paul P. Feeny *Cornell University*
Nancy Field *Bellevue Community College*
Allan Fitzsimmons *University of Kentucky*
George L. Fouke *St. Andrews Presbyterian College*
Lowell L. Getz *University of Illinois at Urbana–Champaign*

Frederick F. Gilbert *Washington State University*
Jay Glassman *Los Angeles Valley College*
Harold Goetz *North Dakota State University*
Jeffery J. Gordon *Bowling Green State University*
Eville Gorham *University of Minnesota*
Ernest M. Gould, Jr. *Harvard University*
Katherine B. Gregg *West Virginia Wesleyan College*
Paul Grogger *University of Colorado*
J. L. Guernsey *Indiana State University*
Ralph Guzman *University of California, Santa Cruz*
Raymond E. Hampton *Central Michigan University*
Ted L. Hanes *California State University, Fullerton*
John P. Harley *Eastern Kentucky University*
Grant A. Harris *Washington State University*
Harry S. Hass *San Jose City College*
Arthur N. Haupt *Population Reference Bureau*
Denis A. Hayes *Environmental consultant*
John G. Hewston *Humboldt State University*
David L. Hicks *Whitworth College*
Eric Hirst *Oak Ridge National Laboratory*
C. S. Holling *University of British Columbia*
Donald Holtgrieve *California State University, Hayward*
Michael H. Horn *University of California, Fullerton*
Marilyn Houck *Pennsylvania State University*
Richard D. Houk *Winthrop College*
Donald Huisingh *North Carolina State University*
Marlene K. Hutt *IBM*
David R. Inglis *University of Massachusetts*
Robert Janiskee *University of South Carolina*
Hugo H. John *University of Connecticut*
David I. Johnson *Michigan State University*
Agnes Kadar *Nassau Community College*
Thomas L. Keefe *Eastern Kentucky University*
Nathan Keyfitz *Harvard University*
Edward J. Kormondy *California State University, Los Angeles*
Judith Kunofsky *Sierra Club*
Theodore Kury *State University College at Buffalo*
Steve Ladochy *University of Winnepeg*
Mark B. Lapping *Kansas State University*
Tom Leege *Idaho Department of Fish and Game*
William S. Lindsay *Monterey Peninsula College*
Valerie A. Liston *University of Minnesota*
Dennis Livingston *Rensselaer Polytechnic Institute*
James P. Lodge *Air pollution consultant*
Ruth Logan *Santa Monica City College*
Robert D. Loring *DePauw University*
T. Lovering *University of California, Santa Barbara*
Amory B. Lovins *Energy consultant*

PART ONE

Humans and Nature: An Overview

It is only in the most recent, and brief, period of their tenure that human beings have developed in sufficient numbers, and acquired enough power, to become one of the most potentially dangerous organisms that the planet has ever hosted.

John McHale

Human despair or default can reach a point where even the most stirring visions lose their regenerating powers. This point, some will say, has already been reached. Not true. It will be reached only when human beings are no longer capable of calling out to one another, when the words in their poetry break up before their eyes, when their faces are frozen toward their young, and when they fail to make pictures in the mind out of clouds racing across the sky. So long as we can do these things, we are capable of indignation about the things we should be indignant about and we can shape our society in a way that does justice to our hopes.

Norman Cousins

1

Population, Resources, Environmental Degradation, and Pollution

GENERAL OBJECTIVES

1. How rapidly is the human population on earth increasing and when might it stabilize?

2. What are the major types of resources and are they in danger of being depleted?

3. What are the major types of environmental degradation and pollution?

4. What are the relationships among human population size, resource use, technology, environmental degradation, and pollution?

5. What are the two major opposing schools of thought about how to solve environmental problems?

We travel together, passengers on a little spaceship, dependent on its vulnerable resources of air, water, and soil . . . preserved from annihilation only by the care, the work, and the love we give our fragile craft.

Adlai E. Stevenson

Today the world is at a critical turning point. We have spent billions to transport a handful of people to the moon, only to learn the importance of protecting the rich diversity of life on the beautiful blue planet that is our home. While technological optimists promise a life of abundance for everyone, environmentalists warn that the earth's life-support systems are being destroyed—or at least severely degraded.

This chapter is devoted to an overview of the interrelated problems of population growth, resource use, pollution of the air and water, and environmental degradation of topsoil, forests, grasslands, and fisheries that support human life. Later chapters will discuss these problems and options for dealing with them in greater depth.

1-1 HUMAN POPULATION GROWTH

The J-Shaped Curve of Human Population Growth If we plot the estimated number of people on earth over time, the resulting curve roughly resembles the shape of the letter *J* (Figure 1-1). This increase in the size of the human population is an example of **exponential growth,** which occurs when some factor—such as population size—grows by a constant percentage of the whole during each unit of time. Although the percentage growth may vary from year to year, the world's population will grow exponentially as long as the number of births exceeds the number of deaths each year.

During the first several million years of human history, when people lived in small groups and survived by hunting wild game and gathering wild plants, the earth's population grew exponentially at an ex-

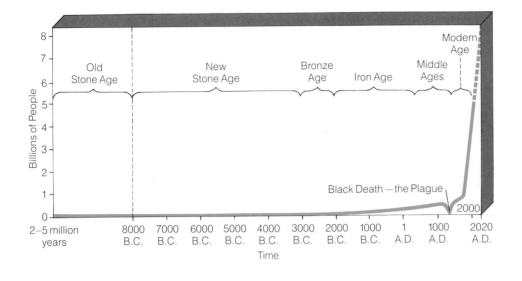

Figure 1-1 J-shaped curve of past exponential world population growth with projections to 2020 A.D. (Data from World Bank and United Nations)

tremely slow average rate of about 0.002% a year. This slow, or lag, phase of exponential growth is represented by the almost horizontal portion of Figure 1-1.

Since then, the average annual exponential growth rate has increased. It reached an all-time high of 2.06% in 1970, before dropping somewhat to 1.7% in 1987. This has led to such a large increase in people that the curve of population growth has rounded the bend of the J and has been heading almost straight up from the horizontal axis (Figure 1-1). This occurs because the size of the population becomes so large that even a small annual percentage increase adds a large number of people.

For example, when a deceptively small annual growth rate of 1.7% acted on a base population of 5.03 billion in 1987, about 86.7 million people were added to the earth's population. At this rate, it takes less than five days to replace the number of people equal to the number of Americans killed in all U.S. wars; slightly more than a year to replace the numerical equivalent of the more than 75 million people killed in the world's largest disaster (the bubonic plague epidemic of the 14th century); and only about two years to replace the numerical equivalent of the estimated 165 million soldiers who died in all wars fought on this planet during the past 200 years.

The 86.7 million new people added in 1987—amounting to an average addition of 1.7 million a week, 238,000 a day, or 9,900 an hour—and the even larger numbers to be added annually for decades need to be fed, clothed, housed, educated, and kept in good health. Each person will use some resources and will add to global pollution and degradation of the earth's life-support systems. Yet this growth is occurring at a time when, according to United Nations estimates, at least half the adults on this planet cannot read or write,

Figure 1-2 One-sixth of the people in the world do not have adequate housing. Lean-to sidewalk shelters like these are homes for many families in Dacca, Bangladesh.

one out of six people is hungry or malnourished and does not have adequate housing (Figure 1-2), one out of four lacks clean drinking water, and one out of three lacks access to adequate sewage disposal or effective health care.

With such exponential growth, it has taken an increasingly smaller number of years to add each additional billion people. It took 2 million to 5 million years to add the first billion people; 130 years to add the

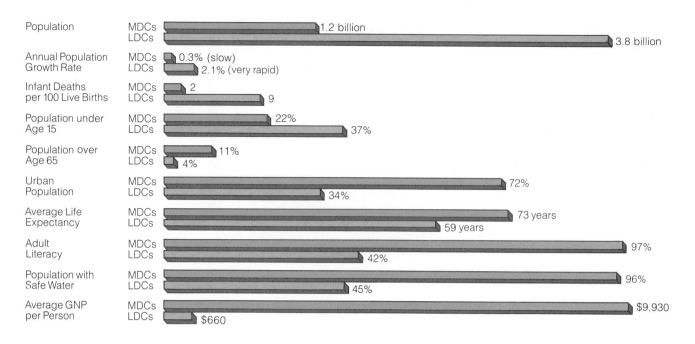

Figure 1-3 Some characteristics of more developed countries (MDCs) and less developed countries (LDCs) in 1986. (Data from United Nations)

second billion; 30 years to add the third billion; 15 years to add the fourth billion; and only 12 years to add the fifth billion, between 1975 and 1987. At present growth rates, the sixth billion will be added during the 12-year period between 1987 and 1999. This is what it means to go around the bend of a J-shaped curve of exponential growth.

By the year 2000, if the annual growth rate of population drops as projected to 1.5%, there will be 6.16 billion people on earth. Because the population will still be growing exponentially and the population base will have increased by more than 1 billion people, 89 million people would be added in 2000—2 million more than the number added in 1987.

Population Growth in the More Developed and Less Developed Countries The world's 166 countries can be divided into two general groups based primarily on the average annual **gross national product,** or **GNP** (the total market value of all goods and services produced per year), per person (Figure 1-3). The 33 **more developed countries (MDCs),** consisting of the United States, the Soviet Union, all European countries, Japan, Australia, and New Zealand, have significant industrialization, have a high average GNP per person, and are located primarily in the Northern Hemisphere with mostly favorable climates and fertile soils.

The 133 **less developed countries (LDCs)** have low to moderate industrialization, have a very low to moderate average GNP per person, and are located pri-

marily in the Southern Hemisphere in Africa, Asia, and Latin America, often with less favorable climates and less fertile soils than those in the Northern Hemisphere (see Spotlight on p. 5). At present, 92% of the world's annual population growth is taking place in the LDCs.

Although the world is feeding more people than ever before, an estimated 750 million people—one out of every six people on earth—are desperately poor, living mostly in low- and very-low-income countries. These people do not have enough fertile land or money to grow their own food in rural areas or enough money to buy the food they need in cities. As a result, between 12 million and 20 million die prematurely each year from starvation, malnutrition (lack of sufficient protein and other nutrients needed for good health), or normally nonfatal diseases such as diarrhea brought on by contaminated drinking water, which for people weakened by malnutrition becomes deadly.

This means that during your lunch hour, 1,400 to 2,300 people died of such causes; by the time you eat lunch tomorrow, 33,000 to 55,000 more will have died; and by this time next week, 231,000 to 385,000. Half are children under the age of 5 (Figure 1-4). This starvation and malnutrition is not classified as famine by most officials because it is spread throughout much of the world (especially rural Africa and Asia) and not confined to one country.

Although dividing the world into MDCs and LDCs is convenient for dramatizing major differences in *average* living conditions, it is an oversimplification. Some countries designated MDCs are richer and more

Life for the poor people who make up at least half the population of the 79 LDCs with low and very low average GNPs per person consists of a harsh daily struggle for survival. In typical rural villages or urban slums, groups of malnourished children sit around wood or dung fires eating breakfasts of bread and coffee. The air is filled with the stench of refuse and open sewers. Children and women carry heavy jars or cans of water, often for long distances, from a muddy, microbe-infested river, canal, or village water faucet. At night people sleep on the street in the open, under makeshift canopies, or on dirt floors in crowded single-room shacks, often made from straw, cardboard, rusting metal, or abandoned sections of drainage pipes. Families consisting of a father, mother, and from seven to nine children consider themselves fortunate to have an annual income of $300—an average of 82 cents a day. The parents, who themselves may die by age 50, know that three or four of their children will probably die of hunger or childhood diseases, such as diarrhea or measles, that rarely kill in affluent countries.

Some consider poor people ignorant for having so many children. To most poor parents, however, having a large number of children, especially boys, makes good sense. It gives them much needed help for work in the fields or begging in the streets and provides a form of social security to help them survive when they reach old age (typically in their forties). For people living near the edge of survival, having too many children may cause problems, but having too few can contribute to premature death.

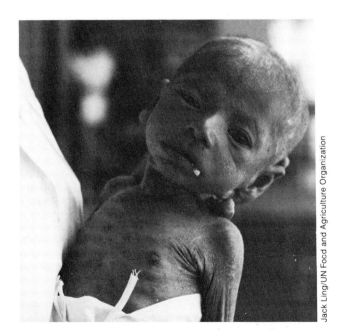

Figure 1-4 This Indonesian child is one of the estimated 750 million people on Earth who suffer malnutrition resulting from a diet insufficient in protein and other nutrients needed for good health.

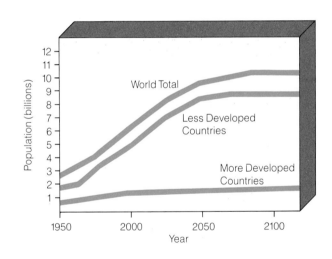

Figure 1-5 Past and projected population size for the more developed countries, less developed countries, and the world, 1950–2120 A.D. (Data from United Nations)

industrialized than others; some LDCs are poorer than others; and some poverty is found in the richest countries. In the United States during 1986, for example, 32 million people—one of every seven, and almost one of every three Hispanic and black Americans—were classified by the government as living below the poverty level.

When Might World Population Growth Come to a Halt? If the annual rate of population growth continues to decrease, we will eventually reach a state termed **zero population growth (ZPG),** in which the annual number of births equals the number of deaths (births per year − deaths per year = 0). Some MDCs in Europe have reached or are approaching ZPG, and some, such as West Germany, Hungary, and Denmark, are even experiencing population declines. In most LDCs, however, population growth will continue for many decades (Figure 1-5). Unless a global nuclear war occurs or famine and disease increase

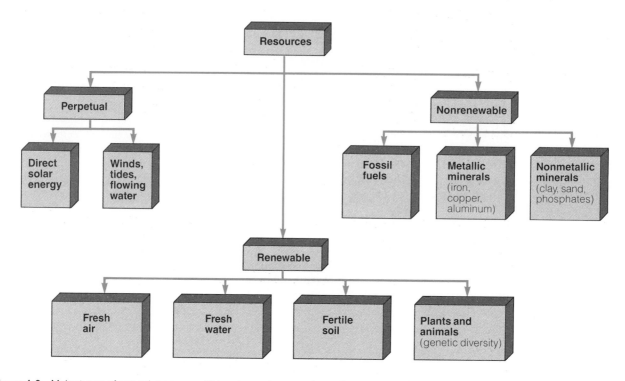

Figure 1-6 Major types of natural resources. This scheme, however, is not fixed; renewable resources can be converted to nonrenewable if used for a prolonged time faster than they are renewed by natural processes.

greatly, UN population experts project that the world is not likely to attain ZPG until around 2100, when it will reach a population of 10.4 billion—more than twice that in 1987—with most of the growth taking place in the LDCs.

1-2 RESOURCES AND ENVIRONMENTAL DEGRADATION

What Is a Resource? A **resource** is usually defined as anything obtained from the physical environment to meet human needs. Some resources, such as fresh air and naturally growing edible plants, are directly available for use. But most resources, such as oil, iron, groundwater, fish, and game animals, are not. They become resources only as a result of *human ingenuity* (using science and technology to find, extract, process, and convert them to usable forms), *economics* (deposits that can be exploited at a reasonable cost), and *cultural beliefs* (made available in acceptable forms). For example, groundwater found deep below the earth's surface was not a resource until we developed the technology for drilling a well and installing pumps to bring it to the surface. Petroleum was a mysterious fluid until people learned how to locate it, extract it, and refine it into gasoline, home heating oil, road tar, and other products at affordable prices.

Types of Resources Resources can be classified as perpetual, nonrenewable, and renewable (Figure 1-6). A **perpetual resource,** such as solar energy, is one that comes from an essentially inexhaustible source and thus will be available in a relatively constant supply regardless of whether or how we use it.

Nonrenewable resources, such as copper and oil, exist in a fixed amount (stock) in various places in the earth's crust and either are not replenished by natural processes (copper) or are replenished much more slowly than they are used (oil). Typically, a nonrenewable resource such as copper or oil is considered depleted from an economic standpoint when 80% of its total estimated supply has been removed and used. To find, extract, and process the remaining 20% generally costs more than it is worth.

Some nonrenewable resources can be recycled or reused to stretch supplies—copper, aluminum, iron, and glass, for example. **Recycling** involves collecting and remelting or reprocessing a resource (aluminum beverage cans), whereas **reuse** involves using a resource over and over in the same form (refillable beverage bottles). But discarded aluminum cans, refillable bottles, and abandoned car hulks can be dispersed so widely that collecting them for reuse or recycling becomes too costly. Other nonrenewable resources, such as fossil fuels (coal, oil, and natural gas), cannot be recycled or reused. Once burned, the fuels lose their high-quality energy forever.

Examples of major types of environmental degradation of potentially renewable resources include **(1)** covering productive land with water, silt, concrete, asphalt, or buildings to such an extent that agricultural productivity declines and places for wildlife to live are lost; **(2)** cultivating land so intensively without proper soil management that agricultural productivity is reduced by soil erosion and depletion of plant nutrients; **(3)** irrigating cropland without sufficient drainage so that excessive accumulation of water (*waterlogging*) or salts (*salinization*) in the soil decreases agricultural productivity; **(4)** removing trees from large areas without adequate replanting (*deforestation*) so that wildlife habitats are destroyed and long-term timber productivity is decreased by flooding and soil erosion; and **(5)** depleting grass on land grazed by livestock (*overgrazing*) so that soil is eroded to the point where productive grasslands are converted into nonproductive deserts (*desertification*).

Environmental degradation in many areas of LDCs is caused by hundreds of millions of poor people trying to survive in the short term, diminishing the normally renewable resource base on which the future economic productivity and growth of these countries depend. Environmental degradation in both LDCs and MDCs takes place when attempts to stimulate short-term economic growth are made with little concern for the future availability of renewable resources that their future economic growth depends on.

Another situation giving rise to environmental degradation involves common, or shared, resources. A **commons** is a resource to which a population has free and unmanaged access—in contrast to a private property resource (accessible only to the owner) and a socialized resource (where access is controlled by elected or appointed managers). A phenomenon known as the **tragedy of the commons** occurs when resources such as clean air, clean water in a river or lake, or fish of the sea are considered common property free to be used by anyone. They are then overharvested (fish) or polluted (air and water) because each user reasons, "If I don't use this resource, someone else will." Individuals continue maximizing their use of such a common resource until the cumulative effect of many people doing the same thing so depletes the usable supply that it is no longer available at an affordable price.

Sometimes a substitute or replacement for a nonrenewable resource that is scarce or too expensive can be found. Although some resource economists argue that we can use ingenuity to find a substitute for any nonrenewable resource, this is not always the case at a particular time or for a particular purpose. Some materials have unique properties that cannot easily be replaced; the would-be replacements are inferior, too costly, or otherwise unsatisfactory. For example, nothing now known can replace steel and concrete in skyscrapers, nuclear power plants, and dams.

A **renewable resource** is one that can be depleted in the short run if used or contaminated too rapidly but normally will be replaced through natural processes in the long run. Examples include trees in forests, grasses in grasslands, animals such as fish and game, fresh surface water in lakes and rivers, most deposits of groundwater, fresh air, and fertile soil.

Classifying something as a renewable resource, however, does not mean that it is inexhaustible and that it will always remain renewable. The highest rate at which a renewable resource can be used without impairing or damaging its ability to be fully renewed is called its **sustained yield.** If this yield is exceeded, the base supply of a renewable resource begins to shrink and can eventually become nonrenewable on a human time scale or in some cases nonexistent—a process known as **environmental degradation** (see Spotlight above). Considerable evidence indicates that in many parts of the world, especially in LDCs, the sustained yields for potentially renewable resources such as topsoil, groundwater, grasslands, forests, fisheries, and wildlife are being exceeded.

Will There Be Enough Resources? Increasing population causes a corresponding rise in resource use, and a rise in the standard of living creates a significant rise in average use of nonrenewable and renewable resources per person. During the past 100 years and especially since 1950, affluent countries have gone around the bend on a J-shaped curve of increasing average consumption per person of renewable and nonrenewable resources. *With only 24% of the world's population, the MDCs use 80% of the world's processed energy and mineral resources. The United States alone, with only 4.8% of the world's population, produces about 21% of all goods and services, uses about one-third of the world's*

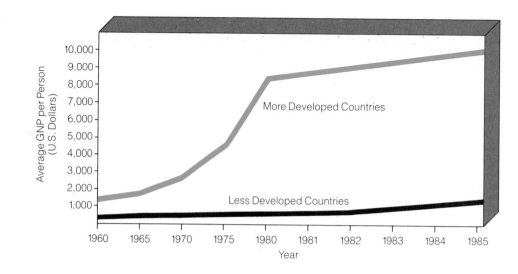

Figure 1-7 The gap in average GNP per person between the more developed and the less developed countries has been widening since 1960, raising fears in many LDCs that they may become never-developed countries. Adjusting for inflation, the average GNP per person in LDCs has actually decreased since 1960. (Data from United Nations)

processed energy and mineral resources, and produces at least one-third of the world's pollution. For example, the average U.S. citizen consumes 50 times more steel, 56 times more energy, 170 times more synthetic rubber and newsprint, 250 times more motor fuel, and 300 times more plastic than the average citizen of India.

Because the gap between the rich and poor countries has been increasing since 1960 (Figure 1-7), an increasing number of leaders and citizens in LDCs fear that rapid depletion of nonrenewable resources and degradation of renewable resources by the MDCs may not leave enough resources for their countries to ever become MDCs. They are calling for the MDCs to waste fewer resources and to provide greatly increased assistance and access to trade and markets to help the LDCs become more self-sufficient and have a more equitable share of the world's resources.

1-3 POLLUTION

What Is Pollution? Any change in the physical, chemical, or biological characteristics of the air, water, or soil that can affect the health, survival, or activities of human beings or other forms of life in an *undesirable* way is called **pollution.** Pollution does not have to cause physical harm; pollutants such as noise and heat may cause injury but more often cause psychological distress, and aesthetic pollution such as foul odors and unpleasant sights offend the senses.

People, however, may differ in what they consider to be a pollutant, on the basis of their assessment of benefits and risks to their health and economic well-being. For example, visible and invisible chemicals spewed into the air or water by an industrial plant might be harmful to people and other forms of life living nearby. However, if the installation of expensive pollution controls forced the plant to shut down, workers who would lose their jobs might feel that the risks from polluted air and water are minor weighed against the benefits of profitable employment. The same level of pollution can also affect two people quite differently—some forms of air pollution might be a slight annoyance to a healthy person but life-threatening to someone with emphysema. As the philosopher Georg Hegel pointed out, the nature of tragedy is not the conflict between right and wrong but between right and right.

Types and Sources of Pollutants As long as they are not overloaded, natural processes or human-engineered systems (such as sewage treatment plants) can biodegrade, or break down, some types of pollutants to an acceptable level or form. Pollutants can be classified as being **rapidly biodegradable** (such as animal and crop wastes), **slowly biodegradable** (such as DDT and PCBs), and **nonbiodegradable** (such as toxic mercury and lead compounds and some radioactive substances).

Polluting substances can enter the environment naturally or through human activities. Most natural pollution is dispersed over a large area and is often diluted or degraded to harmless levels by natural processes. In contrast, the most serious human pollution problems occur in or near urban and industrial areas, where large amounts of pollutants are concentrated in relatively small volumes of air, water, and soil. Furthermore, many pollutants from human activities are synthetic (human-made) chemicals that are slowly biodegradable or nonbiodegradable.

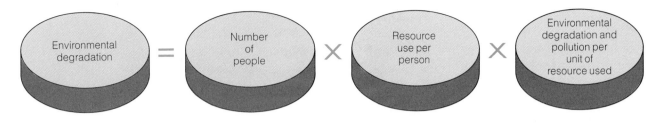

Figure 1-8 Simplified model of how three factors affect overall environmental degradation and pollution or environmental impact.

Determining Harmful Levels of Pollutants Determining the amount of a particular pollutant that can cause a harmful or undesirable effect in human beings or other organisms is a difficult scientific problem. The amount of a chemical or pollutant in a given volume of air, water, or other medium is called its **concentration.** Concentrations of pollutants are often expressed as **parts per million (ppm)** or **parts per billion (ppb)**—the number of parts of a chemical or pollutant found in 1 million or 1 billion parts of air, water, or other medium. Although 1 ppm and 1 ppb represent very small concentrations for some organisms, with some pollutants they represent dangerous levels.

During a lifetime an individual is exposed to many different types and concentrations of potentially harmful pollutants. The scientific evidence correlating a particular harmful effect to a particular pollutant is usually statistical or circumstantial—as is most scientific evidence. For example, so far no one has been able to show which specific chemicals in cigarette smoke cause lung cancer; however, smoking and lung cancer are causally linked by an overwhelming amount of statistical evidence from more than 32,000 studies.

Another complication is that certain pollutants acting together can cause a harmful effect greater than the sum of their individual effects. This phenomenon is called a **synergistic effect.** For example, asbestos workers, already at a higher-than-average risk of lung cancer, greatly increase that risk if they smoke, because of an apparent synergistic effect between tobacco smoke and tiny particles of asbestos inhaled into the lungs. Testing all the possible synergistic interactions among the thousands of possible pollutants in the environment is prohibitively expensive and time-consuming, even for their effects on one type of plant or animal.

Pollution Control Pollution can be controlled in two fundamentally different ways. **Input pollution control** prevents potential pollutants from entering the environment or sharply reduces the amounts released. In this preventive approach, taxes, incentives, or other economic devices are used to make the resource inputs of a process so expensive that these resources will be used more efficiently, thus decreasing the output of waste material.

The other is a "treat-the-disease," or **output pollution control,** approach that deals with wastes after they have been produced. The three major methods of output control are (1) cleaning up polluted air, water, or land by reducing pollutants to harmless levels or by converting them to harmless or less harmful substances; (2) disposing of harmful wastes by burning them, dumping them in the air or water in the hope that they will be diluted to harmless levels, or burying them in the ground and hoping they will remain there; and (3) recycling or reusing materials output from human activities rather than discarding them.

1-4 RELATIONSHIPS AMONG POPULATION, RESOURCE USE, TECHNOLOGY, ENVIRONMENTAL DEGRADATION, AND POLLUTION

The Roots of Environmental Degradation and Pollution According to one simple model, the total environmental degradation and pollution or environmental impact of population in a given area depends on three factors: (1) the number of people; (2) the average amount of resources each person uses; and (3) the environmental degradation and pollution resulting from each unit of resource used (Figure 1-8).

In general, **overpopulation** occurs when the people in a country, a region, or the world are using nonrenewable and renewable resources to such an extent that the resulting degradation or depletion of the resource base and pollution of the air, water, and soil are impairing their life-support systems. Differences in the relative importance of each factor in the model shown in Figure 1-8 have been used to distinguish between two types of overpopulation (Figure 1-9).

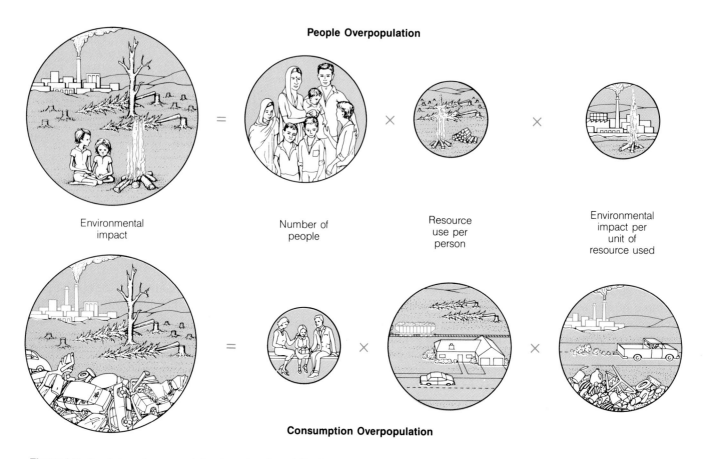

People Overpopulation

Environmental impact = Number of people × Resource use per person × Environmental impact per unit of resource used

Consumption Overpopulation

Figure 1-9 Two types of overpopulation based on the relative importance of the factors in the model shown in Figure 1-8.

The type known as **people overpopulation** exists where there are more people than the available supplies of food, water, and other vital resources can support, or where the rate of population growth so exceeds the rate of economic growth that an increasing number of people are too poor to grow or buy sufficient food, fuel, and other vital resources. In this type of overpopulation, population size and the resulting environmental degradation of potentially renewable soil, grasslands, forests, and fisheries tend to be the most important factors determining the total environmental impact. In the world's poorest LDCs, people overpopulation results in premature death for 12 million to 20 million human beings each year and bare subsistence for hundreds of millions more—a situation that many fear will worsen unless population growth is brought under control and improved resource management is used to restore degraded renewable resource bases.

Affluent and technologically advanced countries such as the United States, the Soviet Union, and Japan are said by some to have a second type of overpopulation, known as **consumption overpopulation.** It is based on the fact that without adequate pollution and land use controls, a small number of people using resources at a high rate produces more pollution and environmental degradation than a much larger number of people using resources at a much lower rate. With this type of overpopulation, high rates of resource use per person and the resulting high levels of pollution per person tend to be the most important factors determining overall environmental impact.

Other Factors   The three-factor model shown in Figure 1-8, though useful, is far too simple. The actual situation is much more complex: an interacting mix of problems and contributing factors shown in simplified form in Figure 1-10. For example, pollution and environmental degradation are intensified not only by population size but also by *population distribution*. The most severe air and water pollution problems usually occur when large numbers of people are concentrated in urban areas. Conversely, spreading people out can have a devastating effect on potentially renewable soil, forest, grassland, and recreational resources. *War* also has a devastating environmental impact.

Some *scientific and technological developments*, such as the automobile and phosphate detergents, create new environmental problems or aggravate existing

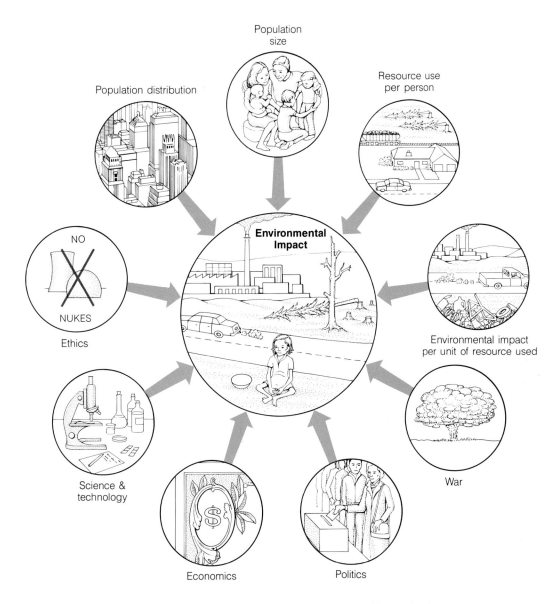

Figure 1-10 Environmental problems are caused by a complex, poorly understood mix of interacting factors, as illustrated by this greatly simplified multiple-factor model.

ones. Other scientific and technological developments can help solve various environmental and resource problems. Substitutes have been developed for many scarce resources. Light bulbs, for example, have replaced whale oil in lamps, thus helping protect the world's rapidly diminishing population of whales from extinction. Unnecessary resource waste has been reduced. For example, more energy is recovered from a ton of coal than in the past, and processes to control and clean up many forms of pollution have been developed.

One major attempt to use technology wisely is the increased global emphasis on appropriate technology. **Appropriate technology** is usually small, simple, decentralized, and inexpensive to build and maintain,

and it usually utilizes locally available materials and labor. The use of huge tractors to plow fields in a poor rural village in India is often cited as an example of inappropriate and destructive technology. In such villages the most plentiful resource is people willing and needing to work on farms. The tractor deprives these people of their only means of survival and forces them to migrate to already overpopulated cities, looking for nonexistent jobs. The wealthier farmers who remain become dependent on industrialized countries for expensive gasoline and parts and find the tractor too complex to be repaired by untrained local people. Instead of a large tractor, an appropriate technology would be a well-designed metal plow, made and repaired by a local blacksmith and pulled by locally available draft animals such as oxen or water buffaloes. Supporters of appropriate technology recognize that it is not a cure for all our environmental problems but believe that its increasing use is an encouraging trend that should be nurtured.

Economic, political, and ethical factors are also involved. We can manipulate the *economic system* to control pollution, environmental degradation, and resource waste by making such practices unprofitable (in free-market economies) or illegal (in centrally controlled economies). We can also use the *political process* to enact and enforce pollution-control and land-use-control laws. However, such economic and political efforts will not be undertaken until an informed and politically active segment of the population (probably 5% to 10%) in countries with free elections and the leaders in other countries realize that to abuse the world's life-support systems for short-term economic gain is both unwise and *unethical*.

1-5 WHAT SHOULD BE DONE? NEO-MALTHUSIANS VERSUS CORNUCOPIANS

Two schools of thought present diametrically opposed views about what the role of people in the world should be, how serious the world's present and projected future environmental and resource problems really are, and what should be done about them. **Neo-Malthusians** (called "gloom-and-doom pessimists" by their opponents) believe that if present trends continue, the world will become more crowded and more polluted, leading to greater political and economic instability and increasing the threat of nuclear war as the rich get richer and the poor get poorer. The term *neo-Malthusian* refers to an updated and greatly expanded version of the hypothesis proposed in 1789 by Thomas Robert Malthus, an English clergyman and economist, that human population growing exponentially will eventually outgrow food supplies and will be reduced in size by starvation, disease, and war.

The opposing group, called **cornucopians** (or unrealistic "technological optimists" by their opponents), believes that if present trends continue, economic growth and technological advances based on human ingenuity will produce a less crowded, less polluted world, in which most people will be healthier, will live longer, and will have greater material wealth. The term *cornucopian* comes from *cornucopia*, the horn of plenty, which symbolizes abundance. (Major differences between these two schools of thought are summarized in the Spotlight on pp. 13–14.)

This debate between cornucopians (most of whom are economists) and neo-Malthusians (most of whom are environmentalists and conservationists) has been going on for decades. But it is much more than an intellectual debate between people who generally use the same data and general trends to reach quite different conclusions. At a more fundamental level it represents radically different views of how the world works—one's *worldview*—and thus how we should operate in the world.

Cornucopians generally have a **frontier** or **throwaway worldview**. They see the earth as a place of unlimited resources where any type of conservation that hampers short-term economic growth is unnecessary. If we pollute or deplete the resources in one area, we will find substitutes, control the pollution through technology, and if necessary obtain additional resources from the moon and asteroids in the "new frontier" of space.

In contrast, neo-Malthusians generally have a **sustainable-earth worldview.*** Seeing the earth as a place with finite room and resources, they believe that ever-increasing production and consumption inevitably put severe stress on the complex, poorly understood natural processes that renew and maintain the air, water, and soil. Some neo-Malthusians have used the term "Spaceship Earth" to help people see the need to protect the earth's life-support systems. However, other neo-Malthusians have criticized this image, believing the spaceship analogy subtly reinforces the arrogant idea that the role of human beings is to dominate and control nature; it encourages us to view the earth merely as a machine that we can manipulate at will and to think that we have essentially complete understanding of how nature works.

As we examine major environmental problems and their possible solutions throughout this book, we should be guided by the motto of philosopher and mathematician Alfred North Whitehead (1861–1947): "Seek simplicity and distrust it," and by writer and social critic H. L. Mencken (1880–1956), who warned: "For every problem there is a solution—simple, neat, and wrong."

What is the use of a house if you don't have a decent planet to put it on?
Henry David Thoreau (1817–1862)

---

*Others have used the terms *sustainable society* and *conserver society* to describe this view. I use the word *earth* to make clear that it is the entire earth system, not just the subsystem of human beings and their societies, that must be sustained.

CORNUCOPIANS	NEO-MALTHUSIANS

Role of Human Beings on Earth

Use our ingenuity to develop technologies for *conquering nature* to promote an economy based on ever-increasing production and consumption of material goods.

Use our ingenuity to develop technologies for *working with nature* to promote kinds of economic growth that sustain rather than deplete and degrade the earth's life-support systems.

Seriousness of Environmental Problems

Exaggerated; can be cured by increased economic growth and technological innovations.

Serious now and could become more serious if many present forms of economic growth and technology are not replaced with forms that place less stress on the environment.

Population Growth and Control

Should not be controlled because people are our most vital source of new technologies that will solve the world's problems.

Should be controlled because adding billions of mostly poor and malnourished people in LDCs (people overpopulation) will not solve the world's problems and will condemn hundreds of millions—perhaps billions—to premature deaths in the future; allowing wasteful types of resource consumption to continue in MDCs (consumption overpopulation) will increase regional and global pollution and environmental degradation.

Economic development will lead to lower birth rates and a stabilized population in LDCs as it has in MDCs since the Industrial Revolution.

Economic development cannot slow population growth in time to avoid greatly increased famine an environmental degradation in the poorer LDCs unless it is coupled with a well-funded, well-planned program for population control.

Many forms of birth control go against some individuals' religious beliefs and are an infringement on people's freedom to have as many children as they want.

People should not be forced to control the number of children they have, but they should have easy access to any form of birth control they find acceptable; people should also be educated to understand that a lower population size in LDCs and MDCs will help reduce resource depletion, pollution, environmental degradation, famine, poverty, social tension, war, and erosion of individual freedom.

Resource Depletion

We will not run out of renewable resources; as resources are degraded, price increases for food, lumber, and other products will lead to better management or a switch to alternatives.

There are no substitutes for potentially renewable resources such as topsoil, grasslands, forests, and fisheries, which are already being overused and converted to nonrenewable resources in many LDCs for survival and in many MDCs for short-term profit.

We will not run out of scarce nonrenewable resources because price increases will stimulate people to invent ways to mine lower-grade deposits or to find substitutes.

Once a nonrenewable resource is about 80% depleted, it is usually uneconomic to use what's left; substitutes for some widely used resources may not be found or may take too long to develop and phase in without causing severe economic hardship.

Continued

Resource Depletion

MDCs are not overdeveloped countries suffering from consumption overpopulation; in the United States and most MDCs since the 1970s, economic growth has produced industrial products, technology, and knowledge, all of which can be used to control pollution, find new resources, help LDCs grow economically, and increase average life expectancy.

As a result of very high rates of resource use and unnecessary waste, MDCs are causing unacceptable regional and global resource depletion, environmental degradation, and pollution, which can eventually wipe out short-term gains in average life expectancy; when adjusted for inflation, the average GNP per person in LDCs has decreased since 1960—not increased as cornucopians have been promising for decades.

Resource Conservation

Renewable resources should be used and managed efficiently to leave them in reasonable shape for future generations, but not at the expense of short-term economic growth by the present generation.

Renewable resources should be used and managed efficiently for people on earth today, but not at the expense of their future sustainability, on which long-term economic productivity depends.

Reducing unnecessary resource waste, while desirable, is not a high priority because human ingenuity can always find a substitute for any scarce resource.

Reducing unnecessary resource waste should receive the highest priority to stretch supplies that are being depleted and to reduce the environmental impact of resource extraction and use; substitutes may not be found or may be inferior or more costly.

Emphasis on recycling and reusing materials can reduce short-term profits and is unnecessary because we can always find a substitute for any scarce resource.

Recycling and reuse of nonrenewable resources are necessary to stretch supplies, allow more time to find and phase in substitutes (if possible), and to reduce the environmental impact of finding, extracting, and processing primary or virgin resources.

Pollution Control

Pollution control should not be increased at the expense of short-term economic growth, which can provide funds for cleaning up the environment as needed.

Insufficient pollution control leads to short- and long-term damages that will reduce long-term economic productivity; costs of most products and services do not reflect their actual harmful environmental, economic, and health impacts, thus misleading consumers about the impact of their lifestyles.

When pollution control is necessary, emphasis should be on *output control* to clean up pollution that has entered the environment or to burn, dump, or bury waste materials.

Emphasis should be on *input control*; preventing pollution or sharply reducing the amount entering the environment is more effective and less costly than attempting to reduce levels of widely dispersed pollutants after they have entered the environment; some pollutants should be viewed not as wastes but as valuable resources that can be recycled or reused.

Technology

Large, centralized technology, such as large electric power plants, provides more profit for large corporations, enhances a country's short-term economic growth, and provides a more reliable source of energy and other goods and services.

Smaller, decentralized forms of technology use and waste fewer resources, increase national security by making a country less vulnerable to attack and natural disaster, and give individuals more control over how they obtain goods and services and over the prices they pay.

DISCUSSION TOPICS

1. Is the world overpopulated? Why or why not? Is the United States suffering from consumption overpopulation? Why or why not?

2. Do you favor zero population growth (ZPG) for the world? For the United States? Explain.

3. Debate the following proposition: High levels of resource use by the United States and other MDCs is beneficial because it allows greater purchases of raw materials from poor countries, providing them with funds to stimulate economic growth, and because increased economic growth in the MDCs can provide funds for higher levels of financial aid to LDCs.

4. Debate the following proposition: The world will never run out of resources because technological innovations will produce substitutes or allow use of lower grades of scarce resources.

5. Should economic growth in the United States and in the world be limited? Why or why not? Is all economic growth bad? Which forms, if any, do you believe should be limited? Which types, if any, should be encouraged?

6. Do your own views more closely resemble those of a neo-Malthusian or a cornucopian? Does your lifestyle indicate that in reality you are acting as if you were a cornucopian or a neo-Malthusian? Compare your views with those of others in your class.

2

Human Impact on the Earth

GENERAL OBJECTIVES

1. What major impacts did early and advanced hunter-gatherer societies have on the environment and what was their primary relationship to nature?

2. What major impacts have early agricultural societies and present-day nonindustrialized agricultural societies had on the environment and what is their primary relationship to nature?

3. What major impacts do present-day industrialized societies have on the environment and what is their primary relationship to nature?

4. What are the major phases in the history of resource exploitation, resource conservation, and environmental protection in the United States?

5. What hopeful signs are there that we may be gradually shifting from today's mix of industrialized and nonindustrialized societies to new, sustainable-earth societies?

A continent ages quickly once we come.
Ernest Hemingway

During the 2 million years human beings have lived on earth, they have been able to gain increasing control over their environment through a series of major cultural changes, each with a greatly increased environmental impact. The J-shaped curves of increasing population, resource use, pollution, and environmental degradation are symptoms of the cultural changes, from the hunter-gatherers who survived nearly 2 million years by hunting wild game and fish and gathering wild plants, to agriculturalists who learned to tame and breed wild animals and cultivate wild plants about 10,000 years ago, to today's rapidly growing population of 5 billion living in a mix of countries ranging from nonindustrialized to highly industrialized.

2-1 HUNTING-AND-GATHERING SOCIETIES

Early Hunter-Gatherers During 99.9% of the time that human beings have lived on earth, they were **hunters and gatherers,** obtaining food by gathering edible wild plants and hunting wild game and fish from the nearby environment. Archaeological findings and studies of the few, rapidly decreasing hunter-gatherer tribes that remain today indicate that early hunter-gatherer societies and the more advanced ones that followed lived in small groups of rarely more than 50 people, who worked together to secure enough food to survive. If a group became so large that its members could not find enough food within reasonable walking distances, it split up to form other groups, which moved to different areas. Many of these widely scattered groups were nomads, moving with the seasons and migrations of game animals to obtain sufficient food and to allow renewal of plant and animal life in the areas they left behind.

The hunters and gatherers were experts in survival. Their intimate knowledge of nature enabled them to predict the weather, find water even in the desert, locate a wide variety of edible plants and animals, and learn which plant and animal parts had medicinal properties. Their material possessions consisted mostly of primitive weapons and tools made by chipping sticks and stones into forms useful for killing animals, fishing, cutting plants, and scraping hides to be used for clothing, shelter, and other purposes.

Studies of today's few remaining hunter-gatherer societies suggest that women and children spend an average of 15 hours a week gathering the food they need, and men hunt for only about a week out of each month or a few hours each day. The studies also show that these "primitive" people enjoy a more diverse diet, more free time and less stress and anxiety than most people in today's LDCs and MDCs.

Although malnutrition and starvation were rare in early hunter-gatherer societies, infant mortality was high, primarily from infectious diseases. This factor, coupled with infanticide (killing the young), led to an average life expectancy of about 30 years and helped keep population size in balance with available food supplies. Population size was also kept down by the natural birth control and spacing of births caused by the suppression of women's ability to ovulate and conceive during the 3 to 4 years they breast-fed each of their children.

Advanced Hunter-Gatherers   Archaeological evidence indicates that hunter-gatherers gradually developed improved tools and hunting weapons, such as spears with sharp-edged stone points mounted on wooden shafts and later the bow and arrow. They also learned to cooperate with members of other groups to hunt herds of reindeer, woolly mammoths, European bison, and other big game; to use fire to flush game from thickets toward hunters lying in wait and to stampede herds of animals into traps or over cliffs; and to burn vegetation to promote the growth of plants that could be gathered for food and that were favored by some of the animal species they hunted.

These practices meant that advanced hunter-gatherers had a greater impact on their environment than their predecessors, especially in converting forests into grasslands. But because of their relatively small numbers, their movement from place to place, and their dependence on their own muscle power to modify the environment, their environmental impact was still fairly small and localized. Both early and advanced hunter-gatherers were examples of *human beings in nature,* who learned to survive by understanding and cooperating with nature and with one another.

2-2   AGRICULTURAL SOCIETIES: THE AGRICULTURAL REVOLUTION

Domestication of Wild Animals and Plants   One of the most significant changes in human history is believed to have begun about 10,000 years ago. During this period, people in several parts of the world independently began learning how to domesticate—herd, tame, and breed—wild game for food, clothing, and carrying loads and to domesticate selected wild food plants, planting and growing them close to home instead of gathering them over a large area.

Archaeological evidence indicates that the first type of plant cultivation, which today we call *horticulture,* began when women in tropical areas discovered they could grow some of their favorite wild food plants such as yam, sweet potato, taro, and arrowroot by digging holes with a stick (a primitive hoe) and placing the roots or tubers of these plants in the holes between tree stumps in small patches of forest. They cleared the ground by **slash-and-burn cultivation**—cutting down trees and other vegetation, leaving the cut vegetation on the ground to dry, and then burning it (Figure 2-1). The resulting wood ashes added plant nutrients to the nutrient-poor soils found in most tropical forest areas, which were chosen because they had fewer roots than grassland soils and thus were easier to cultivate with digging sticks.

These early growers also discovered the principle of **shifting cultivation:** A single plot was harvested and replanted for 2 to 5 years, until further cultivation was no longer worthwhile because of a reduction in soil fertility or because the area had been invaded by a dense growth of vegetation from the surrounding forest. A new plot was then cleared to begin a new cycle of cutting, burning, planting, and harvesting for several years. The growers learned that each abandoned patch had to be left fallow (unplanted) for 10 to 30 years to allow a new growth of trees to become established and the soil to be renewed before it could again be used to grow crops.

These growers practiced **subsistence agriculture,** growing only enough food to feed their families. The dependence of these early subsistence horticulturists on human muscle power and crude stone or stick tools meant that they could cultivate only small plots; thus, they had relatively little impact on their environment.

True *agriculture* (as opposed to horticulture) began around 5000 B.C. with the invention of the metal plow,

Figure 2-1 Slash-and-burn cultivation in tropical forests was probably the first technique used to grow crops. It is still practiced today in some tropical areas by 150 million to 200 million people, as shown in this cleared patch of tropical rainforest in the Tuxtla Mountains, Veracruz, Mexico.

pulled by domesticated animals and steered by the farmer. Animal-pulled plows greatly increased crop productivity by allowing farmers to cultivate larger parcels of cropland and to cultivate fertile grassland soils in spite of their more extensive root systems. In some arid (dry) regions, early farmers further increased crop output by diverting supplies of water into hand-dug ditches and canals to irrigate crops. With this animal- and irrigation-assisted agriculture, families could provide themselves with sufficient food and sometimes have enough left over for sale or for storage to provide food when flooding, prolonged dry spells, insect infestation, or other natural disasters reduced crop productivity.

The Emergence of Agriculture-based Urban Societies The gradually increasing ability of a few farmers to produce enough food to feed their families

plus a surplus that could be traded and used to feed other people had four important effects: (1) Population began to increase because of a larger, more constant supply of food. (2) People cleared increasingly larger areas of land and began to control and shape the surface of the earth to suit their needs. (3) Urbanization began as the need for farming decreased and former farmers moved into villages, many of which gradually grew into towns and eventually into cities. (4) Specialized occupations and long-distance trade developed as former farmers in villages and towns learned crafts such as weaving, toolmaking, and pottery to produce handmade goods that could be exchanged for food.

The trade in food and manufactured goods made possible by the agriculture-based urban societies created wealth and the need for a managerial class to regulate the distribution of goods, services, and land. As ownership of land and water rights became a valuable economic resource, conflict increased. Armies and war leaders rose to power and took over large areas of land. A new class of powerless people, the slaves, minorities, and landless peasants, were forced to do the hard, disagreeable work of producing food.

Impact on the Environment The rise of agriculture-based urban societies created an environmental impact far exceeding that of hunting-and-gathering societies and early subsistence farmers. Forests were cut down and grasslands were plowed up to provide vast areas of cropland and grazing land to feed the growing populations of these emerging civilizations and to provide wood for fuel and for buildings to serve the growing number of city dwellers. Such massive land clearing destroyed and altered the habitats of many forms of plant and animal wildlife, endangering their existence and in some cases causing or hastening their extinction.

Poor management of many of the cleared areas led to greatly increased deforestation, soil erosion, and overgrazing of grasslands by huge herds of sheep, goats, and cattle—helping convert once fertile land to desert. The silt that washed off these denuded areas polluted streams, rivers, lakes, and irrigation canals, making them useless. The concentration of large numbers of people and their wastes in cities helped spread infectious human diseases and parasites. The gradual degradation of the vital resource base of soil, water, forests, grazing land, and wildlife was a major factor in the downfall of many once great civilizations.

The development and gradual spread of agriculture meant that most of the earth's population shifted from life as hunter-gatherers *in nature* to life as shepherds, farmers, and urban dwellers *against nature*. For the first time, human beings saw themselves as distinct and apart from the rest of nature, and they created the concepts of "wild" and "wilderness" to apply to animals, plants, and parts of the earth not under their control.

2-3 INDUSTRIAL SOCIETIES: THE INDUSTRIAL REVOLUTION

Early Industrial Societies The next major cultural change, known as the Industrial Revolution, began in England in the mid-18th century. Human inventiveness led to discoveries of how to use the chemical energy stored in fossil fuels such as coal and later oil and natural gas. The invention of the coal-burning steam engine (1765) was followed by the the steamship (1807), the steam locomotive (1829), and the internal combustion engine (various versions, 1860–92), which allowed carriers running on fossil fuels to replace horse-drawn vehicles and wind-powered ships. Within a few decades, these innovations transformed agriculture-based urban societies in western Europe and North America into even more urbanized early industrial societies.

These societies and the more advanced ones that followed greatly increased the average energy resource use per person. The use of fossil-fuel-powered machines in agriculture greatly reduced the number of people needed to produce food and increased the number of former farmers migrating from rural to urban areas. Many found jobs in the growing number of mechanized factories, where they worked long hours for low pay and were usually subjected to noisy, dirty, and hazardous working conditions.

Advanced Industrial Societies After World War I (1914–18), the greatly increased use of fossil fuels, the development of more efficient machines and techniques for mass production, and advances in science and technology led to the development of today's advanced industrial societies in the United States and other MDCs. These societies have provided a number of important benefits for most people living in them: (1) the creation and mass production of many useful and economically affordable products; (2) significant increases in the standard of living in terms of average GNP per person; (3) a sharp increase in average agricultural productivity per person as a result of advanced industrialized agriculture, in which a small number of individual and corporate farmers produce large outputs of food; (4) a sharp rise in average life expectancy as a result of improvements in sanitation, hygiene, nutrition, medicine, and birth control; and (5) a gradual decline in the rate of population growth from simultaneous decreases in average birth rates and death rates as a result of improvements in health, birth control, education, average income, and old-age security.

Environmental Impact Along with the many benefits, advanced industrialized societies have intensified many existing environmental problems and created a series of new ones, such as air, water, and soil pollution from DDT, radioactive wastes, and a host of hazardous synthetic substances. These and other environmental problems now threaten people's well-being not only at the local level but also at the regional level (acid deposition) and the global level (buildup of carbon dioxide caused by the burning of fossil fuels).

By decreasing the number of people engaged in food production, industrialized agriculture has led to massive shifts of population from rural to urban areas and has helped intensify a variety of social, political, economic, and environmental problems. The combination of industrialized agriculture, increased mining, and urbanization has also led to increased degradation of topsoil, forests, grasslands, and wildlife—the same problems that helped lead to the downfall of earlier civilizations. Because of these numerous, interrelated, harmful effects, industrialized societies must now use an increasingly larger fraction of their financial, human, and natural resources for control of pollution and land use to prevent environmental overload and degradation.

The benefits of the Industrial Revolution are so great that very few people would propose that we abandon most of the technological achievements of the past two and a half centuries. However, by enabling human beings to have much greater control over nature and decreasing the number of people who live close to the land, the Industrial Revolution encouraged people to view themselves as apart from and superior to nature.

2-4 BRIEF HISTORY OF RESOURCE EXPLOITATION, CONSERVATION, AND ENVIRONMENTAL PROTECTION IN THE UNITED STATES

Phase I: Resource Exploitation and Early Conservation Warnings (1607–1870) When European colonists began settling in North America in 1607, they found a vast continent with what appeared to be inexhaustible supplies of timber, fertile soil, water, minerals, and other resources for their own use and for export to Europe. Such abundance, coupled with the frontier view that most of the continent was a hostile wilderness to be conquered, opened up, cleared, and

Figure 2-2 Some early American conservationists.

Henry David Thoreau
1817–1862

George P. Marsh 1801–1882

John Muir 1838–1914

Gifford Pinchot 1865–1946

Theodore Roosevelt 1858–1919

Aldo Leopold 1886–1948

exploited, led to considerable resource waste and little regard for future needs.

By the mid-1800s, 80% of the total land area of the United States was government owned, mostly acquisitions that ignored the rights of members of various Native American tribes, or nations, who had lived on these lands for centuries. By 1900, more than half of this land had been given away or sold at nominal cost to railroad, timber, and mining companies, land developers, states, schools, universities, technical schools, and homesteaders to encourage settlement and economic development of the continent and thus strengthen it against its enemies. By artificially depressing the prices of resources, these land transfers encouraged widespread exploitation, waste, and degradation of the country's forests, grasslands, and minerals.

Between 1830 and 1870, a number of early conservationists such as George Catlin, Horace Greeley, Ralph Waldo Emerson, Henry David Thoreau, Frederick Law Olmsted, Charles W. Eliot, and George Perkins Marsh (Figure 2-2) warned that America's timber and grassland resources were being exploited at an alarming rate through overgrazing, deforestation, and general misuse. They proposed that part of the land be withdrawn from public use and preserved in the form of national parks.

Most of the early conservationists, because they were intellectuals whose writings did not reach popular audiences, had relatively little influence on politicians. In 1864, however, George P. Marsh, a scientist and congressman from Vermont, published a book, *Man and Nature* (see Further Readings), that eventually proved to be influential in the debate over conservation versus exploitation. Marsh questioned the prevailing idea of the inexhaustibility of the country's resources, showed how the rise and fall of past civilizations were linked to their use and misuse of their resource base, and set forth basic conservation principles still used today.

Phase II: Beginnings of the Government's Role in Resource Conservation (1870–1910) In 1872 President Ulysses S. Grant signed an act designating over 8,100 square kilometers (2 million acres) of forest in northeastern Wyoming for preservation as Yellowstone National Park. This action was the world's first instance of large-scale forest preservation in the public interest and marked the beginning of the first wave of resource conservation in the United States.

In 1891 Congress designated Yellowstone Timberland Reserve—the first federal forest reserve—and authorized the president to set aside additional federal lands to ensure future availability of adequate timber. Between 1891 and 1897, Presidents Benjamin Harrison and Grover Cleveland withdrew millions of acres of public land, located mostly in the West, from timber cutting. Powerful political foes—especially Westerners accustomed to using these public lands as they pleased—called these actions undemocratic and un-American.

Effective protection of the national forest reserves did not exist, however, until 1901, when Theodore Roosevelt became president. He tripled the size of the forest reserves and transferred administration of them from the Department of the Interior, which had a reputation for lax enforcement, to the Department of Agriculture. In 1905 Congress created the U.S. Forest Service to manage and protect the forest reserves, and Roosevelt appointed Gifford Pinchot as its first chief (Figure 2-2). Pinchot pioneered efforts to manage these renewable forest resources according to the principles of *sustained yield* (Section 1-2) and *multiple use,* by which forests are to be used for a variety of purposes, including timbering, recreation, grazing, wildlife conservation, and water conservation.

Roosevelt also used the Lacey Act, passed by Congress in 1906, to protect several million acres of public land as national monuments, many of which—the Grand Canyon, for instance—would later become part of the National Park System. He also began the National Wildlife Refuge System in 1903 by designating Pelican Island off the east coast of Florida as a federal wildlife refuge.

Phase III: Preservation versus Scientific Conservation (1911–32) After 1910 the conservation movement split into two schools of thought: preservationists and scientific conservationists. **Preservationists,** with their roots in philosophy and aesthetics, emphasized preserving and protecting public lands from mining, timbering, and other forms of development by establishing parks, wilderness areas, and wildlife refuges whose beauty and wealth could be used and enjoyed by present generations and passed on unspoiled to future generations.

Preservationists were led by California nature writer John Muir, who founded the Sierra Club in 1892, and, after Muir's death in 1914, by forester Aldo Leopold (Figure 2-2). According to Leopold, the role of the human species should be that of a member, citizen, and protector of nature—not its conqueror. Another ardent and effective supporter of wilderness preservation was Robert Marshall, an officer in the U.S. Forest Service, who together with Leopold founded the Wilderness Society in 1935. Others, who have led preservationist efforts in more recent years, include David Brower, former head of the Sierra Club and founder of Friends of the Earth, Ernest Swift, and Stewart L. Udall.

Scientific conservationists saw public lands as resources to be used now to enhance economic growth and national strength and to be protected from depletion by being managed for sustained yield and multiple use. The scientific conservationists were led by Roosevelt, Pinchot, John Wesley Powell, Charles Van Hise, and others. Pinchot angered preservationists, who had been active allies in Roosevelt's conservation efforts, when he stated his principle of the wise use of a resource:

The first great fact about conservation is that it stands for development. There has been a fundamental misconception that conservation means nothing but the husbanding of resources for future generations. There could be no more serious mistake. . . . The first principle of conservation is the use of the natural resources now existing on this continent for the benefit of the people who live here now.

In 1912 Congress created the U.S. National Park System and in 1916 passed legislation declaring that national parks are set aside to conserve and preserve scenery, wildlife, and natural and historic objects for the use, observation, health, and pleasure of people and are to be maintained in a manner that leaves them unimpaired for future generations. The same law established the National Park Service within the Department of the Interior to manage the system, which by then included 16 national parks and 21 national monuments, most of them in the western states. The Park Service's first director, Stephen Mather, recruited a corps of professional park rangers to manage the parks.

Phase IV: Expanding Federal Role in Land Management (1933–60) The second wave of national resource conservation began during the early 1930s,

Figure 2-3 CCC crew in 1937, cleaning up woodlands, removing dead trees, pruning live ones.

as President Franklin D. Roosevelt attempted to get the country out of the Great Depression (1929–41). To provide jobs for 2 million unemployed young men, he established the Civilian Conservation Corps (CCC), which worked on conservation projects such as planting trees, developing parks and recreation areas, restoring silted waterways, providing flood control, controlling soil erosion, and protecting wildlife (Figure 2-3). In 1933 the Soil Erosion Service under the Department of Agriculture was created to correct some of the erosion problems that contributed to the Depression. In 1935 it was renamed the Soil Conservation Service and Hugh H. Bennett became its first director.

For many decades, public lands, especially in the West, where they were subject to periodic drought, had been heavily overgrazed because of ranchers' ignorance and greed. The Taylor Grazing Act of 1934 placed 80 million acres of public land into grazing districts to be managed jointly by the Grazing Service, established within the Department of the Interior and committees of local ranchers. This law also established a system of fees for use of federal grazing lands and placed limits on the number of animals that could be grazed.

From the start, however, ranchers resented government interference with their long-established unregulated use of public land. Since 1934 they have led repeated efforts to have these lands removed from government ownership and turned over to private ranching, mining, timber, and development interests. Until 1976, western congressional delegations kept the Grazing Service (which in 1946 became the Bureau of Land Management, or BLM) so poorly funded and staffed, and without enforcement authority, that many

ranchers and mining and timber companies openly continued to misuse western public lands.

Between 1940 and 1960, there were relatively few new developments in federal conservation policy, because of World War II and the economic recovery from the war.

Phase V: Rise of the Environmental Movement (1960–80) During the early 1960s, a third wave of national resource conservation began during the short administration of John F. Kennedy (1961–63) and expanded under the administration of Lyndon B. Johnson between 1963 and 1969.

In 1962 biologist Rachel Carson published *The Silent Spring* (see Further Readings), a book that described pollution of air, water, and wildlife from the widespread use of persistent pesticides such as DDT. The book helped broaden the concept of resource conservation to include the preservation of the *quality* of the air, water, and soil, which were under assault by a country experiencing rapid economic growth. This marked the beginnings of what is now known as the *environmental movement* in the United States.

In 1964 Congress passed the Wilderness Act, which authorizes the government to protect undeveloped tracts of public land from development as part of the National Wilderness System unless Congress later decides they are needed for the national good. Between 1965 and 1970, extensive media coverage of pollution and ecology and the popular writings of biologists such as Paul Ehrlich, Barry Commoner, and Garrett Hardin helped the public become aware of the interlocking relationships between population growth, resource use, and pollution (Section 1-4). As a result, between 1970 and 1980, sometimes called the "environmental decade," more than two dozen separate pieces of legislation (Appendix 3) were passed to protect the air, water, land, and wildlife.

In 1970 President Richard M. Nixon used his administrative powers to create the Environmental Protection Agency (EPA), empowered to determine environmental standards and enforce federal environmental laws. William D. Ruckelshaus was appointed its first director.

The accomplishments that took place between 1965 and 1980 represent a fourth wave of national resource conservation, expanding from protection of public land resources to protection of shared air and water resources. During this period, citizen-supported environmental organizations such as the Sierra Club, the Wilderness Society, the National Wildlife Federation, Friends of the Earth, the Environmental Defense Fund, and the Natural Resources Defense Council lobbied for better protection and management of public lands. They also began suing the government to secure enforcement of environmental laws. In addition, private organizations such as the Nature Conservancy and the Audubon Society accelerated their efforts to buy and protect unique areas of land threatened by development.

Oil embargoes in the mid-1970s dramatically demonstrated the need for effective conservation of energy resources, especially oil, and mineral resources that are found, extracted, processed, and converted into manufacutured goods by the use of large amounts of fossil fuels. Between 1977 and 1981, President Jimmy Carter created the Department of Energy and appointed a number of experienced administrators to key posts in the EPA and the Department of the Interior, drawing heavily on established environmental organizations for such appointees and for advice on policy. He also created the $1.6 billion Superfund to clean up toxic waste sites. Just before leaving office, Carter tripled the amount of land in the National Wilderness System and doubled the area under the administration of the National Park Service, primarily by adding vast tracts of land in Alaska.

Phase VI: Continuing Controversy and Some Retrenchment (the 1980s) The Federal Land Policy and Management Act of 1976 gave the Bureau of Land Management its first real authority to manage the public lands, mostly in the West, under its control. This angered western ranchers, farmers, miners, users of off-road motorized vehicles, and others, who had been doing pretty much as they pleased on BLM lands.

In the late 1970s, western ranchers, who had been paying low fees for grazing rights that encouraged overgrazing, launched a political campaign that came to be known as the "Sagebrush Rebellion." The goals of this campaign, like earlier attempts since the 1930s, were to transfer most western public lands (including national forests) from federal ownership to state ownership and then to influence state legislatures to sell or lease the resource-rich lands at low prices to ranching, mining, timber, land development, and other corporate interests. Three leaders in this campaign were James G. Watt, Robert Burford, and Anne Gorsuch (who later married Burford), who in 1981 were appointed by President Ronald Reagan to head, respectively, the Department of the Interior, the Bureau of Land Management, and the Environmental Protection Agency.

Environmentalists and conservationists have helped resist this rebellion for decades. They have argued that resources on public lands owned jointly by all Americans should not be sold or leased to private interests at below average market prices, because this provides subsidies and increased profits for a few influential corporations and individuals at the taxpayers' expense. They have also noted that legislatures of the western states have long been dominated by cor-

porations and have some of the country's weakest conservation policies, poorest environmental records, and most corrupt histories in the management of public lands. Indeed, departments and agencies such as the Interior Department, the EPA, and the BLM were set up primarily because most *states* had failed to protect commonly shared land, air, and water resources from exploitation and abuse by private interests.

In 1981 Ronald Reagan, a declared sagebrush rebel, became president. He had won the election by a substantial margin, campaigning as an advocate of a strong national defense, less federal government control, and reduced government spending to lower the national debt and help combat the economic recession that had followed the sharp rises in oil prices during the 1970s. Since 1981 considerable controversy has been generated by his environmental policies (see Spotlight on p. 25).

Because of strong opposition by Congress, public outcry, and legal challenges, many of the Reagan administration's attempts to set back earlier environmental progress were thwarted. Although enforcement had been weakened, the major environmental legislation passed during the 1970s remained intact. Two of Reagan's key environmental appointees, James G. Watt as head of the Department of the Interior and Anne Gorsuch as head of the EPA, were forced to resign before the end of Reagan's first term. Although the full-scale attack on the environmental movement was blunted, most environmentalists and conservationists consider the period between 1981 and 1988 a time of retrenchment in the country's efforts to protect its resource base.

2-5 SOME POSSIBLE FUTURES: THE NEXT CULTURAL CHANGE

A Critical Turning Point Our numbers are now so large and our forms of technology so powerful that we are at a critical turning point in human cultural development—a new hinge of history. Signs are growing that over the next 50 to 75 years we will undergo another major cultural change to one of several possibilities (Figure 2-4): (1) a series of *sustainable-earth societies* throughout the world based on people learning how to work *with* nature to preserve our life-support systems (the Neo-Malthusian dream); (2) a series of *superindustrialized societies* based on major advances in technology that allow people even greater control over nature (the cornucopian dream); or (3) a small number of people in scattered bands trying to survive as *modern hunter-gatherers* in a world polluted and depleted of resources as a result of global nuclear war or excessive industrialization and population

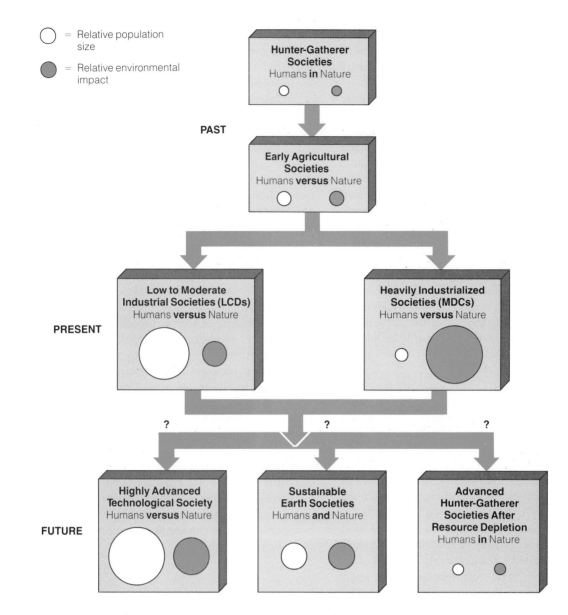

Figure 2-4 Major past cultural changes and some possibilities for the next cultural change.

growth without sufficient resource conservation and environmental protection.

Some Hopeful Signs   Today in the MDCs, awareness of the global problems of population, pollution, and resource depletion is widespread. This knowledge is spreading rapidly to the LDCs. In addition to seeing themselves as citizens of a particular country, a growing number of people also consider themselves *citizens of the earth*, whose primary loyalty is to protect and restore the natural systems that support all life on this planet.

On April 22, 1970, the first annual Earth Day in the United States, 20 million people in more than 2,000 communities took to the streets to demand better environmental quality. Today polls show that 70%

of Americans are more concerned about the environment today than at any time in the past, and 65% indicate they favor greater protection of the environment even if it costs them 10% more in taxes and cost of living.

Today more than 4,000 organizations worldwide are devoted to environmental issues. By 1987, environmental protection agencies had been established in 151 of the world's 166 countries—including 112 LDCs, compared to only 11 LDCs in 1972. The U.S. concept of setting aside land for protection in national parks, wilderness areas, and wildlife refuges has been adopted by more than 100 other countries, including many LDCs.

Most MDCs have passed laws designed to protect the air, water, land, wildlife, and public health, although enforcement is sometimes lax. More than 80 federal

Between 1982 and 1985, several of the country's major environmental groups studied President Reagan's environmental administrative actions and legislative proposals in detail (see Further Readings). Some of their major charges were that the administration

- Appointed people who came from industries or legal firms that opposed existing federal environmental and land-use legislation and policies to key positions in the Interior Department, EPA, and the BLM—something like putting foxes in charge of the henhouse.

- Barred established environmental and conservation organizations and leaders from having any input into such appointments and into the administration's environmental policies.

- Made enforcing existing environmental laws difficult by encouraging drastic budget and staff cuts in environmental programs under the guise of reducing spending—while adding far more to the federal deficit than any president in history.

- Greatly increased energy and mineral development and cutting of timber by private enterprise on public lands, often sold at giveaway prices to large corporations, thus depriving taxpayers of funds that could have been used to reduce the federal deficit or to prevent cuts in environmental programs.

- Gave ranchers excessive influence over how public rangelands should be used and managed.

- Increased the federal budget for nuclear power, which is still not economically feasible even with taxpayer subsidies of over $32 billion and which would never have been developed if it had been forced to compete with other energy alternatives in a free market.

- Drastically cut the budget for energy conservation and the development of renewable energy from the sun and wind with the rationale that these energy alternatives (unlike nuclear energy) should be developed under free-market competition.

The Reagan administration contended that

- It is normal for a newly elected president to appoint key personnel who wish to see the president's policies carried out.

- Drastic budget and staff cuts were necessary to reduce waste and help decrease the mounting national debt.

- Increased energy and mineral exploration and sale of timber on federal lands are necessary to encourage private enterprise, stimulate economic growth, and improve national security by ensuring that the country will have sufficient resources.

- It is better to return decision making about public rangelands to ranchers and to transfer ownership of some public lands to states and private ownership, because federal bureaucrats are inept managers.

- The federal government should continue to subsidize nuclear power, because it is a safe and proven technology that can provide the country with much-needed electricity in the future.

- The budget for energy conservation and alternative renewable energy sources should be decreased so that these emerging technologies can be developed by private enterprise under free-market competition.

What do you think?

protection laws have been enacted in the United States since 1970 (see Appendix 3). By 1987, government and industry in the United States were spending $70 billion a year—an average of $293 a year for each American—to reduce pollution.

These efforts have paid off. Since 1965, as many as 75 U.S. rivers, lakes, and streams have been cleaned up. Restrictions against the dumping of untreated sewage, garbage, and toxic chemicals into lakes, rivers, and oceans have helped arrest a serious decline in the quality of the country's surface water. About 3,600 of the country's 4,000 major industrial water polluters are meeting cleanup deadlines. Twenty major

U.S. cities now have measurably cleaner air than before the passage of the Clean Air Act of 1970, and about 90% of major factories are in compliance with federal air pollution regulations. The overall rate of improvement in air quality, however, did slow down and in some cases deteriorated between 1982 and 1988, presumably because of budget cuts in environmental protection and enforcement since 1981.

Other industrialized countries have also made significant progress in pollution control. Smog in London has decreased sharply since 1952, and the Thames River is returning to life. Japan, once regarded as the most polluted country in the world, has dramatically reduced air pollution in most of its major cities and has upgraded the quality of its waters since passing antipollution laws in 1967. The Japanese environment is still highly degraded, however, partly because the country's small size relative to its large population means that most of the people live in crowded cities, thus creating centers of concentrated pollution.

When energy prices rose in the 1970s, many Americans began to realize that energy conservation is our cheapest, least environmentally harmful energy option and the best and quickest way to reduce excessive dependence on the world's rapidly diminishing supply of oil (Chapters 15 and 17). Since the oil embargo in 1973, the United States has derived 100 times more energy from energy conservation than from oil, coal, and nuclear power combined.

However, there is still a long way to go. A temporary glut of oil in the 1980s has led some drivers to return to bigger cars with poor gas mileage and has encouraged cutbacks in funds and efforts for energy conservation and development of renewable energy alternatives such as solar and wind energy. Most American-made cars still get much poorer gas mileage than most of those made in Japan and Europe. Most homes and buildings are still underinsulated. Although household appliances that use half the electricity of 1970 models are available, relatively few homes have them. Although using nuclear energy to produce electricity is now considered uneconomic by most financial experts, the government still uses taxpayer dollars to provide massive subsidies for nuclear power research and tax breaks to utilities to offset sizable losses from building nuclear plants (see Chapter 16).

With regard to population, the good news is that between 1970 and 1987, the annual growth of the world's population slowed from an all-time high of 2.1% to 1.7% (Section 1-1). Despite this encouraging trend, United Nations population experts project that the world will probably not reach zero population growth until 2100, with a population of 10.4 billion. The rate of population growth in the United States has slowed significantly, with the average number of live children born per woman decreasing from 3.8 in 1957 to 1.8 in 1987. If this trend persists, the United States could reach ZPG by 2020 or sooner, depending on the annual addition of legal and illegal immigrants.

To some, the amazing thing is not the lack of progress in dealing with environmental and resource problems in many parts of the world but that so much has been done since 1965. Nevertheless, we cannot afford to settle back and rest on our progress to date. Envi-

ronmentalists must constantly struggle to see that existing environmental laws (Appendix 3) are enforced and to prevent them from being weakened.

Moreover, many serious environmental problems have only recently been identified: leaking hazardous-waste dumps that can poison wells and groundwater supplies, acid deposition that can kill trees and aquatic life, rapid depletion of the world's remaining tropical forests, depletion of stratospheric ozone that protects life on earth from the sun's harmful ultraviolet radiation, and possible long-term changes in the earth's climate as a result of carbon dioxide and other gases released into the air by human activities.

We found our house—the planet—with drinkable, potable water, with good soil to grow food, with clean air to breathe. We at least must leave it in as good a shape as we found it, if not better.

Rev. Jesse Jackson

DISCUSSION TOPICS

1. Those wishing to avoid dealing with environmental problems sometimes argue, "People have always polluted and despoiled this planet, so why all the fuss over ecology and pollution? We've survived so far." Identify the core of truth in this position and then discuss its serious deficiencies.

2. Make a list of the major benefits and drawbacks of an advanced industrial society such as the United States. Do you feel that the benefits of such a society to its citizens outweigh its drawbacks? Explain. What are the alternatives? Do you feel that the benefits of such a society to people in other parts of the world outweigh its drawbacks? Explain.

3. Explain how various cultural changes in human societies have led to the environmental and resource problems we face today.

4. Some analysts believe that continued economic growth and technological innovation in today's industrial societies offer the best way to solve the environmental problems we face. Others believe that these problems can be dealt with only by changing from a predominantly industrial society to a sustainable-earth society over the next 50 to 75 years. Which position do you support? Why?

5. Do you believe that a cultural change to a sustainable-earth society is possible over the next 50 to 75 years? What changes, if any, have you already made and what changes do you plan to make in your lifestyle to promote such a society?

PART TWO

Basic Concepts

USDI/National Park Service/George Grant

Some Environmental Principles

1. *Everything must go somewhere; there is no away (law of conservation of matter).*

2. *You can't get something for nothing; there is no free lunch (first law of energy, or law of conservation of energy).*

3. *You can't even break even; if you think things are mixed up now, just wait (second law of energy).*

4. *Everything is connected to everything else, but how?*

5. *A thing is right when it tends to preserve the integrity, stability, and beauty of the biotic community. It is wrong when it tends otherwise.*

6. *Natural systems can take a lot of stress and abuse, but there are limits.*

7. *In nature you can never do just one thing, so always expect the unexpected; there are numerous effects, often unpredictable, to everything we do.*

3

Matter and Energy Resources: Types and Concepts

1. What are the major physical and chemical forms of matter and what is matter made of?

2. What physical law governs changes of matter from one physical or chemical form to another?

3. What are the three major types of nuclear changes that matter can undergo and how dangerous to humans is the radioactivity released by such changes?

4. What are the major types of energy?

5. What two physical laws govern changes of energy from one form to another?

6. How can the two physical laws governing changes in energy from one form to another be used to help us evaluate present and future sources of energy?

7. How are all three physical laws governing changes of matter and energy from one form to another related to resource use and environmental disruption?

The laws of thermodynamics control the rise and fall of political systems, the freedom or bondage of nations, the movements of commerce and industry, the origins of wealth and poverty, and the general physical welfare of the human race.

Frederick Soddy, Nobel laureate, chemistry

This book, your hand, the water you drink, and the air you breathe are all samples of *matter*—the stuff all things are made of. The light and heat streaming from a burning lump of coal and the force you must use to lift this book are examples of *energy*—what you and all living things use to move matter around, change its form, or cause a heat transfer between two objects at different temperatures.

All transformations of matter and energy from one form to another are governed by certain natural or physical scientific laws, which, unlike the social laws people enact, cannot be broken. This chapter begins our study of basic concepts with a look at the major types of matter and energy and the scientific laws governing all changes of matter and energy from one form to another. These laws are used throughout this book to help you understand many environmental problems and evaluate proposed solutions.

3-1   MATTER: FORMS AND STRUCTURE

Physical and Chemical Forms of Matter   Anything that has mass (or weight on the earth's surface) and takes up space is **matter.** All matter found in nature can be viewed as being organized in identifiable patterns, or levels of organization (Figure 3-1). This section is devoted to a discussion of the three lowest levels of organization of matter—subatomic particles, atoms, and molecules—that make up the basic components of all higher levels. Chapter 4 discusses the five higher levels of organization of matter—organisms, populations, communities, ecosystems, and the biosphere—that are the major concern of ecology.

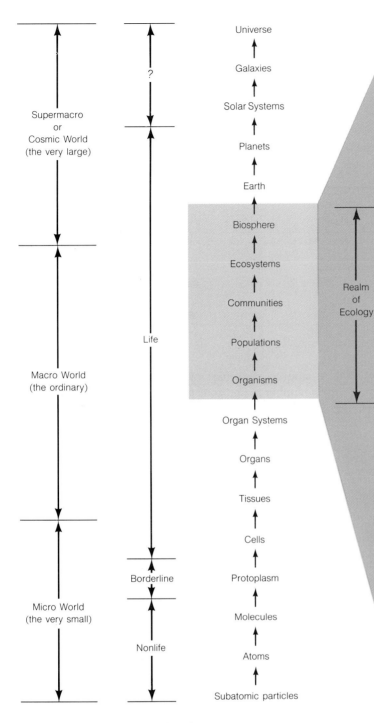

Figure 3-1 Levels of organization of matter.

Any matter, such as water, can be found in three *physical forms:* solid (ice), liquid (liquid water), and gas (water vapor). All matter also consists of *chemical forms:* elements, compounds, or mixtures of elements and compounds. Any element or compound needed in large or small amounts for the survival, growth, and reproduction of a plant or animal is called a **nutrient.**

Elements Distinctive forms of matter that make up every material substance are known as **elements.** Examples of these basic building blocks of all matter include hydrogen (represented by the symbol H), carbon (C), oxygen (O), nitrogen (N), phosphorus (P), sulfur (S), chlorine (Cl), sodium (Na), and uranium (U).

Figure 3-2 Isotopes of hydrogen and uranium.

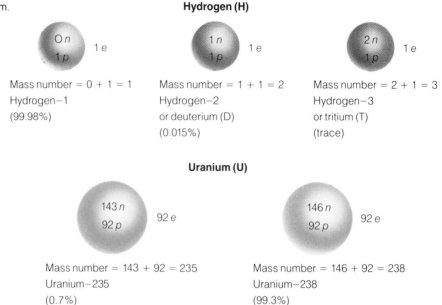

Hydrogen (H)

Mass number = 0 + 1 = 1
Hydrogen−1
(99.98%)

Mass number = 1 + 1 = 2
Hydrogen−2
or deuterium (D)
(0.015%)

Mass number = 2 + 1 = 3
Hydrogen−3
or tritium (T)
(trace)

Uranium (U)

Mass number = 143 + 92 = 235
Uranium−235
(0.7%)

Mass number = 146 + 92 = 238
Uranium−238
(99.3%)

All elements are composed of an incredibly large number of distinctive types of minute particles called **atoms.** Some elements, such as the nitrogen and oxygen gases making up about 99% of the volume of air we breathe, consist of **molecules** formed when two or more atoms of the same element combine in fixed proportions. For example, two atoms of nitrogen (N) can combine to form a nitrogen molecule with the shorthand chemical formula N_2 (read as "N-two"), where the subscript following the symbol for an element indicates the number of atoms of that element in a molecule. Similarly, most of the oxygen gas in the atmosphere exists as O_2 molecules, although a small amount found mostly in the upper atmosphere (stratosphere) exists as ozone molecules with the formula O_3.

All atoms in turn are made up of even smaller **subatomic particles:** protons, neutrons, and electrons. Each atom of an element can be viewed as having a characteristic internal structure consisting of a tiny center, or **nucleus,** containing a certain number of positively charged **protons** (represented by the symbol p) and uncharged **neutrons** (n), and one or more negatively charged **electrons** (e) whizzing around somewhere outside the nucleus. Each atom of the same element always has the same number of positively charged protons inside its nucleus and an equal number of negatively charged electrons outside its nucleus, so that the atom as a whole has no net electrical charge. For example, each atom of the lightest element, hydrogen, has one positively charged proton in its nucleus and one negatively charged electron outside. Each atom of a much heavier element, uranium, has 92 protons and 92 electrons (Figure 3-2).

Because an electron has an almost negligible mass (or weight) compared to a proton and a neutron, the approximate mass of an atom is determined by the number of neutrons plus the number of protons in its nucleus. This number is called its **mass number.** Although atoms of the same element must have the same number of protons and electrons, they may have different numbers of uncharged neutrons in their nuclei, and thus different mass numbers. These different forms of the same element with different mass numbers or numbers of neutrons in their nuclei are called **isotopes.**

Isotopes of the same element are identified by appending the mass number to the name or symbol of the element: hydrogen-1, or H-1; hydrogen-2, or H-2 (common name, deuterium), and hydrogen-3, or H-3 (common name, tritium) (Figure 3-2). A natural sample of an element contains a mixture of its isotopes in a fixed proportion or percent of abundance (Figure 3-2).

Atoms of some elements can lose or gain one or more electrons to form **ions:** atoms or groups of atoms with one or more net positive (+) or negative (−) electrical charges. The charge is shown as a superscript after the symbol for an atom or group of atoms. Examples of positive ions are sodium ions (Na^+) and ammonium ions (NH_4^+). Common negative ions are chloride ions (Cl^-), nitrate ions (NO_3^-), and phosphate ions (PO_4^{3-}).

Compounds Most matter exists as **compounds:** combinations of two or more atoms (*molecular compounds*) or oppositely charged ions (*ionic compounds*) of two or more different elements held together in fixed proportions by chemical bonds. Water, for example, is a molecular compound composed of H_2O (read as "H-two-O") molecules, each consisting of two hydrogen atoms bonded to an oxygen atom. Other molecular compounds you will encounter in this book are nitric oxide (NO); carbon monoxide (CO); carbon dioxide (CO_2); nitrogen dioxide (NO_2); sulfur dioxide (SO_2); ammonia (NH_3); methane (CH_4), the major component of natural gas; and glucose ($C_6H_{12}O_6$), a sugar that most plants and animals break down in their cells to obtain energy. Sodium chloride, or table salt, is an ionic compound consisting of a network of formula units of oppositely charged ions (Na^+Cl^-) held together by the forces of attraction that exist between opposite electric charges.

3-2 LAW OF CONSERVATION OF MATTER AND CHANGES IN MATTER

There Is No "Away" The earth loses some gaseous molecules to space and gains small amounts of matter from space, mostly in the form of meteorites. However, because these overall losses and gains of matter are extremely small compared to earth's total mass, *the earth has essentially all the matter it will ever have.*

Although people talk about consuming or using up material resources, we don't actually consume any matter. We only use some of the earth's resources for a while—taking materials from the earth, carrying them to another part of the globe, processing, using, and then discarding, reusing, or recycling them. In this process we may change various elements and compounds from one physical or chemical form to another, but in every case we neither create from nothing nor destroy to nothingness any measurable amount of matter.

This information, based on many thousands of measurements of matter being changed from one physical or chemical form to another, is expressed in the **law of conservation of matter:** When matter is changed from one physical or chemical form to another, no measurable amount of matter is created or destroyed. In other words, in such changes we can't create or destroy any of the atoms involved. All we can do is rearrange them into different patterns.

This law means that there is no "away." *Everything we think we have thrown away is still here with us, in one form or another.* We can collect dust and soot from the smokestacks of industrial plants, but these solid wastes must then go somewhere. We can collect garbage and remove solid grease and sludge from sewage, but these substances must either be burned (perhaps causing air pollution), dumped into rivers, lakes, and oceans (perhaps causing water pollution), or deposited on the land (perhaps causing soil pollution and water pollution). We can certainly make the environment cleaner

and convert some potentially harmful chemicals to less harmful or even harmless physical or chemical forms. But the law of conservation of matter means that we will always be faced with pollution of some sort.

Physical and Chemical Changes Elements and compounds can undergo **physical changes** in which their chemical composition is not changed and **chemical changes,** or **chemical reactions,** in which they are changed into new combinations of the starting elements or compounds. Any physical or chemical change either gives off energy or requires energy, usually in the form of heat.

Two examples of physical changes are melting and boiling, in which an element or a compound is changed from one physical state to another. For example, when ice is melted or water is boiled, none of the H_2O molecules involved are altered; instead, they are organized in different spatial patterns. In solid ice, the water molecules are close to one another in a highly ordered crystal. In water vapor, the water molecules are relatively far apart and in continuous chaotic motion.

A chemical change can be represented in shorthand form by an equation using the chemical formulas for the elements and compounds involved. Formulas of the original starting chemicals, called *reactants*, are placed to the left and formulas of the new chemicals produced, called *products*, are placed to the right; an arrow ($\rightarrow$) is used to indicate that a chemical change has taken place.

For example, when coal (which is mostly carbon, or C) burns, it combines with oxygen gas (O_2) in the atmosphere to form the gaseous compound carbon dioxide (CO_2). In this case energy is given off, explaining why coal is a useful fuel. This chemical change can be represented in the following manner:

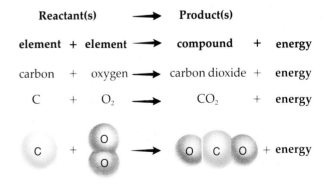

Reactant(s)	$\longrightarrow$	**Product(s)**		
element + element	$\longrightarrow$	compound	+	energy
carbon + oxygen	$\longrightarrow$	carbon dioxide	+	energy
C + O_2	$\longrightarrow$	CO_2	+	energy

A **balanced chemical equation** is one that contains the same number of atoms of each element on each side in accordance with the law of conservation of matter. Notice that the equation for the burning of carbon has one carbon atom and two oxygen atoms

Figure 3-3 The three major types of ionizing radiation emitted by radioactive isotopes vary considerably in their penetrating power.

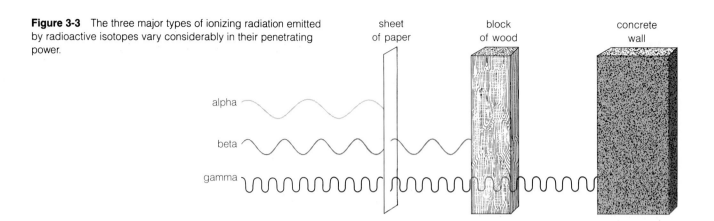

on each side and thus is balanced. This balanced equation shows how the burning of coal or any carbon-containing compound, such as those in wood, natural gas (CH_4), oil, and gasoline, adds carbon dioxide gas to the atmosphere, possibly causing warming of the atmosphere through the greenhouse effect (Section 18-5).

Air contains about 79% nitrogen gas (N_2) and 20% oxygen gas (O_2). When any fuel is burned in air at high temperatures, some of the N_2 and O_2 combine to form molecules of nitric oxide (NO) gas. Without adequate pollution control devices, this oxide of nitrogen spews out of smokestacks, chimneys, and automobile exhaust pipes and is an ingredient in the type of smog found in cities such as Los Angeles.

The equation for the formation of NO from N_2 and O_2 shows that equations are not balanced simply by writing the formulas for the elements and compounds involved:

$$N_2 + O_2 + \textbf{energy} \rightarrow NO \ (\textit{unbalanced})$$

This equation is unbalanced because there are two nitrogen atoms on the left and only one on the right and two oxygen atoms on the left and only one on the right. To balance this equation so that it does not violate the law of conservation of matter, the same number of atoms of each element is needed on each side. This is achieved by forming two molecules of NO. This is indicated by placing the number 2 in front of the formula for NO:

element + element + energy ⟶ compound

$$N_2 \ + \ O_2 \ + \ \textbf{energy} \longrightarrow 2NO \ (\textit{balanced})$$

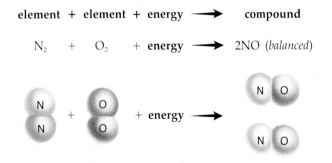

You do not need to know how to balance chemical equations to understand the material in this book. This concept has been shown so that you will understand the shorthand numbers and symbols involved for the small number of other important chemical changes cited in this book and to illustrate the meaning of the law of conservation of matter.

Nuclear Changes and Radioactivity The nuclei of certain isotopes are unstable and can undergo **nuclear changes,** changing into one or more different isotopes by altering the number of neutrons and protons in their nuclei. The three major types of nuclear change are natural radioactivity, nuclear fission, and nuclear fusion.

Natural radioactivity is a nuclear change in which unstable nuclei spontaneously shoot out "chunks" of mass, energy, or both at a fixed rate. An isotope of an atom whose unstable nucleus spontaneously emits fast-moving particles (particulate radiation), high-energy electromagnetic radiation, or both is called a **radioactive isotope,** or **radioisotope.**

Radiation emitted by radioisotopes is called **ionizing radiation** because it has enough energy to dislodge one or more electrons from atoms it hits to form positively charged ions, which can react with and damage living tissue. The two most common types of ionizing particulate radiation are high-speed **alpha particles** (positively charged chunks of matter that consist of two protons and two neutrons) and **beta particles** (negatively charged electrons).

Radio waves, infrared light, and ordinary light are examples of nonionizing electromagnetic radiation, which does not have enough energy to cause ionization of atoms in living tissue. Although X rays are a form of high-energy ionizing radiation that can pass through the body and cause damage, they are not given off by radioisotopes. The most common form of ionizing electromagnetic radiation released from radioisotopes is high-energy **gamma rays,** which are even more penetrating than X rays. Figure 3-3 illustrates the

Scientists agree that exposure to any ionizing radiation can damage cells in the human body. The effects depend on the amount and frequency of exposure, the type of ionizing radiation, its penetrating power (Figure 3-3), and whether it comes from outside or inside the body.

From the outside, alpha particles and beta particles can cause burns at high levels of exposure, but neither can penetrate the skin to cause internal damage. However, if a radioactive isotope that emits alpha or beta particles is inhaled or ingested into the body, the particles can cause considerable damage to nearby vulnerable tissues. Gamma rays and high-energy neutrons are so penetrating that they pass through the body easily and inflict internal cellular damage from outside or inside the body.

Most damage occurs in tissues with rapidly dividing cells, such as the bone marrow (where blood cells are made), spleen, digestive tract (whose lining must be constantly renewed), reproductive organs, and lymph glands. Rapidly growing tissues of the developing embryo are also extremely sensitive, so pregnant women should avoid all exposure to radioactivity and X rays unless they are essential for health or diagnostic purposes.

The two most widely used units for measuring the dose or amount of ionizing radiation in terms of its potential damage to living tissues are the *rem* and the *millirem* (mrem), which is one-thousandth of a rem. Small doses of ionizing radiation over a long period of time cause less damage than the same total dosage given all at once, because our body apparently has some ability to repair itself. Exposure to a large dose of ionizing radiation over a short time, however, can be fatal within a few minutes to a few months later, depending on the dose.

---

relative penetrating power of alpha, beta, and gamma ionizing radiation. Some unstable nuclei emit only one type of ionizing radiation; others emit 2 or 3 types.

The rate at which a particular radioisotope spontaneously emits one or more forms of ionizing radiation is usually expressed in terms of its **half-life:** the length of time it takes for half the nuclei in a sample to decay by emitting one or more types of ionizing radiation and, in the process, change into another nonradioactive or radioactive isotope. Each radioisotope has a unique, characteristic half-life. For example, plutonium-239 (an alpha and gamma emitter), which is produced in nuclear fission power plant reactors and by nuclear fission bombs, has a half-life of 24,000 years. This means that half a given sample of plutonium-239 is still radioactive after 24,000 years and one-fourth is still radioactive after two half-lives, or 48,000 years. When inhaled into the lungs, a small speck of plutonium-239 greatly increases one's chances of developing lung cancer within two or three decades. Any exposure to ionizing radiation can cause potential harm to tissue in the human body (see Spotlight above).

Normally, it takes at least ten half-lives for a sample of a radioisotope to decay to what is considered a safe level of ionizing radiation. Thus, unless it is cleaned up thoroughly (a difficult and expensive procedure), an area contaminated with plutonium-239 by explosion of an atomic bomb or a severe nuclear power plant accident remains dangerously radioactive for at least 10 x 24,000 years, or 240,000 years.

Nuclear fission is a nuclear change in which nuclei of certain heavy isotopes with large mass numbers, such as uranium-235, are split apart into two lighter nuclei, known as *fission fragments*, when struck by slow- or fast-moving neutrons; this process also releases more neutrons and energy (Figure 3-4). Fissions of uranium-235 nuclei, found in small quantities in uranium ore obtained from the earth's crust, can produce any of over 450 different fission fragments or isotopes, most of them radioactive. The two or three neutrons produced by each fission can be used to fission many additional uranium-235 nuclei if enough are present to provide the **critical mass** needed for efficient capture of these neutrons. These multiple fissions taking place within the critical mass represent a **chain reaction** that releases an enormous amount of energy (Figure 3-5).

In an atomic or nuclear fission bomb, a massive amount of energy is released in a fraction of a second in an *uncontrolled* nuclear fission chain reaction. This is normally done by using an explosive charge to suddenly push a mass of fissionable fuel together from all sides to attain the critical mass needed to capture enough neutrons for a massive chain reaction to take place almost instantly.

In the nuclear reactor of a nuclear electric power plant, the rate at which the nuclear fission chain reaction takes place is *controlled* so that, on the average, only one of each two or three neutrons released is used to split another nucleus. In conventional nuclear

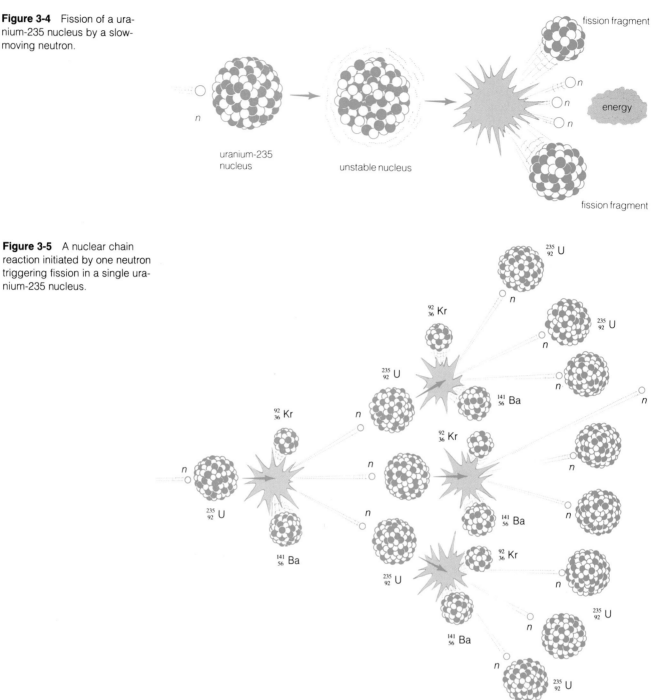

Figure 3-4 Fission of a uranium-235 nucleus by a slow-moving neutron.

uranium-235 nucleus

unstable nucleus

fission fragment

energy

fission fragment

Figure 3-5 A nuclear chain reaction initiated by one neutron triggering fission in a single uranium-235 nucleus.

fission reactors, nuclei of uranium-235 are split apart to produce energy. Another fissionable radioisotope is plutonium-239, which is formed from nonfissionable uranium-238 in breeder nuclear fission reactors, which may be developed in the future to extend supplies of uranium (Section 16-3).

Nuclear fusion is a nuclear change in which two nuclei of isotopes of light elements such as hydrogen are forced together at extremely high temperatures of 100 million to 1 billion degrees Celsius (°C) until they fuse to form a heavier nucleus with the release of energy. Because such high temperatures are needed to force the positively charged nuclei (which strongly repel one another) to join together, fusion is much more difficult to initiate than fission. But once initiated, fusion releases far more energy per gram of fuel than fission. Fusion of hydrogen atoms to form helium atoms is what takes place in the sun and other stars.

After World War II, the principle of uncontrolled nuclear fusion was used to develop extremely pow-

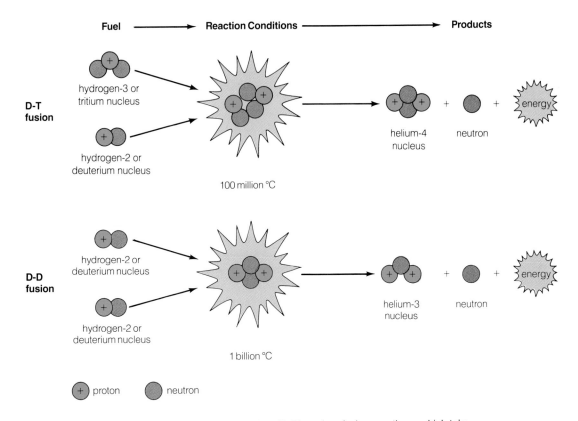

Figure 3-6 The deuterium-tritium (D-T) and deuterium-deuterium (D-D) nuclear fusion reactions, which take place at extremely high temperatures.

erful hydrogen, or thermonuclear, bombs and missile warheads. These weapons involve the D-T fusion reaction, in which a hydrogen-2, or deuterium (D), nucleus and a hydrogen-3, or tritium (T), nucleus are fused to form a larger helium-4 nucleus, a neutron, and energy (Figure 3-6). Scientists have also tried to develop controlled nuclear fusion, in which the D-T reaction is used to produce heat that can be converted into electricity. However, this process is still at the laboratory stage despite 36 years of research. If it ever becomes technologically and economically feasible—a big *if*—it is not projected to be a commercially important source of energy until 2050 or later (Section 16-4).

Exposure to Ionizing Radiation   All living things are exposed to small amounts of ionizing radiation, known as **natural, or background, ionizing radiation** (Table 3-1). Sources include cosmic rays (a high-energy form of ionizing electromagnetic radiation) from outer space; naturally radioactive isotopes, such as radon-222 found in certain types of rock and in bricks, stone, and concrete used in construction; and other natural radioactivity that finds its way into our air, water, and food.

We receive additional exposure to ionizing radiation as a result of various human activities, most from dental and medical X rays and diagnostic tests involving the injection or ingestion of radioactive isotopes. These important tools save many thousands of lives each year and reduce human misery. But some observers contend that many X rays and diagnostic tests involving radioactive isotopes are taken primarily to protect doctors and hospitals from liability suits. If your doctor or dentist proposes an X ray or diagnostic test involving radioisotopes, ask why it is necessary, how it will help find what is wrong and influence possible treatment, and what alternative tests are available with less risk.

The smallest amount of human-caused exposure to ionizing radiation in the United States comes from nuclear power plants and other nuclear facilities—assuming that they are operating normally. Nuclear power critics contend that the greater danger from nuclear power is not from small, routine emissions of radioactivity but from the extremely small but real possibility of accidents that could result in the emission of large quantities of ionizing radiation. Such an accident occurred in the Soviet Union at the Chernobyl nuclear power plant in 1986 (Section 16-2).

Table 3-1 Estimating Your Average Annual Radiation Dose from Background Radiation and Human Activities	
Source of Radiation	Approximate Annual Dose (millirems)
Natural or Background Radiation	
Cosmic rays from space	
At sea level (average)	40
Add 1 mrem for each 30.5 m (100 ft) you live above sea level	_____
Radioactive minerals in rocks and soil: ranges from about 30 to 200 mrem depending on location	55 (U.S. average)
Radioactivity in the human body from air, water, and food: ranges from about 20 to 400 mrem depending on location and water supply	25 (U.S. average)
Radiation from Human Activities	
Medical and dental X rays and tests; to find your total, add 22 mrem for each chest X ray, 500 mrem for each X ray of the lower gastrointestinal tract; 910 mrem for each whole-mouth dental X ray, and 5 million mrem for radiation treatment of a cancer	80 (U.S. average)
Living or working in a stone or brick structure; add 40 mrem for living and an additional 40 mrem for working in such a structure	_____
Smoking a pack of cigarettes a day; add 40 mrem	_____
Nuclear weapons fallout	4 (U.S. average)
Air travel; add 2 mrem for each 2,400 km (1,500 mi) flown that year	_____
TV or computer screens; add 4 mrem per year for each 2 hr of viewing a day	_____
Occupational exposure: 100,000 mrem per year for uranium ore miner, 600 to 800 mrem for nuclear power plant personnel, 300 to 350 mrem for medical X-ray technicians, 50 to 125 mrem for dental X-ray technicians, 140 mrem for jet plane crews	0.8 (U.S. average)
Living next door to a normally operating nuclear plant: boiling water reactor, add 76 mrem; pressurized water reactor, add 4 mrem	_____
Living within 8 km (5 mi) of a normally operating nuclear power plant: add 0.6 mrem	_____
Normal operation of nuclear power plants, nuclear fuel processing, and nuclear research facilities	0.10 (U.S. average)
Miscellaneous: luminous watch dials, smoke detectors, industrial wastes.	2 (U.S. average)
Your annual total	= _____ mrem
Average annual exposure per person in the United States = 230 mrem (with 130 mrem from background radiation and 100 mrem from human activities)	

Each year Americans are exposed to an average of 230 millirems (0.230 rem) of ionizing radiation. According to estimates by the National Academy of Sciences, this exposure over an average lifetime causes about 1% of all fatal cancers and 5% to 6% of all normally encountered genetic defects in the U.S. population.

3-3 ENERGY: TYPES AND CHANGES

Types of Energy Energy is used to grow our food, keep us and other living things alive, move us and other forms of matter from one place to another, change matter from one physical or chemical form to another, and to warm and cool our bodies and the buildings where we work and live. **Energy** is the ability to do work or to cause a heat transfer between two objects at different temperatures. **Work** is what happens when a force is used to push or pull a sample of matter, such as this book, over some distance. Any physical or chemical change in matter either requires an input of energy from the environment (boiling water and combining nitrogen and oxygen gas) or gives off energy to the environment (water freezing and coal burning).

Everything going on in and around us is based on work in which one form of energy is transformed

Type of Energy	Potential	Kinetic
Mechanical	firewood being held above ground	firewood dropped, which does work on experimenter's toe
Chemical	match being held near firewood	energy being released as heat and light from lit fire
Electrical	charged battery	battery being discharged through a wire
Nuclear	Nuclear power plant potential energy in nuclei of certain atoms	electricity produced (kinetic energy)

Figure 3-7 Forms of potential energy and kinetic energy.

into one or more other forms of energy. Scientists classify most forms of energy as either potential energy or kinetic energy (Figure 3-7).

Kinetic energy is the energy that matter has because of its motion and mass. Examples include a moving car, a falling rock, a speeding bullet, and the flow of water or charged particles (electrical energy). **Potential energy** is the energy stored by an object as a result of its position or the position of its parts. A rock held in your hand, a stick of dynamite, still water behind a dam, the nuclear energy stored in the nuclei of atoms, the chemical energy stored in the molecules of gasoline, and the carbohydrates, proteins, and fats in food are all examples of potential energy.

Energy Resources Used by People The direct input of essentially inexhaustible solar energy alone provides 99% of the thermal energy used to heat the earth and all buildings free of charge. Were it not for this perpetual direct input of various forms of electromagnetic energy from the sun, the earth's average temperature would be $-240°C$ ($-400°F$) and life as we know it would not have arisen.

Human ingenuity has developed a number of ways to use various forms of perpetual, renewable, and nonrenewable energy resources to supplement this direct input of solar energy and to provide the remaining 1% of the energy we use on earth (Figure 3-8). This supplemental energy is used primarily to provide us with low-temperature heat for heating buildings (space heating) and water, high-temperature heat for industrial processes and producing electricity, and mechanical energy for propelling vehicles. Conserving energy by decreasing the amount unnecessarily wasted in developing and using any of the supplemental energy resources shown in Figure 3-8 is also a major perpetual source of useful energy.

Most people think of solar energy in terms of direct heat from the sun. But broadly defined, **solar energy** includes not only the perpetual *direct* energy from the sun but also a variety of *indirect* forms of energy produced as a result of the direct input. Major indirect forms of solar energy include perpetual wind and falling and flowing water (hydropower), and renewable biomass (solar energy converted to chemical energy in trees and other plants).

We have learned how to capture some of these direct and indirect forms of solar energy. The direct input of energy from the sun is captured and used to heat water and buildings by *passive solar energy systems,* such as double- or triple-paned windows that face toward the sun, and *active solar energy systems,* such as specially designed roof-mounted collectors that concentrate the energy and then use pumps to transfer this heat to water or the interior of a building (Section 17-2). We have also developed wind turbines and hydroelectric power plants to convert indirect solar energy in the form of wind and falling or flowing water into electricity.

Direct solar energy has also been converted to chemical energy stored in fossil fuels such as natural gas, coal, crude oil, tar sands, and oil shale. But this form of solar energy stored for us hundreds of millions of years ago is being used up at such a rapid rate that it is *nonrenewable* on a human time scale. About 82% of the supplemental energy used throughout the world (91% in the U.S.) is based on burning oil, natural gas, and coal. Supplies of oil (and perhaps natural gas) may become increasingly scarce and expensive within the next few decades (Chapter 15).

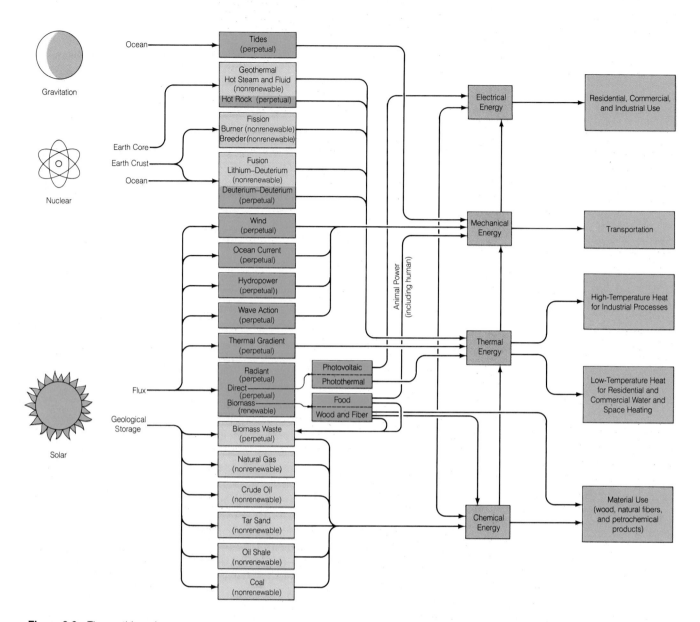

Figure 3-8 The earth's major energy resources and how they can be converted into chemical, thermal, mechanical, and electrical energy for various services. (Adapted from material supplied by the Office of Energy Research and Planning, State of Oregon)

3-4 THE FIRST AND SECOND LAWS OF ENERGY

First Law of Energy: You Can't Get Something for Nothing In studying millions of falling objects, physical and chemical changes, and changes of temperature in living and nonliving systems, scientists have observed and measured energy being transformed from one form to another, but they have never been able to detect any creation or destruction of energy.

This important information about what we find occurring in nature without fail is summarized in the **law of conservation of energy**, also known as the **first law of energy or thermodynamics:** In any physical or chemical change, movement of matter from one place to another, or change in temperature, energy is neither created nor destroyed but merely converted from one form to another. In other words, the energy gained or lost by any living or nonliving *system*—any collection of matter under study—must equal the energy lost or gained by its *surroundings* or *environment*—everything outside the system.

This law means that we can never get more energy out of an energy transformation process than we put in: *Energy input always equals energy output.* For example, the total amount of chemical energy contained in a gallon of gasoline exactly equals the output of energy in the form of mechanical energy and heat when the gasoline is burned. The first energy law also means that in terms of energy quantity, *it always takes energy to get energy; we can't get something for nothing (there is no free lunch).*

Energy Quality   Because the first law of energy states that energy can neither be created nor destroyed, you might think that there will always be enough energy. Yet after filling a car's tank with gasoline and driving around or using a battery until it is too dead to power a flashlight, you have lost something. If it isn't energy, what is it?

The answer involves understanding that energy varies in its *quality* or ability to do useful work—moving matter, changing the physical or chemical form of matter, or altering the temperature of matter. **High-quality energy,** like that in electricity, coal, oil, gasoline, sunlight, wind, nuclei of uranium-235, and high-temperature heat, is concentrated and has great ability to perform useful work. By contrast, **low-quality energy,** like low-temperature heat, is dispersed, or dilute, and has little ability to do useful work. For instance, the total amount of low-temperature heat stored in the Atlantic Ocean is greater than the high-quality chemical energy stored in all the oil deposits in Saudi Arabia. But this low-quality heat is so widely dispersed in the ocean we can't do much with it.

Although sunlight is a form of high-quality energy, it does not melt metals or char our clothes, because only a relatively small amount reaches each square meter of the earth's surface per minute or hour during daylight hours—even though the total amount of solar energy reaching the entire earth is enormous. Wind also is a form of high-quality energy, but to perform large amounts of useful work it must flow into a given area at a fairly high rate. Thus *the overall usefulness of a perpetual energy source such as direct sunlight, flowing water, and wind is determined both by its quality and by its flow rate (flux)—the amount of high-quality energy reaching a given area of the earth per unit of time.*

Unfortunately, many forms of high-quality energy, such as high-temperature heat, electricity, gasoline, hydrogen gas (a useful fuel that can be produced by heating or passing electricity through water), and concentrated sunlight, do not occur naturally. We must use other forms of high-quality energy like fossil, wood, or nuclear fuels to produce, concentrate, and store them, or to upgrade their quality so that they can be used to perform certain tasks.

Second Law of Energy: You Can't Break Even
Millions of measurements by scientists have shown that in any transfer of heat energy to useful work, some of the initial energy input is always degraded to lower-quality, less useful energy, usually low-temperature heat that flows into the environment. This summary of what we always find cccurring in nature is known as the **second law of energy or thermodynamics.** No one has ever found a violation of this funadamental physical law.

Consider three examples of the second energy law in action. First, when a car is driven, only about 10% of the high-quality chemical energy available in its gasoline fuel is converted to mechanical energy used to propel the vehicle. The remaining 90% is degraded to low-quality heat that is released into the environment. Second, when electrical energy flows through the filament wires in an incandescent light bulb, it is converted into a mixture of about 5% useful radiant energy, or light, and 95% low-quality heat. *Much of modern civilization is built around the internal combustion engine and the incandescent light, which, respectively, waste 90% and 95% of their initial energy input.* When oil and other forms of energy are abundant and cheap, such waste has little effect on the future availability of energy resources. But as oil and other nonrenewable energy resources become more scarce and expensive, reducing such unnecessary energy loss becomes quite important.

A third example of the degradation of energy quality takes place when a green plant converts solar energy to high-quality chemical energy stored in molecules of glucose and low-quality heat given off to the environment. When a person eats a plant food, such as an apple, its high-quality chemical energy is transformed within the body to high-quality electrical and mechanical energy (used to move the body and perform other life processes) and low-quality heat. In each of these energy conversions, some of the initial high-quality energy is degraded into lower-quality heat that flows into the environment (Figure 3-9).

According to the first energy law, we will never run out of energy, because energy can neither be created nor destroyed. But according to the second energy law, the overall supply of concentrated, high-quality energy available to us from all sources is being continually depleted and in the process converted to low-quality energy. *Not only can we not get something for nothing in terms of energy quantity (the first energy law), we can't break even in terms of energy quality (the second energy law).*

The second energy law also means that *we can never recycle or reuse high-quality energy* to perform useful work. Once the high-quality energy in a piece of food, a gallon of gasoline, a lump of coal, or a piece of uranium is released, it is degraded and lost forever.

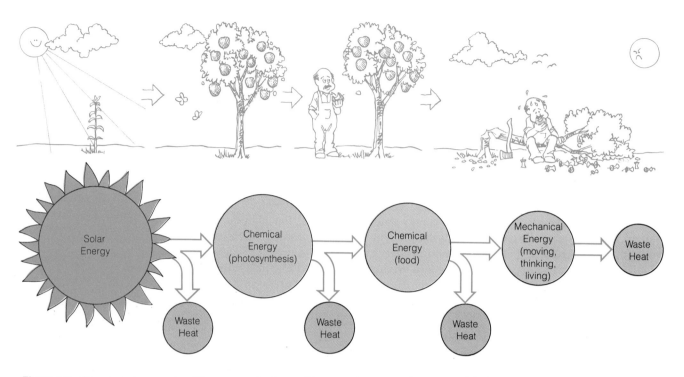

Figure 3-9 The second energy law. When energy is changed from one form to another, some of the initial input of high-quality energy is always degraded, usually to low-quality heat, which is added to the environment.

Second Energy Law and Increasing Environmental Disorder The second energy law can be stated in various ways. For example, since energy tends to flow or change spontaneously from a concentrated and ordered form to a more dispersed and disordered form, the second energy law also can be stated as follows: Heat always flows spontaneously from hot (high-quality energy) to cold (lower-quality energy). You learned this the first time you touched a hot stove.

By observing the spontaneous processes that are going on around us, we might conclude that a system of matter spontaneously tends toward increasing randomness or disorder, often called **entropy.** A vase falls to the floor and shatters into a more disordered state. Your desk and room seem to spontaneously become more disordered after a few weeks of benign neglect. Smoke from a smokestack and exhaust from an automobile disperse spontaneously to a more random or disordered state in the atmosphere and thus help air pollution levels decrease. Similarly, pollutants dumped into a river spread spontaneously throughout the water. Indeed, until we discovered that the atmosphere and water systems could be overloaded, we assumed that such spontaneous dilution was an easy and cheap remedy for air and water pollution.

But the hypothesis that *all* systems tend spontaneously toward increasing disorder or entropy is incorrect. Some systems do and some don't. For example, living systems survive only by maintaining highly ordered (low-entropy) systems of molecules and cells.

You are a walking, talking contradiction of the idea that systems tend spontaneously toward increasing disorder or entropy.

One way out of this seeming dilemma is to look at changes in disorder not just *within* a system but both in the system and in its environment or surroundings. Look at your own body. To form and preserve its highly ordered arrangement of molecules and its organized network of chemical changes, you must continually obtain matter resources and high-quality energy resources from your surroundings, use these resources, and then return more disordered, low-quality heat and waste matter to your surroundings. For example, your body continuously gives off heat equal to that of a 100-watt light bulb—explaining why a closed room full of people gets warm.

Planting, growing, processing, and cooking the foods you eat all require additional use of high-quality energy and matter resources that add low-temperature heat (low-quality energy) and waste materials to the environment. In addition, enormous amounts of low-quality heat (disorder) and waste matter are added to the environment when concentrated deposits of minerals and fuels are extracted from the earth, processed, and used or burned to heat and cool the buildings you use, to transport you, and to make roads, clothes, shelter, and other items you use.

Measurements show that the total amount of disorder, in the form of low-quality heat, added to the environment to keep you (or any living thing) alive

Many cornucopians believe that we will always be able to find enough oil or other fossil fuels so that conserving these resources need not be as high a priority as searching the world for the remaining supplies. Cornucopians believe that if oil becomes scarce, we should convert solid coal—the world's most abundant fossil fuel—to liquid and gaseous synthetic fuels. They also believe that we should shift to greatly increased use of electricity produced in large, centralized, coal-burning and nuclear fission power plants. After the year 2020, they propose shifting from conventional nuclear fission to breeder nuclear fission, which could synthesize nuclear fuel and thus prolong uranium supplies for at least 1,000 years (Chapter 16). After 2050, there would be a gradual shift to almost complete dependence on centralized nuclear fusion power plants, if this energy alternative should prove to be technologically, economically, and environmentally acceptable.

Neo-Malthusians disagree with this approach. They contend that the quickest, cheapest, and most cost-effective way to meet projected energy needs is *energy conservation*—primarily by improving the energy efficiency of houses, cars, and appliances so that less energy is wasted unnecessarily. They point out that reducing waste saves money and decreases the environmental impact of the use of any energy resource (because less is used to achieve the same amount of work). Conservation also extends supplies of fossil fuels, makes countries such as the United States less dependent on oil imports, and eliminates or sharply reduces the need to build additional electric power plants. Meanwhile, it buys time to phase in a diverse and flexible array of decentralized, mostly perpetual energy resources based on direct sunlight, wind, biomass, and falling and flowing water. Neo-Malthusians also point out that the vulnerability of the United States to nuclear attack would be

significantly decreased if we switched from large, centralized power plants (which can be knocked out by a relatively few missiles) to a widely dispersed array of small-scale energy systems based primarily on locally available energy resources.

In addition, most neo-Malthusians believe that all forms of nuclear power should be phased out, because this method for producing electricity is inefficient, uneconomic, unsafe, and unnecessary compared to other available alternatives (Chapter 16). They also view as unacceptable increased reliance on coal and coal-based synthetic fuels favored by the cornucopians; the massive amounts of carbon dioxide released into the atmosphere when these fuels are burned could bring about undesirable long-term changes in global climate patterns (Section 18-5). What do you think?

---

and to provide the items you use is much greater than the order maintained in your body. Thus *all forms of life are tiny pockets of order maintained by creating a sea of disorder around themselves*. The primary characteristic of any advanced industrial society is an ever-increasing flow of high-quality energy to maintain the order in human bodies and the larger pockets of order we call civilization. As a result, today's advanced industrial societies are adding more entropy to the environment than at any other time in human history.

Considering the *system and surroundings as a whole*, experimental measurements always reveal a net increase in entropy with any spontaneous chemical or physical change. Thus our original hypothesis must be modified to include the surroundings: *Any system and its surroundings as a whole spontaneously tend toward increasing randomness, disorder, or entropy*. In other words, if you think things are mixed up now, just wait. This is another way of stating the second energy law, or second law of thermodynamics. In most apparent violations of this law, the observer has failed to include

the greater disorder added to the surroundings when there is an increase in order within the system.

3-5   ENERGY LAWS AND ENERGY RESOURCES

Which Energy Resources Should We Develop?   The history of energy use since the Industrial Revolution has shown that developing and phasing in the widespread use of any new energy resource takes about 50 years. Thus today's energy research and development will largely determine the energy resources available to us 50 years from now. Using large amounts of human ingenuity and limited financial capital to develop the wrong mix of future energy resources could be disastrous for people in both MDCs and LDCs.

Cornucopians and neo-Malthusians tend to disagree over which mix of the energy resources shown in Figure 3-8 should be relied on for most of the energy used over the next 50 years (see Spotlight above). The

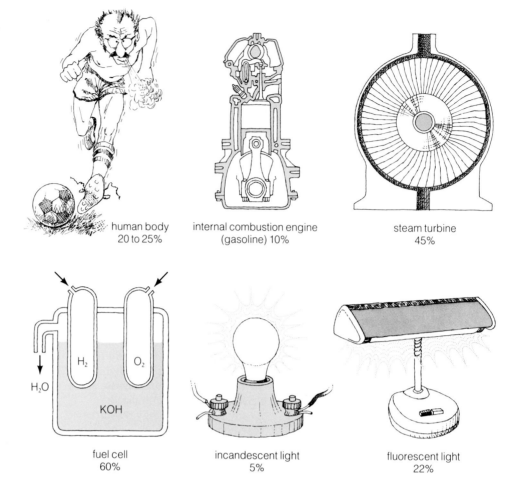

Figure 3-10 Energy efficiency of some common energy conversion devices.

human body
20 to 25%

internal combustion engine
(gasoline) 10%

steam turbine
45%

fuel cell
60%

incandescent light
5%

fluorescent light
22%

two energy laws are important tools in helping us decide how to reduce unnecessary energy waste and in evaluating the usefulness of various present and future energy resources.

Increasing Energy Efficiency One way to cut energy waste and save money, at least in the long run, is to increase **energy efficiency**—the percentage of the total energy input that does useful work and is not converted into low-quality, essentially useless heat in an energy conversion system. The energy conversion devices we use vary considerably in their energy efficiencies (Figure 3-10). We can reduce waste by using the most efficient processes or devices available and by trying to make them more efficient.

We can save energy and money by buying the most energy-efficient home heating systems, water heaters, cars, air conditioners, refrigerators, and other household appliances available. The initial cost of the most energy-efficient models is usually higher, but in the long run they save money. Thus whether an energy conversion device is cost-effective depends on its **lifetime** or **life-cycle cost:** its initial cost plus its lifetime operating cost.

The net efficiency of the entire energy delivery system for a heating system, water heater, or car is determined by finding the energy efficiency of each energy conversion step in the system: extracting the fuel, purifying and upgrading it to a useful form, transporting it, and finally using it. Figure 3-11 shows how net energy efficiencies are determined for heating a well-insulated home (1) passively with an input of direct solar energy through south-facing windows and (2) with electricity produced at a nuclear power plant and transported by wire to the home and converted to heat (electric resistance heating). This analysis reveals that converting high-quality chemical or nuclear energy in nuclear fuel to high-quality heat at several thousand degrees, converting this heat to high-quality electricity, and then using the electricity to provide low-quality heat for warming a house to only about 20°C (68°F) is extremely wasteful of high-quality energy.

According to energy expert Amory Lovins (see Further Readings), such use of high-quality energy to provide low-quality heat "is like using a chain saw to cut butter." By contrast, it is much less wasteful of high-quality energy to use a passive or active solar heating system to obtain low-quality heat from the

Passive Solar

Electricity from Nuclear Power Plant

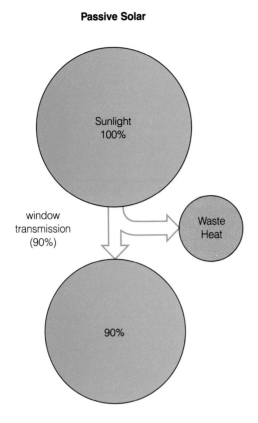
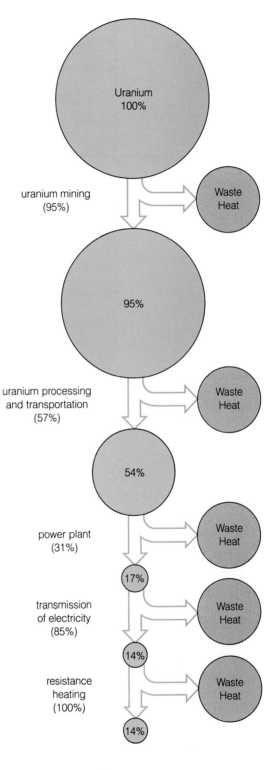

Figure 3-11 Comparison of net energy efficiency for two types of space heating. The cumulative net efficiency is obtained by multiplying the percentage shown inside the circle for each step by the energy efficiency for that step (shown in parentheses).

environment and, if necessary, raise its temperature slightly to provide space heating.

Figure 3-12 lists the net energy efficiencies for a variety of space-heating systems. The cheapest and most energy-efficient way to provide heating is to build a *superinsulated house* that has no need for any type of

conventional heating system. Such a house is so heavily insulated and airtight that even in areas where winter temperatures may average −40°C (−40°F), all of its space heating would be provided by a combination of passive solar gain (about 59%), waste heat from appliances (33%), and body heat from its occupants

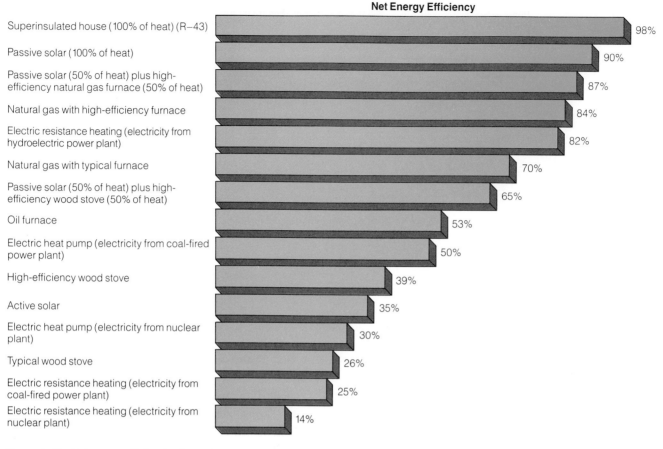

Net Energy Efficiency

System	Efficiency
Superinsulated house (100% of heat) (R–43)	98%
Passive solar (100% of heat)	90%
Passive solar (50% of heat) plus high-efficiency natural gas furnace (50% of heat)	87%
Natural gas with high-efficiency furnace	84%
Electric resistance heating (electricity from hydroelectric power plant)	82%
Natural gas with typical furnace	70%
Passive solar (50% of heat) plus high-efficiency wood stove (50% of heat)	65%
Oil furnace	53%
Electric heat pump (electricity from coal-fired power plant)	50%
High-efficiency wood stove	39%
Active solar	35%
Electric heat pump (electricity from nuclear plant)	30%
Typical wood stove	26%
Electric resistance heating (electricity from coal-fired power plant)	25%
Electric resistance heating (electricity from nuclear plant)	14%

Figure 3-12 Net energy efficiencies for various space-heating systems.

(8%). Passive solar heating is the next most efficient and cheapest method in lifetime cost to heat a house, followed by one of the new, 95% efficient, natural gas furnaces. The least efficient and most expensive way to heat a house is with electricity produced by nuclear power plants. For example, in 1986 the average price of obtaining 250,000 kilocalories (1 million British thermal units, or Btus) for heating space or water in the United States was $5.65 using natural gas, $5.95 using fuel oil, and $21.91 using electricity.

Figure 3-13 gives a similar analysis of net energy efficiency for heating water for washing and bathing. Again the least efficient way is to use electricity produced by nuclear power plants. The most efficient method is to use a tankless, instant water heater fired by natural gas or liquefied petroleum gas. Such heaters fit under a sink and burn fuel only when the hot water faucet is turned on, heating the water instantly as it flows through a small burner chamber and providing hot water only when and as long as it is needed. In contrast, conventional natural gas and electric resistance heaters keep a large tank of water hot all day and night and can run out after a long shower or two. Tankless heaters are widely used in many parts of Europe and are slowly beginning to appear in the United States. A well-insulated, conventional natural gas water heater is also efficient.

Figure 3-14 lists net energy efficiencies for several automobile engine systems. Note that the net energy efficiency for a car powered with a conventional internal combustion engine is only about 10%. In other words, about 90% of the energy in crude oil is wasted by its conversion to gasoline and its subsequent combustion to move a car. An electric engine with batteries recharged by electricity from a hydroelectric power plant has a net efficiency almost three times that of a gasoline-burning internal combustion engine. But this system cannot be widely used in the United States, because most favorable hydroelectric sites are found only in certain areas and have already been developed (Section 17-3). In addition, electric cars will not be cost-effective unless scientists can develop more affordable, longer-lasting batteries. The second most efficient system is a car with a gas turbine engine.

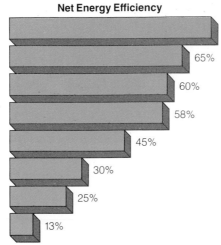

Net Energy Efficiency

Natural gas (tankless instant heater) — 75%

Passive solar (efficient batch heater for 50% plus instant natural gas heater, 50%) — 65%

Natural gas (conventional tank heater) — 60%

Active solar tank (50%) plus instant gas heater (50%) — 58%

Passive solar efficient batch tank heater (100%) — 45%

Active solar tank heater (100%) — 30%

Electric tank heater (electricity from coal-fired plant) — 25%

Electric tank heater (electricity from nuclear plant) — 13%

Figure 3-13 Net energy efficiencies for various water-heating systems.

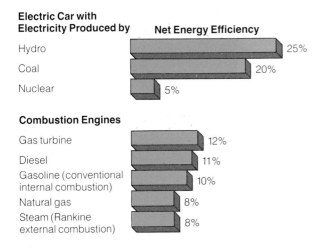

Electric Car with Electricity Produced by   **Net Energy Efficiency**

Hydro — 25%

Coal — 20%

Nuclear — 5%

Combustion Engines

Gas turbine — 12%

Diesel — 11%

Gasoline (conventional internal combustion) — 10%

Natural gas — 8%

Steam (Rankine external combustion) — 8%

Figure 3-14 Net energy efficiencies for automobiles with various engine systems.

American, Japanese, and European car makers have prototype gas turbine engines, but they need more development to determine whether they are cost-effective.

Using Waste Heat Although high-quality energy cannot be recycled, the rate at which the waste heat produced when it is degraded flows into the environment can be slowed. For instance, in cold weather an uninsulated, leaky house loses heat almost as fast as it is produced. By contrast, a well-insulated, airtight house can retain most of its heat for five to ten hours, and a well-designed, superinsulated house can retain most of its heat for up to four days.

In some office buildings, waste heat from lights, computers, and other machines is collected and distributed to reduce heating bills during cold weather, or exhausted in hot weather to reduce cooling bills. Waste heat from industrial plants and electrical power plants can also be distributed through insulated pipes and used as a district heating system for nearby buildings and homes, greenhouses, and fish ponds, as is done in some parts of Europe.

Waste heat from coal-fired and other industrial boilers can be used to produce electricity at half the cost of buying it from a utility company. The electricity can be used by the plant or sold to the local power company for general use. This combined production of high-temperature heat and electricity, known as **cogeneration,** is widely used in industrial plants throughout Europe. If all large industrial boilers in the United States used cogeneration, they could produce electricity equivalent to that of 30 to 200 large nuclear or coal-fired power plants (depending on the technology used) at about half the cost. This would reduce the average price of electricity and essentially eliminate the need to build any large electric power plants through the year 2020.

Net Useful Energy: It Takes Energy to Get Energy Because of the two energy laws, energy is always required to produce useful, high-quality energy. The true value of the energy obtainable from a given quantity of an energy resource is its **net useful energy**— the total energy available from the resource minus the amount of energy used (the first law), automatically wasted (the second law), and unnecessarily wasted in finding, processing, concentrating, and transporting it to a user. For example, if nine units of fossil fuel energy are required to deliver ten units of nuclear, solar, or additional fossil fuel energy (perhaps from a

Space Heating

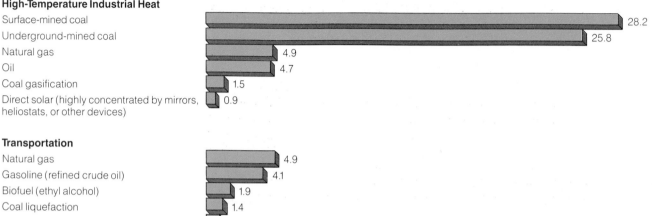

Net Useful Energy Ratio

Passive solar	5.8
Natural gas	4.9
Oil	4.5
Active solar	1.9
Coal gasification	1.5
Electric resistance heating (coal-fired plant)	0.4
Electric resistance heating (natural-gas-fired plant)	0.4
Electric resistance heating (nuclear plant)	0.3

High-Temperature Industrial Heat

Surface-mined coal	28.2
Underground-mined coal	25.8
Natural gas	4.9
Oil	4.7
Coal gasification	1.5
Direct solar (highly concentrated by mirrors, heliostats, or other devices)	0.9

Transportation

Natural gas	4.9
Gasoline (refined crude oil)	4.1
Biofuel (ethyl alcohol)	1.9
Coal liquefaction	1.4
Oil shale	1.2

Figure 3-15 Net useful energy ratios for various energy systems. (Data from Colorado Energy Research Institute, *Net Energy Analysis*, 1976, and Howard T. Odum and Elisabeth C. Odum, *Energy Basis for Man and Nature*, 3rd ed., McGraw-Hill, 1981)

deep well at sea), the net useful energy gain is only one unit.

We can express this as the ratio of useful energy produced to the useful energy used to produce it. In the example above, the net energy ratio would be 10/9, or 1.1. The higher the ratio, the greater the net useful energy yield. When the ratio is less than 1, there is a net energy loss over the lifetime of the system. Figure 3-15 lists estimated net useful energy ratios for various energy alternatives for space heating, high-temperature heat for industrial processes, and gaseous or liquid fuels for vehicles.

Currently, fossil fuels have relatively high net useful energy ratios because they come mainly from rich, accessible deposits. When these sources are depleted, however, the ratios will decline and prices will rise—more money and high-quality fossil fuel will be required to find, process, and deliver new fuel from poorer deposits found deeper in the earth and in remote and hostile areas like the Arctic—far from where the energy is to be used.

Conventional nuclear fission energy has a low net energy ratio, because large amounts of energy are necessary to build and operate power plants and take them apart after their 25 to 30 years of useful life, and

to safely store the resulting highly radioactive wastes. Large-scale solar energy plants for producing electricity or high-temperature heat for industrial processes also have low net useful energy ratios. This is because the small flow of high-quality solar energy in a particular area must be collected and concentrated to provide the necessary high temperatures. Large amounts of money and high-quality energy are necessary to mine, process, and transport the materials used in vast arrays of solar collectors, focusing mirrors, pipes, and other equipment. On the other hand, passive and active solar energy systems for heating individual buildings and for heating water have relatively high net useful energy ratios, because they supply relatively small amounts of heat at moderate temperatures.

3-6 MATTER AND ENERGY LAWS AND ENVIRONMENTAL PROBLEMS

Every Little Bit of Disorder Counts The three physical laws governing the use of matter and energy resources can help us understand how to work with rather than against nature, thus reducing the environ-

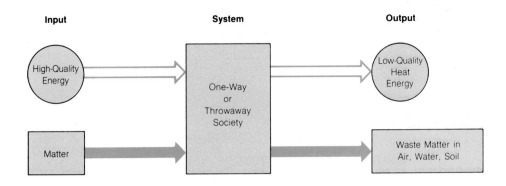

Figure 3-16 The one-way, or throwaway, society in most industrialized countries is based on maximizing the rates of energy flow and matter flow, resulting in the rapid conversion of the world's resources to trash, pollution, and waste heat.

mental impact of human activities. In this chapter we have seen that because of the law of conservation of matter and the second law of energy, the direct and indirect use of resources by each individual person (or any other living thing) automatically adds some waste heat (entropy) and matter to the environment. Your individual use of matter and energy and addition of waste to the environment may seem small and insignificant. But you are either only one of the 1.2 billion individuals in industrialized countries using large quantities of the earth's matter and energy resources at a rapid rate or only one of the 3.8 billion people in the less developed countries who hope to be able to use a greater share of those resources.

Throwaway and Matter-Recycling Societies Today's advanced industrialized countries are **throwaway societies,** sustaining ever-increasing economic growth by maximizing the rate at which matter and energy resources are used and wasted (Figure 3-16). The physical laws of matter and energy tell us that if more and more people continue to use and unnecessarily waste resources at an increasing rate, sooner or later the capacity of the local, regional, and eventually the global environment to absorb heat and waste matter will be exceeded.

A stopgap solution is to convert from a throwaway society to a **matter-recycling society** so that economic growth can continue without depleting matter resources and without producing excessive pollution and environmental disruption. But as we have seen already, there is no free lunch. The two laws of energy tell us that *recycling matter always requires high-quality energy*. However, if a resource is not too widely scattered in its distribution, recycling often requires less high-quality energy than that needed to find, extract, and process virgin or unused resources.

Nevertheless, in the long run, a matter-recycling society based on indefinitely increasing economic growth must have an inexhaustible supply of affordable high-quality energy and an environment with an infinite capacity to absorb waste matter and heat.

Although experts disagree on how much usable high-quality energy we have, supplies of nonrenewable coal, oil, natural gas, and uranium are clearly finite. Increasing evidence indicates that affordable supplies of oil, our most widely used supplementary energy resource, may be used up in several decades (Section 15-3).

"Ah," you say, "but don't we have an essentially infinite supply of high-quality solar energy flowing to the earth?" The problem is that the quantity of sunlight reaching a particular small area of the earth's surface each minute or hour is low, and nonexistent at night. With a proper collection and storage system, using solar energy to provide hot water and to heat a house to moderate temperatures makes good thermodynamic and economic sense. But to provide the high temperatures needed to melt metals or to produce electricity in a power plant, solar energy may not be cost-effective, because it has a very low net useful energy ratio (Figure 3-15).

One promising solar energy technology that may get around this problem is the *solar photovoltaic cell*, which converts solar energy directly to electricity in one simple, nonpolluting step (Section 17-2). If research can lead to continued improvements in the energy efficiency of such cells and decrease their cost, roofs or walls facing the sun could be covered with these cells to meet all household electricity needs. Mass production and transportation of solar cells would require energy and matter resources, but most of the matter would come from widely abundant silicon found in sand and other minerals. The excess electricity produced by the cells during daylight can be stored in deep-cycle batteries (like those used in golf carts and marine vessels) for use when the sun isn't shining or sold to the local utility company.

If scientists and engineers can develop methods for mass-producing such cells at an affordable price (which many believe will happen sometime in the 1990s), most of the large, centralized electric power plants throughout the United States and the world might quickly become obsolete. The first country that develops and sells such a technology could rapidly

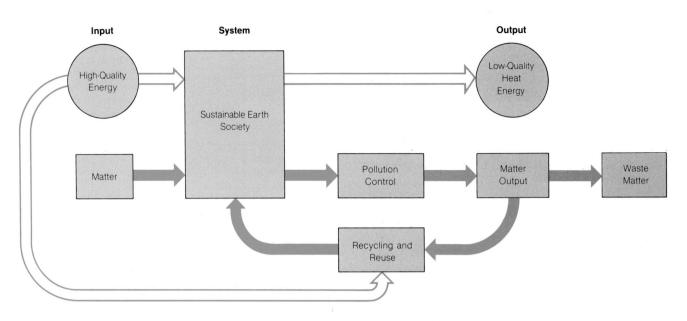

Figure 3-17 A sustainable-earth society, based on energy flow and matter recycling, is characterized by reusing and recycling renewable matter resources.

become the dominant economic force in the world—which helps explain why Japan is increasing its already extensive research efforts in this area. In the United States, government-supported research in photovoltaic cells was significantly expanded during the late 1970s by President Carter, but since 1980 the Reagan administration has sharply reduced this research.

Even if solar cells or some other breakthrough were to provide an essentially infinite supply of affordable useful energy, the first energy law states that we can't increase the universe's available energy supply and the second energy law states that the more energy we use to transform matter into products and to recycle these products, the more low-quality heat and waste matter we add to the environment. Thus the more we order or "conquer" the earth, the greater the disorder or entropy we add to the environment. Although experts argue over how close we are to reaching overload limits, the laws of matter and energy indicate that such limits do exist.

Sustainable-Earth Societies The three physical laws governing matter and energy changes tell us not only what we *cannot* do but, more important, what we *can* do. These laws indicate that the best long-term solution is to shift from a throwaway society based on maximizing matter and energy flow (and in the process wasting an unnecessarily large portion of the earth's resources) to a **sustainable-earth society** (Figure 3-17).

Such a society would go a step further than a matter-recycling society, not only recycling and reusing much of the matter we now discard as trash but also conserving both matter and energy resources by reducing unnecessary waste and by building things that last longer and are easier to recycle, reuse, and repair. Just as important, a sustainable-earth society would cut down on the use of resources by controlling population growth. With such an approach, local, regional, and global limits of the environment to absorb low-quality heat and waste matter would not be exceeded, and the depletion of vital resources would be prevented or at least delayed much further into the future.

The matter and energy laws also indicate why in the long run using input approaches rather than output approaches for controlling pollution makes more sense thermodynamically and economically. For example, removing most of the sulfur dioxide from the emissions of a coal-burning power plant before they leave the smokestack is much easier and cheaper in the long run than trying to remove the gas once it has been widely dispersed in the atmosphere. The three basic physical laws of matter and energy show that, like it or not, we are all interdependent on each other and on the other parts of nature for our survival. In the next chapter, we will apply these laws to living systems and look at some biological principles that can help us work with nature.

The second law of thermodynamics holds, I think, the supreme position among laws of nature. . . . If your theory is found to be against the second law of thermodynamics, I can give you no hope.

Arthur S. Eddington

DISCUSSION TOPICS

1. Explain why we don't really consume anything, and why we can never really throw any form of matter away.

2. A tree grows and increases its mass. Explain why this isn't a violation of the law of conservation of matter.

3. Criticize the statement "Since beta particles and alpha particles can't penetrate skin, they are not harmful to humans."

4. Criticize the statement "Since we are all exposed continuously to small amounts of ionizing radiation from natural sources, there is no need to worry about the small amounts of ionizing radiation released into the environment as a result of human activities."

5. Use the second energy law to explain why a barrel of oil can be used only once as a fuel.

6. Criticize the statement "Any spontaneous process results in an increase in the disorder or entropy of the system."

7. Criticize the statement "Life is an ordering process, and since it goes against the natural tendency for increasing disorder, it breaks the second law of thermodynamics."

8. Explain why most energy analysts urge that improving energy efficiency should form the basis of any individual, corporate, or national energy plan. Does it form a significant portion of your personal energy plan or lifestyle? Why or why not?

9. Explain how using a gas-powered chain saw to cut wood for burning in a wood stove could use more energy than that available from burning the wood. Consider materials used to make the chain saw, fuel, periodic repair, and transportation.

10. You are about to build a house. What energy supply (oil, gas, coal, or other) would you use for space heating, cooking food, refrigerating food, and heating water? Consider long-term economic and environmental impact factors.

11. a. Use the law of conservation of matter to explain why a matter-recycling society will sooner or later be necessary.
    b. Use the first and second laws of energy to explain why in the long run a sustainable-earth society, not just a matter-recycling society, will be necessary.

4

Ecosystems: What Are They and How Do They Work?

GENERAL OBJECTIVES

1. What are the major living and nonliving components of ecosystems?
2. What happens to matter resources in an ecosystem?
3. What happens to energy resources in an ecosystem?
4. How can the role of a particular type of organism in an ecosystem be described?

If we love our children, we must love the earth with tender care and pass it on, diverse and beautiful, so that on a warm spring day 10,000 years hence they can feel peace in a sea of grass, can watch a bee visit a flower, can hear a sandpiper call in the sky, and can find joy in being alive.

Hugh H. Iltis

What plants and animals live in a forest or a pond? How do they get the matter and energy resources needed to stay alive? How do these plants and animals interact with one another and with their physical environment? What changes will this forest or pond undergo through time?

Ecology is the science that attempts to answer such questions. In 1866 German biologist Ernst Haeckel coined the term *ecology* from two Greek words: *oikos*, meaning "house" or "place to live," and *logos*, meaning "study of." Literally, then, ecology is the study of living things in their home. In more formal terms, **ecology** is the study of interactions among organisms and between organisms and the physical and chemical factors making up their environment. This study is usually carried out as the examination of **ecosystems:** forests, deserts, ponds, oceans, or any set of plants and animals interacting with one another and with their nonliving environment. This chapter will consider the major nonliving and living components of ecosystems and how they interact. The next chapter will consider major types of ecosystems and the changes they can undergo as a result of natural events and human activities.

4-1 THE BIOSPHERE AND ECOSYSTEMS

The Earth's Life-Support System What keeps plants and animals alive on this tiny planet as it hurtles through space at a speed of 66,000 miles per hour? The general answer to this question is that life on earth

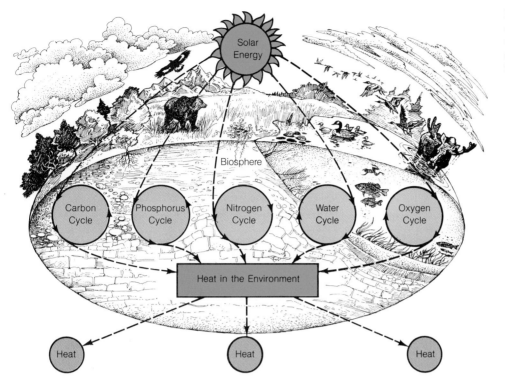

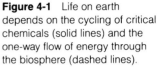

Figure 4-1 Life on earth depends on the cycling of critical chemicals (solid lines) and the one-way flow of energy through the biosphere (dashed lines).

depends on two fundamental processes: *matter cycling* and the *one-way flow of high-quality energy* from the sun, through materials and living things on or near the earth's surface and into space as low-quality heat (Figure 4-1).

All forms of life depend for their existence on the multitude of materials that compose the **(1)** solid **lithosphere,** consisting of the upper surface or crust of the earth, containing soil and deposits of matter and energy resources, and the earth's upper mantle; **(2)** the gaseous **atmosphere** extending above the earth's surface; **(3)** the **hydrosphere,** containing all of the earth's moisture as liquid water, ice, and small amounts of water vapor found in the atmosphere; and **(4)** the **biosphere,** consisting of parts of the lithosphere, atmosphere, and hydrosphere in which living organisms can be found (Figures 4-2 and 4-3).

The biosphere contains all the water, minerals, oxygen, nitrogen, phosphorus, and other nutrients that living things need. For example, your body consists of about 70% water obtained from the hydrosphere, small amounts of nitrogen and oxygen gases continually breathed in from the atmosphere, and various chemicals whose building blocks come mostly from the lithosphere. If the earth were an apple, the biosphere would be no thicker than the apple's skin. Everything in this "skin of life" is interdependent: Air helps purify water and keeps plants and animals alive; water keeps plants and animals alive; plants keep ani-

mals alive and help renew the air and soil; and the soil keeps plants and many animals alive and helps purify water. *The goal of ecology is to find out how everything in the biosphere is related.*

The Realm of Ecology Ecology is primarily concerned with interactions among five of the levels of organization of matter shown in Figure 3-1 (p. 29): organisms, populations, communities, ecosystems, and the biosphere. An **organism** is any form of life. Although biologists classify the earth's organisms in anywhere from 5 to 20 categories, in this book it is only necessary to classify organisms as plants or animals. Plants range from microscopic, one-celled, floating and drifting plants known as phytoplankton to the largest of all living things, the giant sequoia trees of western North America. Animals range in size from floating and drifting zooplankton (which feed on phytoplankton) to the 14-foot-high male African elephant and the 100-foot-long blue whale.

All organisms of the same kind constitute a **species.** For those organisms that reproduce sexually, a species can also be defined as all organisms potentially capable of interbreeding. Worldwide, it is estimated that 5 million to 10 million different plant and animal species exist. Some biologists put the estimate as high as 50 million species, 30 million of them insects. So far, about 1.7 million of the earth's species have

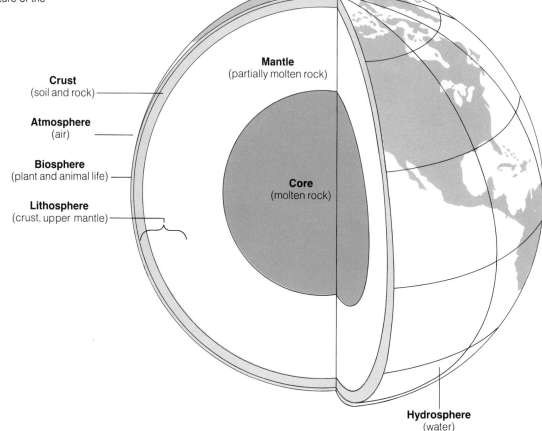

Figure 4-2 Our life-support system: the general structure of the earth.

Crust
(soil and rock)

Atmosphere
(air)

Biosphere
(plant and animal life)

Lithosphere
(crust, upper mantle)

Mantle
(partially molten rock)

Core
(molten rock)

Hydrosphere
(water)

been described and named, and about 10,000 new species are added to the list each year.

Each species is composed of smaller units, known as **populations:** groups of individual organisms of the same species that occupy particular areas at given times. All the striped bass in a pond, gray squirrels in a forest, white oak trees in a forest, people in a country, or people in the world constitute particular populations.

Each organism and population has a **habitat:** the place or type of place where it naturally thrives. A **community,** in ecological terms, is made up of all the populations of plant and animal species living and interacting in a given habitat or area at a particular time. Examples include all the plants and animals found in a forest, a pond, a desert, or an aquarium.

An **ecosystem** is the combination of a community and the chemical and physical factors making up its nonliving environment. An ecosystem can be a tropical rain forest, an ocean, a lake, a desert, a grassland, a field of corn, a fallen log, a terrarium, or a puddle of water, as long as it consists of a self-regulating community of plants and animals interacting with one another and with their nonliving environment. All of

the earth's ecosystems together make up the biosphere (Figure 4-3).

The differences among ecosystems found in various parts of the world are caused by differences in temperature, precipitation, and availability of life-sustaining chemicals or nutrients from the soil, water, and air. Although no two are exactly alike, ecosystems can be classified into general types that contain similar types of plants and animals. Major land ecosystems such as forests, grasslands, and deserts are called **terrestrial ecosystems** or **biomes.** Ponds, lakes, rivers, oceans, and other major ecosystems found in the hydrosphere are called **aquatic ecosystems.**

Whether large or small, ecosystems normally do not have distinct boundaries. Each ecosystem blends into adjacent ones through a transition zone that contains many of the plants and animals and other characteristics found in the adjacent ecosystems.

Components of Ecosystems Ecosystems consist of various nonliving (abiotic) and living (biotic) components. Figures 4-4 and 4-5 are greatly simplified dia-

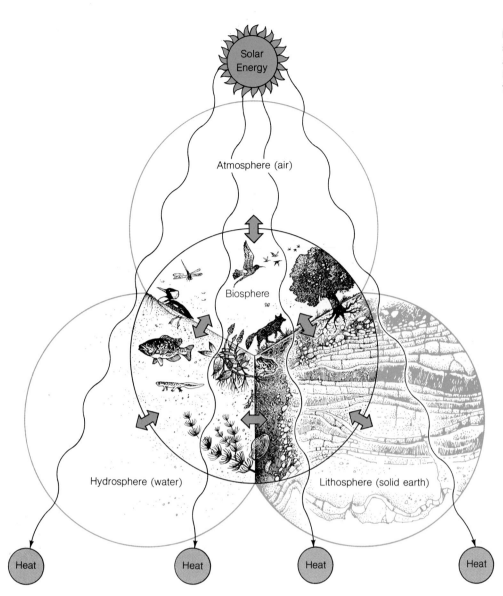

Figure 4-3 The biosphere consists of the parts of the earth's atmosphere, hydrosphere, and lithosphere in which all living things exist and interact.

grams showing a few of the living and nonliving components of ecosystems found in a freshwater pond and in a field.

The nonliving components of an ecosystem include various physical factors (such as sunlight, shade, precipitation, wind, terrain, temperature, and water currents) and chemical factors (all of the elements and compounds in the atmosphere, hydrosphere, and lithosphere that are essential for living organisms).

The different types of plant and animal organisms of an ecosystem are classified as producers, consumers, and decomposers on the basis of their sources of energy and matter resources. **Producers** are plants that can manufacture their own food.

Evergreen plants, such as pines, spruces, and firs, retain some of their leaves or needles throughout the year so that food can be manufactured year round or as soon as climatic conditions become favorable. **Deciduous plants,** such as oak and maple, retain sufficient moisture by losing all their leaves and becoming dormant during winter, when soil moisture may be frozen, or during drought, when it is unavailable. **Succulent plants,** such as desert cacti, survive by having no leaves, thus reducing loss of scarce water; they store water and produce the food they need in the thick, fleshy tissue of their green stems and branches.

Since producer plants alone can manufacture food, all animals must rely on them for nourishment and oxygen: All flesh is grass, so to speak. Organisms, mostly animals, that feed either directly or indirectly on producers are called **consumers.** Depending on their food sources, consumers that feed on living organ-

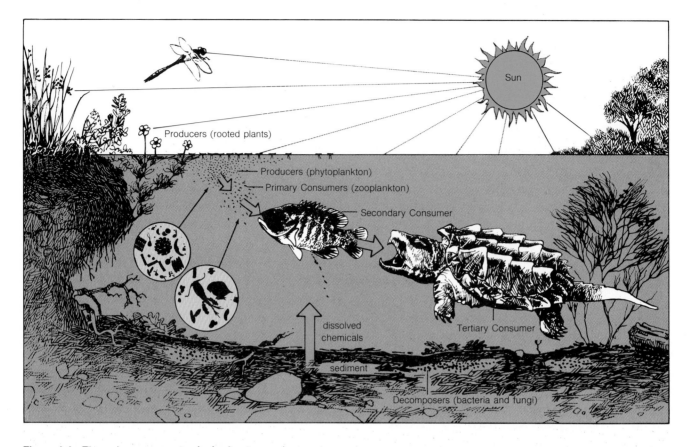

Figure 4-4 The major components of a freshwater pond ecosystem.

isms fall into three major classes: **(1) herbivores** ("plant eaters"), *primary consumers,* which feed directly on all or part of a plant; **(2) carnivores** ("animal eaters"), *secondary and higher-level consumers,* which feed on plant-eating animals; and **(3) omnivores** ("plant and meat eaters"), such as pigs, rats, cockroaches, and humans, which can eat both plants and animals.

In Figure 4-4, the fish that feeds on zooplankton (primary consumers) is a secondary consumer, and the turtle that feeds on this fish is a tertiary consumer. In Figure 4-5, the rabbit that feeds on green plants is a primary consumer (herbivore), and the fox that eats the rabbit is a secondary consumer (carnivore). Higher levels of consumers also exist.

Most consumers devour all or part of the plants or animals they feed on. **Parasites,** however, are primary, secondary, or higher consumers that feed on a plant or animal, known as the **host,** over an extended period of time. Parasites usually feed on their host without killing it—at least not immediately, as most other consumers do—but usually cause harm to it. Parasites such as tapeworms and disease-causing bacteria that live inside their host are called **endoparasites.** Those such as lice and ticks that attach them-

selves to the outside of their host are called **ectoparasites.**

Other consumer organisms in ecosystems feed on dead plant and animal matter, called **detritus,** and are known as **detritus consumers.** There are two major classes of detritus consumers: detritus feeders and decomposers. **Detritus feeders,** such as vultures, termites, earthworms, millipedes, ants, and crayfish, directly feed on dead plant or animal matter.

Much of the detritus in ecosystems, especially dead wood and leaves, is not eaten by detritus feeders. Instead, it undergoes decay, rot, or decomposition, in which its complex molecules are broken into simpler chemicals. Some of these are returned to the soil and water for reuse by producers. This decomposition process is brought about by the feeding activity of another type of detritus consumer, known as **decomposers,** or microconsumers.

Decomposers consist of two classes of organisms: *fungi* (mostly molds and mushrooms) and microscopic, single-celled *bacteria.* The fungi you see attached to a dead tree limb contain an extensive network of microscopic rootlike filaments, called *mycelia.* The mycelia penetrate through the dead wood or other

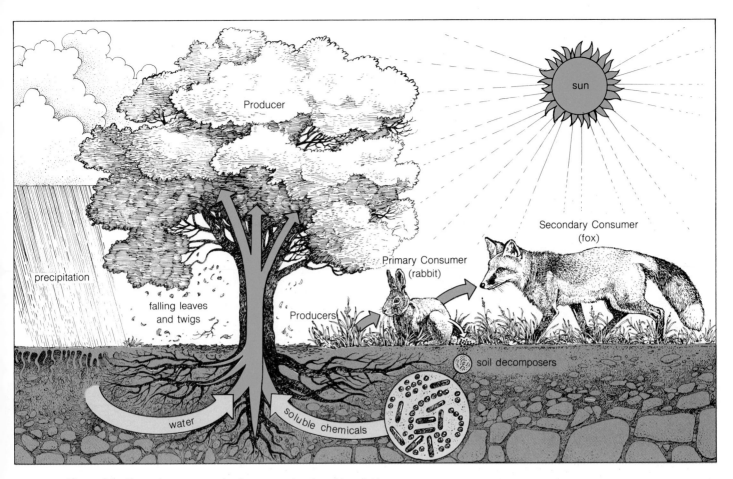

Figure 4-5 The major components of an ecosystem found in a field.

detritus and secrete digestive enzymes that hasten the breakdown into simpler nutrients that are absorbed into the fungal cells. Bacteria obtain the nutrients they need in a similar way. Although a few types of bacteria are parasites that cause diseases in living plants and animals, most species of bacteria are harmless decomposers that feed only on dead organic matter. Bacteria and fungi in turn are an important source of food for organisms such as worms, insects, and protozoans living in the soil and water.

Although the life of an individual organism depends on matter flow and energy flow through its body, the community of plants and animals in an ecosystem survives primarily by a combination of matter cycling and energy flow (Figure 4-6). Figure 4-6 also shows that decomposers are responsible for completing the cycle in which vital chemicals in living organisms are broken down and returned to the soil, water, and air in forms reusable as nutrients.

Although the biosphere as a whole cycles almost all the chemical nutrients needed by living organisms, no individual ecosystem completely cycles all the chemicals it needs. More stable ecosystems such as mature forests recycle most nutrients they need; younger, less stable ecosystems such as open fields, however, recycle only a portion of the elements and compounds they need and thus must obtain some nutrients from other ecosystems.

4-2 MATTER CYCLING IN ECOSYSTEMS: CARBON, OXYGEN, NITROGEN, PHOSPHORUS, AND WATER

Biogeochemical Cycles Of the earth's 92 naturally occurring elements, only 20 to 30 are constituents of living organisms and thus are cycled through the biosphere. In chemical terms, life can almost be summed up in five words: *carbon, oxygen, hydrogen, nitrogen, and phosphorus*. These chemicals as elements and compounds make up 97% of the mass of your body and more than 95% of the mass of all living organisms.

The remaining 15 to 25 elements needed in some form for the survival and good health of plants and animals are required only in relatively small, or trace, amounts. The importance of a particular chemical to

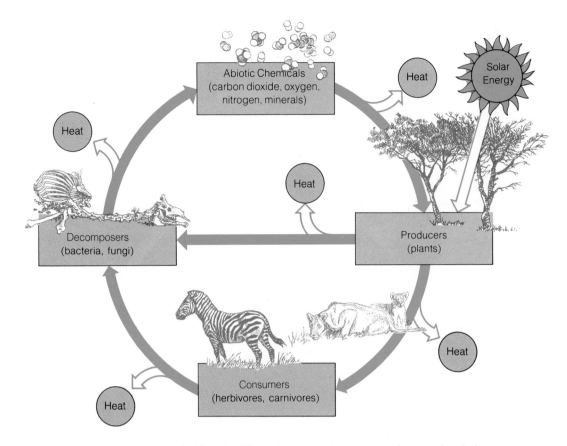

Figure 4-6 Summary of ecosystem structure and function. The major structural components (energy, chemicals, and organisms) of an ecosystem are connected through the functions of energy flow (open arrows) and matter cycling (solid arrows).

a living organism varies with the physical and chemical form and location of the chemical. For example, plants obtain most of their carbon in the form of carbon dioxide gas from the atmosphere or water, and most of their nitrogen and phosphorus as nitrate ions (NO_3^-) and phosphate ions (PO_4^{3-}) from soil water in which compounds containing these ions are dissolved.

Only a small portion of the earth's chemicals exist in forms useful to plants and animals. Fortunately, the essentially fixed supply of elements and compounds needed for life is continuously cycled through the air, water, soil, plants, and animals and converted to useful forms in **biogeochemical cycles** (*bio* meaning "living," *geo* for water, rocks, and soil, and *chemical* for the matter changing from one form to another). These cycles, driven directly or indirectly by incoming energy from the sun, include the carbon, oxygen, nitrogen, phosphorus, and hydrologic cycles (Figure 4-1, p. 51).

Thus a chemical may be part of an organism at one moment and part of its nonliving environment at another moment. This means that one of the oxygen molecules you just inhaled may be one inhaled previously by you, your grandmother, King Tut thousands of years ago, or a dinosaur millions of years ago.

Similarly, some of the carbon atoms in the skin covering your right hand may once have been part of a leaf, a dinosaur hide, or a limestone rock. Without the biogeochemical cycles, the entire world would soon be knee-deep in plant litter, dead animal bodies, animal wastes, and garbage.

Carbon and Oxygen Cycles **Carbon** is the basic building block of the large organic molecules necessary for life, including simple carbohydrates or sugars (such as glucose), complex carbohydrates, fats, proteins, and nucleic acids such as DNA. DNA molecules in the cells of plants and animals carry genetic information and chemical instructions for manufacturing various proteins living organisms need.

Most land plants obtain their carbon by absorbing carbon dioxide gas, which makes up 0.03% of the atmosphere, through pores in their leaves. They obtain the oxygen atoms they need from the oxygen in carbon dioxide and from water molecules in soil or bodies of water. The ocean's microscopic floating plants, known collectively as phytoplankton, get their carbon from atmospheric carbon dioxide that has dissolved in ocean water.

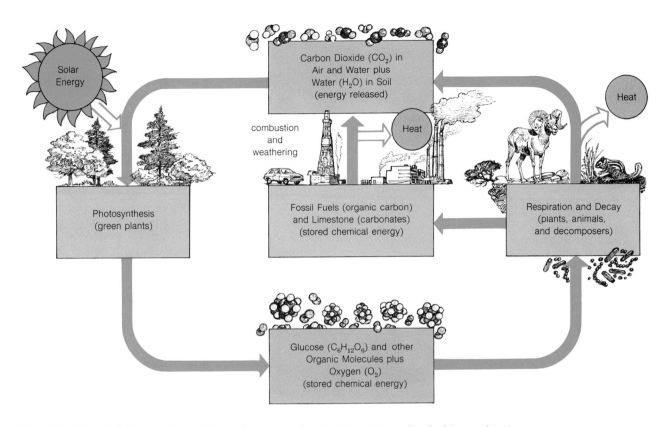

Figure 4-7 Simplified diagram of the carbon and oxygen cycles, showing matter cycling (solid arrows) and one-way energy flow (open arrows).

Chlorophyll molecules and some other pigments in the cells of green plants absorb solar energy and use it to combine carbon dioxide with water to form glucose along with oxygen gas. This complex process in which radiant energy from the sun is converted into chemical energy stored in plant tissue is called **photosynthesis.** Its 80 to 100 different, interconnected chemical changes can be be summarized as follows:

Photosynthesis

$$\text{carbon dioxide} + \text{water} + \textbf{solar energy} \rightarrow \text{glucose} + \text{oxygen}$$

$$6CO_2 + 6H_2O + \textbf{solar energy} \rightarrow C_6H_{12}O_6 + 6O_2$$

The glucose molecules are then converted by the plant itself or by animals eating the plant into more complex sugars, starches, proteins, and fats.

Plants and animals transform a portion of glucose and other, more complex, carbon-containing molecules they synthesize (plants) or eat (consumers) back into carbon dioxide and water by the process of **cellular respiration.** The chemical energy released in this complex process drives the physical and chemical changes needed for plants and animals to survive, grow, and reproduce. The almost 100 interconnected chemical changes involved can be summarized as follows:

Cellular Respiration

$$\text{glucose} + \text{oxygen} \rightarrow \text{carbon dioxide} + \text{water} + \textbf{energy}$$

$$C_6H_{12}O_6 + 6O_2 \rightarrow 6CO_2 + 6H_2O + \textbf{energy}$$

The carbon dioxide released by cellular respiration in all plants and animals is returned to the atmosphere and water for reuse by producers. Although the overall chemical reaction involved in cellular respiration is the reverse of that for photosynthesis, many of the detailed chemical reactions involved in the two processes are different.

Photosynthesis and cellular respiration are the basis of the **carbon** and **oxygen cycles,** shown in greatly simplified form in Figure 4-7, which illustrates some of the ways plants and animals are interdependent. Through these two interconnected cycles, plants produce food and oxygen needed by animals and absorb carbon dioxide given off by animals.

Figure 4-7 also shows that some of the earth's carbon is tied up for long periods in fossil fuels—coal,

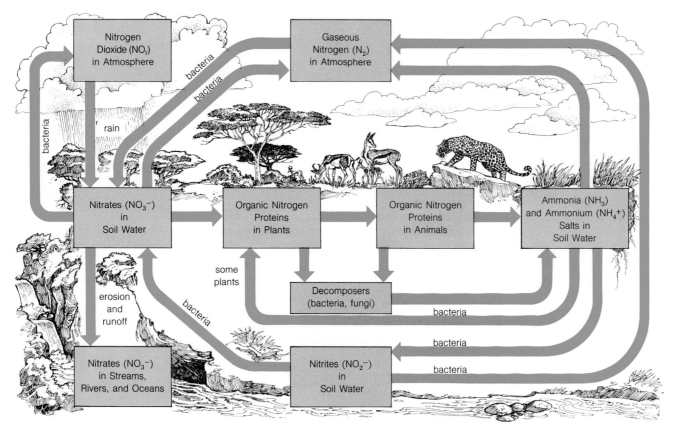

Figure 4-8 The nitrogen cycle (energy flow not shown).

petroleum, natural gas, peat, oil shale, tar sands, and lignite—formed over millions of years in the lithosphere. The carbon in these mineral deposits remains locked deep in the earth's crust until it is released to the atmosphere as carbon dioxide when fossil fuels are extracted and burned. Some of the earth's carbon is also locked for millions of years in deposits of carbonate rocks below the seafloor until movements of the earth's crust expose these rocks as part of an island or a continent. The carbon then reenters the cycle very slowly through erosion and other physical and chemical weathering processes that release it as carbon dioxide into the atmosphere.

Human beings intervene in the carbon and oxygen cycles in two ways that increase the average amount of carbon dioxide in the atmosphere. First, we remove forests and other vegetation without sufficient replanting, so that fewer plants are available worldwide to convert carbon dioxide in the atmosphere to organic nutrients. Second, we burn fossil fuels and wood.

Nitrogen Cycle Living things need nitrogen to manufacture proteins. Thus the growth of many plants can be limited by a lack of nitrogen available from the soil. Too little nitrogen can also cause malnutrition in people, because many of the body's essential functions require nitrogen-containing molecules such as proteins, DNA, and some vitamins.

The nitrogen gas that makes up about 78% of the volume of the earth's atmosphere is useless to most plants and animals. Fortunately, nitrogen gas is converted into water-soluble ionic compounds containing nitrate ions, which are taken up by plant roots as part of the **nitrogen cycle,** shown in simplified form in Figure 4-8. This *nitrogen fixation*—that is, the conversion of atmospheric nitrogen gas into forms useful to plants—is accomplished by **(1)** soil bacteria; **(2)** rhizobium bacteria living in small swellings called nodules on the roots of alfalfa, clover, peas, beans, and other legume plants; **(3)** blue-green algae in water and soil; and **(4)** lightning, which converts nitrogen gas and oxygen gas in the atmosphere to forms that return to the earth as nitrate ions in rainfall and other types of precipitation.

Plants convert nitrates obtained from soil water into large, nitrogen-containing molecules such as the proteins and nucleic acids necessary for life and good health. Animals get most of the proteins and other nitrogen-containing molecules they need by eating plants or other animals that have eaten plants. When

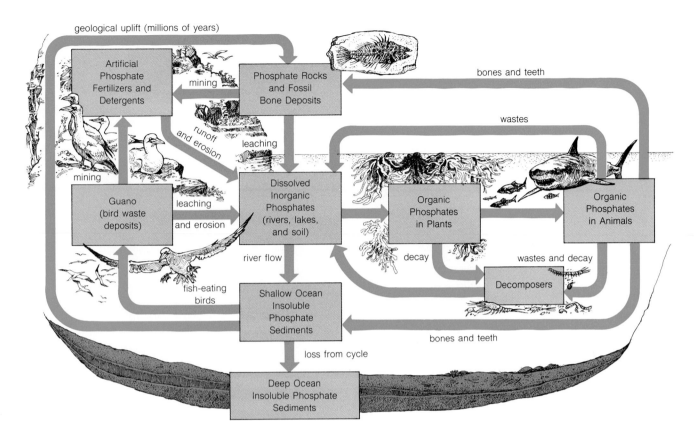

Figure 4-9 The phosphorus cycle (energy flow not shown).

plants and animals die, decomposers break down the nitrogen-containing molecules into ammonia gas and water-soluble salts containing ammonium ions (NH_4^+). Other specialized groups of bacteria then convert these forms of nitrogen back into nitrate ions in the soil and into nitrogen gas, which is released to the atmosphere to begin the cycle again.

Human beings intervene in the nitrogen cycle in several important ways. First, large quantities of NO and NO_2 are added to the atmosphere when fossil fuels are burned in power plants and vehicles. These oxides of nitrogen can react with other chemicals in the atmosphere under the influence of sunlight to form photochemical smog and nitric acid, a major component of acid deposition, commonly known as acid rain (Section 18-2). Second, nitrogen gas and hydrogen gas are converted by an industrial process into ammonia gas ($N_2 + 3H_2 +$ **energy** $\rightarrow 2NH_3$), which is then converted to ammonium compounds used as commercial fertilizer. Third, mineral deposits of compounds containing nitrate ions are mined and used as commercial fertilizers. Fourth, excess nitrate ions are added to aquatic ecosystems via the runoff of animal wastes from livestock feedlots, the runoff of commercial nitrate fertilizers from cropland, and the discharge of untreated and treated municipal sewage. This excess

supply of nitrate plant nutrients can stimulate extremely rapid growth of algae and other aquatic plants, which can deplete the water of dissolved oxygen gas and cause massive fish kills.

Phosphorus Cycle Phosphorus, mainly in the form of phosphate ions, is an essential nutrient of both plants and animals. It is a major constituent of the genetic material coded in DNA molecules and the main component of bones and teeth. It is also used in some commercial fertilizers.

Various forms of phosphorus are cycled through the lower atmosphere, water, soil, and living organisms by the **phosphorus cycle,** shown in Figure 4-9. The major reservoirs of phosphorus are phosphate rock deposits on land and in shallow ocean sediments. Some phosphates released by the slow breakdown of phosphate rock deposits are dissolved in soil water and taken up by plant roots. Animals get their phosphorus by eating plants or animals that have eaten plants. Animal wastes and the decay products of dead animals and plants return much of this phosphorus to the soil, rivers, and eventually to the ocean bottom as insoluble forms of phosphate rock. Some phosphate is returned to the land as *guano*—the phos-

Figure 4-10 "Guano islands" off the coast of Peru are kept as sanctuaries for fish-eating birds whose droppings (guano) are rich in nitrates and phosphates. The birds return some of these chemicals from the sea to the land as part of the biogeochemical cycles.

UN Food and Agriculture Organization/S. Larrain

phate-rich manure from fish-eating birds such as pelicans, gannets, and cormorants (Figure 4-10). This return, however, is small compared to the much larger amounts of phosphate eroding from the land to the oceans each year from natural processes and human activities.

People intervene in the phosphorus cycle in several ways. First, large quantities of phosphate rock are dug up, mostly from shallow ocean deposits, and used primarily to produce commercial fertilizers and detergents. Second, discharge from sewage treatment plants and runoff of commercial fertilizers can overload aquatic ecosystems with phosphate ions. As in the case of nitrate ions, an excessive supply can cause explosive growth of blue-green algae and other aquatic plants that can disrupt life in aquatic ecosystems.

Hydrologic, or Water, Cycle The **hydrologic,** or **water, cycle,** which collects, purifies, and distributes the earth's fixed supply of water, is shown in simplified form in Figure 4-11. Solar energy and gravity continuously move water among the ocean, air, land, and living organisms through evaporation, condensation, precipitation, and runoff back to the sea to begin the cycle again.

When incoming solar energy warms water on or near the surfaces of oceans, rivers, lakes, soil, and plant leaves, the water evaporates and enters the atmosphere as water vapor, leaving behind dissolved impurities. Water and various dissolved compounds in the soil are drawn up through the roots of plants into leaves and other surfaces. When warmed by the sun, water in the exposed parts of plants passes through leaf pores in a process called *transpiration* and then evaporates into the atmosphere. This transfer of liquid water in plant tissue and soil to water vapor in the atmosphere through a combination of transpiration and evaporation is called **evapotranspiration.**

Fresh water removed from the oceans and other bodies of water is returned from the atmosphere to the land and bodies of water as **precipitation** in the forms of rain, sleet, hail, and snow. Some of this fresh water becomes locked in glaciers, and some sinks downward through the soil, where it may remain for hundreds to thousands of years as groundwater in slow-flowing, slowly renewed, undergound reservoirs. When rain falls faster than it can infiltrate downward into the soil, it collects in puddles and ditches and runs off into nearby streams, rivers, and lakes, which carry water back to the oceans, completing the cycle. This runoff of fresh water from the land causes the weathering (slow disintegration) of rock and erosion of soil, which move various chemicals through portions of other biogeochemical cycles.

The hydrologic cycle is influenced by terrestrial organisms in many ways. Plants covering the soil decrease the impact of raindrops and reduce the rate of erosion. Various types of matter in the soil act as sponges to hold water in place for use by plants. When

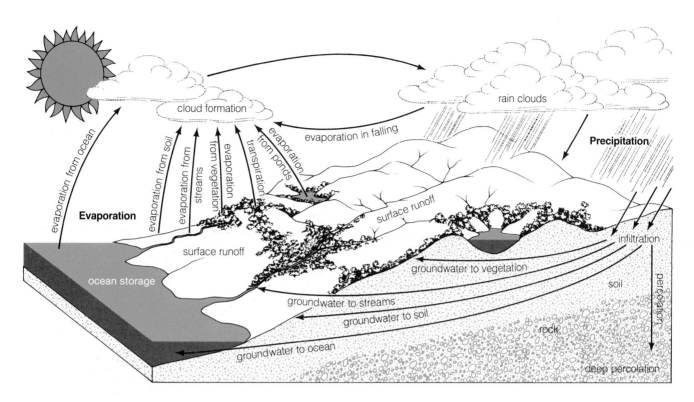

Figure 4-11 The hydrologic cycle. (U.S. Department of Agriculture)

some of the water obtained by plants from the soil is transpired and evaporated into the atmosphere, the water vapor has a cooling effect on climate.

Human beings also intervene in the water cycle. Large quantities of fresh water are withdrawn from rivers and lakes and pumped from underground supplies for irrigation, manufacturing, and domestic uses. In heavily populated or heavily irrigated areas, withdrawals have led to groundwater depletion or intrusion of ocean salt water into underground water supplies. The clearing of land for agriculture, mining, roads, parking lots, construction, and other activities can increase the rate at which water returns from the land to bodies of water, increase soil erosion, reduce the seepage that recharges groundwater supplies, and increase the risk of flooding.

4-3 ENERGY FLOW IN THE BIOSPHERE AND ECOSYSTEMS

The Sun: Source of Energy for Life The source of the radiant energy that sustains all life on earth is the sun. It lights and warms the earth and provides energy used by green plants to synthesize the compounds that keep them alive and serve as food for almost all

other organisms. Solar energy also powers the biogeochemical cycles and drives the climate and weather systems that distribute heat and fresh water over the earth's surface.

The sun is a gigantic gaseous fireball composed mostly of hydrogen and helium gases. Temperatures in its inner core reach 30 million degrees Fahrenheit, and pressures there are so enormous that the hydrogen nuclei are compressed and fused to form helium gas. This thermonuclear, or nuclear fusion, reaction (Section 3-2) taking place at the center of the sun continually releases massive amounts of energy, which pass through a thick zone of hot gases surrounding the inner core and eventually reach the surface. There the energy is radiated into space as a spectrum of heat, light, and other forms of *radiant energy* (Figure 4-12) that travel outward in all directions through space at a speed of 300,000 kilometers (186,000 miles) per hour.

Each type of radiant energy or electromagnetic radiation shown in Figure 4-12 can be viewed as a wave with a different **wavelength:** the distance between the crests of one wave and the next. The longer the wavelength, the lower the energy content of a wave of radiant energy. This explains why the lower-energy, longer-wavelength types of radiant energy shown in the right portion of Figure 4-12 are not harmful to most living organisms, whereas the higher-energy, shorter-wavelength types in the left portion are forms

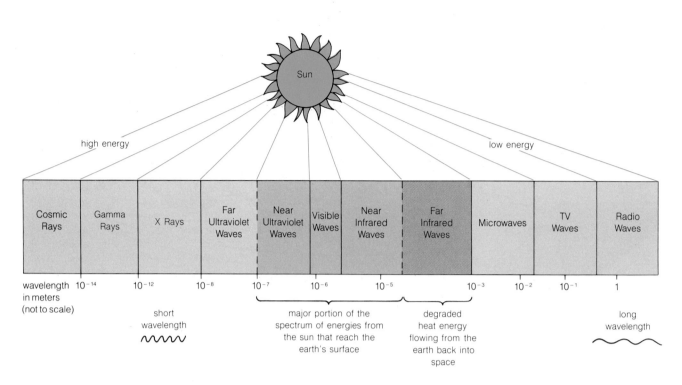

wavelength in meters (not to scale)

10^{-14} 10^{-12} 10^{-8} 10^{-7} 10^{-6} 10^{-5} 10^{-3} 10^{-2} 10^{-1} 1

short wavelength

major portion of the spectrum of energies from the sun that reach the earth's surface

degraded heat energy flowing from the earth back into space

long wavelength

Figure 4-12 The sun radiates a wide range of energies with different wavelengths and energy contents. Since much of this incoming radiant energy is either reflected or absorbed by the atmosphere, little of the harmful shorter-wavelength radiant energy actually reaches the earth's surface.

of ionizing radiation harmful to most organisms (Section 3-2). Fortunately, most of these harmful forms of radiant energy from the sun are absorbed by molecules of ozone (O_3) in the upper atmosphere and water vapor in the lower atmosphere. Without this screening effect, most life on earth could not exist.

Energy Flow in the Biosphere Figure 4-13 shows what happens to the solar radiant energy reaching the earth. About 34% of incoming solar radiation is immediately reflected back to space by clouds, chemicals, and dust in the atmosphere and by the earth's surface. Most of the remaining 66% warms the atmosphere and land, evaporates water and cycles it through the biosphere, and generates winds; a tiny fraction (0.023%) is captured by green plants and used to make glucose essential to life.

Most of the incoming solar radiation not reflected away is degraded into longer-wavelength heat, or far-infrared radiation, in accordance with the second law of energy, and flows into space. The amount of energy returning to space as heat is affected by the presence of molecules such as water, carbon dioxide, methane, and ozone and by some forms of solid particulate matter in the atmosphere. These substances, acting as gatekeepers, allow short-wavelength radiant energy

from the sun to pass through the atmosphere and back into space, but they absorb and reradiate some of the resulting longer-wavelength heat (far-infrared radiant energy) back toward the earth's surface.

Concern is growing that human activities affect global climate patterns by disrupting the rate at which incoming solar energy flows through the biosphere and returns to space as longer-wavelength heat. For example, according to some scientists, increases in the average levels of carbon dioxide in the earth's atmosphere, due primarily to the burning of fossil fuels and land clearing, may trap increasing amounts of far-infrared radiation that otherwise would escape into space, thus raising the average temperature of the atmosphere. These possible effects of human activities on climate are discussed more fully in Section 18-5.

Energy Flow in Ecosystems: Food Chains and Food Webs In general, the study of the flow of energy and the cycling of matter through an ecosystem is the study of what eats or decomposes what. *There is no waste in ecosystems.* All organisms, dead or alive, are potential sources of food for other organisms. A caterpillar eats a leaf; a robin eats the caterpillar; a hawk eats the robin. When the plant, caterpillar, robin, and hawk die, they are in turn consumed by decomposers.

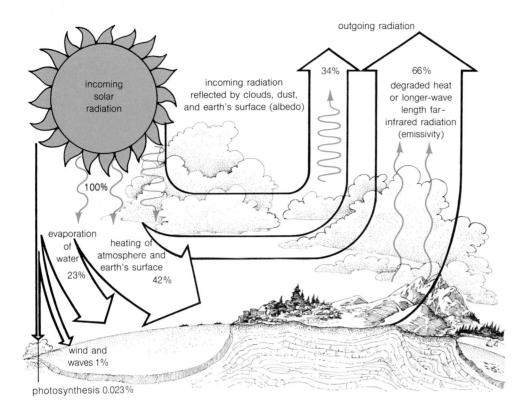

Figure 4-13 The flow of energy to and from the earth.

outgoing radiation

34%

66% degraded heat or longer-wave length far-infrared radiation (emissivity)

incoming solar radiation

incoming radiation reflected by clouds, dust, and earth's surface (albedo)

100%

evaporation of water

heating of atmosphere and earth's surface 42%

23%

wind and waves 1%

photosynthesis 0.023%

A series of organisms, each eating or decomposing the preceding one, is called a **food chain** (Figure 4-14), a one-way flow of chemical energy from producers to consumers and eventually to decomposers. Three other common terrestrial food chains are rice → human; leaves → bacteria (decomposers); and grass → steer → human. Most aquatic food chains, such as phytoplankton → zooplankton → perch → bass → human, involve more energy transfer steps.

All organisms that share the same general types of food in a food chain are said to be at the same **trophic level.** As shown in Figure 4-14, all producers belong to the first trophic level; all primary consumers, whether feeding on living or dead producers, belong to the second trophic level; and so on.

The food chain concept is useful for tracing chemical cycling and energy flow in an ecosystem, but simple food chains like the one shown in Figure 4-14 rarely exist by themselves. Very few herbivores or primary consumers feed on just one kind of plant, nor in turn are they eaten by only one type of carnivore or secondary consumer. In addition, omnivores eat several different kinds of plants and animals at several trophic levels. Thus the organisms in a natural ecosystem are involved in a complex network of many interconnected food chains, called a **food web.** A simplified food web in a terrestrial ecosystem is diagrammed in

Figure 4-15, which shows that trophic levels can be assigned in food webs just as in food chains.

The most obvious form of species interaction in food chains and webs is **predation:** An individual organism of one species, known as the **predator,** captures and feeds on parts or all of an organism of another species, the **prey.** Together, the two organisms involved are said to have a **predator-prey relationship.** We act as predators whenever we eat any plant or animal food. In most cases a predator species has more than one prey species. Likewise, a single prey species may have several different predators.

Food Chains, Food Webs, and the Second Law of Energy At each transfer from one trophic level to another in a food chain or web, work is done, low-quality heat is given off to the environment, and the availability of high-quality energy to organisms at the next trophic level is reduced. This reduction is the result of the inexorable energy-quality tax imposed by the second law of energy.

The percentage of the available high-quality energy transferred from one trophic level to another varies from 2% to 30%, depending on the type of species and the ecosystem in which it is found. However, as an average, only about 10% of the high-quality chem-

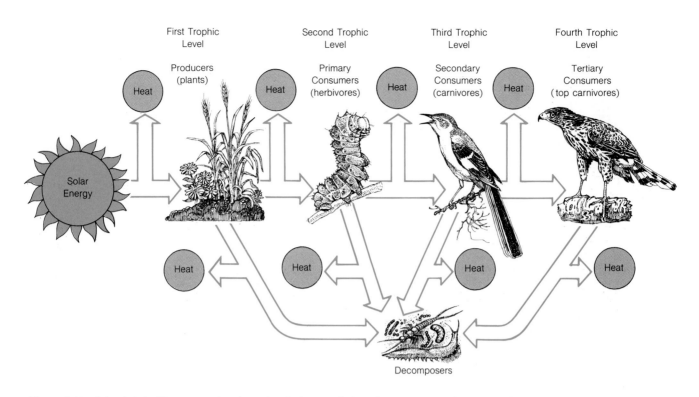

Figure 4-14 A food chain. The arrows show how chemical energy in food flows through various trophic levels, with most of the high-quality chemical energy being degraded to low-quality heat in accordance with the second law of energy.

ical energy available at one trophic level is transferred and stored in usable form as chemical energy in the bodies of the organisms at the next level. The remaining 90% is eventually degraded and lost as low-quality heat to the environment in accordance with the second law of energy.

Figure 4-16 illustrates this loss of usable high-quality energy at each step in a simple food chain. The resulting *pyramids of energy and energy loss* show that the greater the number of steps in a food chain, the greater the cumulative loss of usable high-quality energy. The *pyramid of numbers* shows that in moving from lower to higher trophic levels, the total number of organisms that can be supported usually decreases drastically. For example, a million phytoplankton in a small pond may support 10,000 zooplankton, which in turn may support 100 perch, which might feed one human being for a month or so. The pyramid of numbers helps explain why in a forest or field we find more plants than plant-eating rabbits and more plant-eating rabbits than rabbit-eating foxes.

An important principle affecting the ultimate population size of an omnivorous species such as human beings emerges from a consideration of the loss of available energy at successively higher trophic levels in food chains and webs: *The shorter the food chain, the less the loss of usable energy.* This means that a larger population of people can be supported if people shorten the food chain by eating grains directly (for example, rice → human) rather than eating animals that fed on the grains (grain → steer → human). For good health, such a vegetarian diet must include a variety of plants that provide all the nitrogen-containing molecules needed by the body to form essential proteins.

Net Primary Productivity of Plants The rate at which the plants in a particular ecosystem produce usable chemical energy or food is called **net primary productivity.** It is equal to the rate at which all the plants in an ecosystem convert solar energy to chemical energy stored in plant material through photosynthesis minus the rate at which these plants use some of this chemical energy in cellular respiration to stay alive, grow, and reproduce.

$$\begin{matrix} \textbf{net primary} \\ \textbf{productivity} \end{matrix} = \begin{matrix} \text{rate at which} \\ \text{plants produce} \\ \text{chemical energy} \end{matrix} - \begin{matrix} \text{rate at which} \\ \text{plants use} \\ \text{chemical energy} \end{matrix}$$

Note that net primary productivity is the *rate* at which the plants in an ecosystem produce net useful chem-

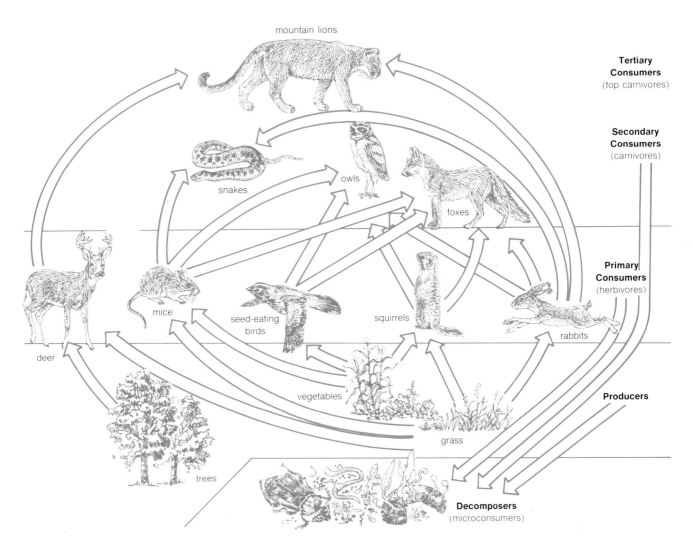

Figure 4-15 Greatly simplified food web for a terrestrial ecosystem.

ical energy—not the total amount of net useful chemical energy produced. Net primary productivity is usually reported as the amount of energy produced by the plant material in a specified area of land over a given time.

Net primary productivity can be used to evaluate the potential of various ecosystems for producing plant material that forms the base of the food supply for human beings and other animals. Farmers attempt to grow the crops that yield the highest net primary productivity in their area of the world. Ecologists have estimated the average annual net primary production per square meter for the major terrestrial and aquatic ecosystems. Ecosystems with the highest net primary productivities are in estuaries (coastal areas where the land meets the sea), swamps and marshes, and tropical rain forests; the lowest are tundra (Arctic grasslands), open ocean, and desert.

You might conclude that we should clear tropical forests to grow crops there and that we should harvest plants growing in estuaries, swamps, and marshes to help feed the growing human population. Such a conclusion is incorrect. One reason is that the plants—mostly grasses—in estuaries, swamps, and marshes are not very useful for direct human consumption, although they are extremely important as food sources and spawning areas for fish, shrimp, and other forms of aquatic life that provide protein for human beings. In tropical forests, most of the nutrients are stored in the trees and other vegetation rather than in the soil. When the trees are cleared, the exposed soil is so infertile that food crops can be grown only for a short time without massive, expensive inputs of commercial fertilizers.

Another reason is that this analysis is not based on how much of each ecosystem is available through-

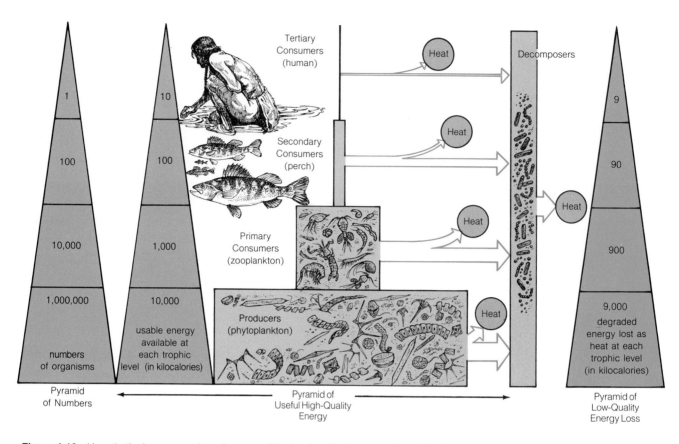

Figure 4-16 Hypothetical energy and number pyramids showing the decrease in usable high-quality energy available at each succeeding trophic level in a food chain.

out the world. Since the world's total area of estuaries is small, it drops way down the list. On the other hand, because about 71% of the earth's surface is covered with oceans, the open ocean ecosystems head the list. But this can be misleading. The world net primary productivity for oceans is high because they cover so much of the globe, not because they have a high productivity per square meter per year. Moreover, harvesting widely dispersed algae and seaweeds from the ocean requires enormous amounts of energy—more than the chemical energy available from the food that would be harvested.

Biological Amplification of Chemicals in Food Chains and Webs A factor affecting the survival of some individual organisms and populations of organisms is **biological amplification**: the phenomenon in which concentrations of certain chemicals soluble in the fatty tissues of organisms feeding at high trophic levels in a food chain or web are drastically higher than the concentrations of the same chemicals found in organisms feeding at lower trophic levels.

Biological amplification plays a devastating role in certain types of pollution. Many pollutants are either diluted to relatively harmless levels in the air or water or degraded to harmless forms by decomposers and other natural processes, as long as the portion of the environment receiving these chemicals is not overloaded. However, some synthetic chemicals, such as the pesticide DDT, some radioactive materials, and some toxic mercury and lead compounds become more concentrated in the fatty tissues of organisms at successively higher trophic levels. This accumulation of potentially harmful chemicals is especially pronounced in aquatic food chains and webs, because they generally consist of four to six trophic levels rather than the two or three levels of most terrestrial food chains and webs. Figure 4-17 illustrates the biological amplification of DDT in a five-step food chain of an estuary ecosystem adjacent to Long Island Sound near New York City.

Such biological amplification depends on three factors: (1) the second law of energy; (2) chemicals that are soluble in fat but insoluble in water; and (3) chemicals that either are not broken down or are broken down slowly in the environment. DDT, for example,

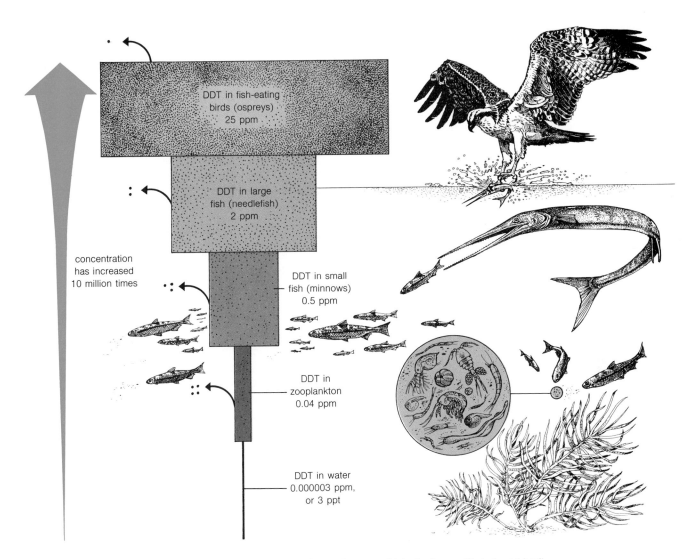

Figure 4-17 The concentration of DDT in the fatty tissues of organisms was biologically amplified about 10 million times in this food chain of an estuary adjacent to Long Island Sound near New York City. Dots represent DDT and arrows show small losses of DDT through respiration and excretion.

is insoluble in water, soluble in fat, and breaks down slowly in the environment. Thus if each phytoplankton concentrates one unit of water-insoluble DDT from the water, a small fish eating thousands of phytoplankton will store thousands of units of DDT in its fatty tissue. Then a large fish that eats ten of the smaller fish will receive and store tens of thousands of units of DDT. A bird or person that feeds on several large fish can ingest hundreds of thousands of units of DDT. The high concentrations of DDT or other chemicals can reduce the populations of such species by directly killing the organisms, by reducing their ability to reproduce, or by weakening them so that they are more susceptible to diseases, parasites, and predators. Biological amplification of certain chemicals helps explain why dilution is not always the answer to some forms of pollution.

4-4 ECOLOGICAL NICHES

The Niche Concept The **ecological niche** (pronounced "nitch") is a description of all the physical, chemical, and biological factors that a species needs to survive, stay healthy, and reproduce in an ecosystem. To describe a species' ecological niche, one must know its *habitat niche* (where it lives); its *food niche* (the species it eats or decomposes and the species it competes with, the species that prey on it, and where it leaves its wastes for decomposition); its *reproductive niche* (how and when it reproduces); and its *physical and chemical niche* (its temperature, shade, slope, humidity, and other requirements, the chemicals it can and cannot tolerate, and its effect on the nonliving parts of its environment).

A common analogy is that an organism's habitat, which is only part of its niche, is its "address" in an ecosystem, while its ecological niche is its "occupation" and "lifestyle." For example, the habitat of a robin includes such areas as woodlands, forests, parks, pasture lands, meadows, orchards, gardens, and yards. Its ecological niche includes such characteristics as nesting and roosting in trees, eating insects, earthworms, and fruit, and dispersing fruit and berry seeds in its droppings. Information about ecological niches helps people to manage species as sources of food or other resources and to predict the effects of either adding or removing a species to or from an ecosystem. Despite its importance, determining the interacting factors that make up an organism's complete ecological niche is very difficult.

Specialist and Generalist Niches The niche of an organism can be classified as specialized or generalized, depending primarily on its major sources of food, the extent of its habitat, and the degree of tolerance it has to temperature and other physical and chemical factors. Most species of plants and animals can tolerate only a narrow range of climatic and other environmental conditions and feed on a limited number of different plants or animals. Such species have a *specialized niche*, which limits them to fairly specific habitats in the biosphere. The giant panda, for example, has a highly specialized niche because it obtains 99% of its food by consuming bamboo plants. The destruction and mass die-off of several species of bamboo in parts of China where the vulnerable panda is found has led to its near extinction. In a tropical rain forest, an incredibly diverse array of plant and animal life survives by occupying a variety of specialized ecological niches in distinct vertical layers of the forest's vegetation (Figure 4-18).

Species with a *generalist niche* are very adaptable, can live on much of the terrestrial portion of the planet, eat a wide variety of foods, tolerate a wide range of environmental conditions, and are usually in much less danger of extinction than species with a specialized niche. Examples of generalist species include flies, cockroaches, mice, rats, and human beings.

Competition between Species As long as commonly used resources are abundant, different species can share them. However, when two or more species in the same ecosystem attempt to use the same scarce resources, they are said to be engaging in **interspecific competition.** The scarce resource may be food, water, carbon dioxide, sunlight, soil nutrients, shelter, or anything needed for survival.

One species gains an advantage over other competing species in the same ecosystem by producing more young, obtaining more food or solar energy, defending itself more effectively, or being able to tolerate a wider range of temperature, light, water salinity, or concentrations of certain poisons. For example, as species of large-leafed trees grow, they slowly create shade to eliminate or reduce competition from other sun-loving plant species. Some plants use chemical warfare to give themselves a competitive advantage: Apple trees do poorly and tomatoes will not grow at all near black walnut trees, which give off a poison through their roots.

Human beings spend a lot of time, energy, and money attempting to reduce populations of various species that compete with us for food. Farmers apply insecticides to kill insects and rodenticides to kill rats and other rodents that compete with people and livestock for grain and other crops. They also apply herbicides to kill weeds, rapidly growing plants that compete with food crops for space, water, and other nutrients.

Experience, however, has shown that in the long run this approach usually increases—not decreases—the numbers of pest species. Species such as insects, rats, and weeds, with short generation times and numerous offspring, can gain a competitive advantage by rapid **genetic adaptation** to a new environmental condition. For example, a few members of any large insect population have a built-in genetic resistance to insecticides such as DDT. When DDT is used to reduce the insect population, these members survive and rapidly breed new populations with a larger number of genetically resistant individuals. The more the insect population is exposed to a particular pesticide, the more resistant it becomes. As one observer put it, "I hope that when the insects take over the world, they will remember that we always took them along on our picnics."

Competitive Exclusion Principle The more similar the ecological niches of two species, the more they will compete for the same food, shelter, space, and other critical resources. According to the **competitive exclusion principle,** no two species in the same ecosystem can occupy exactly the same ecological niche indefinitely. Populations of some animal species can avoid or reduce competition with more dominant species by moving to another area, switching to a less accessible or less readily digestible food source, or hunting for the same food source at different times of the day or in different places. For example, hawks and owls feed on similar prey, but hawks hunt during the day and owls hunt at night. Where lions and leopards occur together, lions take mostly larger animals as prey and leopards take smaller ones.

Two species of fish-eating birds, the common cormorant and the shag cormorant, look alike, fish in a

Figure 4-18 Stratification of plant and animal niches in a tropical rain forest.

similar manner, and live on the same cliffs near the ocean. Careful observations, however, reveal that these two species have different ecological niches. The shag cormorant fishes mainly in shallow water for sprats and sand eels and nests on the lower portion of the cliffs. The common cormorant fishes primarily for shrimp and a few fish farther out to sea and lives nearer the top of the cliffs.

Other Species Interactions We generally think of one species as benefiting from and the other as being harmed by interspecific competition. However, the ecological niches of some species involve interactions in which both species benefit. This phenomenon is known as **mutualism.**

For example, pollinating insects such as butterflies and bees depend on flowering plants for food in the form of pollen and nectar. In turn, the plants ben-

efit by being pollinated—the insects carry plants' male reproductive cells contained in pollen grains to the female flowering parts of other flowers of the same species. Another mutualistic relationship exists in the nitrogen cycle (Figure 4-8) between legume plants (such as peas, clover, beans, and alfalfa) and rhizobium bacteria, which live in nodules on the roots of these plants. Large colonies of bacteria in these nodules "fix," or convert, gaseous nitrogen (N_2) in the atmosphere to forms of nitrogen such as nitrate (NO_3^-) and ammonium (NH_4^+) ions that can be used as nutrients by the plant and by the bacteria themselves.

In another type of interaction between species, called **commensalism,** one species benefits from the association while the other is apparently neither helped nor harmed. For example, in tropical forests rootless plants called epiphytes live high above the ground on tree trunks and branches. These plants are able to use their tree hosts to get access to sunlight in the relatively dark forest. They also use their own leaves and cupped petals to collect water and minerals that drip down from the tops of trees. The tree is neither harmed nor benefited.

This chapter has shown that *the essential feature of the living and nonliving parts of an ecosystem and of the biosphere is interdependence.* The next chapter shows how this interdependence is the key to understanding the earth's major types of terrestrial and aquatic ecosystems and how such ecosystems change in response to natural and human stresses.

We sang the songs that carried in their melodies all the sounds of nature—the running waters, the sighing of winds, and the calls of the animals. Teach these to your children that they may come to love nature as we love it.

Grand Council Fire of American Indians

DISCUSSION TOPICS

1. Distinguish among *ecosystem, biosphere, population,* and *community,* and give an example of each. Rank them by increasing scale and complexity in terms of their level of organization of matter.

2. Distinguish among *herbivores, carnivores,* and *omnivores,* and give two examples of each.

3. a. A bumper sticker asks "Have you thanked a green plant today?" Give two reasons for appreciating a green plant.
   b. Trace the sources of the materials that make up the sticker and see whether the sticker itself represents a sound application of the slogan.

4. a. How would you set up a self-sustaining aquarium for tropical fish?
   b. Suppose you have a balanced aquarium sealed with a transparent glass top. Can life continue in the aquarium indefinitely as long as the sun shines regularly on it?
   c. A friend cleans out your aquarium and removes all the soil and plants, leaving only the fish and water. What will happen?

5. Using the second law of energy, explain why there is such a sharp decrease in high-quality energy along each step of a food chain. Doesn't an energy loss at each step violate the first law of energy? Explain.

6. Using the second law of energy, explain why many people in less developed countries exist primarily on a vegetarian diet.

7. Using the second law of energy, explain why a pound of steak is more expensive than a pound of corn.

8. Why are there fewer lions than mice in an African ecosystem supporting both types of animals?

9. What characteristics must a chemical have before it will be biologically amplified in a food chain or web?

10. Compare the ecological niches of people in a small town and in a large city; in a more developed country and in a less developed country.

5

Ecosystems: What Are the Major Types and What Can Happen to Them?

GENERAL OBJECTIVES

1. How much environmental change can various species of plants and animals tolerate?

2. What are the major types of terrestrial ecosystems (biomes) and how does climate influence the type found in a particular area?

3. What are the major types of aquatic ecosystems and what factors influence the kinds of life they contain?

4. How can populations adapt to natural and human-induced stresses to preserve their overall stability?

5. How can ecosystems adapt to small- and large-scale natural and human-induced stresses to preserve their overall stability?

6. What major impacts can human activities have on ecosystems?

When we try to pick out anything by itself, we find it hitched to everything else in the universe.

John Muir

Organisms have a variety of characteristics that allow them to live in certain environments and adapt to changes in environmental conditions. Some like it hot, some like it cold. Some like it wet, others dry. Some thrive in the sunlight, others in the shade or dark.

The biosphere, in which all organisms are found, is a mosaic of *terrestrial ecosystems,* such as deserts, grasslands, and forests, and *aquatic ecosystems,* such as lakes, reservoirs, ponds, rivers, and oceans. Each of these ecosystems has a characteristic plant and animal community adapted to certain environmental conditions. These ecosystems, however, are dynamic, not static. Their plant and animal populations are always changing and adapting in response to major and minor changes in environmental conditions. This ability to adapt and yet sustain themselves is truly a remarkable feature of ecosystems.

5-1 RANGES OF TOLERANCE AND LIMITING FACTORS

Tolerance Range Each species and each individual organism of a species has a particular **range of tolerance** to variations in chemical and physical factors in its environment, such as temperature (Figure 5-1). This tolerance range includes an *optimum range* of values that allow populations of a species to thrive and operate most efficiently and values slightly above or below the optimal level of each abiotic factor that usually support a smaller population size. Once values exceed the upper or lower limits of tolerance, few if any organisms of a particular species survive.

These observations are summarized in the **law of tolerance:** The existence, abundance, and distribution of a species are determined by whether the levels of one or more physical or chemical factors fall above or

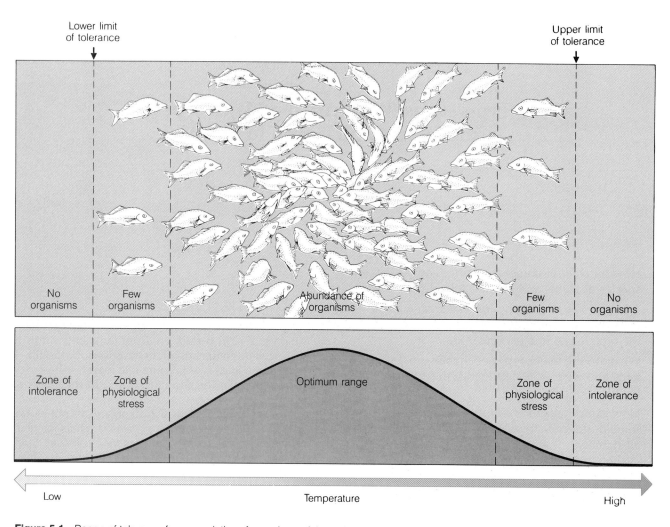

Figure 5-1 Range of tolerance for a population of organisms of the same species to an environmental factor—in this case, temperature.

below the levels tolerated by the species. Although organisms of the same species share the same range of tolerance to various abiotic factors, individual organisms within a large population of a species may have slightly different tolerance ranges because of small differences in their genetic makeup. For example, it may take a little more heat or a little more of a poisonous chemical to kill one frog or one human being than another. This is why the tolerance curve shown in Figure 5-1 represents the response of a population composed of many individuals of the same species rather than an individual organism to variations in some environmental factor such as temperature.

Usually the range of tolerance to a particular stress also varies with the physical condition and life cycle of the individuals making up a species. Individuals already weakened by fatigue or disease are usually more sensitive to stresses than healthy ones. For most animal species, tolerance levels are much lower in juveniles (where body defense mechanisms may not be fully developed) than in adults.

Organisms of most species have a better chance of adjusting or acclimating to an environmental change that takes place gradually. For example, we can tolerate a higher water temperature by getting into a tub of fairly hot water and then slowly adding hotter and hotter water. This ability to adapt slowly to new conditions is a useful protective device, but it can also be dangerous. With each change, the organism comes closer to its limit of tolerance until suddenly, without any warning signals, the next small change triggers a harmful or even fatal effect—much like adding the single straw that breaks an already overloaded camel's back. This **threshold effect** partly explains why so many environmental problems, such as acid deposition, which can kill certain species of trees and aquatic life, or hazardous wastes in groundwater supplies, seem to pop up, even though they have actually been building or accumulating for a long time.

Different species of plants and animals have varying ranges of tolerance for a particular environmental factor such as temperature (Figure 5-2). Those with

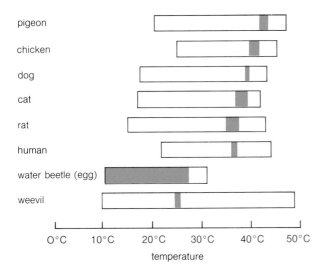

Figure 5-2 Ranges of tolerance for temperature vary among different species. Shaded areas represent the optimum temperature range.

Table 5-1	Some Effects of Environmental Stress

Organism Level

Physiological and biochemical changes
Psychological disorders
Behavioral changes
Fewer or no offspring
Genetic defects in offspring (mutagenic effects)
Birth defects (teratogenic effects)
Cancers (carcinogenic effects)
Death

Population Level

Population increase or decrease
Change in age structure (old, young, and weak may die)
Survival of strains genetically resistant to a stress
Loss of genetic diversity and adaptability
Extinction

Community—Ecosystem Level

Disruption of energy flow
   Decrease or increase in solar energy input
   Changes in heat output
   Changes in trophic structure in food chains and food webs
Disruption of chemical cycles
   Depletion of essential nutrients
   Excessive addition of nutrients
Simplification
   Reduction in species diversity
   Reduction or elimination of habitats and filled ecological niches
   Less complex food webs
   Possibility of lowered stability
   Possibility of ecosystem collapse

the widest ranges of tolerance can be found in many different types of environments and generally are less susceptible to extinction than those with narrow tolerance ranges.

Limiting Factors Another related ecological principle is the **limiting factor principle:** Too much or too little of any single abiotic factor can limit or prevent growth of the populations of a particular plant and animal species in an ecosystem even if all other factors are at or near the optimum range of tolerance for the species. The single factor that is found to be limiting growth of the population of a particular organism is called the **limiting factor.**

For example, suppose a farmer plants corn in a field containing too little phosphorus. Even if the corn's requirements for water, nitrogen, potassium, and other nutrients are met, the corn will stop growing when it has used up the available phosphorus. In this case, availability of phosphorus is the limiting factor that determines how much corn will grow in the field. Similarly, even with plenty of food and air, you will die in a relatively short time without sufficient water. Growth can also be limited by the presence of too much of a particular abiotic factor. For example, plants can be killed by being overwatered or overfertilized.

Effects of Environmental Stress Table 5-1 summarizes what can happen to organisms, populations, and ecosystems if one or more limits of tolerance are exceeded. The stresses that can cause the changes shown in Table 5-1 may result from natural hazards (earthquakes, volcanic eruptions, hurricanes, drought, floods, and fires) or from human activities (industrialization, warfare, transportation, and agriculture).

Organisms, populations, communities, and ecosystems all have some ability to withstand or recover from externally imposed changes or stresses—provided these external stresses are not too severe. In other words, they have some degree of *stability*. This stability is maintained, however, only by constant change. Although an organism maintains a fairly stable structure over its life span, it is continually gaining and losing matter and energy. Similarly, in a mature tropical rain forest ecosystem, some trees will die, others will take their place. Some species may disappear, and the number of individual species in the forest may change. But unless it is cut, burned, or blown down, you will recognize it as a tropical rain forest 50 years from now.

It is useful to distinguish between two aspects of stability in living systems. **Inertia stability,** or **persistence,** is the ability of a living system to resist being disturbed or altered. **Resilience stability** is the ability of a living system to restore itself to an original condition if the outside disturbance is not too drastic. Nature is remarkably resilient. For example, human societies survive natural disasters and wars; the genetic structure of insect populations is altered to survive massive doses of deadly pesticides and ionizing radiation; and plants eventually recolonize areas devastated by volcanoes, retreating glaciers, mining, and abandoned farmlands—although such natural restoration usually takes a long time on a human time scale.

5-2 MAJOR TYPES OF TERRESTRIAL ECOSYSTEMS

Climate and Terrestrial Ecosystems   Why is one area of the earth a desert, another a grassland, and another a forest? Why are there different types of deserts, grasslands, and forests? What determines the variations of plant and animal life among these types? In general, the answer to all these questions is differences in climate. Though the climate of an area is affected by a number of variables, the two most important are its *average temperature* and *average precipitation.*

Ecologists find it useful to classify ecosystems according to their general similarities in structure. The three major types of large terrestrial ecosystems, or biomes, are deserts, grasslands, and forests, each with a characteristic set of plants, animals, climatic conditions, and, often, general soil type. The map inside the front cover shows the general distribution of the major types of desert, grassland, and forest biomes throughout the world, and the color landscape photos inside the front and back covers illustrate most of these types.

Effects of Precipitation and Temperature on Plant Types   With respect to plants, *precipitation is the limiting factor that determines whether the biomes of most of the world's land areas are desert, grassland, or forest.* A region will normally be a **desert,** containing little vegetation or widely spaced, mostly low vegetation, if its average amount of precipitation is less than 25 centimeters (10 inches) a year (Figure 5-3). This is true regardless of the average temperature and amount of sunlight and no matter how high the concentrations of essential plant nutrients in the soil.

Grasslands are found in regions where moderate average precipitation, ranging from 25 to 75 centimeters (10 to 30 inches) a year, is great enough to allow

grass to prosper yet so erratic that periodic drought and fire prevent large stands of trees from growing. Undisturbed areas with an average precipitation of 75 centimeters (30 inches) or more a year tend to be covered with **forest,** consisting of various species of trees and smaller forms of vegetation.

The combination of average precipitation and average temperature determines the particular type of desert, grassland, or forest found in a given area. Acting together, these two factors lead to *tropical, temperate* (mid-latitude), and *polar* (high-latitude) deserts, grasslands, and forests (Figure 5-3).

Major Types of Deserts   In combination with low average precipitation, different average temperatures give rise to three types of desert: (1) *tropical deserts,* such as the Sahara, where daytime temperatures are hot year-round and nights are cold because of the lack of sufficient vegetation to moderate temperature declines; (2) *temperate (mid-latitude) deserts,* where daily temperatures are warm most of the year (see photo inside the back cover); and (3) *cold (high-latitude) deserts,* such as the Gobi, where winters are cold and summers are hot (Figure 5-3).

Tropical deserts, making up about one-fifth of the world's desert area, consist primarily of barren dunes covered with rock or sand. Temperate and cold deserts contain widely scattered thorny bushes and shrubs (acacia, mesquite, sagebrush, greasewood, creosote bush), succulents such as cacti, and small, fast-growing wildflowers that bloom in spring or after a rare, brief, drenching rain. Wide spacing between plants reduces competition for scarce water. The nutrient-poor soils, slow growth rate of plants, and lack of water make deserts fragile biomes; vegetation destroyed by human activities such as motorcycling and other off-road driving may take decades to grow back.

Major Types of Grasslands   *Tropical grasslands,* or *savannahs,* are found in areas with high average temperatures, very dry seasons about half of the year, and abundant rain the rest of the year. They are located in a wide belt on either side of the equator between the tropics (see map inside the front cover). Some of these biomes, such as Africa's Serengeti Plain, consist of open plains covered with low or high grasses; others contain grasses along with varying numbers of widely spaced, small, mostly deciduous trees and shrubs, which shed their leaves during the dry season to avoid excessive water loss (see photo inside the back cover).

Temperate (mid-latitude) grasslands, located in the large interior areas of continents, have moderate average temperatures, more even distribution of precipitation throughout the year than in tropical grasslands, cold winters with snow covering the ground at times,

Figure 5-3 Average precipitation and average temperature act together over a period of 30 years or more as limiting factors that determine the type of desert, grassland, or forest ecosystem found in a particular area.

hot and dry summers, and winds blowing almost continuously. *Polar (high-latitude) grasslands* or *Arctic tundra* in areas just below the Arctic region of perpetual ice and snow have bitter winter cold, icy galelike winds, fairly low average annual precipitation occurring primarily during a brief summer period, and long winter darkness. The wet Arctic tundra is covered with a thick, spongy mat of low-growing plants such as lichens, sedges (grasslike plants often growing in dense tufts in marshy places), mosses, grasses, and low woody shrubs (see photo inside the back cover). Although the ground is frozen in winter, snowfall is rare.

One effect of the extreme cold of this biome is the presence of **permafrost**—water permanently frozen year-round in thick undergound layers of soil. During the brief summer, when sunlight persists almost around the clock, the surface layer of soil thaws and the biome is turned into a soggy landscape dotted with shallow lakes, marshes, bogs, and ponds. The low rate of decomposition, the shallow soil, and the slow growth rate of plants make the Arctic tundra perhaps the earth's most fragile biome. Vegetation destroyed by human activities can take decades to grow back. Unless buildings, roads, pipelines, and railroads are built over bedrock, on insulating layers of gravel, or on deep-seated pilings, they melt the upper layer of permafrost and tilt or crack as the land beneath them shifts and settles (Figure 5-4).

Major Types of Forests   *Tropical rain forests* are found near the equator in areas with a warm but not hot annual mean temperature that varies little either daily or seasonally, high humidity, and heavy rainfall almost daily. These biomes are dominated by broadleaf evergreen trees that cannot tolerate freezing temperatures (Figure 4-18, p. 69). The almost unchanging climate in such forests means that neither water nor temper-

Figure 5-4 Subsidence from thawing of permafrost during construction made this railroad track near Valdez, Alaska, useless.

U.S. Geological Society/O. J. Ferrains

ature is a limiting factor as they are in other biomes. Because of this year-round growing season and lack of major limiting factors, a mature tropical rain forest has a greater diversity of plant and animal species per unit of area than any other biome. More species of animals can be found in a single tree in such a forest than in an entire forest at higher latitudes.

Most of the nutrients in this biome are found in the vegetation, not in the upper layers of soil as in most other biomes. Once the vegetation is removed, the thin soils rapidly lose the few nutrients they have and cannot grow crops for more than a few years without large-scale use of commercial fertilizers. Furthermore, when vegetation is cleared, the heavy rainfall washes away most of the thin layer of topsoil; thus regeneration of a mature rain forest on large cleared areas is next to impossible.

Temperate (mid-latitude) deciduous forests occur in areas with moderate average temperatures that change significantly during four distinct seasons. They have long summers, not very severe winters, and abundant precipitation spread fairly evenly throughout the year. They are dominated by a few species of broadleaf deciduous trees such as oak, hickory, maple, poplar, sycamore, and beech (see photo inside the front cover) that can survive during the winter by dropping their leaves and going into a dormant state.

Cold (high-latitude) northern coniferous forests, also called *boreal forests* and *taiga,* are found in regions with a subarctic climate, where winters are long and dry with only light snowfall, temperatures range from cool to extremely cold, and summers are very brief with mild to warm temperatures. These forests, which form an almost unbroken belt across North America and northern Eurasia (see map inside the front cover), are dominated by a few species of coniferous evergreen trees such as spruce, fir, larch, and pine (see photo inside the front cover). Their needle-shaped, waxy-coated leaves conserve heat and water during the long, cold, dry winters. Beneath the dense stands of trees, a carpet of fallen needles and leaf litter covers the nutrient-poor soil, which remains waterlogged during the brief summers. Plant diversity is low in the northern forests, because few species can survive the long, cold winters, when soil moisture is frozen.

Effects of Other Abiotic Factors   Other abiotic factors such as soil type (especially those low in nutrients), light, and fire can act as limiting factors that cause variations in the plant life (and thus the animal life) of major biomes. For example, within the deciduous forests found in the eastern United States, beeches and maples predominate on nutrient-rich soils, oaks and hickories on rocky, nutrient-poor, well-drained soils.

Light is often the limiting factor determining the plants and animals in various layers of vegetation in forest ecosystems. For example, in a tropical rain forest, the dense overhead canopy of trees (Figure 4-18, p. 69) allows relatively little sunlight to reach the forest floor. Hence relatively little vegetation is found on the forest floor. In a deciduous forest, the lack of cover during the early spring and late fall allows enough light to support a number of plant species on the forest floor. Floods, natural and deliberately set fires, and land clearing for human activities can also drastically modify the general type of biome found in an area.

5-3   AQUATIC ECOSYSTEMS

Limiting Factors of Aquatic Ecosystems   The main factors affecting the types and numbers of organisms found in aquatic ecosystems are salinity (the concen-

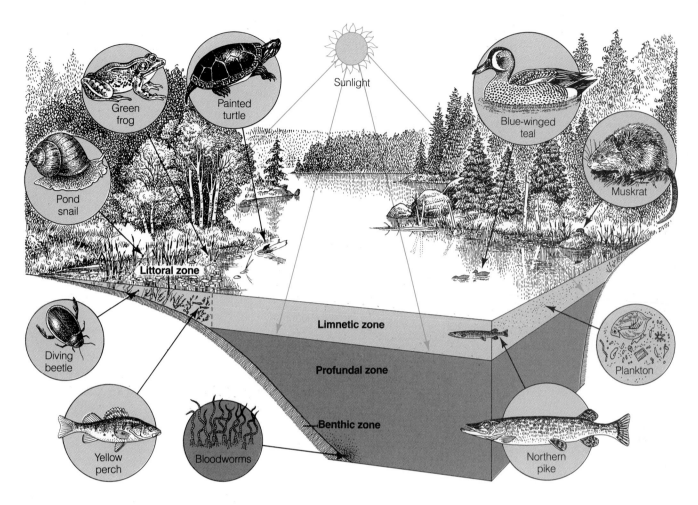

Figure 5-5 Four major zones of life in a lake.

tration of dissolved salts, especially sodium chloride, in a body of water), depth to which sunlight penetrates, amount of dissolved oxygen, and water temperature. Salinity levels are used to divide aquatic ecosystems into two major classes: *freshwater ecosystems*, consisting of inland bodies of standing water (lakes, reservoirs, ponds, and wetlands) and flowing water (streams and rivers) with low salinity, and *marine* or *saltwater ecosystems*, such as oceans, estuaries (where fresh water from rivers and streams mixes with seawater), coastal wetlands, and coral reefs with high to very high salinity levels.

Freshwater Lakes and Reservoirs Large natural bodies of standing fresh water formed when precipitation, land runoff, or groundwater flows filled depressions in the earth created by glaciation (the Great Lakes of North America), earthquakes (Lake Nyasa in East Africa), volcanic activity (Lake Kivu in Africa), and crashes of giant meteorites are called **lakes.**

Lakes normally consist of four distinct zones (Fig-

ure 5-5) that provide a variety of ecological niches for different species of plant and animal life. The **littoral zone,** the shallow, nutrient-rich waters near the shore, contains rooted aquatic plants and an abundance of other forms of aquatic life. The **limnetic zone,** the open water surface layer, receives sufficient sunlight for photosynthesis and contains varying amounts of floating phytoplankton, plant-eating zooplankton, and fish, depending on the availability of plant nutrients. The **profundal zone** of deep water not penetrated by sunlight is inhabited mostly by fish, such as bass and trout, that are adapted to its cooler, darker water and lower levels of dissolved oxygen. The **benthic zone** at the bottom of the lake is inhabited primarily by large numbers of bacteria, fungi, bloodworms, and other decomposers, which live on dead plant debris, animal remains, and animal wastes that float down from above.

Eutrophication is the process in which lakes receive inputs of plant nutrients (mostly nitrates and phosphates) from the surrounding land basin as a result of natural erosion and runoff. This process gradually fills lakes with sediment over thousands to millions of years. However, near urban or agricultural centers,

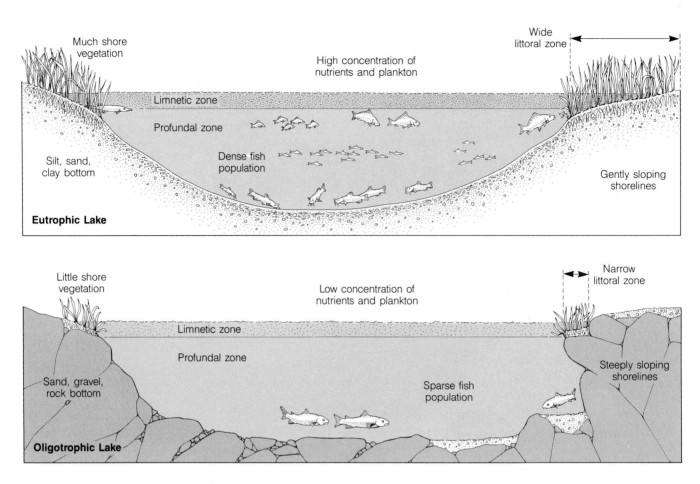

Eutrophic Lake

Much shore vegetation

High concentration of nutrients and plankton

Wide littoral zone

Limnetic zone

Profundal zone

Dense fish population

Silt, sand, clay bottom

Gently sloping shorelines

Oligotrophic Lake

Little shore vegetation

Low concentration of nutrients and plankton

Narrow littoral zone

Limnetic zone

Profundal zone

Sand, gravel, rock bottom

Sparse fish population

Steeply sloping shorelines

Figure 5-6   Eutrophic, or nutrient-rich, lake and oligotrophic, or nutrient-poor, lake.

Figure 5-7   Reservoir formed behind Shasta Dam on the Sacramento River north of Redding, California.

U.S. Department of Interior/Bureau of Reclamation/J. C. Dahilig

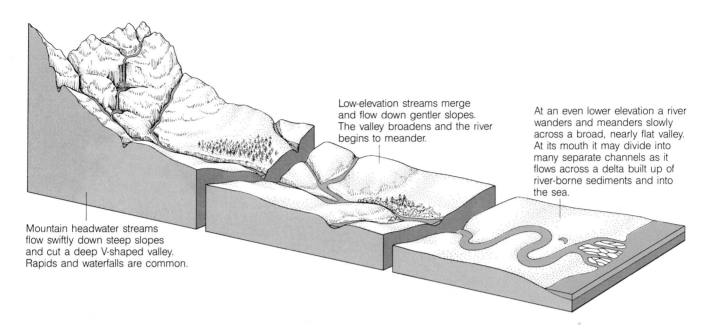

Mountain headwater streams flow swiftly down steep slopes and cut a deep V-shaped valley. Rapids and waterfalls are common.

Low-elevation streams merge and flow down gentler slopes. The valley broadens and the river begins to meander.

At an even lower elevation a river wanders and meanders slowly across a broad, nearly flat valley. At its mouth it may divide into many separate channels as it flows across a delta built up of river-borne sediments and into the sea.

Figure 5-8 Three phases in the flow of water downhill from mountain headwater streams to wider, lower-elevation streams to rivers, which empty into the ocean.

the input of nutrients can be greatly accelerated as a result of human activities. This **accelerated, or cultural, eutrophication** is caused by effluents from sewage treatment plants, runoff of fertilizers and animal wastes, and soil erosion.

Lakes can be grouped into two major types: eutrophic and oligotrophic (Figure 5-6). A lake with a large or excessive supply of plant nutrients is called a **eutrophic lake.** This type of lake is usually shallow and has cloudy, warm water, large populations of phytoplankton (especially algae) and zooplankton, and diverse populations of fish. A lake with a low supply of plant nutrients is called an **oligotrophic lake.** This type of lake is usually deep and has crystal-clear water, cool to cold temperatures, and relatively small populations of phytoplankton and fish. Many lakes fall somewhere between these two extremes of nutrient enrichment and are called **mesotrophic lakes.**

Sometimes incorrectly called lakes, **reservoirs** are fairly large and deep, human-created bodies of standing fresh water often built behind a dam (Figure 5-7). Reservoirs are built primarily for water storage. Unlike lakes, the volume of water they contain is determined by the amount being used for hydroelectric power production, irrigation, or domestic consumption.

Freshwater Streams and Rivers Precipitation that does not infiltrate into the ground or evaporate remains on the earth's surface as surface water. This water becomes runoff that flows into streams and rivers and eventually downhill to the oceans for reuse in the

hydrologic cycle (Figure 4-11, p. 61). The entire land area that delivers the water, sediment, and dissolved substances via streams to a major river, and ultimately to the sea, is called a **watershed** or **drainage basin.**

The downward flow of water from mountain highlands to the sea takes place in three phases (Figure 5-8). In the first phase, narrow headwater or mountain highland streams with cold, clear water rush down steep slopes. As this turbulent water flows and tumbles downward over waterfalls and rapids, it dissolves large amounts of oxygen from the air. Thus most fish that thrive in this environment are cold-water fish, such as trout, which require a high level of dissolved oxygen.

In the second phase, various headwater streams merge to form wider, deeper, lower-elevation streams that flow down gentler slopes and meander through wider valleys. Here the water is warmer and usually less turbulent and can support a variety of cold-water and warm-water fish species with slightly lower oxygen requirements. Gradually these streams coalesce into wider and deeper rivers that meander across broad, flat valleys. The main channels of these rivers support a distinctive variety of fish, whereas their backwaters support species similar to those found in lakes. At its mouth a river may divide into many channels as it flows across a **delta**—a built-up deposit of river-borne sediments—before reaching the ocean.

Inland Wetlands Land that remains flooded all or part of the year with fresh or salt water is called a

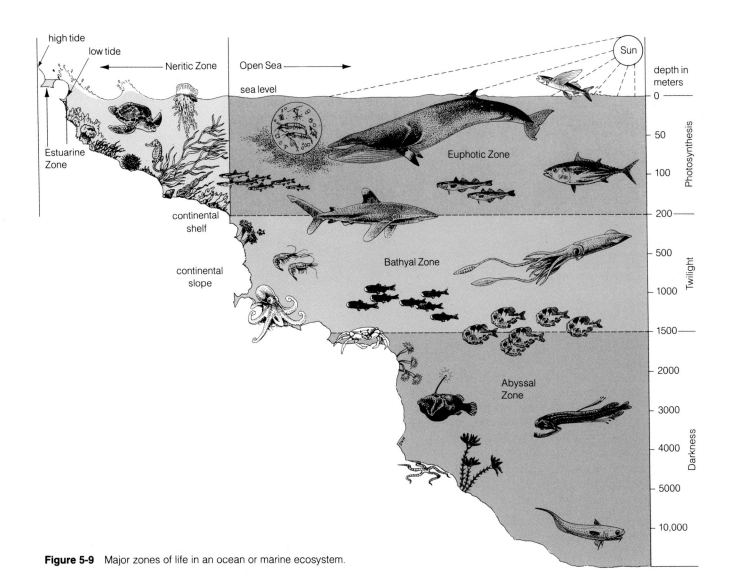

Figure 5-9 Major zones of life in an ocean or marine ecosystem.

wetland. Bogs, marshes, swamps, and river-overflow lands that are covered with fresh water and found inland are called **inland wetlands.** Those found near the coast and covered with salt water are known as **coastal wetlands.**

In the United States, inland wetlands found in the lower 48 states are roughly equivalent in total area to the state of California. Almost twice this area of inland wetlands is found in Alaska. These wetlands include a variety of ecosystems: red-maple swamps and black-spruce bogs in the northern states, bottomland hard-wood forests in the Southeast, prairie potholes in the Midwest, and wet tundra in Alaska.

In addition to providing habitats for a variety of fish and wildlife, inland wetlands store and regulate stream flow, reducing flooding frequency and down-stream peak flood levels. They also improve water quality by trapping stream sediments and reducing levels of many toxic pollutants through uptake by veg-etation and by degradation to less harmful sub-

stances. By holding water, they also allow increased infiltration, thus helping recharge groundwater sup-plies. They are also used to grow important crops such as blueberries, cranberries, and wild rice.

Because people are often unaware of their ecolog-ical importance, inland wetlands are dredged or filled in and used as croplands, garbage dumps, and as sites for urban and industrial development. Over half of the country's original wetlands have been destroyed, and at least an additional 120,000 hectares (300,000 acres) are destroyed each year.

Marine Aquatic Ecosystems As landlubbers, we tend to think of the earth in terms of land. It is more accurately described, however, as the "water planet," because 71% of its surface is covered by water, and 97% of that water is in the oceans. These oceans play key roles in the survival of life on earth. As the ulti-mate receptacle for terrestrial water, they dilute many

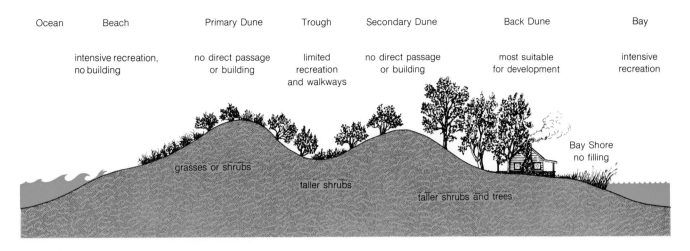

Ocean	Beach	Primary Dune	Trough	Secondary Dune	Back Dune	Bay
	intensive recreation, no building	no direct passage or building	limited recreation and walkways	no direct passage or building	most suitable for development	intensive recreation

grasses or shrubs

taller shrubs

taller shrubs and trees

Bay Shore
no filling

Figure 5-10 Primary and secondary dunes on a barrier beach. Ideally, construction and development should be allowed only behind the second strip of dunes, with walkways to the beach built over the dunes to keep them intact.

human-produced wastes to less harmful or harmless levels. They also help redistribute solar energy as heat through currents and evaporation as part of the hydrologic cycle, and they figure significantly in other major biogeochemical cycles. In addition, they serve as a gigantic reservoir of dissolved oxygen and carbon dioxide, which, respectively, help regulate the composition of the air we breathe and the temperature of the atmosphere. Oceans also provide ecological niches for about 250,000 species of marine plants and animals, which provide food for many organisms, including human beings, and serve as a source of iron, sand, gravel, phosphates, oil, natural gas, and many other valuable resources.

Coastal Zone Each of the world's oceans can be divided into two major zones: neritic, or coastal, and open ocean (Figure 5-9). The **neritic, or coastal, zone** is the relatively warm, nutrient-rich, shallow water that extends from the high-tide mark on land to the edge of a shelflike extension of continental landmassses known as the *continental shelf. The neritic zone, representing less than 10% of the total ocean area, contains 90% of all ocean plant and animal life and is the site of most of the major commercial marine fisheries.*

The neritic zone includes a number of different habitats. Some coasts have gently sloping **barrier beaches** at the water's edge. If not destroyed by human activities, the two rows of sand dunes on such beaches serve as the first line of defense against the ravages of the sea (Figure 5-10). For more effective flood protection, buildings should be placed behind the secondary dunes; walkways should be built over both dunes to keep them intact. When coastal developers remove the dunes or build behind the first set of dunes,

minor hurricanes and sea storms can flood and even sweep away houses and other buildings. Other coasts contain steep **rocky shores** pounded by waves; many organisms live in the numerous intertidal pools in the rocks.

Estuaries, found along coastlines where fresh water from rivers mixes with salty oceanic waters, provide aquatic habitats with a lower average salinity than the waters of the open ocean. Extending inland from the estuaries are **coastal wetlands.** In temperate areas, coastal wetlands usually consist of a mix of bays, lagoons, and salt marshes where grasses are the dominant vegetation (see photo inside the front cover); in tropical areas, we find mangrove swamps dominated by mangrove trees.

The coastal zones of warm tropical and subtropical oceans often contain **coral reefs** (see photo inside the front cover), consisting mostly of calcium carbonate secreted by photosynthesizing red and green algae and small coral animals. These reefs support at least one-third of all marine fish species as well as numerous other marine organisms. Along some steep coastal areas, such as those of Peru, winds blow the surface water away from the shore, allowing cold, nutrient-rich bottom waters, or **upwellings,** to rise to the surface and support large populations of plankton, fish, and fish-eating seabirds (Figure 5-11).

Strings of thin **barrier islands** in some coastal areas (such as portions of North America's Atlantic and Gulf coasts) help protect beaches, estuaries, and wetlands by dissipating the energy of approaching storm waves. Human structures built on these slender ribbons of sand with water on all sides will sooner or later be damaged or destroyed by flooding erosion as a result of storms and hurricanes and by the slow but continual movement of the islands toward the mainland

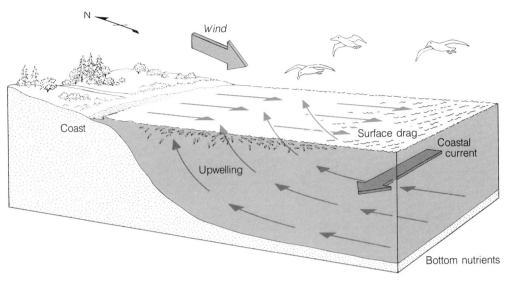

Figure 5-11 An upwelling is caused when deep, cool, nutrient-rich waters are drawn up to replace surface water moved away from the steep coast by wind-driven currents. Such areas support large populations of plankton, fish such as anchovies, and fish-eating seabirds.

N

Wind

Coast

Surface drag

Coastal current

Upwelling

Bottom nutrients

under the influence of currents and winds. As they move, the islands relocate with no net loss of beach, but human structures built near the original high-tide line are gradually undermined.

Many people view estuaries and coastal wetlands as desolate, mosquito-infested, worthless lands that should be drained, dredged, filled in, built on, or used as depositories for human-generated pollutants and waste materials. Nothing could be further from the truth. These areas provide us and many other species with a number of vital services. They supply food and serve as spawning and nursery grounds for many species of marine fish and shellfish; they are feeding, nesting, and rearing sites for millions of birds and waterfowl. Three-fourths of the commercially important aquatic animal species in the United States spend all or part of their life in estuaries and coastal wetlands. In the United States, the $15-billion-a-year commercial and recreational marine fishing industry provides jobs for millions of people.

The coastal areas also dilute and filter out large amounts of waterborne pollutants, helping protect the quality of waters used for swimming, fishing, and wildlife habitats. It is estimated that one acre of tidal estuary substitutes for a $75,000 waste treatment plant and is worth a total of $83,000 when its production of fish for food and recreation is included.

In addition, estuaries, coastal wetlands, and the sand dunes of barrier beaches help protect coastal areas by absorbing damaging waves caused by violent storms and hurricanes and serving as giant sponges to absorb floodwaters. In many heavily populated coastal areas, human activities are increasingly threatening the abundance of plant and animal life in estuaries and coastal wetlands and destroying some of the important services these ecosystems provide (see Spotlight on p. 84).

Open Sea The sharp increase in the depth of the water at the edge of the continental shelf marks the separation of the neritic zone from the **open sea** (Figure 5-9). This marine zone contains about 90% the total surface area of the ocean but only about 10% of its plant and animal life. The open sea is divided into three vertical zones. The surface layer, through which enough sunlight can penetrate for photosynthesis, is called the **euphotic zone.** It supports scattered populations of phytoplankton fed upon by zooplankton, which in turn support commercially important herrings, sardines, anchovies, and other small fish that feed at the surface and their larger predators such as tuna, mackerel, and swordfish.

Below this zone is the **bathyal zone,** a colder, darker layer where some sunlight penetrates but not enough to support photosynthesis. Going deeper we find the **abyssal zone,** a layer of deep, pitch-dark, usually near-freezing water, and the ocean bottom. About 98% of the ocean's different species (many of them decomposer bacteria) are found in the abyssal zone. Most survive by feeding on dead plants and animals and their waste products, which sink down from the surface waters, and by making daily migrations (usually near dusk) to surface waters to feed.

In 1977 a highly productive aquatic ecosystem was discovered deep within the abyssal zone around vents in the ocean floor that spew forth large amounts of very hot, rotten-egg-smelling hydrogen sulfide gas. In this pitch-dark, abnormally warm environment, teeming clouds of specialized bacteria convert hydrogen sulfide into chemical energy without the presence of sunlight, through a process called **chemosynthesis.** These bacteria in turn support a variety of abnormally large, strange-looking worms, clams, blind white crabs, and other animals. This recent discovery reminds us how little we know about life in the deep ocean.

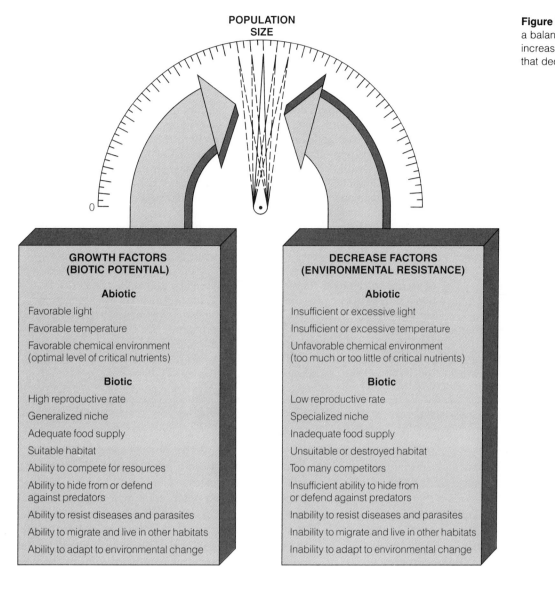

POPULATION SIZE

Figure 5-12 Population size is a balance between factors that increase numbers and factors that decrease numbers.

GROWTH FACTORS (BIOTIC POTENTIAL)

Abiotic

Favorable light

Favorable temperature

Favorable chemical environment (optimal level of critical nutrients)

Biotic

High reproductive rate

Generalized niche

Adequate food supply

Suitable habitat

Ability to compete for resources

Ability to hide from or defend against predators

Ability to resist diseases and parasites

Ability to migrate and live in other habitats

Ability to adapt to environmental change

DECREASE FACTORS (ENVIRONMENTAL RESISTANCE)

Abiotic

Insufficient or excessive light

Insufficient or excessive temperature

Unfavorable chemical environment (too much or too little of critical nutrients)

Biotic

Low reproductive rate

Specialized niche

Inadequate food supply

Unsuitable or destroyed habitat

Too many competitors

Insufficient ability to hide from or defend against predators

Inability to resist diseases and parasites

Inability to migrate and live in other habitats

Inability to adapt to environmental change

5-4 POPULATION RESPONSES TO STRESS

Changes in Population Size Populations of plants and animals that make up ecosystems respond in various ways to changes in environmental conditions such as an excess or shortage of food or other critical nutrients. The birth rate may increase to allow the population to take advantage of an increase in food or other resources, or the death rate may increase to reduce the population to a size that can be supported by available resources. Also, the structure of the population may change. The old, very young, and weak members may die, leaving the population more capable of surviving such stresses as a more severe climate, an increase in predators, or an increase in disease organisms. The major abiotic and biotic factors that tend to increase or decrease the population size of a given species are summarized in Figure 5-12.

The response of the population of a species to a change in resource availability or to environmental stress usually can be represented by two simple curves: a J-shaped curve and an S-shaped curve (Figure 5-13). With unlimited resources and ideal environmental conditions, a species can produce offspring at its maximum rate, called its **biotic potential.** Such growth starts off slowly and then increases rapidly to produce an exponential, or J-shaped, curve of population growth. Species such as bacteria, insects, and mice, which can produce a large number of offspring in a short time, have high biotic potentials; larger species such as elephants and humans, which take a much longer time to produce only a few offspring, have low biotic potentials. Since environmental conditions usually are not ideal, a population rarely reproduces at its biotic potential.

About 55% of the U.S. population lives along the coastlines of the Atlantic Ocean, the Pacific Ocean, and the Great Lakes, and two out of three Americans live within 80 kilometers (50 miles) of these shorelines. By the end of this century, three out of four will live in or near the coastal zone. Nine of the country's largest cities, most major ports, about 40% of the manufacturing plants, and two out of three nuclear and coal-fired power plants are located in coastal counties. The coasts are also the sites of large numbers of motels, hotels, condominiums, beach cottages, and other developments.

Because of these multiple uses and stresses, nearly 50% of the estu- aries and coastal wetlands in the United States have been destroyed or damaged, primarily by dredging and filling and contamination by wastes. Fortunately, about half of the country's estuaries and coastal wetlands remain undeveloped, but each year additional areas are developed. Some coastal areas have been purchased by federal and state governments and by private conservation agencies, which protect them from development and allow most of them to be enjoyed for recreational purposes as parks and wildlife habitats.

The National Coastal Zone Management Acts of 1972 and 1980 provided federal aid to the 37 coastal and Great Lakes states and territories to help them develop voluntary programs for protecting and managing coastlines not under federal protection. By the end of 1987, more than 90% of the country's coastal areas in all but six of the eligible states fell under federally approved state coastal management plans. These plans, however, are voluntary, and many are vague and do not provide sufficient enforcement authority; since 1980 their implementation has also been hindered by federal budget cutbacks. California and North Carolina are considered to have the strongest programs, but developers and other interests make continuing efforts to weaken such programs.

The maximum population size of each species that an ecosystem can support *indefinitely* under a given set of environmental conditions is called that ecosystem's **carrying capacity.** All the limiting factors that reduce the growth rate of a population are called the population's **environmental resistance.** These factors include predation, interspecific competition for resources, food shortage, disease, adverse climatic conditions, and lack of suitable habitat. As a population encounters environmental resistance, the J-shaped curve of population growth bends away from its steep incline and eventually levels off at a size that typically fluctuates above and below the ecosystem's carrying capacity. In other words, environmental resistance lowers the biotic potential of a population and converts a J-shaped curve into an S-shaped curve (Figure 5-13).

Some populations (especially rapidly reproducing ones such as insects, bacteria, and algae) may surpass the carrying capacity and then undergo a rapid decrease in size, known as a **population crash** (Figure 5-14). Some population crashes involve sharp increases in the death rate; others involve combinations of a rise in the death rate coupled with emigration of large numbers of individuals to other areas. A population crash can also occur when a change in environmental conditions lowers the carrying capacity of an ecosystem.

Crashes have occurred in the human populations of various countries throughout history. Ireland, for example, experienced a population crash after a fungus infection destroyed the potato crop in 1845. Dependent on the potato for a major portion of their diet, by 1900 half of Ireland's 8 million people had died of starvation or emigrated to other countries. In spite of such local and regional disasters, the overall human population on earth has continued to grow, because technological, social, and other cultural changes have extended the earth's carrying capacity for human beings. In essence, human beings have been able to alter their ecological niche by increasing food production, controlling disease, and using large amounts of energy and matter resources to make normally uninhabitable areas of the earth inhabitable. Figure 5-15 shows that after each major technological change—except the industrial-scientific revolution we are still experiencing—the population grew rapidly and then leveled off.* The dashed lines in Figure 5-15 show three projections of future population growth based on our present understanding. No one, of course, knows what the present or future worldwide or regional carrying capacity for the human population is or what

---

*The curve shown in Figure 5-15 is plotted by a different mathematical method (a plot of the logarithm of population size versus the logarithm of time) from the method (a plot of population size versus time) used in Figure 1-1 on p. 3.

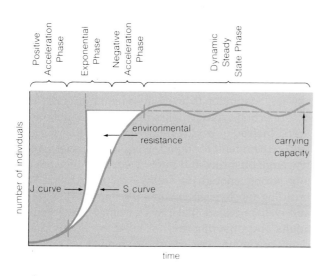

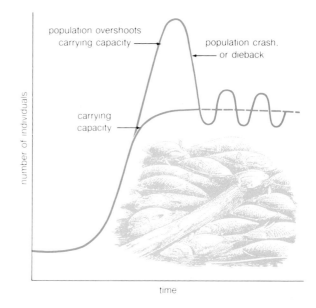

Figure 5-13 The J-shaped curve of population growth of a species is converted to an S-shaped curve when the population encounters environmental resistance and exceeds one or more limiting factors.

Figure 5-14 A population crash can occur when a population temporarily overshoots its carrying capacity or a change in environmental conditions lowers the carrying capacity of its environment.

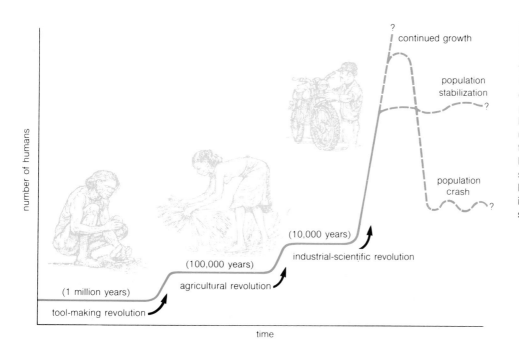

Figure 5-15 Human beings have expanded the earth's carrying capacity for their species through technological innovation, leading to several major cultural changes. Dashed lines represent possible future changes in human population size: continued growth, population stabilization, and continued growth followed by a crash and stabilization at a much lower level. (These curves are generalized log-log plots not drawn to scale.)

the limiting factor or factors might be—food, air, water, or capacity of the environment to absorb pollution and recover from environmental degradation.

Biological Evolution and Natural Selection In addition to changes in population size, a population of a particular species can also adapt to a change in environmental conditions by changing its genetic composition (*gene pool*). All individuals of a population do not have exactly the same genes. These genetic differences (*genotypes*) mean that some individuals in the population have more tolerance to a particular environmental change than others. When subjected to a stress, more of these individuals survive than those without such traits and pass their favorable genetic traits to their offspring. As a result, the gene pool of the population is made more adaptable to a particular environmental stress. Such a change in the genetic composition of a population, when individuals with one kind of genetic endowment regularly outproduce individuals with other genetic endowments as the population is exposed to an altered environment, is called **biological evolution** or **evolution.**

The process by which some genes and gene combinations in a population are reproduced more than others is called **natural selection.** Charles Darwin, who in 1858 first proposed this idea, described natural selection as meaning "survival of the fittest." This has often been misinterpreted to mean survival of the strongest, biggest, or most aggressive. Instead, survival of the fittest means that the most fit genotype in a population is the one that on the average produces the most offspring in a given generation.

Today we have a much better understanding of how natural selection takes place. We now know that hereditary traits of individual organisms are carried in the genes and that this information is coded in the sequence of certain groups of atoms in molecules of DNA found in genes. Individuals in a population of a species don't have exactly the same genes because of **mutations,** inheritable changes in the structure of their genes. Some mutations result from exposure to various external environmental factors, such as ionizing radiation, heat, and certain chemicals that subject DNA molecules to changes in their chemical makeup. Mutations can also occur internally when a cell divides (asexual reproduction) and when two cells, sperm and egg, fuse (sexual reproduction).

Some of these inheritable mutations are harmful and others are helpful. If a new genetic variation interferes with survival and reproduction, organisms with this trait die prematurely. With each successive generation, more of these harmful genes are eliminated from the gene pool of the population. However, if a particular mutation enhances survival and reproduction of an organism, this variation is passed on to greater numbers of offspring in succeeding generations; thus natural selection has taken place.

Species differ widely in how rapidly they can undergo evolution through natural selection. Some species have many offspring and short generation times, others have few offspring and long generation times. Those that can quickly produce a large number of tiny offspring with short average life spans (weeds, insects, and rodents) can adapt to a change in environmental conditions through natural selection in a relatively short time. Most of the offspring of such species die before reproducing because after birth they are usually left to fend for themselves. However, such a large number of offspring are produced that the chances of a few individuals surviving, reproducing, and passing survival-enhancing genes to the next generation are ensured. For example, in only a few years many species of mosquitoes have become genetically

resistant to DDT and other pesticides, and many species of bacteria have become genetically resistant to widely used antibiotics, such as penicillin.

Other species, such as elephants, horses, tigers, white sharks, and humans, have long generation times and a small number in each litter and thus cannot reproduce a large number of offspring rapidly. For such species, adaptation to an environmental stress by natural selection typically takes thousands to millions of years. Instead of having numerous tiny offspring and letting them fend for themselves, such species have a few large offspring and nurse them until they are big and strong enough to survive on their own. This breeding mechanism ensures the species' long-term survival, provided that most or all of the breeding adults and offspring are not killed by disease, accident, predation, starvation, or some other environmental stress before they can reproduce. If that happens, the species becomes extinct and the genetic line ends.

Not only has the human species used its intelligence to develop cultural mechanisms for controlling and adapting to environmental stresses, it has also learned to bring about genetic change in other species at a fast rate—first through crossbreeding and recently through genetic engineering.

Speciation Genetic changes in existing species can occur not only within a single genetic line but also by the splitting of lines of descent into new species through **speciation:** the gradual formation of two or more species from one in response to new environmental conditions. Speciation has led to the estimated 5 million to 10 million different species found on earth today. Although in some rapidly producing organisms speciation may take place in thousands or even hundreds of years, in most cases it takes from tens of thousands to millions of years.

Exactly how speciation takes place is not fully understood, partly because it takes place so slowly that biologists have been able to observe only part of the process. In general, most speciation is believed to occur when a population of a particular species becomes distributed over areas with quite different environments for long periods. For example, different populations of the same species may be geographically isolated when floods, hurricanes, earthquakes, or other geological processes break up a single land mass into separate islands. Populations also split up when part of the group migrates in search of food (Figure 5-16).

If different populations of a single species are geographically isolated from one another over a great many generations (typically 1,000 to 100,000 generations), they begin to diverge genetically through natural selection in response to their different climates, food sources, soils, and other environmental factors. Even-

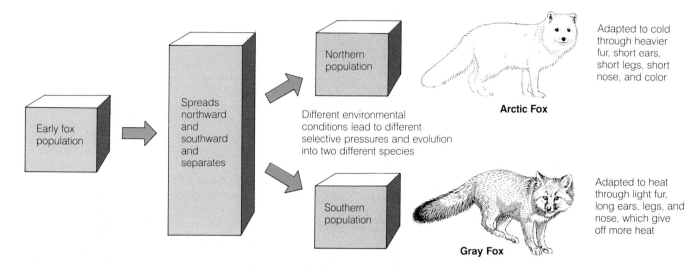

Northern population

Adapted to cold through heavier fur, short ears, short legs, short nose, and color

Arctic Fox

Different environmental conditions lead to different selective pressures and evolution into two different species

Southern population

Adapted to heat through light fur, long ears, legs, and nose, which give off more heat

Gray Fox

Early fox population

Spreads northward and southward and separates

Figure 5-16 Speciation of an early species of fox into two different species as a result of migration of portions of the original fox population into areas with different climates.

tually, the geographically isolated populations become so genetically different that they are no longer capable of successfully interbreeding should they subsequently come to occupy the same area again. Thus they have become different species—speciation has taken place.

The incredible genetic diversity created over billions of years within populations of the same species and among different species is nature's "insurance policy" against known and unforeseen disasters. Although many more species than currently exist have appeared and disappeared throughout earth's history, every species here today represents stored genetic information that allows the species to adapt to certain changes in environmental conditions. *This genetic information contained in living species and the cultural information passed from one human generation to the next are the most valuable resources on this planet.*

5-5 ECOSYSTEM RESPONSES TO STRESS

Responses to Small and Moderate Stress Ecosystems are so complex and variable that ecologists have little understanding of how they maintain their inertia stability and resilience stability. One major problem is the difficulty of conducting controlled experiments. Identifying and observing even a tiny fraction of the interacting variables found in simple ecosystems are virtually impossible. In addition, ecologists cannot run long-term experiments in which only one variable in a natural ecosystem is allowed to change. Greatly simplified ecosystems can be set up and observed under laboratory conditions, but to extrap-

olate the results of such experiments to more complex, natural ecosystems is difficult if not essentially impossible.

Numerous observations of laboratory and natural ecosystems, however, have led ecologists to the conclusion that one factor affecting the inertia stability, or persistence, of *some* ecosystems under small or moderate environmental stress is **species diversity**—the number of different species and their relative abundances in a given area. High species diversity tends to increase long-term persistence, because with so many different species, ecological niches, and linkages between them, risk is spread more widely. Because it does not "put all its eggs in one basket," an ecosystem with a diversity of species has more ways available to respond to most environmental stresses.

For example, the loss or drastic reduction of one species in a diverse ecosystem with complex food webs usually does not threaten the existence of others, because most predators have several alternative food supplies. In contrast, a highly simplified agricultural ecosystem planted with a single crop (monoculture) such as corn or wheat is highly vulnerable to destruction from a single plant disease or insect. The situation is parallel to that of a city that has a number of diverse economic interests and businesses and therefore is more likely to have long-term persistence than a city that depends primarily on only one business.

However, some research indicates that exceptions to this intuitively appealing idea may exist. Coastal salt marshes, for example, are not very diverse ecosystems but have persistence if not disturbed too severely, whereas the intertidal ecosystems of rocky seashores have a very high species diversity but can be upset by a single change in their species composition—especially loss of the starfish population.

Controversies have arisen, however, over the interpretation of these and other studies of the relationships between species diversity and ecosystem inertia stability—primarily because of differences in how the terms *diversity* and *stability* are defined for experimental purposes. For example, should species diversity be confined to counts of just the number of different species in an area, or should the relative abundance of each species also be considered? Should the characteristics of the plant and animal species also be considered? Is an ecosystem containing 100 deer and 1,000 rabbits as diverse as one containing 100 deer and 1,000 grasshoppers? Is an ecosystem considered stable as long as its species composition undergoes little change (persistence) or as long as the population sizes of its species remain relatively constant (constancy)?

Does an ecosystem need both high inertia stability and high resilience stability to be considered stable? Evidence indicates that some ecosystems have one type of stability but not the other. For example, California redwood forests and tropical rain forests have high species diversity and high inertia stability and are thus hard to alter significantly or destroy through natural processes. However, once large tracts of these diverse ecosystems are completely cleared, they have such low resilience stability that restoring them is nearly impossible. On the other hand, grasslands, much lower in species diversity, can burn easily and thus have low inertia stability, but because most of their plant matter consists of roots beneath the ground surface, these ecosystems have high resilience stability that allows them to recover quickly. The grassland can be destroyed only if its roots are plowed up and the soil is used to plant wheat or some other crop.

Clearly, we have a long way to go in understanding how the factors involved in natural ecosystems interact. We don't know for certain that high species diversity always contributes to either the inertia or resilience stability, or both, of a particular ecosystem. But considerable evidence indicates that simplifying an ecosystem by the intentional or accidental removal of a species often has unpredictable short- and long-term harmful effects (see Spotlight above).

Responses to Large-Scale Stress: Ecological Succession Most ecosystems can adapt not only to small and moderate changes in environmental conditions but also to quite severe changes. Sometimes, for example, little vegetation and soil are left as a result of a natural environmental change (retreating glaciers, fires, floods, volcanic eruptions, earthquakes) or a human-induced change (fires, land clearing, surface mining, flooding to create a pond or reservoir, pollution).

After such a large-scale disturbance, life usually

begins to recolonize a site in a series of stages. First, a few hardy pioneer species invade the environment and start creating soil or, in aquatic ecosystems, sediment. Eventually these pioneer species change the soil or bottom sediments and other conditions so much that the area is less suitable for them and more suitable for a new group of plants and animals with different ecological niche requirements. Gradually these new invaders alter the local environment still more by changing the soil, providing shade, and creating a greater variety of ecological niches, paving the way for invasion by a third wave of plants and animals. Each successive invasion makes the local environment more suitable for future invaders and less suitable for previous communities. This process, in which communities of plant and animal species are replaced over time by a series of different and usually more complex communities, is called **ecological succession.**

exposed rocks

lichens
and mosses

small herbs
and shrubs

heath mat

jack pine,
black spruce,
and aspen

balsam fir,
paper birch, and
white spruce
climax community

Time ⟶

Figure 5-17 Primary ecological succession of plant communities over several hundred years on a patch of exposed rock on Isle Royale in northern Lake Superior. Succession of animal communities is not shown.

If not severely disrupted, ecological succession usually continues until the community becomes much more self-sustaining and stable than the preceding ones and what ecologists call the **climax ecosystem** or **climax community** occupies the site. Depending primarily on the climate, climax terrestrial ecosystems may be various types of mature grasslands, forests, or deserts (Figure 5-3, p. 75).

Ecologists recognize two types of ecological succession: primary and secondary. Which type takes place depends on the conditions at a particular site at the beginning of the process. **Primary succession** is the sequential development of communities in a bare or soilless area. Examples of such areas include the rock or mud exposed by a retreating glacier or mudslide, cooled volcanic lava, a new sandbar deposited by a shift in ocean currents, and surface-mined areas from which all topsoil has been removed. On such barren surfaces, primary succession from bare rock to a mature forest may take hundreds to thousands of years.

Figure 5-17 shows the stages of primary succession from bare rock exposed by retreating glaciers to a climax natural community of balsam fir, paper birch,

and white spruce on Isle Royale in northern Lake Superior. First, retreating glaciers exposed bare rock. Wind, rain, and frost weathered the rock surfaces to form tiny cracks and holes. Water collecting in these depressions slowly dissolved minerals out of the rock's surface. The minerals—the inorganic basis of a pioneer soil layer—supported hardy pioneer plants, such as lichens (two plants, a fungus and an alga, living together in a mutualistic relationship). Gradually covering the rock surface and secreting a weak acid (carbonic acid), which dissolved additional minerals from the rock, the scaly or crusty lichens were replaced by mosses. Decomposer organisms moved in to feed on the dead lichens and mosses and were followed by a few small animals such as ants, mites, and spiders. This first successfully integrated set of plants, animals, and decomposers is called the **pioneer community**.

After several decades, the pioneer community had built up enough organic matter in its thin soil to support the roots of small herbs and shrubs. These newcomers slowed the loss of moisture and provided food and cover for new plants, animals, and decomposers. Under these modified conditions, the species in the

annual
weeds

perennial weeds
and grasses

shrubs

young pine forest

mature oak forest

canopy

lower
canopy trees

tall shrub
understory

Time ⟶

Figure 5-18 Secondary ecological succession of plant communities on an abandoned farm field in North Carolina over about 150 years. Succession of animal communities is not shown.

pioneer community were crowded out and gradually replaced with a different community.

As this new community thrived, it too added organic matter to the slowly thickening upper layers of soil, leading to the next stage of succession, a compact layer of vegetation called a *heath mat*. This mat, in turn, provided the thicker and richer soil needed for the successful germination and growth of trees such as jack pine, black spruce, and aspen. Over several decades these trees increased in height and density, crowding out the compact plants of the heath mat. Shade and other new conditions created by these trees allowed the successful germination and growth of balsam fir, paper birch, white spruce, and other shade-tolerant climax species. As these taller trees created a canopy overhead, most of the earlier, shade-intolerant species were eliminated, because they could no longer reproduce. After several centuries, what was once bare rock had become a mature, or climax, ecosystem.

The more common type of succession is **secondary succession,** the sequential development of communities in an area where the natural vegetation has been removed or destroyed but the soil or bottom sediment is not destroyed. Examples of areas that can undergo secondary succession include abandoned farmlands, burned or cut forests, land stripped of veg-

etation for surface mining, heavily polluted streams, and land that has been flooded naturally or to produce a reservoir or pond. Because some soil or sediment is present, new vegetation can usually sprout within only a few weeks.

In the central (Piedmont) region of North Carolina, European settlers cleared away the native oak-hickory climax forests and planted the land in crops. Figure 5-18 shows how this abandoned farmland, covered with a thick layer of soil, has undergone secondary succession over a period of about 150 years until the area is again covered with a mature oak-hickory forest. Newly created lakes, reservoirs, and ponds also undergo secondary succession: As they gradually fill up with bottom sediments, they are eventually converted to terrestrial ecosystems, which then undergo terrestrial ecological succession.

Comparison of Immature and Mature Ecosystems Pioneer, or immature, ecosystems and climax, or mature, ecosystems have strikingly different characteristics, as summarized in Table 5-2. Immature communities at the early stages of ecological succession have only a few species (low species diversity) and fairly simple food webs, made up mostly of producers fed upon by herbivores and relatively few

Table 5-2 Ecosystem Characteristics at Immature and Mature Stages of Ecological Succession

Characteristic	Immature Ecosystem	Mature Ecosystem
Ecosystem Structure		
Plant size	Small	Large
Species diversity	Low	High
Quantity of living matter	Small	Large
Quantity of nonliving matter	Small	Large
Trophic structure	Mostly producers, few decomposers	Mixture of producers, consumers, and decomposers
Ecological niches	Few, mostly generalized	Many, mostly specialized
Community organization (number of interconnecting links)	Low	High
Ecosystem Function		
Plant growth rate	Rapid	Slow
Food chains	Simple, mostly plant → herbivore with few decomposers	Complex, dominated by decomposers
Nutrient chemical cycles	Mostly open	Mostly closed
Efficiency of nutrient recycling	Low	High
Efficiency of energy use	Low	High

decomposers. Most of the plants are small annuals that grow close to the ground and expend most of their energy in producing large numbers of small seeds for reproduction rather than in developing large root, stem, and leaf systems. They are partially open systems that receive some matter resources from other ecosystems, because they are too simple to hold and recycle many of the nutrients they receive.

In contrast, the community in a mature ecosystem is characterized by high species diversity, relatively stable populations, and complex food webs dominated by decomposers that feed on the large amount of dead vegetation and animal wastes. Most plants in mature ecosystems are larger perennial herbs and trees that produce a small number of large seeds and expend most of their energy and matter resources in maintaining their large root, trunk, and leaf systems rather than in producing large numbers of new plants. They tend to be closed systems, because they have the complexity needed to entrap, hold, and recycle most of their nutrients.

5-6 HUMAN IMPACTS ON ECOSYSTEMS

Human Beings and Ecosystems In modifying ecosystems for our use, we simplify them. For example, we bulldoze and plow grasslands and forests and replace their thousands of interrelated plant and ani-

mal species with a greatly simplified, single-crop, or monoculture, ecosystem or with structures such as buildings, highways, and parking lots.

Modern agriculture is based on deliberately keeping ecosystems in early stages of succession, where net primary productivity of one or a few plant species (such as corn or wheat) is high. But such simplified ecosystems are highly vulnerable. A major problem is the continual invasion of crop fields by unwanted pioneer species, which we call *weeds* if they are plants, *pests* if they are insects or other animals, and *disease* if they are harmful microorganisms such as bacteria, fungi, and viruses. Weeds, pests, or disease can wipe out an entire crop unless it is artificially protected with pesticides such as insecticides (insect-killing chemicals) and herbicides (weed-killing chemicals). When quickly breeding species develop genetic resistance to these chemicals, farmers must use ever-stronger doses or switch to a new product. Persistent, broad-spectrum insect poisons kill not only the pests but also species that prey on the pests. This further simplifies the ecosystem and allows pest populations to expand to even larger sizes.

Thus in the long run, every pesticide increases the rate of natural selection of the pests to the point that the effectiveness of the chemical is eventually doomed to failure. This illustrates biologist Garret Hardin's **first law of ecology:** We can never do merely one thing. Any intrusion into nature has numerous effects, many of which are unpredictable.

Cultivation is not the only factor that simplifies ecosystems. Ranchers who don't want bison or prairie

Table 5-3	Comparison of a Natural Ecosystem and a Simplified Human System	
Natural Ecosystem (marsh, grassland, forest)	**Simplified Human System (cornfield, factory, house)**
Captures, converts, and stores energy from the sun	Consumes energy from fossil or nuclear fuels
Produces oxygen and consumes carbon dioxide	Consumes oxygen and produces carbon dioxide from the burning of fossil fuels
Creates fertile soil	Depletes or covers fertile soil
Stores, purifies, and releases water gradually	Often uses and contaminates water and releases it rapidly
Provides wildlife habitats	Destroys some wildlife habitats
Filters and detoxifies pollutants and waste products free of charge	Produces pollutants and waste, which must be cleaned up at our expense
Usually capable of self-maintenance and self-renewal	Requires continual maintenance and renewal at great cost

dogs competing with sheep for grass eradicate these species from grasslands, as well as wolves, coyotes, eagles, and other predators that occasionally kill sheep. Far too often, ranchers allow livestock to overgraze grasslands until excessive soil erosion helps convert these ecosystems to simpler and less productive deserts. The cutting of vast areas of diverse tropical rain forests is causing an irreversible loss of many of their plant and animal species. Human beings also tend to overfish and overhunt some species to extinction or near extinction, further simplifying ecosystems. The burning of fossil fuels in industrial plants, homes, and vehicles creates atmospheric pollutants that fall to the earth as acidic compounds, which simplify forest ecosystems by killing trees and aquatic ecosystems by killing fish. To environmentalists, these are all signs that we have already gone too far in simplifying the world's ecosystems.

It is becoming increasingly clear that the price we pay for simplifying, maintaining, and protecting such stripped-down ecosystems is high: It includes time, money, increased use of matter and energy resources, loss of genetic diversity, and loss of natural landscape (Table 5-3). There is also the danger that as the human population grows, we will convert too many of the world's mature ecosystems to simple, young, productive, but highly vulnerable forms. The challenge is to maintain a balance between simplified, human ecosystems and the neighboring, more complex, natural ecosystems our simplified systems depend on.

Some Lessons from Ecology   It should be clear from the brief discussion of ecological principles in this and the preceding two chapters that ecology forces us to recognize six major features of living systems: *interdependence, diversity, resilience, adaptability, unpredictability,* and *limits*. Ecology's message is not to avoid growing food, building cities, and making other changes that affect the earth's plant and animal communities, but to recognize that such human-induced changes have far-reaching and unpredictable consequences. Ecology is a call for wisdom, care, and restraint as we alter the biosphere.

What has gone wrong, probably, is that we have failed to see ourselves as part of a large and indivisible whole. For too long we have based our lives on a primitive feeling that our "God-given" role was to have "dominion over the fish of the sea and over the fowl of the air and over every living thing that moveth upon the earth." We have failed to understand that the earth does not belong to us, but we to the earth.

Rolf Edberg

DISCUSSION TOPICS

1. List a probable limiting factor for each of the following ecosystems: (a) a desert, (b) the surface layer of the open sea, (c) the Arctic tundra, (d) the floor of a tropical rain forest, and (e) the bottom of a deep lake.

2. What bodies of water, if any, in your area have suffered from cultural eutrophication and what is being done to correct this situation?

3. Why are the oceans so important to life on this planet?

4. Since the deep oceans are vast, self-sustaining ecosystems located far away from human habitats, why not use them as a depository for essentially all of our radioactive and other hazardous wastes? Give your reasons for agreeing or disagreeing with this proposal.

5. Draw up a list of rules and regulations designed to protect beaches, estuaries, and wetlands while still allowing them to be used for recreation and ecologically sound development.

6. Someone tells you not to worry about air pollution because the human species through natural selection will develop lungs that can detoxify pollutants. How would you reply?

7. Are human beings or insects such as flies and mosquitoes better able to adapt to environmental change? Defend your choice and indicate the major way each of these species can adapt to environmental change.

8. Explain how a species can change local conditions so that the species becomes extinct in a given ecosystem. Could human beings do this to themselves? Explain.

9. Explain why a simplified ecosystem such as a cornfield is much more susceptible to harm from insects, plant diseases, and fungi than a more complex, natural ecosystem. Why are natural ecosystems less susceptible?

6

Risk-Benefit and Cost-Benefit Analysis

We cannot command nature except by obeying her.
Sir Francis Bacon

GENERAL OBJECTIVES

1. How are the short- and long-term risks associated with using a particular technology or product determined?

2. What is risk-benefit analysis and what are its limitations?

3. How can government or other agencies manage risks to protect the public?

4. What lifestyle factors can increase the risk of contracting cancer?

5. What are cost-benefit and cost-effectiveness analysis and what are their limitations?

Almost everything we do and every form of technology involves some degree of direct or indirect risk to human beings and other species. The key question is whether the risks outweigh the benefits. In addition, each type of technology, product, or form of development has some economic benefits and costs associated with it. In this case, the question is whether its economic benefits outweigh its economic costs.

6-1 RISK ASSESSMENT

Hazards and Risks A **hazard** is something that can cause injury, disease, death, economic loss, or environmental deterioration. **Risk** is the probability that something undesirable will happen from deliberate or accidental exposure to a hazard. Risks involve individuals, groups (such as the workers in a factory), or society as a whole. Risks may also be local, regional, or global and can last a short or a long time. Expecting or demanding that any activity have zero risk is unrealistic, because everything we do involves some degree of risk from one or more types of hazards.

Determining Risks The process of determining the short- and long-term adverse consequences to individuals or groups from the use of a particular technology in a particular area is known as **risk assessment.** Calculating the hazardous risk of a particular type of technology is difficult. Probabilities based on past experience are used to estimate risks from older technologies. For new technologies, however, much less accurate statistical probabilities, based on models rather than actual experience, must be calculated. Engineers and systems analysts try to identify everything that could go wrong, the probability of each of

Figure 6-1 Control room of a nuclear power plant. Watching these indicators is such a boring job that government investigators have found some operators asleep and in one case found no one in the control room. Most nuclear power plant accidents have resulted primarily from human errors.

Department of Energy

these failures occurring, and then the probabilities of various combinations of such events taking place.

The more complex the system, the more difficult it is to make realistic calculations based on statistical probabilities. The total reliability of any system is the product of two factors:

$$\frac{\text{system}}{\text{reliability}} = \frac{\text{technology}}{\text{reliability}} \times \frac{\text{human}}{\text{reliability}}$$

With careful design, quality control, maintenance, and monitoring, a high degree of technology reliability can usually be obtained in complex systems such as a nuclear power plant, space shuttle, or early warning system for nuclear attack. However, human reliability is almost always much lower; to be human is to err, and human behavior is highly unpredictable. Workers who carry out maintenance or who monitor warning panels in complex systems such as the control rooms of nuclear power plants (Figure 6-1) become bored and inattentive, because most of the time nothing goes wrong. They may fall asleep while on duty (as has happened at several U.S. nuclear plants); they may falsify maintenance records, because they believe that the system is safe without their help; they may be distracted by personal problems or illness; they may be told by managers to take shortcuts to enhance short-term profits or to make the managers look more efficient and productive.

For example, assuming that the technology reliability of a system such as a nuclear power plant is 95% (0.95) and the human reliability is 65% (0.65), then overall system reliability is only 62%, or 0.62 (0.95 x 0.65 = 0.62). Even if we could increase the technology reliability to 100% (1.0), the overall system reliability would still be only 65%, or 0.65 (1.0 x 0.65 = 0.65). This crucial dependence of even the most carefully designed systems on human reliability helps explain the occurrence of such extremely unlikely events as the Three Mile Island and Chernobyl nuclear power plant accidents, the tragic explosion of the space shuttle *Challenger,* and the far too frequent false alarms given by early warning defense systems on which the fate of the entire world depends.

One way to improve system reliability is to move more of the potentially fallible elements from the human side to the technical side, making the system more foolproof or "fail-safe." But chance events such as a lightning bolt can knock out automatic control systems, and no machine can replace all the skillful human actions and decisions involved in seeing that a complex system operates properly and safely. Furthermore, the parts in any automated control system are manufactured, assembled, tested, certified, and maintained by fallible human beings, who often are underpaid, have little knowledge of the importance of their work, and are unaware of how their work fits into the overall system.

6-2 RISK-BENEFIT ANALYSIS

Desirability Quotients The real question we face is whether the estimated short- and long-term benefits of using a particular technology outweigh the estimated short- and long-term risks compared to other alternatives. One method for making such evaluations is **risk-benefit anaylsis.** It involves calculating the short- and long-term societal benefits and risks involved and then dividing the benefits by the risks to find a **desirability quotient:**

$$\text{desirability quotient} = \frac{\text{societal benefits}}{\text{societal risks}}$$

Examples of Risk-Benefit Analysis Assuming that accurate calculations of benefits and risks can be made (often a big assumption), here are several possibilities:

1. $\text{large desirability quotient} = \dfrac{\text{large societal benefits}}{\text{small societal risks}}$

Example: *X rays.* Use of ionizing radiation in the form of X rays to detect bone fractures and other medical problems has a large desirability quotient—provided they are not overused merely to protect doctors from liability suits, the dose is no larger than needed, and less harmful alternatives are not available. Ionizing radiation from X rays is not persistent and can cause mutations only during use, unlike that from a radioactive isotope, which can cause cellular damage for at least ten times the half-life of the isotope (Section 3-2). Other examples with large societal benefits and whose societal risks are relatively small or could be reduced at acceptable cost include mining, most dams, and airplane travel.

2. $\begin{array}{l}\text{very small}\\\text{desirability}\\\text{quotient}\end{array} = \dfrac{\text{very small societal benefits}}{\text{very large societal risks}}$

Example: *Nuclear war.* Global nuclear war has essentially no societal benefits (except the short-term profits made by companies making weapons and weapons defense systems) and involves totally unacceptable risks to most of the earth's present human population and to many future generations. Although the interlocking problems of population growth, resource depletion, pollution, and environmental degradation are serious, the single greatest human and environmental threat to the earth's life-support systems is war— especially global nuclear war (see Spotlight on p. 96).

Although most people prefer not to think about it, many experts are convinced that the threat of global nuclear war continues to grow as the number of nuclear weapons and warheads and the number of countries with the knowledge to produce nuclear weapons continue to increase. We live in a world with enough atomic weapons to kill everyone on earth 67 times. Together these nuclear weapons have an explosive power equal to 6,667 times that of all the explosives detonated during World War II. The total firepower of the nuclear missiles carried on a single U.S. Trident submarine is equivalent to over six times all the explosives detonated in World War II.

By 1987, five countries—the United States, the Soviet Union, Great Britain, France, and China—had built and tested nuclear weapons, and many other countries are believed to have the knowledge and materials needed to build such weapons. It is projected that by the end of this century, 60 countries— one of every three in the world—will have the knowledge and capability to build nuclear weapons. Most of the knowledge and materials enabling these countries to have a nuclear capability have been provided by the United States, the Soviet Union, France, Italy, and West Germany, which have given or sold commercial nuclear power plants and small, research-oriented nuclear fission reactors to other countries.

In addition to the ecological threats, the buildup of nuclear and conventional weapons drains funds and creativity that could be used to solve most of the world's food, health, and environmental problems. For example, about 40% of the world's research and development expenditures and 60% of its physical scientists and engineers are devoted to developing weapons to improve our ability to kill one another.

3. $\begin{array}{l}\text{small desirability}\\\text{quotient}\end{array} = \dfrac{\text{large societal benefits}}{\text{much larger societal risks}}$

Example: *Nuclear and coal-burning electric power plants.* Nuclear and coal-burning power plants provide society with electricity—a highly desirable benefit. Although proponents of these technologies disagree, many analysts contend that the short- and long-term societal risks from widespread use of these technologies outweigh the benefits and that many other economically acceptable alternatives exist for producing electricity with less severe societal risks.

The probability of a catastrophic release of ionizing radiation from a nuclear reactor is extremely small, but the immediate and long-lasting harm of such a release and the large population exposed make the overall societal risks very high (Section 16-2). For example, if a "worst-case" accident did occur, as many

Evaluation of previously overlooked calculations has suggested that even a limited nuclear war could kill 2 billion to 4 billion people (40% to 80% of the world's population). An estimated 1 billion people, mostly in the Northern Hemisphere, where a nuclear exchange is most likely to take place, would die immediately or within a few weeks as a result of the direct effects of the explosions. Tens of millions more would suffer serious burns (Figure 6-2) and other injuries.

Within the next two years, another 1 billion to 3 billion might die from starvation caused by disruption of world agricultural production, first in the Northern Hemisphere and later in the Southern Hemisphere. Some models indicate that the mid-latitudes of the Northern Hemisphere, where most of the world's food is grown, would experience what is called the **nuclear winter effect**: Massive amounts of smoke, soot, dust, and other debris, lifted into the atmosphere as a result of the nuclear explosions and subsequent fires, would coalesce into huge smoke clouds. Within two weeks, these dense clouds would cover large portions of the Northern Hemisphere and prevent 50% to 90% of all sunlight from reaching these areas. During the weeks or months the clouds persisted, these models suggest, the regions under the clouds would be subjected to varying periods of darkness or semidarkness, day and night, and drops in temperature to as low as 4.4°C (40°F), regardless of the season. The resulting sharp decline in food production would cause widespread starvation in the Northern Hemisphere and in African and Asian countries dependent on food imports from countries such as the United States and Canada.

Food production in the Southern Hemisphere would also be affected as the smoke clouds gradually became less dense and spread southward. Although drops in temperature and sunlight would be less severe than in the Northern Hemisphere, even a slight reduction in temperature could be disastrous to agriculture in tropical and subtropical forest areas and could lead to the extinction of numerous plant and animal species.

The projected nuclear winter effect would not be the only cause of food scarcity. Plagues of rapidly producing insects and rodents—the animal life forms best equipped to survive nuclear war—would damage stored food (and spread disease). In areas where crops could still be grown, farmers would be isolated from supplies of seeds, fertilizer, pesticides, and fuel. People hoping to subsist on seafood would find many surviving aquatic species contaminated with radioactivity, silt, runoff from ruptured tanks of industrial liquids, and oil pouring out of damaged offshore rigs.

All of these projected environmental effects of limited nuclear war are based on calculations and computer models. Some scientists have questioned whether the effects would be quite as great as those projected by these initial studies. Instead of a nuclear winter, they say that we would have a **nuclear autumn**—temperatures and light levels would drop, but not so severely. However, even these milder effects could still lead to widespread disruption of agricultural productivity for at least one growing season and starvation for at

---

as 100,000 people could die within several months from exposure to high levels of ionizing radiation. Half a million or more other people would have a high risk of developing thyroid disorders and various types of cancer from 5 to 50 years after the accident from exposure to moderate to low levels of radiation. There would also be tens to hundreds of billions of dollars in economic damage. Investors in the Three Mile Island facility lost over $1 billion, even though no immediate deaths and injuries occurred. The moderately severe nuclear accident at the Chernobyl nuclear power plant in the Soviet Union in 1986 cost an estimated $7 billion to $14 billion, killed 31 people within a few months,

and exposed at least 100,000 people to potentially damaging levels of ionizing radiation.

Widespread use of coal-fired plants also poses high short- and long-term societal risks (Section 15-5). Air pollutants released by coal-fired power plants in the United States are estimated to cause annually about 10,000 premature deaths, 100,000 cases of respiratory illness, and several billion dollars in losses from damage to trees, aquatic life, crops, and buildings. They are also major contributors to possible long-term changes in the global climate resulting from the greenhouse effect (Section 18-5). With much stricter air pollution controls, the short-term societal risks from coal-

least 1 billion people. For example, even an average temperature drop of only 4°C (7.2°F) during a growing season can cut the yields of major crops such as wheat or corn by 50% or more; an average drop of 5°C (9°F) can wipe out such crops. A one- or two-day nonfreezing cold spell at a sensitive phase of the growing season is enough to destroy rice crops. In view of even these less drastic projections, a country launching a so-called limited nuclear attack would be killing much of its own population—even if not a single missile were fired in return.

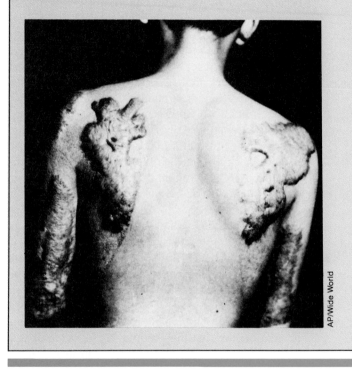

Figure 6-2 Burn scars of a survivor of the explosion of a single nuclear fission atomic bomb over Hiroshima, Japan, by the United States in August 1945. Another atomic bomb was detonated over Nagasaki, Japan. These two bombs—which were quite small compared to today's nuclear fission and fusion bombs and warheads—killed about 120,000 people immediately. Within a year, another 100,000 had died, mostly from exposure to radioactive fallout.

AP/Wide World

burning plants could be sharply reduced. But the cost of electricity from such plants would increase, and no way is presently known to reduce their potential long-term effect on global climate.

4. uncertain desirability quotient $= \dfrac{\text{large benefits}}{\text{large risks}}$

Example: *Genetic engineering.* For many decades, human beings have intervened in and shortened natural selection processes by crossbreeding genetic varieties of plants and animals to develop new varieties with certain desired qualities as human food sources. Today

"genetic engineers" have learned how to splice genes and recombine sequences of DNA molecules to produce DNA with new genetic characteristics (recombinant DNA). This new biotechnology, which is just beginning to emerge from the laboratory, may in the near future allow human beings greatly increased control over the course of natural selection for the earth's living species.

Some see this as a way to increase crop and livestock yields and to produce plants and livestock that have greater resistance to diseases, pests, frost, and drought and that provide greater quantities of nutrients such as proteins. They also hope to develop bacteria that can destroy oil spills and degrade toxic wastes and new vaccines, drugs, and therapeutic hormones like insulin. Gene therapy would also be used to eliminate certain genetic diseases and other human genetic afflictions.

Others are horrified by this prospect. Most of these critics recognize that stopping the development of genetic engineering is virtually impossible, but they believe that this technology should be kept under strict control. These critics do not believe that people have enough understanding of how nature works to be trusted with such great control over the genetic characteristics of human beings and other species.

Critics also fear that unregulated biotechnology could lead to the development of "superorganisms," which, if released deliberately or accidentally into the environment, could cause many unpredictable, possibly harmful effects. For example, genetically altered bacteria designed to clean up ocean oil spills by degrading the oil might multiply rapidly and eventually degrade the world's remaining oil supplies—including the oil in our cars. The risks of this or other catastrophic events resulting from biotechnology are quite small, but critics contend that biotechnology is a potential source of such enormous profits that without strict controls, greed—not ecological wisdom and restraint—will take over.

Genetic biologists answer that it is highly unlikely that the release of genetically engineered species would cause serious and widespread ecological problems. To have a serious effect, such organisms would have to be outstanding competitors and relatively resistant to predation. In addition, they would have to be capable of becoming dominant in ecosystems and in the biosphere, which are already populated with a vast diversity of organisms that act as checks and balances against dominance by any single species (except humans so far). Some proponents of biotechnology, however, acknowledge that some genetically engineered organisms might contribute to localized or temporary ecological disasters and join the critics in calling for strict control over this emerging technology.

Limitations of Risk-Benefit Analysis Although it appears quite simple and straightforward, calculation of desirability quotients is an extremely difficult and controversial undertaking. For example, many people—especially those who make short-term economic profits from the technologies involved—would disagree with the examples just given.

Short- and long-term benefits such as kilowatts of electricity produced, time saved, jobs created, and potential profits from use of a particular technology or product are usually fairly easy to estimate. Determining the risks, however, is usually much more difficult. Sufficient data to analyze particular risks are usually not available, either because such information has not been collected in the past or is not available for new technologies. Typically, we do not know how many people will be exposed to a particular hazard, how much exposure they might get, whether the hazard will cause damage at any level of exposure or only above a certain threshold level, and how the hazard interacts with other factors, which might enhance or reduce the hazard.

Other problems in calculating societal risks arise because some technologies benefit one group of people (population A) while imposing a risk on another (population B) and because some risk-benefit analysts emphasize short-term risks while others put more weight on long-term risks. There is also the problem of who will carry out a particular risk-benefit analysis. Should it be the corporation or government agency involved in developing or managing the technology, or some independent laboratory or panel of scientists? If it involves outside evaluation, who chooses which persons to do the study? Who pays the bill and thus has the potential to influence the outcome by refusing to give the lab or agency future business? Once the study is done, who reviews the results—a government agency, independent scientists, the general public—and what influence will outside criticism have on the final decisions? Clearly, politics, economics, and value judgments that can be biased in either direction are involved at every step of the risk-benefit analysis process.

The difficulty in making risk-benefit assessments does not mean that they should not be made or that they are not useful. But scientists, politicians, and the general public who must evaluate such analyses and make decisions based on them should be aware that at best they can only be expressed as a range of probabilities based on different assumptions—not the precise bottom-line numbers that decision makers want.

6-3 RISK MANAGEMENT

Deciding How and What Risks to Manage The process of **risk management** encompasses all of the administrative, political, and economic actions taken to decide how, and if, a particular societal risk is to be reduced to a certain level and at what cost. Risk management involves deciding (1) which of the vast number of risks facing society should be evaluated and managed with the limited funds available; (2) the sequence or priority in which these risks should be evaluated and managed; (3) the reliability of the risk-benefit analysis carried out for each risk; (4) how much risk is acceptable; (5) how much money will be needed to reduce each risk to an acceptable level; (6) what level of risk reduction will be attained if sufficient funds are not available, as is usually the case; and (7) how the risk management plan will be monitored and enforced.

Each step in this process involves value judgments and trade-offs to find some reasonable compromise between conflicting political and economic interests. For example, deciding that a pollutant should be reduced to a level of zero risk would bankrupt almost any company producing the chemical and would not take into account that nature often provides some degree of pollution control for many harmful chemicals free of charge. On the other hand, it would be wrong for risk managers not to provide society with a reasonable degree of protection on the grounds that such protection reduces private profits or exceeds rigidly imposed administrative budget cuts. Managing risks is a high-wire juggling act that is almost bound to be criticized from all sides regardless of the final decision.

True and Perceived Risks Risk management involves comparing the estimated true risk and harm of a particular technology or product with the risk and harm perceived by the general public. This comparison helps the risk manager develop a management program that is acceptable to the public but that does not cause economic ruin of the companies involved or require prohibitively high expenditures of tax dollars. The public generally perceives that a technology has a greater risk than its actual estimated risk when it (1) is a relatively new technology (genetic engineering) rather than a familiar one (dams, automobiles); (2) is involuntary (nuclear power plants, nuclear weapons) instead of voluntary (smoking, drinking alcohol); and (3) involves a large number of deaths and injuries from a single catastrophic accident (nuclear power plant accident or plane crash) rather than the same number of deaths spread out over a longer time (coal-burning power plants and automobiles).

For example, U.S. citizens tolerate 45,000 deaths from automobile accidents each year—equivalent to a fully loaded passenger jet crashing with no survivors every day—because these deaths are distributed in space and time. If they all occurred at one place and at the same time, like a plane crash, they would be considered a national catastrophe and would not be tolerated.

Some risk-benefit analysts decry this irrational evaluation of risks by the general public. But other observers contend that when it comes to evaluation of large-scale, complex technologies, the public often is better at seeing the forest than the risk-benefit specialists, who look primarily at the trees. This commonsense wisdom does not usually depend on understanding the details of scientific risk-benefit analysis. Instead, it is based on the average person's understanding that science has limits and that the people designing, building, running, monitoring, and maintaining potentially hazardous technological systems and products are fallible just like everyone else.

6-4 RISK FACTORS AND CANCER

What Is Cancer? Tumors are growths of cells that enlarge and reproduce at higher than normal rates. A benign tumor is one that remains within the tissue where it develops. **Cancer** is the name for a group of more than 120 different diseases—one for essentially each major cell type in the human body—all characterized by a tumor in which cells multiply uncontrollably and invade the surrounding tissue. If not detected and treated in time, many cancerous tumors undergo **metastasis;** that is, they release malignant (cancerous) cells that travel in body fluids to various parts of the body, making treatment much more difficult.

Evidence suggests that human cancer is the result of a genetic error or mutation in one or more of the normal genes found in each human cell. Because most types of cancer are not inherited, genes are usually mutated after birth spontaneously or from exposure to an environmental agent like radiation (X rays, radioactivity, and the sun's ultraviolet rays) or one or more chemicals called **carcinogens,** such as several substances in tobacco smoke.

Cancer is usually a latent disease, requiring a lag of 15 to 40 years between the initial cause and the appearance of symptoms. The long latency period, limited knowledge about the causes of different cancer types, and our lifetime exposure to many potential carcinogens make identifying what causes a particular cancer extremely difficult. This lag also deters many people from taking simple precautions that would greatly decrease their chances of getting the disease. For instance, healthy high school students and young adults have difficulty accepting the fact that their smoking, drinking, and eating habits *today* will be major influences on whether they will die prematurely from cancer.

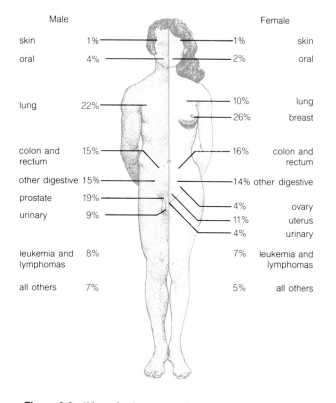

Male			Female
skin	1%	1%	skin
oral	4%	2%	oral
lung	22%	10%	lung
		26%	breast
colon and rectum	15%	16%	colon and rectum
other digestive	15%	14%	other digestive
prostate	19%		
		4%	ovary
urinary	9%	11%	uterus
		4%	urinary
leukemia and lymphomas	8%	7%	leukemia and lymphomas
all others	7%	5%	all others

Figure 6-3 Where fatal cancer strikes; percentages of all U.S. cancer deaths in 1982. (Data from the National Cancer Institute)

Figure 6-4 Deep-seated cancerous tumors can be treated with gamma rays emitted by cobalt-60.

American men and women have different incidences and death rates from different types of cancer (Figure 6-3). An average of one person every 66 seconds dies from cancer in the United States, and one of every three Americans now living will eventually have some type of cancer (see Spotlight on p. 103). But the good news is that almost half of all Americans who get cancer can now be cured, because of a combination of early detection and improved use of surgery, radiation (Figure 6-4), and drug treatments, compared to only a 25% cure rate 30 years ago. Survival rates for many types of cancer now range from 66% to 88% (Table 6-1). On the other hand, little improvement has been registered in the survival rates for people with cancer of the pancreas, esophagus, lung, stomach, and brain.

Cancer Risk Factors Hereditary factors are involved in an estimated 10% to 30% of all cancers. Examples of genetically transferable cancer risks include leukemia in children with Down's syndrome; breast cancer; ovarian cancer; malignant melanoma (which may also be caused by excessive exposure to ultraviolet rays); retinoblastoma (a rare form of eye cancer); and lung cancer (although heredity is not nearly so important as smoking).

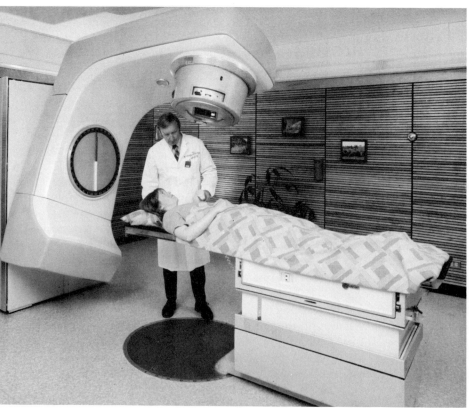

Varian Associates, Palo Alto

Environmental factors—including such lifestyle elements as diet, smoking, and environmental pollution—are believed to contribute to or directly cause the remaining 70% to 90% of all cancers. For example, studies show that U.S. Mormons, who do not smoke or drink alcohol, have one-fourth the incidence of lung, esophagus, and larynx cancers compared to the average U.S. white population. Higher than normal incidences of certain types of cancer in various parts of the world indicate the effects of diet, industrialization, and other environmental factors such as soil and water contamination on cancer rates (Figure 6-5). In the United States, cancer death rates for men are much higher in large cities and in the heavily urbanized and industrialized Northeast, Great Lakes region, and Gulf Coast.

Thus the risks of developing cancer can be greatly reduced by working in a less hazardous environment, by not smoking, by drinking in moderation (no more than two beers or drinks a day) or not at all, by adhering to a healthful diet (see Spotlight on p. 104), and by shielding oneself from the sun. According to experts, 60% of all cancers could be prevented by such lifestyle changes.

Cancer and Smoking   Tobacco is the cause of more death and suffering by far among adults than any other environmental factor (Figure 6-6). Over a billion people—one of every five—throughout the world now smoke. Worldwide, between 2 million and 2.5 million smokers die prematurely each year from heart disease, lung cancer, bronchitis, and emphysema—all related to smoking. In the United States, tobacco is estimated to cause 375,000 deaths each year—one out of of every five deaths—far more than all the deaths caused by alcohol, automobile accidents, hard drugs, suicide, and homicide.

It is estimated that every cigarette smoked reduces one's average life span by five and a half minutes. Overwhelming statistical evidence from more than 30,000 studies shows that smoking causes an estimated one-third of all cancer deaths in the United States, 30% of all heart disease deaths, and three-fourths of all lung cancer deaths in American men. People who smoke two packs of cigarettes a day increase their risk of getting lung cancer 15 to 25 times over nonsmokers. Tobacco use also contributes to cancer of the bladder, lip, mouth, pancreas, esophagus, and pharynx, although alcohol also plays a significant role in the last two types. Fires caused by cigarettes kill between 2,000 and 4,000 Americans each year.

The nicotine in tobacco is a highly addictive drug that, like heroin and crack, can quickly hook its victims. A British government study showed that ado-

Table 6-1   Chances (in Percentages) of Surviving Cancer for Five Years or More in the United States*

Type of Cancer	Diagnosed 1960–1963	Diagnosed 1977–1982	Type of Cancer	Diagnosed 1960–1963	Diagnosed 1977–1982
Among Adults			**Among Adults (continued)**		
Testes	63	88	Brain	18	23
Lining of uterus	73	84	Stomach	11	16
Skin (melanoma)	60	80	Lung	8	13
Bladder	53	76	Esophagus	4	6
Breast	63	74	Pancreas	1	2
Hodgkin's disease	40	73	**Among Children**		
Prostate	50	71	Hodgkin's disease	52	86
Uterine cervix	58	66	Wilm's tumor	57	81
Colon	43	53	Acute lymphocytic leukemia	4	68
Rectum	38	50	Brain and central nervous system	35	52
Kidney	37	48	Neuroblastoma	25	51
Non-Hodgkin's lymphoma	36	48	Non-Hodgkin's lymphoma	18	50
Ovary	32	38	Bone	20	47
Leukemia	14	33	Acute granulocytic leukemia	3	25

*Data from the National Cancer Institute.

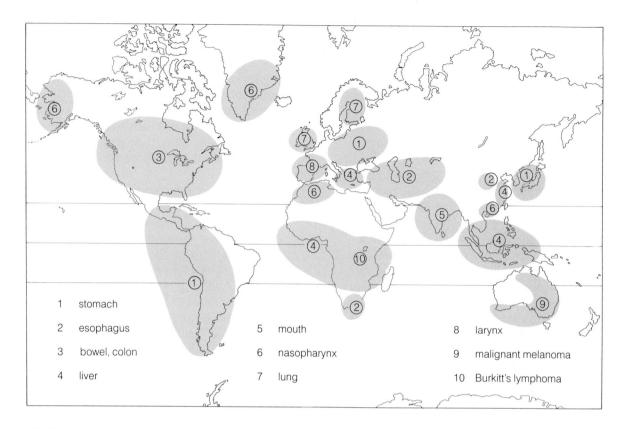

Figure 6-5 Areas around the world where incidence rates of certain types of cancer are much higher than normal. These differences are believed to be caused primarily by environmental factors such as diet, smoking, and pollution. (Data from the American Cancer Society)

1	stomach				
2	esophagus	5	mouth	8	larynx
3	bowel, colon	6	nasopharynx	9	malignant melanoma
4	liver	7	lung	10	Burkitt's lymphoma

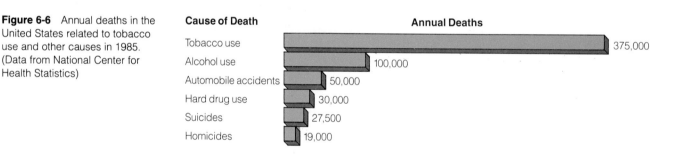

Figure 6-6 Annual deaths in the United States related to tobacco use and other causes in 1985. (Data from National Center for Health Statistics)

Cause of Death — **Annual Deaths**

Cause of Death	Annual Deaths
Tobacco use	375,000
Alcohol use	100,000
Automobile accidents	50,000
Hard drug use	30,000
Suicides	27,500
Homicides	19,000

lescents who smoke more than one cigarette have an 85% chance of becoming smokers. Other studies have shown that a child is about twice as likely to become a smoker if either parent smokes and that 75% of smokers who quit start smoking again within six months.

Smoke from cigarettes and other tobacco products doesn't just harm the smoker. About 86% of nonsmoking Americans involuntarily inhale smoke from other people's cigarettes, amounting to an average of about 1 cigarette a day. People working in smoky bars or living with a chainsmoker passively smoke the

equivalent of 14 cigarettes a day. A 1985 study by the EPA indicated that passive smoke kills from 500 to 5,000 Americans a year. Pregnant women who smoke 10 or more cigarettes a day give birth to underweight babies twice as often as nonsmokers, and their children have reduced lung capacity, higher rates of respiratory illness, and slower intellectual development.

There is some good news, however. After about one year, ex-smokers' chances of developing heart disease are about the same as nonsmokers', assuming all other heart disease risk factors are equal. Studies also show that 10 to 15 years after smokers quit, they have

Skin cancer is by far the most common form of cancer; about one in seven Americans get it sooner or later. Cumulative sun exposure over a number of years is the major cause of basal-cell and squamous-cell skin cancers. These two types can be cured if detected early enough, although their removal may leave disfiguring scars.

However, evidence suggests that just one severe, blistering burn as a child or teenager is enough to double the risk of contracting deadly malignant melanoma later in life, regardless of the skin type or the amount of cumulative exposure to the sun. This form of skin cancer spreads rapidly to other organs and can kill its victims. Between 1950 and 1986, the incidence of malignant melanoma in Americans rose by 700%—from 1 in 600 to 1 in 150.

Although anyone can get skin cancer, those with very fair and freckled skin run the highest risk. White Americans who spend long hours in the sun or under sunlamps (which are even more hazardous than direct exposure to the sun) greatly increase their chances of developing skin cancer and also tend to have wrinkled, dry skin by age 40. Blacks with darker skin are almost immune to sunburn but do get skin cancer, although at a rate ten times lower than whites'.

Clouds are dangerously deceptive: They admit as much as 80% of the sun's harmful ultraviolet radiation but allow people to stay out in the sun longer because they have a cooling effect. A dark suntan also doesn't prevent skin cancer. Outdoor workers are particularly susceptible to cancer of the exposed skin on the face, hands, and arms.

Sunbathers and outdoor workers can reduce this risk greatly and still get a tan (although more slowly) by using lotions containing sunscreen agents to block out the most harmful ultraviolet rays given off by the sun and sunlamps. People should stay away from tanning booths and sunlamps, wear wide-brimmed hats, and avoid prolonged exposure to the sun, especially between 10 A.M. and 3 P.M.

Doctors advise people to check themselves frequently for early signs of skin cancer, using a full-length mirror and a hand mirror for checking the back and hard-to-see places. The warning signs of skin cancer are a change in the size, shape, or color of a mole or wart (the major sign of malignant melanoma, which needs to be treated quickly), sudden appearance of dark spots on the skin, or a sore that keeps oozing, bleeding, and crusting over but does not heal. People should also be on the watch for precancerous growths that appear as reddish-brown spots with a scaly crust. If any of these signs are observed, one should immediately consult a doctor.

about the same risk of dying from lung cancer as those who never smoked. Evidence also indicates that very high doses of vitamin B-12 (half a milligram a day) and folic acid (10 milligrams a day) given to people for four months after they quit smoking reduced the number of premalignant cells in their lungs.

In 1986 the American Medical Association passed a resolution calling for a total ban on cigarette advertising in the United States, the prohibition of the sale of cigarettes and other tobacco products to anyone under 21, and a ban on cigarette vending machines. At present the U.S. tobacco industry spends almost $2 billion a year on advertising designed to create the impression that smokers are young, attractive, sophisticated, healthy, and sexy. By comparison, the federal government's Office on Smoking and Health has a yearly budget of about $3.5 million.

In the United States, smoking costs from $38 billion to $95 billion a year in increased health care and insurance costs, lost work because of illness, and other economic losses. This amounts to an average cost to society of $1.25 to $3.15 per pack of cigarettes sold, excluding the cost of the tobacco and packaging. By one estimate, every nonsmoking American adult pays at least $100 a year in taxes and increased insurance premiums to help cover the health costs of smokers.

Smokers also cost their employers at least $650 a year more in health insurance, lost work time, and cleanup costs. To reduce these losses and protect nonsmokers, an increasing number of firms are prohibiting smoking on the job, and a few refuse to hire smokers. On a strictly economic basis, tobacco's costs to society exceed its benefits in terms of the livelihood of tobacco farmers and employees of tobacco companies by more than two to one.

By 1987, 17 states and hundreds of cities had outlawed smoking in offices and other workplaces; 35% of all U.S. companies restrict smoking, and 2% ban it

The National Academy of Sciences and the American Heart Association advise that the risk of certain types of cancer—lung, stomach, colon, and esophagus cancer—and heart disease can be significantly reduced by a daily diet that cuts down on certain foods and includes others. Such a diet limits **(1)** total fat intake to 30% of total calories, with no more than 10% saturated fat (compared to the 40% fat diet of the average American); **(2)** protein (particularly meat protein) to 15% of total calories; **(3)** alcohol consumption to 15% of total caloric intake—no more than two drinks or beers a day;

(4) cholesterol consumption to no more than 300 milligrams a day; and **(5)** sodium intake to no more than 3 grams a day. We should avoid salt-cured, nitrate-cured, and smoked ham, bacon, hot dogs, sausages, bologna, salami, corned beef, and fish.

The diet should include fruits (especially vitamin C–rich oranges, grapefruit, and strawberries), minimally cooked orange, yellow, and green leafy vegetables such as spinach and carrots, and cabbage-family vegetables such as cauliflower, cabbage, kale, brussels sprouts, and broccoli. It should also incorporate

10 to 15 grams of whole-grain fiber a day from raw bran (the cheapest source), bran in cereals, and fibers in vegetables and fruits, and a daily intake of selenium not exceeding 200 micrograms. Recent preliminary evidence also suggests that eating cold-water fish such as bluefish, salmon, herring, and sardines, which are rich in certain types of fish oils, two or three times a week may help prevent heart disease and help prevent and arrest the growth of breast, colon, and prostate cancers.

outright. Designating nonsmoking areas in buildings, public buses, and planes is only a partial solution, because air conditioning and heating systems recycle the smoke. A 1986 Gallup poll found that 87% of both smokers and nonsmokers in the United States favored a ban on smoking in the workplace or creation of nonsmoking areas, and 94% believed that smoking is a health hazard.

It has also been suggested that all financial subsidies to U.S. tobacco farmers and tobacco companies be eliminated; instead, aid and subsidies would be provided to allow farmers to grow more healthful crops. Others have suggested that cigarettes be taxed at $1.25 to $3.15 a pack to discourage smoking and to make smokers—not nonsmokers—pay for the health and productivity losses now borne by society as a whole.

Cancer and Diet   A second major cause of cancer is improper diet, causing an estimated 35% of all cancer deaths. However, evidence linking specific dietary habits to specific types of cancer is difficult to obtain and controversial. The major factors—especially in cancers of the breast, bowel (colon and rectum), liver, kidney, stomach, and prostate—seem to be fats, nitrosamines, and nitrites. The incidence of cancers of the colon, rectum, and female breast is about five times higher in Americans than in Japanese, who have low-fat diets. Third-generation offspring of Japanese immigrants, however, have about the same incidence of these types of cancer as other Americans. A high-

fat, high-protein diet may also be a factor in cancers of the breast, prostate, testis, ovary, pancreas, and kidney.

High levels of nitrate and nitrite food preservatives, found in smoked and cured meats and in some beers, may increase the risk of stomach cancer because the body converts them to **carcinogenic,** or cancer-causing, compounds known as nitrosamines. These compounds have been implicated in the very high incidence of stomach cancer in Japan, where large amounts of dried, salted, pickled, and smoked fish are consumed. Despite the uncertainty in linking specific cancers to excessive consumption of specific foods, enough evidence exists to suggest a diet that should reduce cancer risks (see Spotlight above).

Cancer and the Workplace   The third major cause of cancer, occupational exposure to carcinogens and radiation, accounts for about 5% to 20% of cancer deaths, according to health scientists. Roughly one-fourth of U.S. workers run the risk of some type of illness from routine exposure to one or more toxic compounds. The National Institute for Occupational Safety and Health estimates that as many as 100,000 deaths a year—with at least half from cancer—are linked to workplace diseases in the United States (Table 6-2). Most work-related deaths could be prevented by stricter laws and enforcement of existing laws governing exposure of workers to radiation and dangerous chemicals. If enforced, the Occupational Safety and

Table 6-2 Cancer Risks in the Workplace in the United States*

Substance	Workers Exposed (millions)	Industries	Cancer Risks
Asbestos	2.5	Asbestos, textiles, insulation, mining	Lung, larynx, mesothelioma, bowel, stomach
Vinyl chloride	3.5	Vinyl chloride and vinyl plastic	Liver, brain, breast
Benzene	3.0	Tire, shoe, paint, cement, glue, varnish, chemicals	Leukemia
Arsenic	1.5	Pesticides, copper, leather tanning, mining, vineyards	Lung, skin, liver, testis, lymphatic system
Chromium	1.5	Bleaching, glass, pottery, batteries, linoleum	Lung, nasal
Nickel	1.4	Nickel refiners	Lung, nasal, larynx
Cadmium	1.4	Electrical, paint, metal alloys	Prostate, renal, respiratory
Carbon tetrachloride	1.4	Dry cleaning, machinists	Liver
Formaldehyde	1.3	Wood finishing, plastics, synthetic resins	Nasal

*Data from Occupational Safety and Health Administration and the AFL-CIO

Health Act of 1970 and the Toxic Substances Control Act of 1975 could establish such controls. However, political pressure by industry officials has hindered effective enforcement of these laws.

Cancer and Pollution A fourth cause of cancer is air and water pollution, estimated to contribute from 1% to 5% of cancer deaths in the United States. This contribution may be more significant, however, for residents of airtight, energy-efficient housing without air-to-air heat exchangers, because of abnormally high levels of indoor air pollution; for nonsmokers who work or live in an environment that exposes them to cigarette smoke; and for residents of cities whose drinking water is contaminated with one or more toxic metals or carcinogenic organic compounds.

6-5 COST-BENEFIT AND COST-EFFECTIVENESS ANALYSIS

Cost-Benefit Analysis The process of estimating and comparing the expected *costs* or losses associated with a particular project or degree of pollution control with the expected *benefits* or gains over a given period of time is known as **cost-benefit analysis.** If the economic benefits exceed the costs, the project or activity is usually considered to be worthwhile.

For example, in 1987 the EPA proposed that the amount of lead permitted in drinking water be reduced from 50 parts per billion (ppb) to 20 ppb. An estimated 40 million Americans drink water that exceeds the 50 ppb standard. According to cost-benefit analysis, achieving the 20 ppb goal would cost from $115 million to $145 million a year but would yield from $800 million to $1 billion in health benefits. Thus the projected benefits exceed the projected costs by about 7 to 1.

Cost-Effectiveness Analysis Evaluation of alternative methods of achieving a desired goal while incurring the least cost is done by using **cost-effectiveness analysis.** The goal of cost-effectiveness analysis is to minimize the total costs of pollution control and still reduce harmful environmental effects to a reasonable or acceptable level (Figure 6-7). Reducing pollution below this level might mean that the high costs would outweigh the economic benefits. Likewise, not reducing a given pollutant to the acceptable level means that the manufacturer is passing on hidden costs to taxpayers.

The cost of pollution control climbs with each additional increment of control and rises very steeply for removing the last few percent. For example, reducing the pollutants emitted by a coal-burning power plant or an industrial plant by 90% might cost $20 million. However, to reduce emissions by 95% might cost $30 million, and a 99% reduction might cost $75 million.

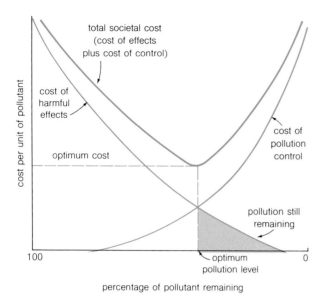

Figure 6-7 Cost-effectiveness analysis involves comparing the costs from the harmful effects of pollution with the costs of pollution control; the total costs of pollution control are minimized while still reducing the pollution to an acceptable level. The shaded area shows that some harmful effects remain, but removing these residual damages would make the costs of pollution control too high.

Problems with Cost-Benefit and Cost-Effectiveness Analysis The considerable controversy over the widespread use of cost-benefit and cost-effectiveness analysis involves several key questions: Who decides what costs and benefits are considered in these calculations? Who should evaluate whether such analyses are valid and not distorted in favor of a particular position? Who gets the benefits and runs the risks? How do you put monetary values on such things as a human life, a wilderness area, whooping cranes?

Environmentalists contend that cost-benefit and cost-effectiveness analyses often distort economic reality by exaggerating regulatory costs and underestimating benefits. They believe that such analyses are often used to weaken environmental regulation. Thus it matters a great deal who carries out the analyses. Industries favoring a particular project or degree of pollution control have a strong financial incentive to exaggerate present costs of pollution control, to downgrade the economic value of a future benefit, and thus to confine their calculations only to the near future.

Similarly, environmentalists tend to place a much higher value on clean air and water and other benefits and to put more emphasis on long-term benefits and costs. Environmentalists also argue that all such evaluations should be open to review and challenge during administrative deliberations. Furthermore, regulatory agencies should be required to consider several different evaluations, including those from parties for and against a particular action as well as from an impartial party.

Another problem is that those who receive the benefits often don't run the risks, and those who run the risks often receive the fewest benefits. For example, in the chemical industry, workers and those living in the plant vicinity often bear the risks of not reduc-

---

Spotlight   How Much Is Your Life Worth?

When asked to put a price tag on their life, most people would either call it infinite or contend that to attempt to do so would be impossible or even immoral. Yet the average American has been allowing industry and government officials to do so for decades, at least informally. Since 1980, the Reagan administration, under the urging of industry, has been proposing that such calculations become a mandatory part of all environmental and health decisions.

Values assigned to a human life in various cost-benefit and cost-effectiveness studies are plucked out of the air and vary from nothing to about $2 million; the most frequently assigned values range from $200,000 to $300,000. The callousness of this purely economic approach was revealed by an oil company representative who protested clean air standards affecting his firm's operations in Montana: "Some of the people who will die from air pollution are unemployed, and therefore have *no* economic value." An evaluation of a number of cost-benefit and cost-effectiveness studies also reveals that from a purely economic perspective, women are less valued than men, housewives less than women working outside the home, retired people less than workers, the sick less than the healthy, low-paid workers less than high-paid workers, and the unemployed less than the employed. What value, if any, do you believe should be placed on your life? Do you think that cost-benefit, risk-benefit, and cost-effectiveness analysis should be made a part of all environmental and health decisions?

ing air pollution emissions below a certain level, while those living far away reap the benefits of being able to purchase less expensive products. Similarly, the poor, who are the most likely to live and work under hazardous and unhealthful conditions, bear the risks of unnecessarily low workplace and general environmental standards. For example, congressional investigators found that 75% of the South's worst hazardous-waste dumps are located near low-income black communities.

The most serious objection to the automatic use of cost-benefit and cost-effectiveness analysis is that many things we value cannot be reduced to dollars and cents. For example, some of the costs of air pollution—such as extra laundry bills, house repainting, and ruined crops—are fairly easy to estimate. But how do we put a price tag on human life (see Spotlight on p. 106), clean air and water, beautiful scenery, and the ability of natural ecosystems to degrade and recycle our wastes? Such quantifications involve political, social, and ethical issues as well as economic ones.

For example, when the EPA did a cost-benefit analysis in 1984 on the proposed revision of the Clean Air Act, it concluded that the net benefits (after deducting the projected costs) ranged from a loss of $1.4 billion to a gain of as much as $110 billion—depending primarily on the monetary value assigned to things such as human life, human health, and a cleaner environment. From this calculation, we can see why cost-benefit and cost-effectiveness analyses are so vulnerable to partisan manipulation and value judgments.

Earth and water, if not too blatantly abused, can be made to produce again and again for the benefit of all. The key is wise stewardship.

Stewart L. Udall

DISCUSSION TOPICS

1. Do you agree or disagree that most of the survivors of a global nuclear war would envy the dead? Why? Would you want to be one of the survivors? Why or why not?

2. Considering the benefits and risks involved, do you believe that
   a. Nuclear power plants should be controlled more rigidly and gradually phased out.
   b. Coal-burning power plants should be controlled more rigidly and gradually phased out.
   c. Genetic engineering using recombinant DNA should be prohibited.
   d. Genetic engineering using recombinant DNA should be very rigidly controlled.
   e. People not wearing seat belts in vehicles should be heavily fined.
   f. Air bags or other automatic passenger protection systems should be made mandatory on all vehicles.
   g. Federal disaster insurance for people choosing to live in areas with high risk of floods, hurricanes, or earthquakes should be prohibited or greatly increased in cost.

   In each case defend your position.

3. Why are risk-benefit analysis and risk management so important? What are their major limitations? Does this mean that these processes are useless? Why or why not?

4. Analyze your lifestyle and diet to determine the relative risks of developing some form of cancer before you reach age 55. Which type of cancer are you most likely to get? How could you significantly reduce your chances of getting this cancer?

5. Give your reasons for agreeing or disagreeing with each of the following proposals:
   a. All advertising of cigarettes and other tobacco products should be banned.
   b. All smoking should be banned in public buildings and commercial airplanes, buses, subways, and trains.
   c. All government subsidies to tobacco farmers and the tobacco industry should be eliminated.
   d. Cigarettes should be taxed at three dollars a pack so that smokers—not nonsmokers—pay for the health and productivity losses now borne by society as a whole.

6. Assume you have been appointed to a technology risk-benefit-assessment board. What major environmental and health risks and benefits would you list for the following: (a) intrauterine devices (IUDs) for birth control, (b) television sets, (c) computers, (d) nuclear power plants, (e) a drug that a woman could take at home to cause a medically safe abortion (presently being tested in France), (f) effective sex stimulants, (g) drugs that would retard the aging process, (h) drugs that would enable people to get high but are physiologically and psychologically harmless, (i) electrical or chemical methods that would stimulate the brain to eliminate anxiety, fear, unhappiness, and aggression, (j) genetic engineering (manipulation of human genes) that would produce people with superior intelligence, strength, and other attributes? In each case, would you recommend that the technology be introduced, and if so, what restrictions would you apply?

7. What obligations concerning the environment do we have to future generations? Try to list the major beneficial and harmful aspects of the environment that were passed on to you during the past 50 years by the last two generations.

PART THREE

Population

Owen Franken/Stock, Boston

We need that size of population in which human beings can fulfill their potentialities; in my opinion we are already overpopulated from that point of view, not just in places like India and China and Puerto Rico, but also in the United States and in Western Europe.

George Wald, Nobel laureate, biology

7

Population Dynamics and Distribution

GENERAL OBJECTIVES

1. How are changes in population size affected by birth rates and death rates?

2. How do migration rates affect the population size of a particular country or area?

3. How are changes in population size affected by the average number of children women have during their reproductive years (total fertility rate)?

4. How are changes in population size affected by the percentage of men and women at each age level in the population of the world or a given country (age structure)?

5. How is the world's population distributed between rural and urban areas?

6. How do transportation systems affect population distribution?

The present extended period of rapid population growth in the world is unique when seen from a long-range perspective; it has never occurred before and is unlikely to occur again.

Jonas Salk and Jonathan Salk

By 1987 the size of the human population was 5.03 billion and was growing by 86.7 million a year (Figure 1-1, p. 3). Judging by present trends, United Nations population experts estimate that world population size will probably reach 6.2 billion by the year 2000 and 7.8 billion by 2020, and perhaps level off around 2100 at about 10.4 billion—more than twice the number of people on earth in 1987.

What are the major factors affecting these dramatic changes in the size and distribution of the human population? How can the size and growth rate of the human population be controlled? The first question is discussed in this chapter and the second one in Chapter 8.

7-1 BIRTH RATE, DEATH RATE, AND NET POPULATION CHANGE

Net Population Change Calculating the difference between the total number of live births and the total number of deaths throughout the world during a given period of time (usually a year) yields the **global net population change** over that period:

$$
\begin{array}{lll}
\text{global net} & & \text{number of} \quad \text{number of} \\
\text{population} & = & \text{live births} \; - \; \text{deaths} \\
\text{change per year} & & \text{per year} \quad\;\; \text{per year}
\end{array}
$$

If more live births than deaths occur, the world's population will increase.

There were about 2.6 births for each death throughout the world in 1987. This amounted to a global net population increase of 86.7 million people—or 165 additional people a minute. Words like *million* and *billion* often make little impression on us. But suppose you decide to take one second to say hello to each of the 86.7 million persons added to the world's

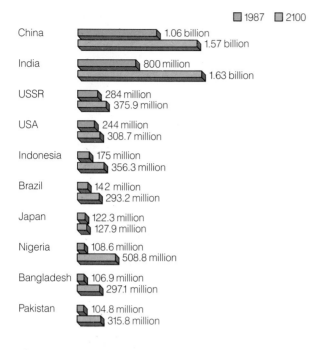

Figure 7-1 The world's ten most populous countries in 1987, with projections of their population size in 2100. (Data from Population Reference Bureau)

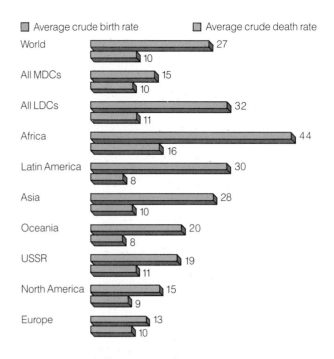

Figure 7-2 Average crude birth rates and crude death rates of various groups of countries in 1987. (Data from Population Reference Bureau)

population during 1987. Working 24 hours a day, you would need about 2.7 years to greet them, and during that time 234 million more people would have been added.

About 92% of these 86.7 million people were born in less developed countries of Africa, Asia, and Latin America, which in 1987 already contained 79% of the world's people. The large difference in population size between LDCs and MDCs is an important factor in the widening gap in average personal income and life quality between countries in the Southern and Northern Hemispheres (Section 1-1 and Figure 1-5, p. 5). This difference between LDC and MDC population sizes is projected to increase sharply from 1987 to 2100, when LDCs are expected to have 86% of the world's population (Figure 1-5, p. 5).

Figure 7-1 shows the world's ten most populous countries in 1987 and their projected population size by 2100. Because six of these countries are in Asia, it is not surprising that Asia is by far the most populous continent, containing 58% of the people on earth.

Crude Birth Rates and Death Rates Demographers, or population specialists, normally use the **crude birth rate** and **crude death rate** rather than total live births and deaths to describe population change. The crude rates give the number of live births and deaths per 1,000 persons at the midpoint of a given

year (July 1), since this should represent the average population for that year:

$$\text{crude birth rate} = \frac{\text{live births per year}}{\text{midyear population}} \times 1,000$$

$$\text{crude death rate} = \frac{\text{deaths per year}}{\text{midyear population}} \times 1,000$$

Figure 7-2 shows the crude birth rates and death rates for the world and various groups of countries in 1987. *The rapid growth of the world's population over the past 100 years is not the result of a rise in birth rates, as might be assumed; rather, it is due largely to a decline in death rates—especially in the LDCs* (Figure 7-3). The interrelated reasons for this general decline in death rates include better nutrition because of increased food production and better distribution; reduction of the incidence and spread of infectious diseases because of improved personal hygiene and improved sanitation and water supplies; and improvements in medical and public health technology through the use of antibiotics, immunization, and insecticides.

Life Expectancy and Infant Mortality Two useful indicators of overall health in a country or region are the average life expectancy at birth and the mortality rate of infants during their first year of life (Figure

Figure 7-3 Changes in crude birth and death rates for the more developed and less developed countries between 1775 and 1987 and projected rates (dashed lines) to 2000. (Data from Population Reference Bureau and United Nations)

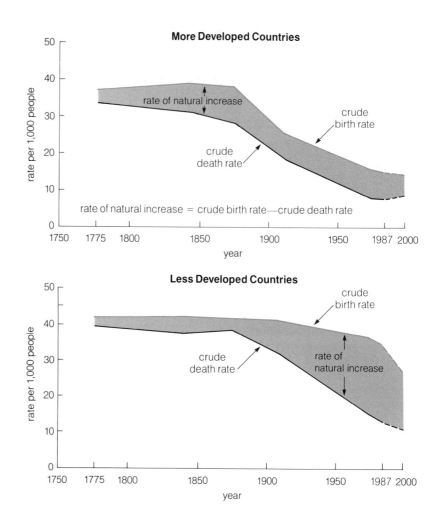

rate of natural increase = crude birth rate—crude death rate

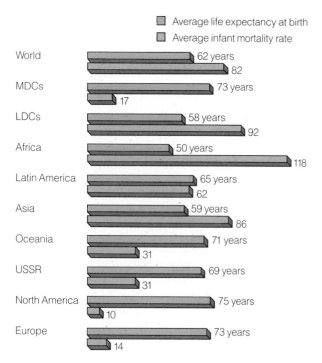

Figure 7-4 Life expectancy at birth and average infant mortality rate for various groups of countries in 1987. (Data from Population Reference Bureau)

7-4). **Life expectancy** is the average number of years a newborn can be expected to live. Increased average life expectancy is a result of better sanitation, nutrition, health care, and living conditions. It is not surprising that people in the world's MDCs have a higher average life expectancy than those in the LDCs. In 1987 average life expectancy at birth ranged from a low of 35 years in Sierra Leone in western Africa to a high of 77 years in Japan, Iceland, and Sweden.

Fortunately, major health improvements can be made in LDCs with preventive and primary measures at a relatively low cost. One important goal should be to provide better nutrition and birth assistance for pregnant women in LDCs, where half of all babies are delivered without any assistance from a trained midwife or doctor. Another would be to provide family planning and improved child care (including the promotion of breast-feeding) to reduce the infant mortality rate. Finally, infant mortality rates can be significantly lowered and average life expectancy rates can be extended by providing clean drinking water and sanitation facilities to the third of the world's population that lacks them. According to the World Health Organization, about 80% of all infectious disease in LDCs is caused by unsafe drinking water and inade-

Excessive alcohol consumption and alcoholism are major forms of personal pollution that have damaging effects on individuals, friends and family members, and society as a whole. An estimated 21 million Americans are alcoholics (people so addicted to alcohol that they have lost control of their drinking) or problem drinkers (people in the early stages of alcoholism). This includes 4 million teenagers age 13 to 17.

Each year alcohol kills at least 100,000 Americans, more than twice the number of soldiers who were killed in the nine-year Vietnam War. In 1985 alcoholism cost Americans about $120 million in lost work time, medical bills, property damage, rehabilitation programs, and other expenses—costs borne by everyone, not just those who abuse alcohol.

Excessive alcohol consumption also damages the liver (cirrhosis or buildup of fatty tissue), esophagus, and digestive tract. The family members and loved ones of millions of alcoholics are psychologically injured each year. Women who have three to nine alcoholic drinks a week may face a 30% higher chance of developing breast cancer than women who don't drink. Women who consume more than nine drinks a week have a 60% higher risk. Pregnant women who have as little as two drinks or beers a day pass alcohol into the bloodstream of the fetus. Babies born to alcoholic mothers can be physically addicted to alcohol, and some suffer from heart problems, reduced motor skills, and mental retardation.

At least half of the 45,000 deaths and hundreds of thousands of serious injuries from car accidents in the United States each year are related to alcohol. Analysts argue that these alcohol-related deaths could be reduced sharply by requiring all passengers to use seat belts, raising the drinking age to 21, significantly increasing the enforce- ment and penalties for driving under the influence of alcohol, and discouraging happy hours, chug-a-lug contests, and other practices that foster excessive alcohol consumption over a short period of time. It has also been suggested that bars and people hosting parties should have simple blood alcohol detection devices available and should provide transportation for those who can't drive safely.

Although millions took to the streets to protest the Vietnam War, few protest the death and destruction caused by alcohol, because it is a socially acceptable drug. Two-thirds of all Americans drink to some degree, and one-third of them are moderate to heavy drinkers. Because alcohol is so widely used in the United States and many other countries, reducing its destructive effects is very difficult. What do you think should be done?

---

quate sanitation. Extending such primary health care to all the world's people would cost an additional $10 billion a year, one-twenty-fifth as much as the world spends each year on cigarettes.

Between 1900 and 1987, average life expectancy at birth increased sharply in the United States, from about 42 to 75. Despite this increase, people in nine countries (Japan, Iceland, Sweden, Norway, the Netherlands, Switzerland, Spain, Australia, and Canada) had an average life expectancy at birth one to two years higher than people in the United States in 1987.

In most MDCs, safe water supplies, public sanitation, adequate nutrition, and immunization have nearly stamped out many infectious diseases. In 1900 pneumonia, influenza, tuberculosis, and diarrhea were the leading causes of death in MDCs. By contrast, the four leading causes of death in MDCs today are heart disease and stroke (48%), cancer (21%), respiratory infections (8%), and accidents, especially car accidents (7%). These deaths are largely a result of environmental and lifestyle factors rather than infectious agents invading the body. Except for car accidents, these deaths are largely attributable to the area in which we live (urban or rural), our work environment, our diet, whether we smoke, and the amount of alcohol we consume (see Spotlight above).

Average life expectancy is influenced by the **infant mortality rate,** the number of deaths of persons under 1 year of age per 1,000 live births. A high infant mortality rate in a country normally decreases the life expectancy at birth of its population. It usually indicates a lack of adequate food, poor nutrition, and a high incidence of infectious diseases (usually from contaminated drinking water). Infant mortality rates differ significantly among groups of countries (Figure 7-4) and among individual countries. In 1987 infant mortality rates ranged from a low of 5.5 deaths per 1,000 live births in Japan to a high of 183 deaths per 1,000 live births in East Timor in Southeast Asia.

In 1987, 19 industrialized countries had lower infant mortality rates than the 10.5 deaths per 1,000 live births in the United States—suggesting a need for further improvement in prenatal and infant health care and nutrition in the United States. Every year in the United

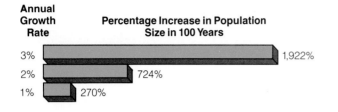

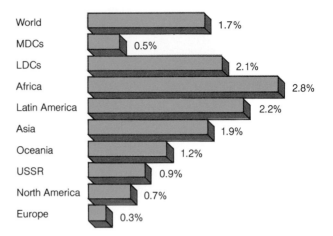

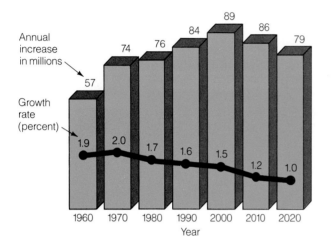

Figure 7-5 Effect of various annual percentage growth rates on population size over 100 years.

Figure 7-6 Actual and projected changes in the annual growth rate and annual increase for the world's population between 1960 and 2020.

Figure 7-7 Average annual population growth rate in various groups of countries in 1987. (Data from Population Reference Bureau)

States, 40,000 infants enter the world and then die before their first birthday. Since 1955 the United States has fallen from sixth to a tie for last place among 20 industrialized countries' infant mortality rates.

Annual Percentage Rates of Population Change

The **natural change rate,** also known as the **annual population change rate** (in percent), indicates how fast a population is growing or decreasing. This rate can be calculated using the following formula:

$$\text{natural change or annual population change rate} = \frac{\text{crude birth rate} - \text{crude death rate}}{10}$$

If the crude birth rate in a particular country or collection of countries is higher than the crude death rate, the population is growing exponentially by a certain percentage each year. If the crude death rate exceeds the crude birth rate, the population is decreasing by a certain percentage each year. Using the crude birth and death rates in Figure 7-2, we can calculate that in 1987 the population grew at a rate of 1.7% [(27 − 10)/ 10 = 17/10 = 1.7%].

A population growth rate of 3% or less a year may seem small, but it leads to enormous increases in population size over a 100-year period (Figure 7-5). For example, Nigeria in western Africa, with a population of 108.6 million and a 2.8% growth rate in 1987, is projected to have a population of 274 million by 2020 and eventually 623 million, more people than now live on the entire continent of Africa.

In the late 1970s, a series of newspaper headlines such as "Population Time Bomb Fizzles," "Another Non-Crisis," and "Population Growth May Have Turned Historic Corner" falsely implied that world population growth had almost stopped. What had actually happened was not a halt in net population growth but a slowing of the annual rate at which the world's population was growing, from a high of 2.0% in 1965 to 1.7% by the mid-1980s (Figure 7-6). Despite this encouraging slowdown in the *annual population growth rate,* the world's *annual net population growth* increased from 57 million in 1960 to 86.7 million in 1987. It is projected that by the year 2000, the world's annual population growth rate will have declined to about 1.5%, but annual net population growth will increase to 89 million persons a year.

Figure 7-7 shows differences in the average annual population change rates in major parts of the world. In 1987 population change rates ranged from a *growth* rate of 3.9% in Kenya in eastern Africa to a decline rate of −0.2% in West Germany and Hungary.

Doubling Time   Another indication of the rate at which a population is growing is called **doubling time:** the time it takes for a population to double in size if present annual population growth continues unchanged. The approximate doubling time in years can

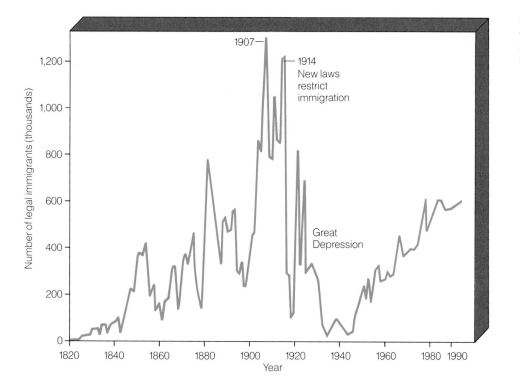

Figure 7-8 Legal immigration to the United States: 1820–1987. (Data from U.S. Immigration and Naturalization Service)

Chart labels:
- 1907
- 1914 New laws restrict immigration
- Great Depression
- y-axis: Number of legal immigrants (thousands)
- x-axis: Year

be found by using the *rule of 70*—that is, by dividing the annual percentage growth rate into 70:

$$\text{doubling time (years)} = \frac{70}{\text{annual percent growth rate}}$$

In 1987 the doubling time for the world's population was 41 years (70/1.7 = 41), compared to a doubling time of 128 years for the MDCs and 33 years for the LDCs. Doubling time is only a crude estimate of future population growth, because it is based on the often incorrect assumption that a population will have the same growth rate over several decades.

7-2 MIGRATION

Net Migration Rate   The rate at which the size of the world's population changes is based only on the difference between crude birth rates and death rates. The annual rate of population change for a particular country, however, is also affected by the net migration or movement of people into *(immigration)* and out of *(emigration)* that country during the year. The **immigration rate** is the number of people migrating into a country each year per 1,000 people in its population. It is usually based only on legal immigrants, because of the difficulty in counting illegal immigrants. The **emigration rate** is the number of people migrating out of a country each year per 1,000 people in its population. A country's **net migration rate** per year is the

difference between its immigration rate and emigration rate. If more persons immigrate than emigrate, the annual net migration rate is positive. Conversely, if more persons leave than enter, it is negative.

Thus the annual rate of population change for a country is the difference between its crude birth rate and death rate plus its net migration rate:

annual rate of population change for a country = (crude birth rate − crude death rate) + net migration rate

Migration also takes place within countries, especially from rural to urban areas, as discussed in Section 7-5.

Immigration and Population Growth in the United States   The United States, founded by immigrants and their children, has admitted a larger number of immigrants and refugees than any other country in the world. Indeed, the 53.7 million legal immigrants admitted to the United States from various parts of the world between 1820 and 1987 is almost twice the number received by all other countries combined during this same period. Between 1820 and 1960, most legal immigrants admitted to the United States came from Europe, but since then most have come from Asia and Latin America.

The number of legal immigrants entering the United States since 1820 has varied during different periods as a result of changes in immigration laws and economic growth (Figure 7-8). Between 1960 and 1987,

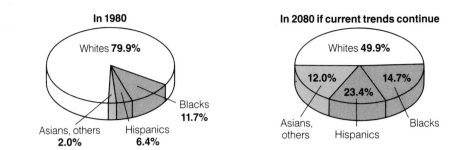

Figure 7-9 Projected changes in the ethnic composition of the U.S. population between 1980 and 2080. (Data from Population Reference Bureau)

the number of legal immigrants admitted to the United States more than doubled, from 250,000 to around 600,000 (Figure 7-8). By 1987 the Census Bureau estimated that the United States also had 4.9 million illegal immigrants, half in California and additional large populations in Texas, Illinois, New York, and Florida. According to the Census Bureau, 200,000 to 500,000 new illegal immigrants—three-fourths from Mexico and other Latin American countries—now enter and remain in the United States each year. Thus in 1987 the total number of legal and illegal immigrants increased the U.S. population size by 800,000–1.1 million people.

Excluding legal and illegal immigration, about 1.7 million people were added to the U.S. population of 244 million in 1987. But if we include the 800,000 to 1.1 million legal and illegal immigrants, the net population growth in 1987 was 2.5 million to 2.8 million—an average increase of about 5 people a minute, 285 an hour, 6,849 a day. Legal and illegal immigration now accounts for 32% to 39% of the annual population growth in the United States. This means that the United States is adding the equivalent of another Los Angeles to its population every year and a new California every decade. If birth and death rates and legal and illegal immigration rates continue at present levels, a dramatic change in the ethnic composition of the U.S. population will take place between 1980 and 2080 (Figure 7-9).

7-3 FERTILITY

Replacement Level Fertility and Total Fertility Rate A key factor affecting the future growth of a population is **fertility,** the average number of live babies born to women in the population during their normal childbearing years (ages 15–44).

Replacement-level fertility is the number of children a couple must have to replace themselves. You might think that two parents would have to have only two children to replace themselves. The actual average replacement-level fertility, however, is slightly higher,

primarily because some children die before reaching their reproductive years. In MDCs, average replacement-level fertility is 2.1 children per couple or woman. In some LDCs with high infant mortality rates, the replacement level may be as high as 2.5 children per couple.

The most useful measure of fertility for projecting future population change is the **total fertility rate (TFR)**—an estimate of the number of children the average woman will bear during her reproductive years, assuming she lives to age 44. In 1987 the average total fertility rate was 3.6 children per woman for the world as a whole, 2.0 in MDCs, and 4.2 in LDCs—ranging from a low of 1.3 in West Germany to a high of 8.0 in Kenya. Figure 7-10 shows changes in TFR in major groups of countries between 1960 and 1985. A significant part of the large drop in Asia was the result of massive family-planning efforts by China, as discussed in Section 8-3.

Fertility and Marriage Age One factor that can affect the total fertility rate is the median age of women at first marriage, or, more precisely, the median age at which women give birth to their first child. Studies indicate that significantly lower fertility tends to occur in countries where the median marriage age of women is at least 25, which reduces potential childbearing years (ages 15–44) by ten years. Even more important, this cuts the *prime* reproductive period from ages 20–29, when most women have children, by about half.

Although data are lacking for some countries, median age at first marriage for women is around 18 in Africa, 20–21 in Asia and Latin America, 22 in Oceania, and 23 in Europe. Note that this correlates closely with the TFRs for these continents (Figure 7-10).

In the United States, the median marriage age for women has increased from 20.3 in 1950 to 23.1 in 1986. During this same period, the median marriage age for U.S. men has risen from 22.8 to 25.8. This has helped reduce the TFR in the United States, although the reduction is offset somewhat by the increase in unmarried young women (including teenagers) having children. Every year, 1 million American teenagers

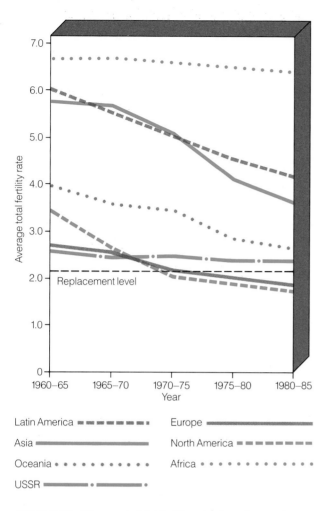

Figure 7-10 Changes in total fertility rates in various regions of the world between 1960 and 1985. (Data from United Nations)

Latin America ━━━━━━━━
Asia ━━━━━━━━━
Oceania ● ● ● ● ● ● ● ●
USSR ━━●━━●━━●━
Europe ━━━━━━━━━
North America ━━ ━━ ━━
Africa ● ● ● ● ● ● ● ● ● ●

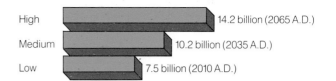

Figure 7-11 Three projections of stabilized world population levels based on different assumptions about the year (shown in parentheses) when the average world total fertility rate drops to replacement level. (Data from United Nations)

High 14.2 billion (2065 A.D.)
Medium 10.2 billion (2035 A.D.)
Low 7.5 billion (2010 A.D.)

become pregnant. About 400,000 will have abortions, accounting for almost one of every three U.S. abortions.

In addition to affecting the U.S. average total fertility rate, the increase in births to teenagers contributes to an infant mortality rate for the United States that is higher than that in 19 other industrialized countries. Studies show that out-of-wedlock births by teenagers are more likely to result in low birth weight (under 5.5 pounds), mental retardation, or death during the first year of life. It is estimated that each baby born in 1987 as the first child of a teenage mother will cost taxpayers an average of $15,600 over the next 20 years. This helps explain why teenage pregnancy is everyone's problem, not merely that of the mother and her family.

Using TFR to Project World Population Stabilization Many people falsely equate a TFR at or below replacement-level fertility with zero population growth (ZPG), by which population size remains constant.

Although achieving a replacement-level TFR is one of the first steps necessary for achieving ZPG, a population can have a TFR at or below 2.1 and still be growing because the number of women entering their childbearing period is still rising.

The larger the number of girls under age 15, the longer it takes to reach population stabilization or ZPG after the TFR reaches 2.1 (discussed further in Section 7-4). ZPG occurs at the global level when the world's birth and death rates are equal. For a particular country or group of countries, ZPG occurs when the birth rate and death rate are equal and net migration is zero.

Figure 7-11 shows UN projections for world population growth and eventual stabilization based on different assumptions about when the average fertility rate will drop to replacement-level fertility of 2.1. Figure 7-12 projects population size and year of stabilization for different groups of countries, using the medium UN projection (usually taken as the most likely) shown in Figure 7-11.

No one knows whether any of these projections will prove accurate. All are based not only on assumptions about TFRs but also on the assumption of adequate supplies of food, energy, and other natural resources. If such supplies are inadequate or if global nuclear war occurs, the resulting sharp increase in death rates could lead to population stabilization at a much lower level.

U.S. Population Stabilization Figure 7-13 shows that the total fertility rate in the United States has oscillated wildly. At the peak of the post–World War II baby boom (1945–63) in 1957, the average TFR reached 3.7 children per woman. Since then the average TFR has generally declined and has been at or below replacement level since 1972. This drop was probably caused by various factors, including reduction in unwanted and mistimed births through widespread use of effective birth control methods, availability of legal abortions, social attitudes in favor of smaller families, greater social acceptance of childless couples, rising costs of raising a family ($175,000 to raise one child born in 1986 to age 18), and an increasing number of women

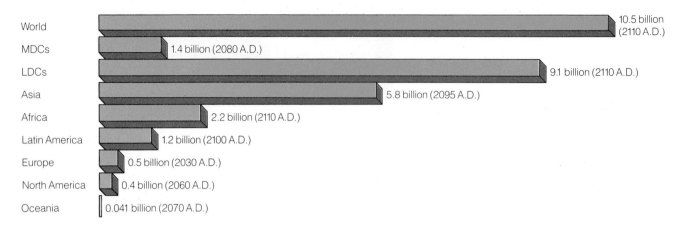

Figure 7-12 United Nations medium projections for stable population size and year of stabilization (shown in parentheses) of various groups of countries.

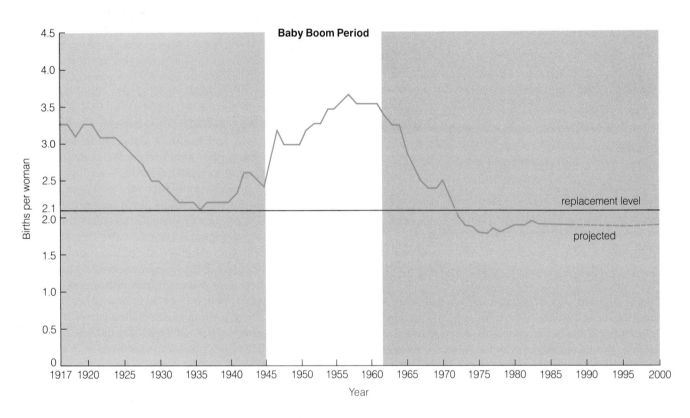

Figure 7-13 Total fertility rate for the United States between 1917 and 1987 and projected rate (dashed line) to 2000. (Data from Population Reference Bureau and U.S. Bureau of the Census)

working outside the home. For example, by 1987 more than 70% of women of childbearing age worked—up from 40% in 1955—and had a childbearing rate one-third of those not in the labor force.

The United States has not reached ZPG in spite of the dramatic drop in average TFR below the replacement level because of the large number of women still moving through their childbearing years and because of the country's high levels of annual legal and illegal immigration. In 1986 the Census Bureau made various projections of future U.S. population growth, assuming different average TFRs, life expectancies, and annual net legal immigration rates. The medium projection, assuming an average annual TFR of 1.9 and an annual net legal immigration of 500,000 persons, is that U.S. population will grow from 244 million in 1987 to 309 million in 2050 and then begin to slow down, reaching ZPG by 2080 with a population of 311 million.

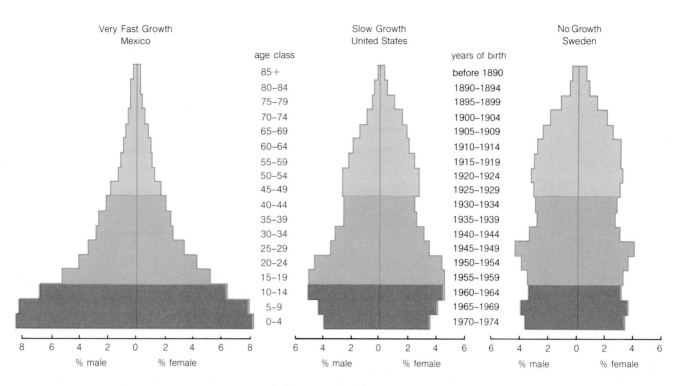

Figure 7-14 Population age structure diagrams for countries with rapid, slow, and zero population growth rates. Dark portions represent preproductive years (0–14), shaded portions represent reproductive years (15–44), and clear portions are postproductive years (45–85+). (Data from Population Reference Bureau)

Some demographers project a somewhat lower growth, based on the assumption that the present lower TFR of 1.8 will be maintained. If this happens, legal and illegal immigration will account for all U.S. population growth by the 2030s. Others project a larger stable population size than 311 million, because they believe that an annual net migration rate (including both legal and illegal immigration) of 800,000 to 1.1 million is more likely than the 500,000 figure used by the Census Bureau.

Still others project a higher population size because of a future rise to a TFR above 1.9, primarily from an increase in the nonwhite population (mostly Hispanic and black), which historically has had fertility rates higher than the U.S. average (Figure 7-9). Each of us will play a role in determining which of these and other demographic possibilities becomes a reality.

7-4 AGE STRUCTURE

Age Structure Diagrams Why will world population most likely keep growing for at least 100 years, even after the average world TFR has reached or dropped below replacement-level fertility of 2.1? Why

do some demographers expect the U.S. birth rate to rise between now and 1994, even though the TFR may stay well below 2.1?

The answer to these questions lies in an understanding of another important factor of population dynamics, the **age structure** of a population: the percentage of the population or the number of people of each sex at each age level in a population. A population age structure diagram is obtained by plotting the percentages of males and females in the total population in three age categories: *preproductive* (ages 0–14), *reproductive* (ages 15–44), and *postproductive* (ages 45–85+).

Figure 7-14 shows the age structure diagrams for countries with rapid, slow, and zero growth rates. Mexico and most LDCs with rapidly growing populations have pyramid-shaped age structure diagrams, indicating a high ratio of children under age 15 (an average of 37% in 1987 for all LDCs) to adults over age 65 (4% in 1987). In contrast, the age structure diagrams for the United States, Sweden, and most MDCs undergoing slow or no population growth have a narrower base, indicating a much smaller percentage of their population under age 15 (an average of 22% in 1987) and a larger percentage above 65 (11% in 1987). Some MDCs—Sweden, for example, which has achieved ZPG—have roughly equal numbers of people at each age level.

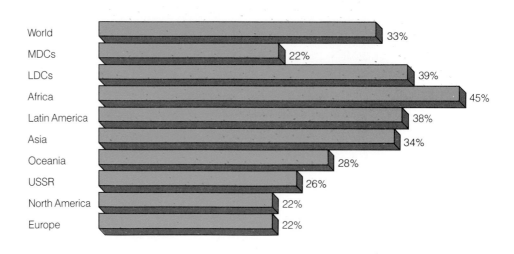

Figure 7-15 Percentage of people under age 15 in various groups of countries in 1987. (Data from Population Reference Bureau)

World 33%
MDCs 22%
LDCs 39%
Africa 45%
Latin America 38%
Asia 34%
Oceania 28%
USSR 26%
North America 22%
Europe 22%

Population Momentum and Age Structure Any country whose population contains a large number of people below age 29, and especially below age 15, has a powerful built-in momentum of population growth. These are the potential parents of the next generation. This momentum exists even if women have only one or two children, simply because the number who can have children is so large. Moreover, the population of a country with a large number of people under 29 continues to expand for approximately one average lifetime—roughly 60 to 70 years—after its average TFR has dropped to replacement level.

In 1987 about 33% of the people on this planet were under 15 years of age. In LDCs the number is even higher, 37%, compared to 22% in MDCs (Figure 7-15). This youth-heavy age structure explains why population will continue to grow, especially in LDCs, long after replacement-level fertility rates are reached—unless death rates rise sharply.

Population age structure also explains why it will probably take 50 to 70 years for the United States to reach ZPG, even if fertility rates remain below the replacement level. Although many couples are now having smaller families, the number of births could easily rise for the next several years because there are more women to have babies. The 37 million women born during the baby boom will affect U.S. population growth through 1994—when women born at the end of the baby boom turn 30 and move out of their prime reproductive years. As mentioned previously, U.S. population growth is also maintained by a large annual net migration rate.

Making Projections from Age Structure Diagrams Figure 7-16 shows that the U.S. baby boom caused a bulge in the age structure. This bulge will move through the prime reproductive ages of 20–29 between 1970 and 1994. This helps explain why the 1960s and 1970s have been called the "youth genera-

tion." Similarly, the period between 1975 and 1990 could be called the "age of young adults," the period between 1990 and 2009 the "age of middle-aged adults," and between 2010 and 2058 the "age of senior citizens."

The diagrams in Figure 7-16 can be used to project some of the social and economic changes that may occur in the United States in coming decades. In the 1970s and early 1980s, large numbers of "baby boomers" flooded the job market, causing high unemployment rates for teenagers and adults under age 29. This situation probably won't begin to ease until after 1993, when the last of the people born during the baby boom are over age 29.

Most baby boomers are having to work harder than the generation that preceded them just to stay even. Many are falling behind and facing financial sacrifice and lowered expectations. Between 1973 and 1983, the real (adjusted for inflation) after-tax income of households headed by a 25-to-34-year-old baby boomer decreased by nearly 19%—even though the proportion of wives in this group working outside the household more than doubled, from 29% to 62%, during the same period. In 1983 the average 30-year-old male homeowner spent 44% of his earnings on mortgage payments, compared to his 1949 counterparts' 14%. This also means that most baby boomers are not able to put very much aside for their retirement. Since the 1970s, for the first time in U.S. history, the economic value of a college degree when adjusted for inflation has declined. It is estimated that between 1980 and 1990, nearly one-quarter of college graduates will be overeducated for the jobs they will be able to get.

By 1999 all of the baby boom generation will have reached middle age (35–64). Many of these adults may find little opportunity for professional advancement unless large numbers of their elders decide to retire early.

Between 2012 and 2029, all baby boomers will reach age 65, and many will live at least another 20 years. Assuming that death rates don't rise, the number of

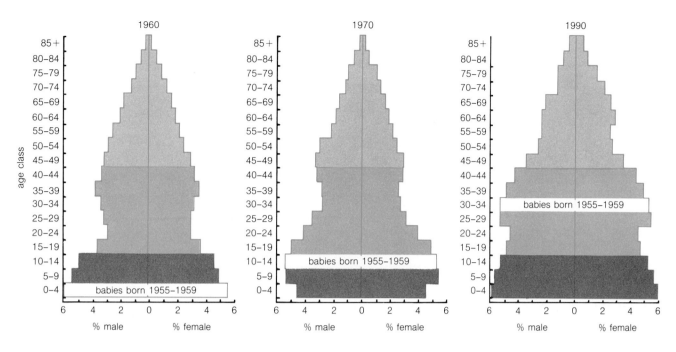

Figure 7-16 Age structure of the U.S. population in 1960, 1970, and 1990 (projection). The population bulge of babies born between 1955 and 1959 is slowly moving up. (Data from Population Reference Bureau)

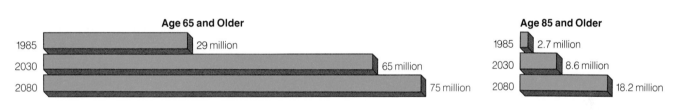

Figure 7-17 Projected growth in the number of people age 65 and older in the United States between 1985 and 2080. (Data from U.S. Bureau of the Census)

people 65 and older will increase dramatically (Figure 7-17). The burden of supporting so many retired people will be on the "baby bust" generation—the much smaller group of people born in the 1970s and 1980s, when average fertility rates were below replacement level.

This large increase in retired citizens will put a severe strain on Medicare and the Social Security system, which is already near bankruptcy. Between 1937 and 1980, the number of workers paying Social Security taxes for each beneficiary dropped from ten to three. By 2030 it is projected that there will be only two workers per beneficiary. Thus many baby boomers may face harsh times after retirement because of lower Social Security benefits, as well as less personal savings.

The baby bust generation should have a much easier time in many respects than the preceding baby boom generation. Much smaller numbers will be competing for education, jobs, and services. Labor shortages should drive up wages for the baby bust generation; in fact, the United States might have to relax current immigration laws to bring in new workers. With a shortage of young adults, the armed forces would be hard pressed to meet recruiting levels, and there may be pressure to reinstate the draft.

On the other hand, the baby bust group may find job promotions elusive as they reach middle age, because most upper-level positions will be occupied by the much larger baby boom group. They also will probably face much higher income taxes and Social Security taxes (to help pay for retired baby boomers). From these few projections, we see that any bulge or indentation in the age structure of a population creates a number of social and economic changes that ripple through a society for decades.

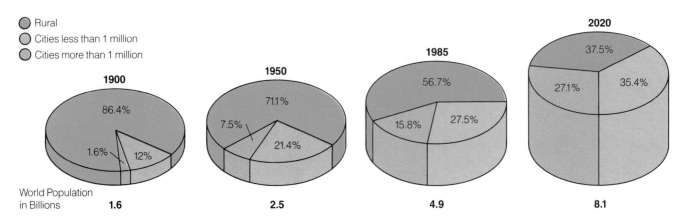

Figure 7-18 Patterns of world urbanization from 1900 to 1985 with projections to 2020. (Data from United Nations and Population Reference Bureau)

7-5 POPULATION DISTRIBUTION: URBANIZATION AND URBAN GROWTH

The World Situation Economic activities, environmental degradation, pollution, and social interactions and problems are affected not only by population growth but also by how population is distributed. Since the Agricultural Revolution about 12,000 years ago, people have flooded into cities to find jobs. Those who get good jobs and do well financially often are enthusiastic about living in cities, which provide the greatest variety of goods and services and the most exciting diversity of social and cultural activities. In addition, urban social and cultural activities enrich the lives of people who live far from city boundaries. Thus it is not surprising that the percentage of the world's people living in cities continues to increase.

A country's **degree of urbanization** is the percentage of its population living in areas with a population of more than 2,500 people. **Urban growth** is the rate of growth of urban populations. Between 1900 and 1985, the percentage of the world's population living in cities increased from 14% to 43% (Figure 7-18). By the year 2000, one-half of the world's population will be living in cities (one-quarter of city dwellers will be homeless), and by 2020 almost two out of three people on earth are expected to be living in cities.

The percentage of people living in cities with a population of more than 1 million is also increasing. Accommodating the 4.2 billion additional people projected to be living in urban areas by 2020 will be a monumental task.

Unprecedented urban growth in MDCs and LDCs has given rise to a new concept: the "supercity," an urban area with a population of more than 10 million. In 1985 there were 10 supercities, most of them located in LDCs (Table 7-1). The United Nations projects that by 2000 there will be 25 supercities, most of them in LDCs.

The degree of urbanization varies in different parts of the world. From 50% to 75% of the population in North America, Latin America, and Europe live in urban areas, whereas about 70% of the people in Africa and Asia live in rural areas and try to make their living from farming.

Although all countries are experiencing urbanization, the rate of urban growth in LDCs surpassed that in MDCs around 1970 and is expected to increase more rapidly in the future (Figure 7-19). LDCs are simultaneously experiencing high rates of natural population increase and rapid internal migration of people from rural to urban areas—each factor contributing about equally to urban growth. In LDCs more than 20 million rural people migrate to cities each year to escape wretched living conditions in the countryside; in the process, they overwhelm already inadequate city services.

For most of these migrants, the city becomes a poverty trap, not an oasis of economic opportunity and cultural diversity. Those few fortunate enough to get a job must work long hours for low wages and are usually exposed to dust, hazardous chemicals, excessive noise, and dangerous machinery. With official unemployment levels of 20% to 30%, most of the urban poor are forced to live on the streets or to crowd into slums and shantytowns, made from corrugated iron, plastic sheets, and packing boxes, which ring the outskirts of most cities in these countries. Because many of these settlements spring up illegally on unoccupied land, their occupants live in constant fear of eviction or of having their makeshift shelters destroyed by bulldozers. Many shantytowns are located on land prone to landslides, floods, or tidal waves or in the most polluted districts of inner cities. Fires are common because most residents use kerosene stoves or fuelwood for heating and cooking.

Table 7-1 The Ten Largest Urban Areas in the World in 1985 and 2000 (Data from United Nations)

1985		2000	
Urban Area	Population (millions)	Urban Area	Projected Population (millions)
Mexico City	18.1	Mexico City	26.3
Tokyo-Yokohama	17.2	Sao Paulo	24.0
Sao Paulo	15.9	Tokyo-Yokohama	17.1
New York–N.E. New Jersey	15.3	Calcutta	16.6
Shanghai	11.8	Greater Bombay	16.0
Calcutta	11.0	New York–N.E. New Jersey	15.5
Greater Buenos Aires	10.9	Seoul	13.5
Rio de Janeiro	10.4	Shanghai	13.5
Seoul	10.2	Rio de Janeiro	13.3
Greater Bombay	10.1	Delhi	13.3

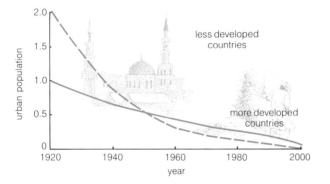

Figure 7-19 Actual and projected urban population growth in billions in MDCs and LDCs between 1920 and 2000. (Data from United Nations)

In most large cities in LDCs, shantytown populations double every five to seven years, four or five times the population growth rate of the entire city. Shantytown and "overnight" squatter settlements now hold up to 30% of the population of many urban centers in LDCs. Because an accurate count of the number of squatters is impossible to get, they are not included in urban population estimates like those given in Table 7-1. Most cities refuse to provide shantytowns and slums with adequate drinking water, sanitation, food, health care, housing, schools, and jobs, because of a lack of money and the fear that such improvements will atttract even more of the rural poor.

Despite joblessness and squalor, shantytown residents cling to life with resourcefulness, tenacity, and hope. Most of them are convinced that the city offers, possibly for themselves and certainly for their children, the only chance of a better life. On balance, most do have more opportunities and are often better off than the rural poor they left behind. They also tend to have fewer children, because there is no room for them and because of better access to family planning programs.

The U.S. Situation In 1987 more than three out of four (76%) Americans lived in urban areas, and two out of three lived in the country's 28 largest urban regions (Figure 7-20). Since 1800 several major internal population shifts have taken place in the United States. The major shift has been from rural to urban areas as the country has industrialized and needed fewer and fewer farmers to produce sufficient food (Figure 7-21). Since 1970, however, many people have moved from large cities to suburbs and to smaller cities and rural areas, primarily because of the creation of large numbers of new jobs in such areas. Since 1980 about 90% of the population increase in the United States has occurred in the South and West, much of this the result of migration from the North and East.

7-6 POPULATION DISTRIBUTION AND TRANSPORTATION SYSTEMS

Compact and Dispersed Cities If suitable rural land is not available for conversion to urban land, a city tends to grow upward rather than outward, occupy a relatively small area, and have a high population density. People living in such compact cities tend to walk or use energy-efficient mass transit and tend to live in multistoried apartment buildings with shared walls that have an insulating effect, reducing heating and cooling costs. Because of the lack of suitable rural land, many European cities are compact and tend to be more energy efficient than the dispersed cities found throughout the United States.

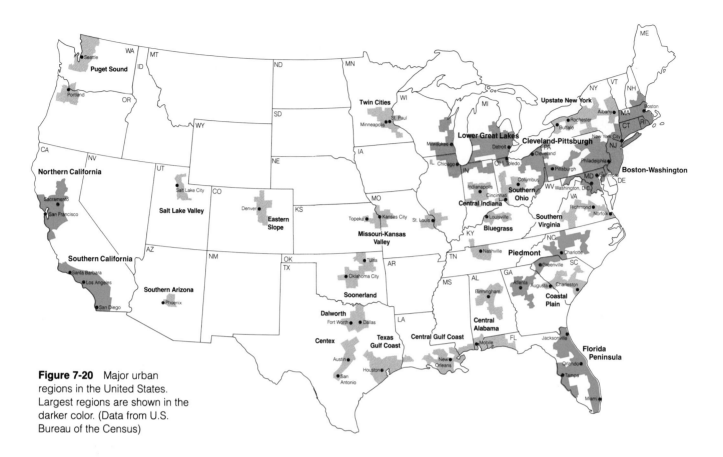

Figure 7-20 Major urban regions in the United States. Largest regions are shown in the darker color. (Data from U.S. Bureau of the Census)

A combination of cheap gasoline and a large supply of rural land suitable for urban development tends to result in a dispersed city with a low population density. Most people living in such cities rely on cars with low energy efficiencies for transportation within the central city and to and from its suburbs. Most live in single-family houses whose unshared walls lose and gain heat rapidly unless they are well insulated. Sharp rises in energy prices (especially gasoline) will stimulate many suburbanites to move back into the city to cut down on transportation, heating, and cooling costs.

Motor Vehicles In the United States the car is now used for about 98% of all urban transportation and 85% of all travel between cities. In 1985 almost two-thirds of all working Americans traveled to and from work alone in their own car with an average round trip of 37 kilometers (23 miles) a day at an average annual cost of $1,300 a person (Figure 7-22). No wonder British author J. B. Priestly remarked, "In America, the cars have become the people."

The automobile provides many advantages. Above all, it offers people privacy, security, and unparalleled freedom to go where they want to go when they want to go there. In addition, much of the U.S. economy is built around the automobile. One out of every six dollars spent and one out of every six nonfarm jobs are connected to the automobile or related industries such as oil, steel, rubber, plastics, automobile services, and highway construction. This industrial complex accounts for 20% of the annual GNP and provides about 18% of all federal taxes.

In spite of their advantages, cars and trucks have harmful effects on human lives and on air, water, and land resources. By providing almost unlimited mobility, automobiles and highways have been a major factor in urban sprawl, stimulating most U.S. cities to become decentralized and dispersed.

The world's 490 million cars and trucks also kill an average of 170,000 people, maim 500,000, and injure 10 million each year. This is equivalent to a death about *every 3 minutes* and a disabling injury *every 20 seconds* of every day. In the United States about 25 million motor vehicle accidents each year kill about 45,000 people and injure about 5 million, at a cost of about $60 billion annually in lost income, insurance, and administrative and legal expenses. Since the automobile was introduced, almost 2 million Americans have been killed on the highways—about twice the number of Americans killed in all U.S. wars.

Large areas of land are also utilized by motor vehicles. Roads and parking space take up 65% percent of the total land area in Los Angeles, more than half of Dallas (see photo on page 1), and more than one-third of New York City and the nation's capital.

Instead of reducing automobile congestion, the construction of thousands of miles of roads has

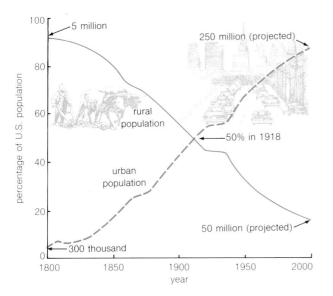

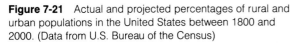

Figure 7-21 Actual and projected percentages of rural and urban populations in the United States between 1800 and 2000. (Data from U.S. Bureau of the Census)

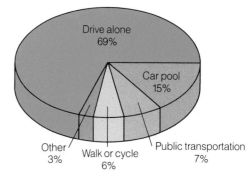

Figure 7-22 How people in the United States get to and from work. (Data from U.S. Bureau of the Census)

encouraged more automobiles and travel, causing even more congestion. In 1907 the average speed of horse-drawn vehicles through the borough of Manhattan was measured at 11.5 miles per hour. Today, cross-town Manhattan traffic—cars and trucks with the potential power of 100 to 300 horses each—creeps along at an average speed of 5.2 miles per hour. In Los Angeles, traffic on the Hollywood Freeway slows to 20 miles per hour for about 14 hours every day. By 2000 the average speed is projected to drop to about 7 miles per hour.

Mass Transit Even though total population has increased significantly, the number of riders on all forms of mass transit (heavy-rail subways and trains, light-rail trolleys, and buses) in the United States has declined drastically, from 24 million in 1945 to about 9 million in 1986. This decline generally parallels the increased use of the automobile and the resulting development of increasingly dispersed cities. These interrelated social changes were stimulated by cheap gasoline and the provision of funds from federal gasoline taxes to build highways. At the same time relatively little federal support was given mass transit.

Mass transit advocates argue that the country needs to increase, not decrease, its support of mass transit and Amtrak to help reduce dependence on imported oil, unnecessary oil waste, highway congestion, and air pollution. They also argue that mass transit systems stimulate new business development and revitalization in central cities and thus offset the federal, state, and local subsidies usually needed for their construction and operation. In 1986 the American Public

Transit Association estimated that every $100 billion spent on mass transit projects results in a $327 million increase in urban business revenues and supports 8,000 jobs.

Heavy-Rail Mass Transit Some analysts see the building of new, fixed heavy-rail mass transit systems and the improvement of existing subway and above-ground urban railroad systems as the key to solving transportation problems in most large cities. Others argue that such systems are useful only where many people live along a narrow corridor, and even then their high construction and operating costs may outweigh their benefits.

Some fixed-rail rapid transit systems have been successful—others have not. Since its opening in 1972, San Francisco's $1.7 billion, computer-controlled Bay Area Rapid Transit (BART) system has suffered from breakdowns, fires, brake problems, computer failures, massive financial losses, and too few riders. The Metro system of Washington, D.C., is better planned than BART and has the advantage of serving a concentrated urban area. But this efficient mass transit system cost about $90 million a mile to build, compared to $22.5 million a mile for BART. Built entirely at federal expense, the Metro has had large annual operating deficits. Critics blame many of the problems of large-scale fixed-rail transit systems on overdependence on federal aid and shortsighted planning that overestimated the number of riders and underestimated the costs of building and operating such systems.

Some cities, however, have built successful fixed-rail systems. Since Atlanta's system opened in 1979, it has steadily added riders and opened new stations. Baltimore's state-owned Metro System is also increasing service, routes, and ridership. Pittsburgh has cleaner air and renewed business vitality partly because of its new subway system, opened in 1985.

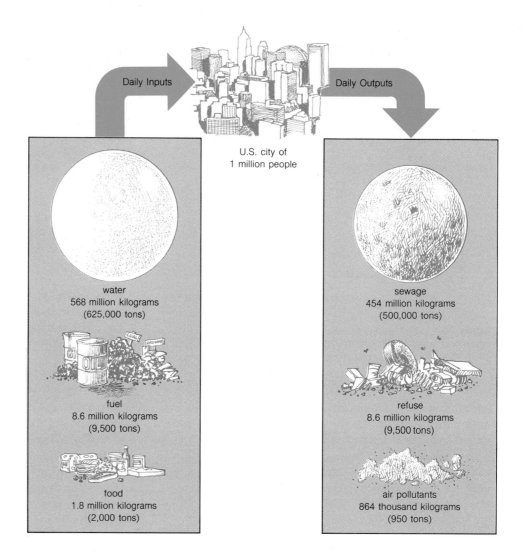

Figure 7-23 Typical daily input and output of matter and energy for a U.S. city of 1 million people.

Daily Inputs

U.S. city of 1 million people

Daily Outputs

water
568 million kilograms
(625,000 tons)

fuel
8.6 million kilograms
(9,500 tons)

food
1.8 million kilograms
(2,000 tons)

sewage
454 million kilograms
(500,000 tons)

refuse
8.6 million kilograms
(9,500 tons)

air pollutants
864 thousand kilograms
(950 tons)

Light-Rail Mass Transit Other cities—many of them medium-size—such as San Diego, California; Buffalo, New York; Portland, Oregon; and Toronto, Canada, have built fixed light-rail trolley systems. These are modernized versions of the streetcar systems found in most major U.S. cities in the 1930s and 1940s, before they were purchased and torn up by General Motors and tire, oil, truck, and bus companies to increase sales of motor vehicles.

Light-rail systems are much less costly per mile to build than heavy-rail systems. Although the start-up cost for a light-rail system is higher than for a bus system carrying a comparable number of passengers, the operating costs are much lower. Light-rail systems are also cleaner and quieter than buses. The biggest barrier to development of new light-rail systems is lack of federal money raised through taxes on gasoline—most of which is used for highway construction.

Bus Mass Transit Buses are cheaper and more flexible than rail systems. They can be routed to almost any area in widely dispersed cities. They also require less capital and have lower operating costs than most rail systems. But by offering low fares to attract riders, they usually lose money. To make up for losses, bus companies tend to cut service and maintenance and seek federal, state, and local subsidies. At any one time, almost half of the city buses in Houston, Texas, are idled by breakdowns, with riders often waiting an hour for service. Philadelphia's aging fleet of buses averages ten accidents a day, often because of improperly done repairs.

Paratransit Because full-size buses are cost-effective only when full, they are being supplemented by car pools, van pools, jitneys, and dial-a-ride systems. These paratransit methods attempt to combine the advantages of the door-to-door service of a private automobile or taxi with the economy of a ten-passenger van or minibus. They represent a practical solution to some of the transportation problems of today's dispersed urban areas.

Dial-a-ride systems operate in an increasing number of American cities. Passengers call for a van, minibus, or tax-subsidized taxi, which comes by to pick them up at the doorstep, usually in about 20 to 50 minutes. Efficiency can be increased by the use of two-way radios and computerized routing. Dial-a-ride systems are fairly expensive to operate. But compared with most large-scale mass transit systems, they are a bargain, and each vehicle is usually filled with passengers.

In cities such as Mexico City; Caracas, Venezuela; and Cairo, large fleets of *jitneys*, small vans or minibuses that travel relatively fixed routes but stop on demand—carry millions of passengers each day. After laws banning jitney service were repealed in 1979 despite objections by taxi and transit companies, privately owned jitney service has flourished in San Diego, San Francisco, and Los Angeles and may spread to other cities. Analysts argue that deregulation of taxi fares and public transport fares would greatly increase the number of private individuals and companies operating jitneys in major U.S. cities.

7-7 MAKING CITIES MORE SUSTAINABLE

Vulnerability of Urban Areas Unlike natural ecosystems, cities do not have enough producers—that is, green plants—to support their human inhabitants. As one observer remarked, "Cities are places where they cut down the trees and then name the streets after them." This scarcity of vegetation is unfortunate, because urban plants, grasses, and trees absorb air pollutants, give off oxygen, help cool the air as water is evaporated from their leaves, muffle noise, and satisfy important psychological needs of city dwellers.

Urban systems survive only by importing food from external, plant-growing ecosystems. Cities also obtain most of their fresh air, water, minerals, and energy resources from external ecosystems. Instead of being recycled, most of the solid, liquid, and gaseous wastes of cities are discharged to ecosystems outside their boundaries (Figure 7-23).

Making Urban Areas More Self-Sufficient Some planners have proposed guidelines and models for building compact towns and cities that waste less matter and energy resources and are more self-reliant than conventional municipalities. Such self-sufficient cities would be surrounded by farms, greenbelts, and community gardens. Homes and marketplaces would be close together, and most local transportation would be by bus, bicycle, and foot. Buildings would be cooled and heated by sun and wind, and wastes would be recycled. Food would be grown locally, and there would be few huge factories.

Although no such cities have been started from scratch, some existing cities have also begun efforts to become more self-sufficient in certain energy resources. Erie, Pennsylvania, for example, drilled two producing oil wells between 1978 and 1980, and Palo Alto, California, has rezoned almost 5,000 acres within the city as open space for agriculture. Fort Collins, Colorado, runs much of its transportation system on methane gas generated from its sewage plant and uses nutrient-rich sludge from the sewage plant to fertilize 600 acres of city-owned land to grow corn, which might be converted to alcohol to fuel more city cars. In New York City, Consolidated Edison has drilled into a large city landfill and has retrieved enough methane from decaying garbage to heat tens of thousands of homes. St. Paul, Minnesota, is planning to build the country's first system that will heat all major downtown buildings with waste heat now being dumped into the Mississippi River by an electric utility.

The population of most less developed countries is doubling every twenty to thirty years. Trying to develop into a modern industrial state under these conditions is like trying to work out the choreography for a new ballet in a crowded subway car.

Garrett Hardin

DISCUSSION TOPICS

1. Why are falling birth rates not necessarily a reliable indicator of future population growth trends?

2. Explain the difference between achieving replacement-level fertility and zero population growth (ZPG).

3. What must happen to the total fertility rate if the United States is to attain ZPG in 40 to 60 years? Why will it take so long?

4. Project what your own life may be like at ages 25, 45, and 65 on the basis of the present age structure of the population of the country in which you live. What changes, if any, do such projections make in your career choice and in your plans for marriage and children?

5. Do you think the world is more likely to reach the high (14.2 billion), medium (10.2 billion), or low (7.5 billion) population size projected by the United Nations? Explain.

6. List the advantages and disadvantages of living in (a) the downtown area of a large city, (b) suburbia, (c) a small town in a rural area, (d) a small town near a large city, and (e) a rural area. Which would you prefer to live in? Which will you probably end up living in? Why?

7. What conditions, if any, would encourage you to rely less on the automobile? Would you regularly travel to school or work in a car pool, on a bicycle or moped, on foot, or by mass transit? Explain.

8. What life-support resources in your community are the most vulnerable to interruption or destruction? What alternate or backup resources, if any, exist?

8

Population Control

GENERAL OBJECTIVES

1. What methods can be used for controlling the size and rate of growth of the human population?

2. What are the major methods of birth control?

3. What success have the world's two most populous countries, China and India, had in trying to control the rate of growth of their populations?

Short of thermonuclear war itself, rampant population growth is the gravest issue the world faces over the decades immediately ahead.

Robert S. McNamara

Some consider rapid population growth one of the most serious problems facing the world today. Others believe it is a problem in most LDCs but not in MDCs such as the United States. A small number of analysts contend that population growth is good and should not be discouraged. How population growth might be controlled is the complex and highly controversial issue discussed in this chapter.

8-1 METHODS FOR CONTROLLING POPULATION GROWTH

Controlling Births, Deaths, and Migration A government can alter the size and growth rate of its population by encouraging a change in any of the three basic demographic variables: births, deaths, and migration. Most MDCs now have relatively low birth and death rates, while most LDCs have relatively low death rates but high birth rates (Figure 7-3, p. 112). Governments of most countries in the world achieve some degree of population control by allowing relatively little immigration from other countries and in some cases by encouraging emigration to other countries to reduce population pressures. Only a few countries, chiefly the United States, Canada, and Australia, allow large annual increases in their population from immigration.

Throughout the world, decreasing birth rate is the focus of most efforts to control population growth. By 1987 countries containing 91% of the people living in LDCs had programs to reduce their fertility rates, although the effectiveness and funding of such programs vary widely from country to country. In 1987 countries containing only 3% of the people living in LDCs considered their fertility rates too low.

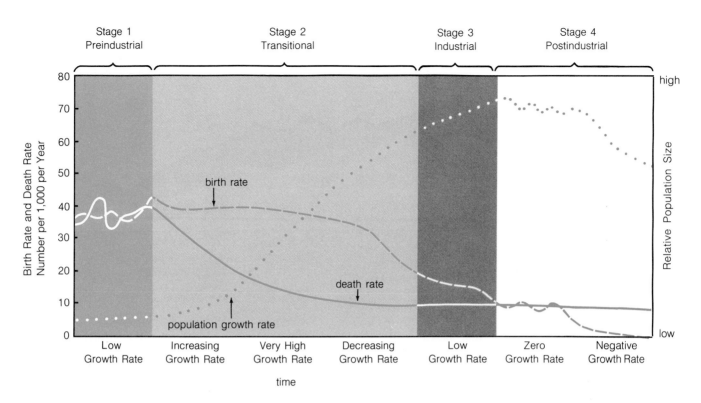

Figure 8-1 Generalized model of the demographic transition.

Two general approaches to decreasing birth rates are *economic development* and *family planning*. Economic development may reduce the number of children a couple desires by bringing about better education, providing more economic security, and reducing the need to consider children a substitute for old age social security. Family planning methods help people regulate the number of children they have and when they have them. Although controversy still exists over which approach is better, increasing evidence shows that a combination of both approaches offers a country the best way to reduce its birth rate and thus its rate of population growth.

Economic Development and the Demographic Transition After examining birth and death rates in western European countries that became industrialized during the 19th century, demographers formed a model of population change and control known as the **demographic transition.** The basic idea of this model is that as western European countries became industrialized, they had declines in death rates followed by declines in birth rates, eventually resulting in decreased population growth.

This transition takes place in four distinct phases (Figure 8-1). In the *preindustrial stage*, harsh living conditions lead to high birth rates (to compensate for high

infant mortality) and high death rates, resulting in little population growth. The second, or *transitional, stage* begins shortly after industrialization is initiated. In this phase, death rates drop, primarily as a result of increased food production and improved sanitation and health. Because birth rates remain high, population growth accelerates and continues at a high rate for a prolonged period (typically about 2.5% to 3% a year). Later in this phase, population growth levels off somewhat as industrialization spreads and living conditions improve.

In the *industrial stage*, industrialization is widespread. Crude birth rates drop and eventually begin to approach crude death rates as better-educated and more affluent couples (who have moved to cities to obtain jobs) become aware that children are expensive to raise and that having too many hinders them from taking advantage of job opportunities in an expanding economy. Population growth continues but at a slower and perhaps fluctuating rate, depending on economic conditions. The United States, Japan, the Soviet Union, Canada, Australia, New Zealand, and most of the industrialized western European countries are now in this third phase. A fourth phase, the *postindustrial stage*, takes place when birth rates decline even further to equal death rates, thus reaching ZPG, and then continue to fall so that total population size begins slowly to decrease.

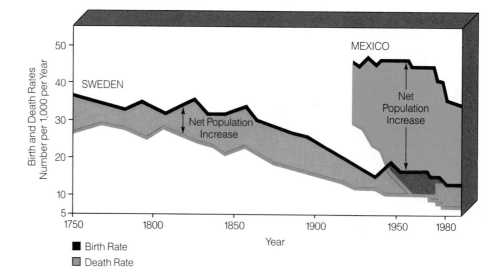

Figure 8-2 Comparison of Sweden, which has undergone the demographic transition, and Mexico, which has not. (Data from Population Reference Bureau.)

By 1987, 14 countries—Austria, Belgium, Bulgaria, Denmark, East Germany, Greece, Hungary, Italy, Luxembourg, Norway, Sweden, Switzerland, the United Kingdom, and West Germany—had reached or were close to ZPG. Together these countries have a population of 264 million, representing 5% of the world's population. Four of these countries—Austria, Denmark, Hungary, and West Germany—were experiencing population declines.

If West Germany maintains its present total fertility rate of 1.3 and does not allow significant immigration, its population will decrease by half, from 61 million in 1987 to 30 million by 2030. With a small percentage of its population under age 15, West Germany has built-in momentum for population decline over the next 50 to 70 years. West German leaders are concerned about this projected population decrease and would like to see birth rates rise. They fear that there will be too few workers to sustain economic growth and to pay taxes to support the increasing portion of population over age 65. Other MDCs entering the postindustrial stage also may become concerned with increasing their birth and fertility rates to reduce their rate of population decline.

In most LDCs today, death rates have fallen but not birth rates. In other words, these LDCs are still in the transitional phase halfway up the economic ladder, with fairly high to high population growth rates. Figure 8-2 compares Sweden, which has entered the postindustrial phase, with Mexico, still in the transitional phase. Mexico, with a population of 82 million in 1987 and a growth rate of 2.5%, is projected to have a population of 138 million by 2020.

Cornucopians believe that today's LDCs will make the demographic transition, becoming MDCs with low population growth rates over the next few decades without increased family planning efforts. However, neo-Malthusians point out that an increase in average annual income per person occurs only when a country's average annual rate of economic growth exceeds its annual rate of population growth. For example, over the last 100 years, Japan's rate of economic growth has averaged 4% a year, while its population growth rate has averaged 1.1% a year. It is the resulting 2.9% average annual growth in income per person that has made Japan one of the world's wealthiest countries. Similarly, over the past 100 years the rate of economic growth in the United States has averaged 3.3% a year, while population has grown at an average rate of 1.5% a year. As a result, the United States has averaged a 1.8% annual growth in average income per person.

Neo-Malthusians fear that the rate of economic growth in many LDCs will not exceed their high rates of population growth. Many LDCs, such as Mexico, could become stuck in the transitional stage of the demographic transition, because some of the conditions that allowed today's MDCs to become developed are not available for today's LDCs. For example, even with large and growing populations, many LDCs do not have enough skilled workers to produce the high-technology products needed to compete in today's economic environment.

Most low- and middle-income LDCs also lack the capital and resources needed for rapid economic development, and the amount of money being given or lent to LDCs—already struggling under tremendous debt burdens—has been decreasing. LDCs face stiff competition from MDCs and recently modernized LDCs in selling products on which their economic growth depends. In addition, energy experts project that cheap supplies of fossil fuel energy, which enabled today's MDCs to make the demographic transition, will decrease in coming decades, and the prices of most other sources of energy will be too high for poor countries.

Family Planning Recent evidence suggests that improved and expanded family planning programs may bring about a more rapid decline in the birth rate and at a lower cost than economic development alone. **Family planning** programs provide educational and clinical services that help couples choose how many children to have and when to have them. Such programs vary from culture to culture but usually provide sex education and information on methods of birth control, distribute contraceptives, and in some cases perform abortions and sterilizations, often without charge or at low rates. With the exception of China, family planning programs in most countries have steered clear of trying to convince or coerce couples to have fewer children.

Family planning services were first introduced in LDCs in the 1940s and 1950s by private doctors and women's groups. Since that time, organizations such as the International Planned Parenthood Federation, the Planned Parenthood Federation of America, the United Nations Fund for Population Activities, the U.S. Agency for International Development, the Ford Foundation, and the World Bank have been helping countries carry out family planning by providing technical assistance, funding, or both.

Between 1970 and 1987, birth rates have dropped, sometimes quite rapidly, in more than 30 LDCs. Family planning has been a major factor in reducing birth and fertility rates in highly populated China and Indonesia, and in some LDCs with relatively small populations, such as Singapore, Hong Kong, Sri Lanka, Barbados, Taiwan, Cuba, Mauritius, Thailand, Colombia, Costa Rica, South Korea, Fiji, and Jamaica. The common denominators in these successful programs are committed leadership, local implementation, and wide availability of contraceptive services.

However, only moderate to poor results have been claimed in more populous LDCs like India, Brazil, Bangladesh, Pakistan, and Nigeria, and in 79 less populous LDCs—especially in Africa and Latin America, where population growth rates are usually very high. Only 3% to 4% of the couples in most African countries use contraception. However, Brazil, the most populous country in Latin America, has recently launched a program to give all women information on birth control methods and a free supply of birth control pills. Mexico has also stepped up a public education program to publicize the importance of family planning.

The Population Crisis Committee estimates that between 1978 and 1983, family planning programs reduced world population by 130 million and saved at least $175 billion in government expenditures for food, shelter, clothing, education, and health care. Despite these efforts, the delivery of family planning services in much of the less developed world is still woefully inadequate, particularly in rural areas.

Between 1975 and 1985, nearly half a million women in 61 LDCs were interviewed in the World Fertility Survey and Contraceptive Prevalence Survey conducted by the United Nations. On the basis of these surveys, the UN concluded that about 400 million women in LDCs desire to limit the size and determine the spacing of their children but lack access to family planning services.

Expanding family planning services to reach these women and those who will soon be entering their reproductive years could prevent an estimated 5.8 million births a year and more than 130,000 abortions a day and help bring down the world's currently projected population in the year 2100 from 10.2 billion (the UN medium projection) to 7.5 billion (the UN low projection) by preventing unwanted births. This would mean 2.7 billion fewer people needing food, water, shelter, and health services.

Family planning could be provided in LDCs to all couples who want it for about $8 billion a year. Currently only about $2 billion is being spent; about one-fifth ($400 million) is provided as foreign aid by MDCs and private organizations such as the World Bank. If MDCs provided 50% of the $8 billion, it would cost each person in the MDCs an average of only 34 cents a year (compared to the 3 cents now being given) to help reduce world population by 2.7 billion.

But even the present inadequate level of expenditure for family planning is decreasing, as a result of the sharp drop of funds provided to international family planning agencies by the United States since 1985. Efforts are also increasing, mostly by antiabortion activists, to have federal assistance to U.S. hospitals, health departments, and community clinics for contraceptive counseling, pregnancy testing, breast exams, and screening for sexually transmitted diseases sharply reduced or eliminated. These centers serve about 5 million poor women and teenagers each year, most of whom cannot afford private health care. None of these facilities spend any federal money to terminate even a single pregnancy. Instead, they inform women of their options for preventing future pregnancies, terminating an unwanted pregnancy, obtaining prenatal and postnatal health services for themselves and their babies, and putting a baby up for adoption.

This federal assistance program helps prevent pregnancies that would lead to about 282,000 additional births annually and reduces the number of legal abortions in the United States by about 433,000 a year. For every $1 invested, this program saves taxpayers $2 to $3 the following year in health and welfare costs.

Restricting Immigration   Most countries control their population growth to some extent by limiting immigration. A major exception is the United States, which throughout its history has admitted large numbers of immigrants and refugees. Today the United States admits between 600,000 and 750,000 legal immigrants each year—about twice as many as all other countries combined. In addition, at least 200,000 illegal immigrants are added each year to the Census Bureau's estimate of 4.9 million illegal aliens already in the country. The Immigration and Naturalization Service (INS), however, puts the estimated number of illegal aliens at 12 million.

In recent years, pressure to reduce illegal immigration into the United States has been growing. Some analysts have also called for an annual ceiling of no more than 450,000 for all categories of legal immigration, including refugees, to help reduce the intensity of some of the country's social, economic, and environmental problems and reach zero population growth sooner. In polls taken in 1985, half of those surveyed favored lower legal immigration levels, up from 33% in 1965.

In 1986 Congress passed a new immigration law designed to control illegal immigration. Illegal immigrants who entered the United States before January 1, 1982, and who can provide evidence that they have lived here continuously since then may become temporary residents. After 18 months in that status, they can apply to become permanent residents and can apply for citizenship 5 years later. Illegal immigrants who worked in American agriculture for at least 90 days between May 1, 1985, and May 1, 1986, may also become temporary residents and then apply for citizenship after 7 years. The federal government is providing $1 billion a year between 1987 and 1991 to reimburse state and local governments for the cost of supplying public assistance or other benefits to these immigrants. After 5 years, newly legal immigrants will be eligible for federally funded public assistance.

The bill prohibits the hiring of illegal immigrants. Employers must examine the identity documents of all new employees; those who knowingly employ illegal aliens will be subject to fines of $250 to $10,000 per violation, and repeat offenders can be sentenced to prison for up to six months. The bill also increases the budget for the INS and requires that the service provide a telephone verification system to help employees identify illegal immigrants and increase the border patrol staff by 50%. Critics of the law contend that unless the government introduces a new, tamperproof Social Security card, many illegals can circumvent the law with readily available fake documents.

Socioeconomic Methods   Some population experts argue that family planning, even coupled with economic development, cannot lower birth and fertility rates fast enough because, as surveys show, most couples in LDCs want 3 or 4 children—well above the 2.1 fertility rate needed to bring about eventual population stabilization. These experts call for increased emphasis on socioeconomic methods, especially discouraging births by means of economic rewards and penalties and reducing fertility by increasing rights, education, and work opportunities for women.

About 20 countries offer small payments to individuals who agree to use contraceptives or become sterilized, and payments to doctors and family planning workers for each sterilization they perform and each IUD they insert. For example, in India a person receives about $15 for being sterilized, the equivalent of about two weeks' pay for an agricultural worker. Such payments, however, are most likely to attract people who already have all the children they want. Although payments are not physically coercive, they have been criticized as being psychologically coercive, because in some cases the poor feel they have to accept them in order to survive.

Some countries, such as China, also penalize couples who have more than a certain number of children—usually one or two. Penalties may include extra taxes and other costs, or not allowing income tax deductions for a couple's third child (used in Singapore, Hong Kong, Ghana, and Malaysia). Families who have more children than the desired limit may also suffer reduced free health care, decreased food allotments, and loss of job choice. Like economic rewards, economic penalties can be psychologically coercive for the poor. Programs that withhold food or increase the cost of raising children are considered by most analysts to be unjust and inhumane, punishing innocent children for actions by their parents.

Experience has shown that economic rewards and penalties designed to reduce fertility work best if they nudge rather than push people to have fewer children, if they reinforce existing customs and trends toward smaller families, if they do not penalize people who produced large families before the programs were established, and if they increase a poor family's income or land ownership.

In 1987 the world's women did almost all of the world's domestic work and child care mostly without pay, provided more health care with little or no pay than all of the world's organized health services put together, and did more than half the work associated with growing food, gathering fuelwood, and hauling water. At the same time, women made up about 60% of the world's almost 900 million adults who can neither read nor write, and they suffered the most malnutrition, because men and children are fed first where food supplies are limited. Making up one-third of the world's paid labor force in 1987, women were concentrated in the lowest-paid occupations and earned about 75% less than men who did similar work (31% less in the United States). They also worked longer, had a lower income, and owned much less of the world's land than men.

Numerous studies have shown that increased education is a strong factor leading women to have fewer children. Educated women are more likely than uneducated women to be employed outside the home rather than to stay home and raise children, marry later (thus reducing the number of their prime reproductive years), and lose fewer infants to death—a major factor in reducing fertility rates.

However, offering more of the world's women the opportunity to become educated and to express their lives in meaningful, paid work and social roles outside the home will require some major social changes. Such changes include eliminating laws and practices that discriminate against women in education and the workplace, providing free or low-cost day-care centers for children of working women, and changing laws that prohibit or discourage women to own land and thus to obtain credit. Making these changes will be difficult because of long-standing political and economic domination by men throughout the world. In addition, competition between men and women for already scarce jobs in many countries should become even more intense by 2000, when another *billion* people will be looking for work.

8-2 METHODS OF BIRTH CONTROL

Present Methods   The ideal form of birth control would be effective, safe, inexpensive, convenient, free of side effects, and compatible with one's cultural, religious, and sexual attitudes. However, no single method ever has been, and probably no method ever will be, accepted universally because of cultural differences among the world's societies and changes in preferences during different phases of a couple's reproductive years.

About 200 million of the world's 800 million couples of reproductive age use some form of birth control. Available methods either *prevent pregnancy* or *ter-*

Table 8-1   Methods of Birth Control

Pregnancy Prevention by Keeping Sperm and Egg from Uniting

Abstention (no intercourse)

Male Sterilization (vasectomy: tubes carrying sperm to penis are cut and tied)

Female Sterilization (fallopian tubes, which carry eggs from ovary to uterus, are either plugged or cut and tied)

Rhythm Method (periodic abstention; that is, no intercourse during woman's fertile period)

Coitus Interruptus (withdrawal of penis from vagina before ejaculation)

Condom (protective sheath worn over penis; only method that helps prevent sexually transmitted diseases)

Spermicide (sperm-killing chemicals inserted into vagina before intercourse)

Sponge Saturated with Spermicide (inserted into vagina before intercourse to absorb and kill sperm and partially block cervix)

Diaphragm (flexible, dome-shaped rubber device placed over cervical opening before intercourse and usually used with a spermicide)

Douche (rinsing of vagina with spermicide immediately after intercourse)

Pregnancy Prevention by Suppressing Release of Egg

Breast-feeding (prevents ovulation for up to 29 months after birth)

Oral Contraceptive—"the Pill" (synthetic female hormones taken daily by mouth to prevent ovulation)

Pregnancy Termination

Intrautrine Device—IUD (plastic or plastic and metal device inserted and left in uterus prevents fertilized egg from implanting on uterine wall)

Abortion by Suction Aspiration (suction device used to remove embryo and placenta during the first 12 weeks of pregnancy without stay in hospital)

Abortion by Dilation and Evacuation (fetus is dismembered by forceps and then removed by a suction device between the 13th and 28th weeks of pregnancy; may require overnight hospital stay)

Abortion by Induced Labor (injection of salt solution into uterus to induce labor; used only after 16 weeks of pregnancy; usually requires overnight hospital stay)

"Month-After" Pill (induces abortion when taken within ten days after a missed period; being tested in France in 1987)

minate pregnancy before birth (Table 8-1). These methods vary widely in their typical effectiveness (Figure 8-3) and amount of use throughout the world and the United States (Figure 8-4).

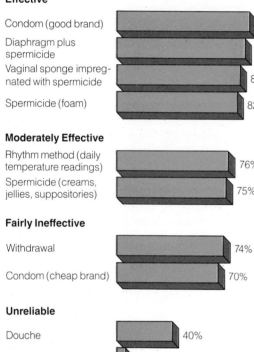

Extremely Effective

Total abstinence — 100%

Abortion — 100%

Sterilization — 99.6%

Highly Effective

Oral contraceptive — 98%

IUD with slow-release hormones — 98%

IUD plus spermicide — 98%

IUD — 95%

Condom (good brand) plus spermicide — 95%

Effective

Condom (good brand) — 90%

Diaphragm plus spermicide — 87%

Vaginal sponge impregnated with spermicide — 83%

Spermicide (foam) — 82%

Moderately Effective

Rhythm method (daily temperature readings) — 76%

Spermicide (creams, jellies, suppositories) — 75%

Fairly Ineffective

Withdrawal — 74%

Condom (cheap brand) — 70%

Unreliable

Douche — 40%

Chance (no method) — 10%

Figure 8-3 Typical effectiveness of birth control methods in the United States.

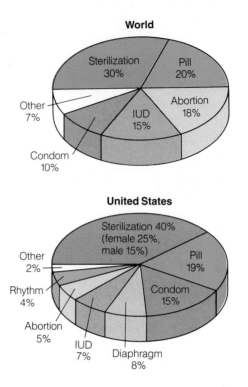

World

Sterilization 30%
Pill 20%
Abortion 18%
IUD 15%
Condom 10%
Other 7%

United States

Sterilization 40% (female 25%, male 15%)
Pill 19%
Condom 15%
Diaphragm 8%
IUD 7%
Abortion 5%
Rhythm 4%
Other 2%

Figure 8-4 Estimated use of various birth control methods in the world and the United States in 1985. (Data from UN Population Division and U.S. National Center for Health Statistics)

It helps reduce infant mortality rates, because mother's milk provides a baby with antibodies to help prevent disease and is usually the most nutritious food available for infants in poor families. It also helps a mother's weight return to normal faster.

Unfortunately, breast-feeding is declining in many LDCs, primarily because large U.S.-based international companies have promoted the use of infant formulas as an alternative to mother's milk. Buying infant formula when free breast milk is available is an unnecessary expense for a poor family struggling to survive. In addition, formula can lead to increases in infant illnesses and mortality, because poor people, lacking fuel, often prepare it with unboiled, contaminated water and use unsterilized bottles.

Terminating Pregnancy: Abortion   Only sterilization and oral contraceptives are more widely used for birth control than abortion. Various types of abortion procedures can be performed during each trimester, or three-month period, of pregnancy. Each year an estimated 50 million legal and illegal abortions—an average of 137,000 a day—are performed in the world, most in Latin America, Asia, and Africa. This means that approximately 1 out of 4 pregnancies end in abor-

Breast-feeding also plays an important role in preventing pregnancy; women who are lactating (producing milk) usually do not ovulate and thus cannot conceive. However, because the number of nursing women is difficult to count, breast-feeding does not show up in official birth control statistics shown in Figure 8-4. Breast-feeding has several other benefits:

Abortion is a highly emotional issue that does not lend itself to compromise or cool debate. On one side are antiabortionists, often calling themselves "pro-lifers," who believe that abortion is an act of murder and should be illegal under all circumstances. On the other side are "pro-choicers," who are not pro-abortion (despite the use of this label by antiabortionists) but who believe that a woman should have the freedom to decide whether to have a legal abortion, especially during her first trimester of pregnancy.

The abortion issue is complicated by inconclusive arguments about when life begins. Most scientists agree that science cannot say when life begins and that this is essentially a moral or religious issue. Most ardent antiabortionists belong to religious groups that hold that life begins at the moment of conception. Other religions make a distinction between an embryo and a "viable" fetus, arguing that until the fetus can survive outside the womb, it is not a true person. Members of many religious and secular groups are not in favor of abortion, but at the same time they are not in favor of having the religious views of a minority forced on them and all Americans. Clearly, the abortion issue will not go away.

tion. It is estimated that at least half of all abortions could be prevented at little cost by making family planning services readily available to all people.

Three-fourths of the world's people live in countries where abortion is legal at least for health reasons; more than half live in areas that have few if any restrictions on obtaining a legal abortion during the first trimester (12 weeks) of pregnancy. Even though abortion is legal in most countries, hundreds of millions of women in such countries are either too poor to pay for an abortion or live in rural areas where such services are not available.

Where legal abortion is not available, especially in Latin America, illegal abortion is widespread. An estimated half of all abortions worldwide are performed illegally and are a leading cause of death among women of childbearing age. Often causing infection and hemorrhage, an illegal abortion is 50 to 100 times more hazardous than a legal one. Perhaps half of all illegal abortions are self-induced: Women commonly

swallow dangerous doses of chemicals sold as home remedies; should this method fail, they often attempt to abort themselves with sharpened sticks or wire coat hangers.

In 1973 the U.S. Supreme Court ruled that during the first trimester of pregnancy, abortion cannot be banned or regulated by states. During this period, the decision to have an abortion must be left entirely to a woman and her doctor, but the procedure must be performed by a licensed doctor. About 91% of the approximately 1.5 million legal abortions that take place in the United States each year are performed during the first trimester. Regarding the second trimester of pregnancy, states can establish regulations on abortion to protect the health of women but cannot prevent a woman from obtaining an abortion. About 8% of legal abortions in the United States are performed between weeks 13 and 20 of the second trimester.

During the third trimester, when the fetus has the potential to live outside the womb, with or without artificial aid, states can ban abortion except in cases where the woman's life or health is threatened. Only about 1%, or 15,000, of the annual abortions in the U.S. take place after the 20th week of pregnancy (near the middle of the second trimester).

Despite the sharp increase in legal abortions since 1973, surveys show that at least one out of four American women who want a legal abortion are unable to obtain it, primarily because 80% of U.S. counties, mostly in rural areas, are without a clinic or hospital that regularly performs abortions. In addition, since 1977, federal law has banned the use of Medicaid funds to pay for abortions, thus making abortions more difficult for poor women to get.

Since 1980, there have been increased efforts on behalf of the estimated 19% of U.S. adults who believe abortion should be illegal to have the 1973 Supreme Court decision overturned—a position publicly supported by President Reagan. Many antiabortionists also favor a constitutional amendment declaring that life begins at the moment of conception, in effect making abortion and use of IUDs and possibly some other forms of birth control acts of criminal homicide (see Spotlight above).

A few antiabortionists have tried to put abortion and family planning clinics out of business by picketing, harassing employees and clients, firebombing, and arson. Between 1982 and 1986, for example, 46 U.S. abortion clinics were firebombed. To cover increased security and liability insurance costs (which have risen up to tenfold), clinics have had to raise their prices, and some have gone out of business. Since 1985, antiabortionists have set up 2,000 free clinics, which outwardly look like nearby abortion or family planning clinics; women seeking an abortion or coun-

| Table 8-2 | Projected New or Improved Technologies for Controlling Fertility |

Likely to Be Available by 1990

Improved IUDs

Improved spermicides for women

Vaginal ring (plastic ring, placed in vagina, that releases contraceptive hormones)

Hormone injections that prevent pregnancy for 1–3 months (such as Depo-Provera, approved for use in 90 countries, and Noristerat, approved for use in 40 countries)

Skin implants of slow-release hormones that prevent pregnancy for 3–5 years (such as Norplant, being tested in 14 countries)

Cervical cap (fits over cervix to block sperm; can be left in place for up to three days)

Physician-administered or self-administered drugs that induce abortion when taken within ten days after a missed period (so-called month-after or abortion pills being tested in France)

Could Be Available by 1990

Monthly steroid-based contraceptive pill

New types of drug-releasing IUDs

Antipregnancy vaccine for women

Sperm-suppression contraceptives for men

Simplified male and female sterilization

Reversible female sterilization

Could Be Available by 2000

Antifertility vaccine for men

Antisperm drugs for men

Better ovulation prediction techniques for more effective use of rhythm method

Reversible male sterilization

Drugs for the sterilization of men and women

U.S. Office of Technology Assessment, *World Population and Fertility Planning Technologies: The Next Twenty Years*, Washington, D.C.: Government Printing Office, 1982.

seling are lured into these clinics and then subjected to scare tactics to jolt them out of considering an abortion. Antiabortion groups are also pressuring the president and Congress to be sure that future appointees to the Supreme Court and other federal courts oppose abortion.

Future Methods of Birth Control Researchers throughout the world are at work trying to develop new and better methods of fertility control (Table 8-2). Despite the importance of population control, annual worldwide expenditures on reproductive research and contraceptive development are minuscule, totaling less than $200 million—an average of 25 cents per person.

In the United States, contraceptive research and development by private drug firms has decreased sharply, because it typically takes 10 to 15 years and up to $50 million before a new product receives approval by the Food and Drug Administration (FDA). Between 1979 and 1985, federal funding for research has also decreased by 25% (adjusted for inflation), because of budget cuts and pressure from antiabortion groups. Most population experts fear that unless annual government and private funding for contraceptive research is at least doubled, few, if any, of the possible new and improved forms of birth control shown in Table 8-2 will be available in the United States.

Highly effective forms of birth control such as Norplant and Depo-Provera, which have been tested and approved for use throughout much of the world, probably won't be available in the United States even with FDA approval. U.S. drug firms consider the financial risks too great because of the threat of liability lawsuits and the high cost of liability insurance. In addition, these companies make much higher profits by selling 28 pills a month than by selling four antifertility shots a year or an implant that lasts five years.

8-3 EFFORTS AT HUMAN POPULATION CONTROL: INDIA AND CHINA

Population Policies In l960 only two countries, India and Pakistan, had official policies to reduce their birth rates. By 1986 about 93% of the world's population and 91% of the people in LDCs lived in countries with some type of family planning program. Few governments, however, spend more than 1% of the national budget on family planning services. To get some idea of the results of population control programs, let's look at what has happened in the world's most populous countries, India and the People's Republic of China.

India India started the world's first national family planning program in 1952, when its population was nearly 400 million, with a doubling time of 53 years. In 1987, after 35 years of population control effort, India was the world's second most populous country, with a population of 800 million and a doubling time of 33 years. In 1952 it was adding 5 million people to its population each year. In 1987 it added 17 million, with only about 25% of Indian couples of childbearing age currently practicing some form of birth control.

The population is projected to more than double to 1.6 billion before leveling off sometime after 2100.

In 1987 at least one-third of India's population had an annual income per person of less than $70 a year, and the overall average income per person was only $260. To add to the problem, nearly half of India's labor force is unemployed or can find only occasional work. Each *week*, 100,000 more people enter the job market, and for most of them jobs do not exist.

Without its long-standing national family planning program, India's numbers would be growing even faster. But the program has yielded disappointing results, partly because of poor planning, bureaucratic inefficiency, low status of women (despite constitutional guarantees of equality), extreme poverty, and lack of sufficient administrative and financial support.

But the roots of the problem are deeper. More than 3 out of every 4 people in India live in 560,000 rural villages, where crude birth rates are still close to 40 per thousand. The overwhelming economic and administrative task of delivering contraceptive services and education to the mostly rural population is complicated by an illiteracy rate of about 71%, with 80% to 90% of the illiterate people being rural women. Although for years the government has provided information about the advantages of small families, Indian women still have an average of 4.3 children, because most couples remain convinced that they need many children as a source of cheap labor and old-age survival insurance. This belief is reinforced by the fact that almost one-third of all Indian children die before age 5. Population control is also hindered by India's diversity: 14 major languages, more than 200 dialects, many social castes, and 11 major religions.

To improve the effectiveness of the effort, in 1976 Indira Gandhi's government instituted a mass sterilization program, primarily for men in the civil service who already had two or more children. The program was supposed to be voluntary, based on financial incentives alone. But officials allegedly used coercion to meet sterilization quotas in a few rural areas. The resulting backlash played a role in Gandhi's election defeat in 1977. In 1978 the government took a new approach, raising the legal minimum age for marriage from 18 to 21 for men and from 15 to 18 for women. After the 1981 census showed that the population growth rate since 1971 was no lower than that of the previous decade, the government vowed to increase family planning efforts and funding, with the goal of achieving replacement-level fertility by 2000. Whether such efforts will succeed remains to be seen.

China Between 1958 and 1962, an estimated 30 million people died from famine in China. Since 1970,

however, the People's Republic of China has made impressive efforts to feed its people and bring its population growth under control. Food production in 1985 was 2.5 times the level of 1960. In 1987 China had enough grain both to export and to feed its population of 1.06 billion. Between 1972 and 1985, China achieved a remarkable drop in its crude birth rate, from 32 to 18, and its total fertility rate had reached 2.1. By 1987, however, its birth rate had risen slightly, to 21, its TFR to 2.4, and its population was growing by 1.3% a year, compared to 2.4% for all other LDCs. At this rate, China's population grew by about 13.5 million in 1987.

China's leaders have a goal of reaching ZPG by 2000 with a population at 1.2 billion, followed by a slow decline to a population between 600 million and 1 billion by 2100. Achieving this goal will be difficult, because 34% of the Chinese people are under age 15. As a result, the United Nations projects that the population of China may be around 1.4 billion by 2020.

To accomplish its sharp drop in fertility, China has established the most extensive and strictest population control program in the world. Its major features include: (1) strongly encouraging couples to postpone marriage; (2) expanding educational opportunities; (3) providing married couples with free, easy access to sterilization, contraceptives, and abortion; (4) offering couples who sign pledges to have no more than one child economic rewards such as salary bonuses, extra food, larger old-age pensions, better housing, free medical care and school tuition for their child, and preferential treatment in employment when the child grows up; (5) requiring those who break the pledge to return all benefits; (6) exerting intense peer pressure on women pregnant with a third child to have an abortion; (7) requiring one of the parents in a two-child family to be sterilized; (8) using mobile units and paramedics to bring sterilization, family planning, health care, and education to rural areas; (9) training local people to carry on the family planning program; and (10) expecting all leaders to set an example with their own family size.

Most countries cannot or do not want to use the coercive elements (especially items 6 and 7) of China's program. Other elements of this program, however, could be used in many LDCs. Especially useful is the practice of localizing the program, rather than asking the people to go to distant centers. Perhaps the best lesson that other countries can learn from China's experience is not to wait to curb population growth until the choice is between mass starvation and coercive measures. Even at that point, coercion can cause a backlash of public resentment and runs a high risk of failure. China's population control program has been successful so far. But, as noted previously, between 1985 and 1987, its birth rate and total fertility rate rose, primarily because of the large number of women moving into their childbearing years coupled with some relaxation of the government's stringent policies.

Population programs aren't simply a matter of promoting smaller families. They also mean guaranteeing that our children are given the fullest opportunities to be educated, to get good health care, and to have access to the jobs and careers they eventually want. It is really a matter of increasing the value of every birth, of expanding the potential of every child to the fullest, and of improving the life of a community.

Pranay Gupte

DISCUSSION TOPICS

1. Should world population growth be controlled? Why or why not?

2. Debate the following resolution: The United States has a serious population problem and should adopt an official policy designed to stabilize its population.

3. Describe the demographic transition hypothesis and give reasons why it may or may not apply to LDCs today.

4. Should federal and state funds be used to provide free or low-cost family planning for the poor in the United States? Explain.

5. Do you believe that each woman should be free to use legalized abortion as a means of birth control or that abortion should be illegal? Give reasons for your position.

6. Should federal and state funds be used to provide free or low-cost abortions for the poor? Explain.

7. Should the number of legal immigrants allowed into the United States each year be sharply reduced? Explain.

8. Should illegal immigration into the United States be sharply decreased? Explain. If so, how would you go about achieving this?

9. What are some ways that women are deliberately and nondeliberately discriminated against in the United States? On your campus?

10. Why has China been more successful than India in reducing its rate of population growth? Do you agree with China's present population control policies? Explain. What alternatives, if any, would you suggest?

PART FOUR

Resources

Kennecott Copper Corporation

Our entire society rests upon—and is dependent upon—our water, our land, our forests, and our minerals. How we use these resources influences our health, security, economy, and well-being.

John F. Kennedy

9

Soil Resources

GENERAL OBJECTIVES

1. Why is soil so important and what are its major components?

2. How is soil formed and what key properties make a soil best suited for growing crops?

3. What are the major differences between grassland, forest, and desert soils?

4. How serious is the problem of erosion in the world and in the United States?

5. What are the major methods for reducing unnecessary soil erosion and depletion of plant nutrients in topsoil?

Below that thin layer comprising the delicate organism known as the soil is a planet as lifeless as the moon.

G. Y. Jacks and R. O. Whyte

Despite its importance in providing us with food, fiber, wood, gravel, and many other materials vital to our existence, soil has been one of the most abused resources. Evidence exists that entire civilizations collapsed because they mismanaged the topsoil that supported their populations, treating this potentially renewable resource as though it could not be depleted by overuse. Unless we wish to relearn the harsh lessons of soil abuse, everyone—not just farmers—needs to be concerned with reducing human-accelerated soil erosion.

9-1 SOIL: USES, COMPONENTS, AND PROFILES

The Base of Life Pick up a handful of soil and notice how it feels and looks. The **soil** you hold in your hand is a complex mixture of inorganic minerals (mostly clay, silt, and sand), decaying organic matter, water, air, and living organisms. The earth's thin layer of soil—at most only a meter or two thick—provides nutrients for plants, which directly or indirectly provide the food we and other animals use to stay alive and healthy. Soil also serves as the base for producing the natural fibers, lumber, and paper we use in large quantities.

Soil Layers and Components The components of soils are arranged in a series of layers called **soil horizons** (Figure 9-1). Each horizon has a distinct thickness, color, texture, and composition that varies with different types of soils. A cross-sectional view of the horizons in a soil is called a **soil profile.** Most mature soils have at least three or four of the six possible horizons.

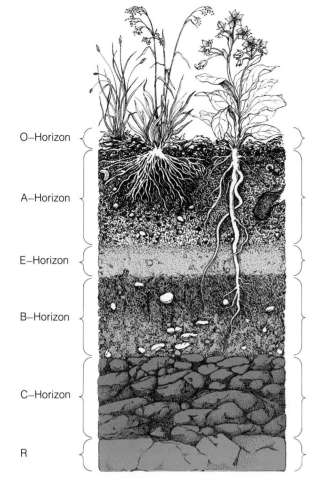

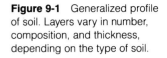

Figure 9-1 Generalized profile of soil. Layers vary in number, composition, and thickness, depending on the type of soil.

O–Horizon

A–Horizon

E–Horizon

B–Horizon

C–Horizon

R

Surface litter:
Freshly fallen leaves and organic debris
and partially decomposed organic matter

Topsoil:
Partially decomposed organic matter (humus),
living organisms, and some inorganic minerals

Zone of leaching:
Area through which dissolved or suspended
materials move downward

Subsoil:
Accumulation of iron, aluminum, and
humic compounds, and clay leached
down from above layers

Parent material:
Partially broken-down
inorganic materials

Bedrock:
Impenetrable layer

The top *surface-litter layer* (*O-horizon*) consists mostly of fresh-fallen and partially decomposed leaves, twigs, animal waste, and other organic debris. It usually has a dark, rich color. The underlying *topsoil layer* (*A-horizon*) is usually a porous mixture of partially decomposed organic matter (humus), living organisms, and some inorganic mineral particles. It is usually darker and looser than deeper layers. Roots of most plants grow in these two upper soil layers (Figure 9-1). Most of a soil's organic matter, which makes up from 1% to 7% of the volume of most soil samples, is found in the first two layers.

The two top layers of most soils are also teeming with bacteria, fungi, molds, earthworms, and small insects, and are home to larger, burrowing animals such as moles and gophers—all of which interact in complex food webs (Figure 9-2). Most of these organisms are bacteria and other decomposer microorganisms—billions in every handful of soil. They partially or completely break down some of the complex inorganic and organic compounds in the upper layers of soil into simpler, water-soluble compounds. Soil water carrying these dissolved plant nutrients and other

nutrients is drawn up by roots and transported through stems and into leaves (Figure 9-3).

Other organic compounds are broken down slowly and form a dark-colored mixture of organic matter called **humus.** This partially decomposed, water-insoluble material remains in the topsoil layer and helps retain water and water-soluble plant nutrients so they can be taken up by plant roots. A fertile soil useful for growing high yields of crops has a thick topsoil layer containing a high content of humus.

We can learn a lot about the suitability of a soil for growing crops by noting the color of its topsoil layer. Dark brown or black topsoil contains a large amount of organic matter and is highly fertile. Gray, bright yellow, or red topsoils are low in organic matter and will require fertilizers and perhaps acidity adjustment to increase their fertility.

The spaces or pores between the solid organic and inorganic particles in the upper and lower soil layers contain varying amounts of two other key inorganic components: air and water. Gaseous oxygen in the air in topsoil is used by the cells in plant roots to carry out cellular respiration. Some of the rain falling on the

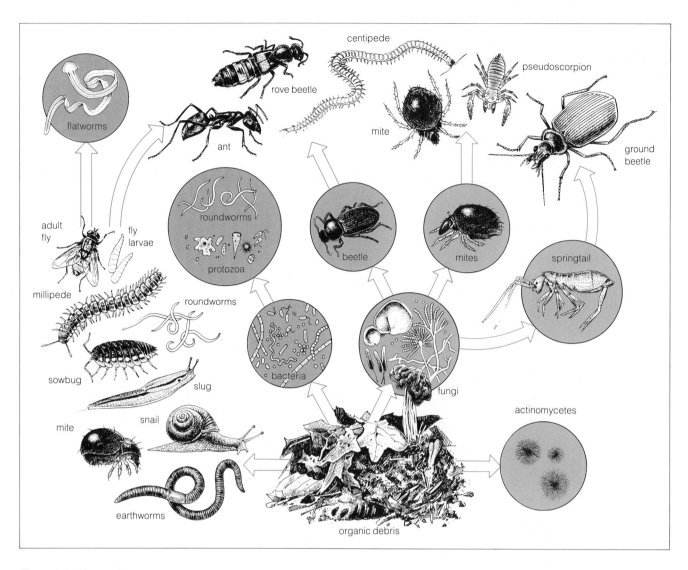

Figure 9-2 Food web of some living organisms found in soil.

soil surface infiltrates, or percolates through, the soil layers and occupies some of the pores. As this water seeps downward, it dissolves and picks up various soil components in upper layers and carries them to lower layers—a process called **leaching.**

Major Types of Soil Mature soils in different biomes of the world vary widely in color, properties such as porosity and acidity, and depth. These differences can be used to classify soils throughout the world into ten major types or orders. Five important soil orders are mollisols, alfisols, spodosols, oxisols, and aridisols, each with a distinct soil profile (Figure 9-4). Most of the world's crops are grown on grassland mollisols and on alfisols exposed when deciduous forests are cleared.

9-2 SOIL: FORMATION, POROSITY, AND ACIDITY

Formation Most soil begins as bedrock, which is gradually broken through an abrasion process called **weathering** into small bits and pieces that make up most of the soil's inorganic material. Other soils develop from the weathering of sediments that have been deposited on the bedrock by wind, water (alluvial soils), volcanic eruptions, or melting glaciers.

Physical weathering involves the breaking down of rock primarily by temperature changes and the physical action of moving ice, water, and wind. For example, in the arid climates of desert biomes, repeated exposure to very high temperatures during the day followed by low temperatures at night causes rocks to

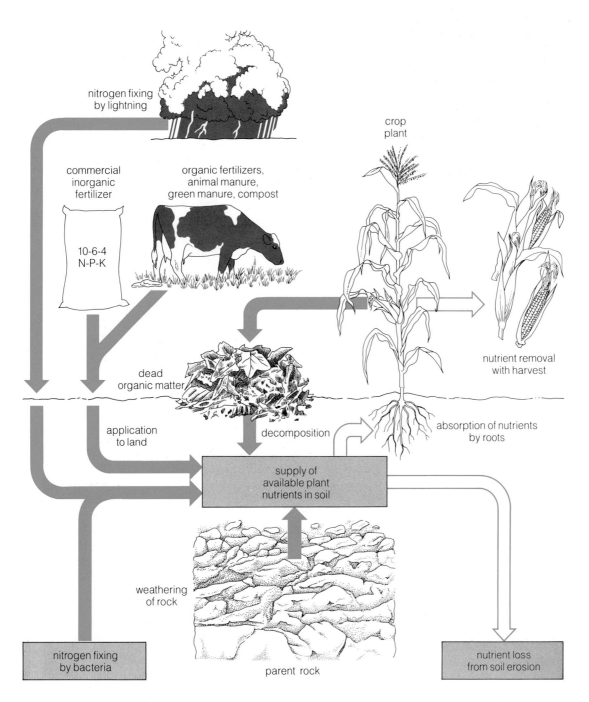

Figure 9-3 Addition and loss of plant nutrients in soils.

expand and contract and eventually to crack and shatter. In cold or temperate climates, rock can crack and break as a result of repeated cycles of expansion of water in rock pores and cracks during freezing and contraction when the ice melts. Growing roots can exert enough pressure to enlarge cracks in solid rock, eventually splitting the rock. Plants such as mosses and lichens also penetrate into rock and loosen particles.

Chemical weathering involves chemical attack and dissolution of rock, primarily through exposure to oxygen gas in the atmosphere (oxidation), rainwater (which becomes slightly acidic by dissolving small amounts of carbon dioxide gas from the atmosphere), and acidic secretions of bacteria, fungi, and lichens. Where average temperatures are high, these chemical reactions take place relatively rapidly; in colder regions they take place at a slower rate.

Figure 9-4 Soil profiles for the major soil orders typically found in five different biomes.

Grassland Soil
(Mollisol)
Semiarid climate

alkaline, dark, and rich in humus

accumulation of clay and calcium compounds

Deciduous Forest Soil
(Alfisol)
Humid mild climate

forest litter
leaf mold
humus-mineral mixture
light, grayish-brown, silt loam
dark brown firm clay
calcareous loam glacial till

Coniferous Forest Soil
(Spodosol)
Humid cold climate

acid litter and humus
light-colored and acidic
humus and iron and aluminum compounds

Desert
(Aridisol)
Hot dry climate

desert pavement
weak humus-mineral mixture
dry, brown to reddish-brown with variable accumulations of clay, calcium carbonate and soluble salts
old alluvium from eroded uplands

Tropical Rain Forest Soil
(Oxisol)
Humid tropical climate

acidic light-colored humus
iron and aluminum compounds mixed with clay

The slope of the land also has an important effect on the type of soil and the rate at which it forms. When the slope is steep, the action of wind, flowing water, and gravity tends to erode the soil constantly. This explains why soils on steep slopes often are thin and never accumulate to a depth sufficient to support plant growth. In contrast, valley soils, receiving mineral particles, nutrients, water, and organic matter from adjacent slopes, are often fertile and highly productive if they are not too wet.

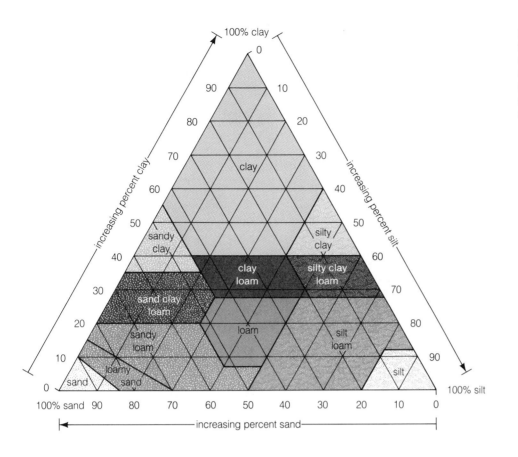

Figure 9-5 Soil porosity and texture depend on the percentages of clay, silt, and sand particles in the soil. Loams are the soils best for growing most crops. (Source: USDA Soil Conservation Service)

Porosity Soils vary in their content of clay (very fine particles), silt (fine particles), sand (coarse particles), and gravel (large particles). The relative amounts of different size particles determine **soil porosity**—the number of pores it has and the distances between them. Porosity is a major factor controlling the amount of water and air the soil can hold and the rate at which water moves downward through it.

Soils can be classified by their clay, silt, and sand content, a useful indicator of their suitability for agriculture (Figure 9-5). Soils consisting of almost equal amounts of sand and silt and somewhat less clay—known as **loams**—are the best for growing most crops. Of medium porosity, loam has sufficient space to provide ample oxygen for plant root cells and retains enough water for roots to absorb without being waterlogged.

Sandy soils have relatively large pores between particles, and the particles have little tendency to adhere to form large clumps. Although sandy soils have good aeration and are easy to work, they retain almost no water and are useful primarily for growing crops without large water requirements, such as peanuts and strawberries. At the other extreme are **clay soils,** in which particles are very small, easily packed together, and when wet can form large clumps—the reason wet clay is so easy to mold into bricks and pottery. Because of their low porosity, clay soils allow little water infil-

tration, especially when compacted, and are poorly aerated and hard to cultivate. Water that does not penetrate drains poorly, so that clay soils are often too waterlogged to grow crops other than onions, potatoes, and celery.

To get a general idea of a soil's porosity, take a small amount of topsoil, moisten it, and rub it between your fingers and thumb. Topsoil that contains too much sand for crops will feel gritty and will have little clumping. Topsoil that contains too much clay is sticky when wet and you can easily roll it up into a single clump or group of large clumps. When it dries, it will harden. A loam topsoil best suited for plant growth has a texture between the other two extremes. It has a crumbly, spongy feeling, and many of its particles are clumped loosely together.

Acidity The acidity or alkalinity of a soil is another important factor determining which types of crops it can support. Acidity and alkalinity are measures of the relative concentrations of hydrogen ions (H^+) and hydroxide ions (OH^-) in a water solution. A volume of solution containing more hydrogen ions than hydroxide ions is an **acid solution;** a volume of solution containing more hydroxide ions than hydrogen ions is a **basic,** or **alkaline, solution.** A solution with an equal number of the two types of ions is called a

neutral solution. Scientists use **pH** as a simple measure of the degree of acidity or alkalinity of a solution. An acid solution has a pH less than 7. The lower the pH below 7, the greater the acidity of a solution. A solution with a pH of 7 is neutral, and one with a pH greater than 7 is basic, or alkaline. Figure 9-6 shows the general pH values for water solutions containing various substances.

Crops vary widely in the pH ranges they can tolerate. Many common food plants, such as wheat, spinach, peas, corn, and tomatoes, grow best in slightly acidic soils; potatoes and berries do best in very acidic soils, and alfalfa and asparagus in neutral soils. When soils are too acidic for the desired crops, the acids can be partially neutralized by adding an alkaline substance such as lime (calcium oxide produced from limestone). But adding lime speeds up the undesirable decomposition of organic matter in the soil, so manure or other organic fertilizer should be added as well. Otherwise, the initial years of good crop yields will be followed by poor yields. Adding organic matter to the soil also helps stabilize pH.

In areas of low rainfall, such as the semiarid valleys in the western and southwestern United States, calcium and other alkaline compounds are not leached away, and soils may be too alkaline (pH above 7.5) for some desired crops. If drainage is good, the alkalinity of these soils can be reduced by leaching the alkaline compounds away with irrigation water. Soil alkalinity can also be reduced by adding sulfur, an abundant and cheap element, which is gradually converted to sulfuric acid (H_2SO_4) by bacteria.

9-3 SOIL EROSION

Natural and Human-Accelerated Soil Erosion
Soil does not remain in one place indefinitely. **Soil erosion** is the movement of soil components, especially topsoil, from one place to another. The two main forces causing soil erosion are wind (Figure 9-7) and flowing water (Figure 9-8). Some soil erosion always takes place as a result of these and other natural processes. But the roots of plants generally protect soil from excessive erosion. Agriculture, logging, construction, and other human activities that remove plant cover greatly accelerate the rate at which soil erodes. Excessive erosion not only reduces soil fertility, the resulting sediment clogs irrigation ditches, navigable waterways, and reservoirs used to generate electric power and provide drinking water for urban areas.

If the average rate of topsoil erosion exceeds the rate of topsoil formation on a piece of land, the topsoil on that land becomes a nonrenewable resource that is

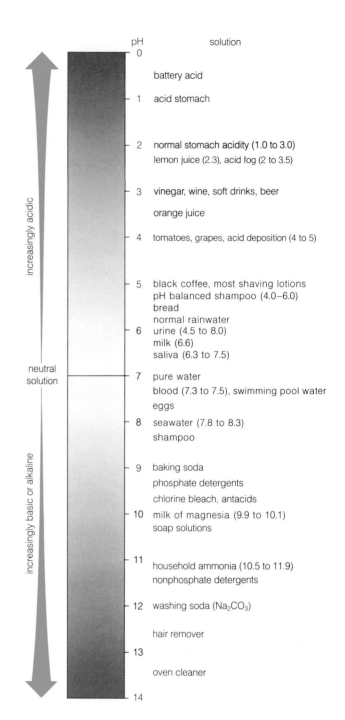

Figure 9-6 Scale of pH, used to measure acidity and alkalinity of water solutions. Values shown are approximate.

being depleted. In tropical and temperate areas, the renewal of 2.5 cm (1 inch) of soil takes an average of 500 years (with a range of 220 to 1,000 years). Worldwide annual erosion rates for agricultural land are 18 to 100 times this natural renewal rate.

Soil erosion also occurs in forestland but is not as severe as that in the more exposed soil of agricultural land. However, soil erosion in managed forests is a major concern, because the soil reformation rate in

Figure 9-7 Wind blowing soil off Iowa farmland in 1930. If grass had been planted between crops, most of this loss of valuable topsoil could have been prevented.

Figure 9-8 Extensive soil erosion and gully formation caused by flowing water. Good farming practices could have prevented the loss of most of this topsoil.

forests is about two to three times slower than that in cropland.

The World Situation Today topsoil is eroding faster than it forms on about 35% of the world's croplands. The amount of topsoil washing and blowing into the world's rivers, lakes, and oceans each year would fill a train of freight cars long enough to encircle the planet 150 times. At this rate, the world is losing about 7% of its topsoil from potential cropland each decade. This situation, which is undermining agricultural productivity in many parts of the world, is worsening as cultivation is being extended into areas poorly suited for agriculture in order to feed the world's growing population.

The ninefold increase in fertilizer use and the tripling of the world's irrigated cropland between 1950 and 1987 have temporarily masked the effects of soil erosion on crop productivity. However, as we will see in Section 9-4, commercial fertilizer is not a complete substitute for naturally fertile topsoil and merely hides for a time the gradual depletion of this vital resource.

Erosion rates vary in different regions because of topography, rainfall, wind intensity, and the type of agricultural practices used. Accelerated erosion from human activities is most widespread in India, China, the Soviet Union, and the United States, which together account for over half the world's food production and contain almost half the world's people. In China, for example, the average annual soil loss is reported to be more than 40 tons per hectare. At least 34% of the country's land is severely eroded, and river siltation is now a nationally recognized threat. Soil erosion and river siltation are also major problems in India, with erosion affecting one-quarter of the country's land area. The Worldwatch Institute estimates that the Soviet Union, which has the world's largest cropland area, may be losing more topsoil than any other country. Increasingly, poor people in many LDCs must either grow crops on marginal land and use cultivation methods that will increase soil erosion or face starvation.

The U.S. Situation According to surveys by the Soil Conservation Service, about one-third of the original topsoil on U.S. croplands in use today has already been washed or blown into rivers, lakes, and oceans. While the average soil erosion rate in the United States is about 18 tons per hectare, in states like Iowa and Missouri average rates are greater than 35 tons per hectare. Some of the country's richest agricultural lands, such as those in Iowa, have lost about half their topsoil.

Enough topsoil erodes away *each day* in the United States to fill a line of dump trucks 5,600 kilometers (3,500 miles) long, with the bulk of this coming from less than one-fourth of the country's cropland. The plant nutrient losses from this erosion are worth at least $18 billion a year. Unless soil conservation efforts are increased, projected soil erosion may destroy productivity on U.S. cropland acreage equal to the combined areas of the states of New York, New Jersey, Maine, New Hampshire, Massachusetts, and Connecticut over the next 50 years.

Erosion from U.S. cropland is not the only problem. Improper use and management of forest, range, and pasture lands, as well as activities such as mining and construction account for about 40% of the soil eroded from land in the United States. In addition to reduced crop productivity, runoff of sediment into U.S. waterways causes about $6 billion in damages a year as a result of loss of water storage capacity in reservoirs, destruction of wildlife, and filling of navigable waterways.

Some analysts, however, estimate that continuation of 1977 rates of soil erosion at an average of 18 tons per hectare for 50 years would reduce national average crop yields by only 2% to 5%, a deficit they believe can easily be made up by advances in agricultural technology and application of commercial inorganic fertilizers. However, using an evaluation of all the ecological effects caused by erosion, including a reduction in soil depth, reduced water availability for crops, and reduction in soil organic matter and nutrients, agronomists and ecologists project a 7% to 15% reduction in average crop yields over the next 50 years if soil continues to erode each year at an average of 18 tons per hectare.

These analysts also point out that using average soil erosion and crop productivity figures is misleading, because much of the most severe erosion occurs on some of the country's prime farmland. For example, over half of U.S. cropland soil erosion occurs in the corn belt and the northern plains, among the most productive agricultural areas in the world. Without increased use of well-known but little-used soil conservation methods, these fertile Midwest plains are subject to increased erosion from continuous high winds and periodic prolonged drought (see Spotlight on p. 149).

9-4 SOIL CONSERVATION AND LAND-USE PLANNING

Importance of Soil Conservation The practice of **soil conservation** involves using various management methods to reduce soil erosion by holding the soil in place, to prevent depletion of soil nutrients, and to restore nutrients already lost by erosion, leaching, and excessive crop harvesting. Soil conservation usually does not receive high priority among many govern-

ments and farmers because erosion usually takes place at such a slow rate that it may take decades for its cumulative effects to become apparent. For example, the removal of 1 millimeter (1/254 inch) of soil, an amount easily lost during a rain or wind storm, goes undetected; but the accumulated soil loss at this rate over a 25-year period would amount to 25 mm (1 inch)—an amount that would take about 500 years to replace by natural processes.

Conservation Tillage The principal methods of controlling soil erosion and its accompanying runoff of sediment involve maintaining adequate vegetative cover on soils. One important factor is the degree to which soil is disturbed during the tilling or planting of crops.

In conventional, intensive-tillage farming, land is plowed, disked several times, and smoothed to make a planting surface. If plowed in the fall so that crops can be planted early in the spring, the soil is left bare during the winter and early spring months—a practice that increases its susceptibility to erosion.

To lower labor costs, save energy, and reduce soil erosion, an increasing number of U.S. farmers are replacing this approach with **conservation-tillage farming,** also known as minimum tillage or no-till farming, depending on the degree to which the soil is disturbed. Farmers using this method disturb the soil as little as possible when crops are planted. For the minimum-tillage method, special subsurface tillers are used to break up and loosen the subsurface soil without turning over the topsoil, the previous crop residues, and any cover vegetation. In the no-till version of this approach, even subsurface tillage is eliminated. Special planters are used to drill a hole in the soil for each plant and to inject seeds, fertilizers, and weed killers (herbicides) into unplowed soil (Figure 9-10).

In addition to reducing soil erosion, conservation tillage reduces fuel and tillage costs, water loss from the soil, and soil compaction and increases the number of crops that can be grown during a season (multiple cropping). It also usually allows crops to be planted, treated with herbicide, and harvested when conventional tilled fields would be too muddy to enter. Yields are as high as or higher than yields from conventional tillage. Conservation tillage, however, requires increased use of herbicides to control weeds that compete with crops for soil nutrients, and it cannot be used on all soils (especially not those in the northern corn belt). Depending on the soil, this approach can be used for three to seven years before more extensive soil cultivation is needed to prevent crop yields from declining.

The Great Plains of the United States stretch through ten states, from Texas through Montana and the Dakotas. The region is normally dry and very windy and periodically experiences long, severe droughts. Before settlers began grazing livestock and planting crops in the 1870s, the extensive root systems of prairie grasses held the rich mollisol (Figure 9-4). When the land was planted, these perennial grasses were replaced by annual crops with less extensive root systems. In addition, the land was plowed up after each harvest and left bare part of the year. Overgrazing also destroyed large areas of grass, leaving the ground bare. The stage was set for crop failures during prolonged droughts, followed by severe wind erosion.

The droughts arrived in 1890 and 1910 and again with even greater severity between 1926 and 1934. In 1934 hot, dry windstorms created dust clouds thick enough to cause darkness at midday in some areas (Figure 9-9); the danger of breathing the dust-laden air was revealed by dead rabbits and birds left in its wake. During May 1934, the entire eastern half of the United States was blanketed with a massive dust cloud of rich topsoil blown off the Great Plains from as far as 2,415 kilometers (1,500 miles) away. Ships 320 kilometers (200 miles) out in the Atlantic Ocean received deposits of topsoil. Thus the Great Plains acquired a tragic new name: the Dust Bowl.

About 9 million acres of cropland were destroyed and 80 million acres severely damaged. Thousands of displaced farm families from Oklahoma, Texas, Kansas, and other states migrated westward toward California or to the industrial cities of the Midwest and East. Upon arriving at their destinations, most found no jobs, because the country

Figure 9-9 Dust storm approaching Prowers County, Colorado, 1934.

USDA/Soil Conservation Service

was in the midst of the Great Depression. The migrants joined massive numbers of unemployed people waiting in line for enough free food to keep themselves and their families alive.

In May 1934, Hugh Bennett of the U.S. Department of Agriculture (USDA) addressed a congressional hearing, pleading for new programs to protect the country's topsoil. Lawmakers in Washington took action when dust blown from the Great Plains began seeping into the hearing room. In 1935 the United States established the Soil Conservation Service (SCS) as part of the USDA. With Bennett as its first head, the SCS began promoting good conservation practices in the Great Plains and later in every state, establishing local soil conservation districts and providing technical assistance to farmers and ranchers.

These efforts, however, did not completely solve the erosion problems of the Great Plains. From both economic and ecological viewpoints,

the climate of much of the region makes it better suited for grazing than for farming—a lesson its farmers have relearned several times since the 1930s. For example, because of severe drought and soil erosion in the 1950s, the federal government had to provide emergency relief funds to many Great Plains farmers.

In 1975 the Council of Agricultural Science and Technology warned that severe drought could again create a dust bowl in the Great Plains, pointing out that despite large expenditures for soil erosion control, topsoil losses were 2.5% worse than in the 1930s. So far, these warnings have not been heeded. Great Plains farmers, many of them debt-ridden because of low crop prices, have continued to stave off bankruptcy by maximizing production and minimizing expenditures for soil conservation. What do you think should be done about this situation?

Figure 9-10 No-till farming. A specially designed machine plants seeds and adds fertilizers and weed killers at the same time with almost no disturbance of the soil.

Figure 9-11 On this gently sloping land, contoured rows planted with alternating crops (strip cropping) reduce soil erosion.

Increased use of conservation tillage is the most hopeful sign of progress toward reducing soil erosion from U.S. croplands. By 1987 conservation tillage was used on about 38% of the country's croplands and is projected to be used on over half by 2000. The USDA estimates that using conservation tillage on 80% of U.S. cropland would reduce soil erosion by at least half. So far the practice is not widely used in other parts of the world.

Contour Farming, Terracing, and Strip Cropping

Soil erosion can be reduced 30% to 50% on gently sloping land through **contour farming**—plowing and planting along rather than up and down the sloped contour of the land (Figure 9-11). Each row planted at a right angle to the slope of the land acts as a small dam to help hold soil and slow the runoff of water.

Terracing can be used on steeper slopes. The slope is converted into a series of broad, nearly level terraces with short vertical drops from one to another, following the slope of the land (Figure 9-12). Water running down the permanently vegetated slope is retained and delayed by each terrace. This provides water for crops at all levels and decreases soil erosion by reducing the amount and speed of water runoff. In areas of high rainfall, diversion ditches must be built behind each terrace to permit adequate drainage.

In **strip cropping,** a wide row of one crop, such as corn or soybeans, is planted; the next row is planted with a crop such as alfalfa, which forms a complete cover and thus reduces erosion. The alternating rows of ground cover also reduce water runoff, help prevent the spread of pests and plant diseases from one row to another, and help restore soil fertility, especially when nitrogen-rich legumes such as soybeans are planted in some of the rows. On sloping land, strip cropping can reduce soil losses by up to 75% when combined with either terracing or contour farming (Figure 9-11).

Gully Reclamation and Windbreaks

On sloping land not covered by vegetation, deep gullies can be created quickly by water runoff (Figure 9-8). Thus **gully reclamation** is an important form of soil conservation. Small gullies can be seeded with quick-growing plants such as oats, barley, and wheat to reduce erosion. For severe gullies, small dams can be built to collect silt and gradually fill in the channels. The soil can then be stabilized by planting rapidly growing shrubs, vines, and trees and providing channels to divert water away from the gully.

Erosion caused by exposure of cultivated lands to high winds can be reduced by using **windbreaks,** or **shelterbelts**—long rows of trees planted in a north-to-south direction to partially block wind flow over cropland (Figure 9-13). They are especially effective if land not under cultivation is kept covered with vegetation. Windbreaks also provide habitats for birds, pest-eating and pollinating insects, and other animals. Unfortunately, many of the windbreaks planted in the

Figure 9-12 Terracing to reduce soil erosion. These terraces in Pisac, Peru, help reduce soil erosion.

Figure 9-13 Shelterbelts, or windbreaks, reduce erosion on this farm in Trail County, North Dakota.

upper Great Plains following the Dust Bowl disaster of the 1930s have been destroyed to make way for large irrigation systems and farm machinery.

Appropriate Land Use and Land-Use Planning To encourage wise land use and reduce erosion, the SCS has set up the classification system summarized in Table 9-1 and illustrated in Figure 9-14. An obvious land-use approach to reducing erosion is to avoid planting crops on or clearing vegetation from marginal land (classes V through VIII in Table 9-1), which, because of slope, soil structure, high winds, periodic drought, or other factors, is subject to high rates of erosion.

Most urban areas and some rural areas practice some form of **land-use planning** to decide the best present and future use for each parcel of land in the area. Most land-use planning is based on extrapolating from existing trends or reacting to crisis; both methods are ineffective in anticipating and preventing long-term conflicts and problems. **Ecological land-use planning,** in which all major variables are considered and integrated into a model designed to anticipate present and future needs, is a much better method (see Spotlight on p. 155). However, it is not widely used because of political conflicts over valuable pieces of land, reluctance of officials seeking reelection every few years to become concerned with long-term problems, and lack of cooperative planning between various municipalities in the same general area.

Since World War II, the typical pattern of suburban housing development in the United States has been to bulldoze a patch (or tract) of woods or farmland and build rows of houses, each standard house on a standard lot (Figure 9-15). By removing most vegetation, this approach increases soil erosion during and after construction. In recent years, a new pattern, known as *cluster development* or *planned unit development* (PUD), has been used with increasing success to preserve medium-size blocks of open space and natural vegetation, which also helps reduce erosion. Houses, townhouses, condominiums, and garden apartments are built on a relatively small portion of land, with the rest of the area left as open space, either in its natural state or modified for recreation (Figure 9-16, p. 154).

Maintaining and Restoring Soil Fertility Organic fertilizers and commercial inorganic fertilizers can be applied to soil to restore and maintain plant nutrients lost from the soil by erosion, leaching, and crop harvesting and to increase crop yields. Three major types of **organic fertilizer** are animal manure, green manure, and compost. **Animal manure** includes the dung and urine of cattle, horses, poultry, and other farm animals. Application of animal manure improves soil structure, increases organic nitrogen content, and stimulates the growth and reproduction of soil bacteria and fungi. It is particularly useful on crops such as corn, cotton, potatoes, cabbage, and tobacco.

Table 9-1 Land Capability Classification According to the Soil Conservation Service

Land Class	Characteristics	Primary Uses	Secondary Uses	Conservation Measures
Land Suitable for Cultivation				
I	Excellent flat, well-drained land	Agriculture	Recreation Wildlife Pasture	None
II	Good land, has minor limitations such as slight slope, sandy soil, or poor drainage	Agriculture Pasture	Recreation Wildlife	Strip cropping Contour farming
III	Moderately good land with important limitations of soil, slope, or drainage	Agriculture Pasture Watershed	Recreation Wildlife Urban industry	Contour farming Strip cropping Waterways Terraces
IV	Fair land, severe limitations of soil, slope, or drainage	Pasture Orchards Limited agriculture Urban industry	Pasture Wildlife	Farming on a limited basis Contour farming Strip cropping Waterways Terraces
Land Not Suitable for Cultivation				
V	Use for grazing and forestry slightly limited by rockiness; shallow soil, wetness, or slope prevents farming	Grazing Forestry Watershed	Recreation Wildlife	No special precautions if properly grazed or logged; must not be plowed
VI	Moderate limitations for grazing and forestry	Grazing Forestry Watershed Urban industry	Recreation Wildlife	Grazing or logging should be limited at times
VII	Severe limitations for grazing and forestry	Grazing Forestry Watershed Recreation-Aesthetics Wildlife Urban industry		Careful management required when used for grazing or logging
VIII	Unsuitable for grazing and forestry because of steep slope, shallow soil, lack of water, too much water	Recreation-Aesthetics Watershed Wildlife Urban industry		Not to be used for grazing or logging; steep slope and lack of soil present problems

Despite its usefulness, the use of animal manure in the United States has decreased. One reason is that separate farms for growing crops and animals (feedlots) have replaced most of the mixed animal- and crop-farming operations in the United States. Although large amounts of animal manure are available at feedlots (normally located near urban areas), collecting and transporting it to distant rural crop-growing areas usually costs too much. In addition, tractors and other forms of motorized farm machinery have largely replaced horses and other draft animals that naturally added manure to the soil.

Green manure is fresh or still growing green vegetation plowed into the soil to increase the organic matter and humus available to the next crop. It may consist of weeds in an uncultivated field, grasses and clover in a field previously used for pasture, or legumes such as alfalfa or soybeans grown for use as

Beneath the photograph, the following table appears:

LAND CAPABILITY CLASSES				
SUITABLE FOR CULTIVATION		NO CULTIVATION - PASTURE, HAY, WOODLAND AND WILDLIFE		
I	REQUIRES GOOD SOIL MANAGEMENT PRACTICES ONLY	V	NO RESTRICTIONS IN USE	
II	MODERATE CONSERVATION PRACTICES NECESSARY	VI	MODERATE RESTRICTIONS IN USE	
III	INTENSIVE CONSERVATION PRACTICES NECESSARY	VII	SEVERE RESTRICTIONS IN USE	
IV	PERENNIAL VEGETATION - INFREQUENT CULTIVATION	VIII	BEST SUITED FOR WILDLIFE AND RECREATION	

Figure 9-14 Classification of land according to capability; see Table 9-1 for description of each class.

fertilizer to build up soil nitrogen. The effects of green manure on the soil are similar to those of animal manure. **Compost** is a rich natural fertilizer; farmers produce it by piling up alternating layers of carbohydrate-rich plant wastes (such as cuttings and leaves), animal manure, and topsoil, providing a home for microorganisms that aid the decomposition of the plant and animal manure layers.

Today the fertility of most soils, especially in the United States and other industrialized countries, is partially restored and maintained by the application of **commercial inorganic fertilizers.** The most common plant nutrients in these products are nitrogen (as nitrate), phosphorus (as phosphate), and potassium. These nutrients are designated by numbers; for exam-

ple, a 6-12-12 fertilizer contains 6% nitrogen, 12% phosphorus, and 12% potassium. Other plant nutrients may also be present. Soil and harvested crops can be chemically analyzed to determine the exact mix of nutrients that should be added. Use of inorganic commercial fertilizers throughout the world increased more than ninefold between 1950 and 1987, because they are a concentrated source of nutrients that can be easily transported, stored, and applied. By 1987 the additional food they helped produce fed about one out of every three persons in the world.

However, commercial inorganic fertilizers do have disadvantages. They do not add any humus to the soil. If not supplemented by organic fertilizers, continued use of inorganic fertilizers causes the soil to

Figure 9-15 A suburban housing tract laid out in conventional row-by-row pattern.

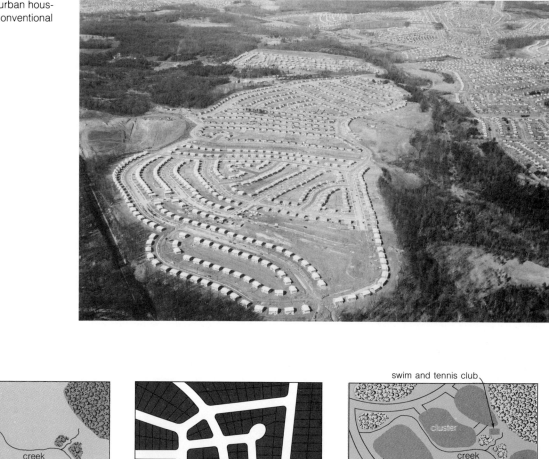

swim and tennis club

creek

marsh

cluster

cluster

creek

pond

Figure 9-16 Conventional and cluster development as they would appear if constructed on the same land area.

become compacted and less suitable for crop growth and reduces its natural ability to produce nitrogen in forms usable by plants. Inorganic fertilizers also reduce the oxygen content of soil by altering its porosity, so that any added fertilizer is not taken up as efficiently. In addition, most commercial fertilizers do not contain many of the nutrients needed in trace amounts by plants.

Water pollution is another problem resulting from extensive use of commercial inorganic fertilizers, especially on sloped land near rivers, streams, and lakes. Some of the plant nutrients in inorganic fertilizers are washed into nearby bodies of water, where they can cause excessive growth of algae, oxygen

depletion, and fish kills. Rainwater seeping through soil can leach the nitrates in commercial fertilizers into groundwater supplies, where excessive levels of nitrate ions can make drinking water toxic, especially for infants.

A third method for preventing depletion of soil nutrients is **crop rotation.** Crops such as corn, tobacco, and cotton remove large amounts of nutrients (especially nitrogen) from the soil and can deplete the topsoil nutrients if planted on the same land several years in a row. Through crop rotation, farmers reduce nutrient depletion by planting areas or strips with corn, tobacco, and cotton one year and planting the same areas the next year with legumes, which add nitrogen to the

1. *Making an environmental and social inventory:* Appropriate experts make a comprehensive survey of geologic variables (such as soil types and water availability), ecological variables (such as forest types and quality, wildlife habitats, stream quality, and pollution), economic variables (such as housing and industrial development), and health and social variables (such as disease and crime rates, ethnic distribution, and illiteracy).

2. *Determination of goals and their relative importance:* Experts, public officials, and the general public decide on goals and rank them in order of importance. For example, is the primary goal to encourage or to discourage further economic development and population growth? To preserve prime cropland from development? To reduce soil erosion? This is a very important and difficult planning step, which should also set guidelines for settling conflicts over the use of various parcels of land.

3. *Production of individual and composite maps:* Data for each variable obtained in step 1 are plotted on separate transparent plastic maps. The transparencies are superimposed on one another or combined by computer to give three composite maps—one each for geologic variables, ecological variables, and socioeconomic variables.

4. *Development of a comprehensive plan:* The three composite maps are combined to form a master composite, which shows the suitability of various areas for different types of land use.

5. *Evaluation of the comprehensive plan:* The comprehensive plan (or series of alternative comprehensive plans) is evaluated by experts, public officials, and the general public, and a final comprehensive plan is drawn up and approved.

6. *Implementation of the comprehensive plan:* The plan is set in motion and monitored by the appropriate governmental, legal, environmental, and social agencies.

Civilization can survive the exhaustion of oil reserves, but not the continuing wholesale loss of topsoil.

Lester R. Brown

DISCUSSION TOPICS

1. Why should everyone, not just farmers, be concerned with soil conservation?

2. Explain how a plant can have ample supplies of nitrogen, phosphorus, potassium, and other essential nutrients and still have stunted growth.

3. List the following soils in order of increasing porosity to water: loam, clay, sand, and sandy loam.

4. What are the key properties of a soil that is good for growing most crops?

5. Describe briefly the Dust Bowl phenomenon of the 1930s and explain how and where it could happen again. How would you prevent a recurrence?

6. Distinguish among contour farming, terracing, strip cropping, and no-tillage farming, and explain how each can reduce soil erosion.

7. What is crop rotation and how can it be used to help restore soil fertility?

8. Visit rural or relatively undeveloped areas near your campus and classify the lands according to the system shown in Figure 9-14 and Table 9-1. Look for examples of land being used for purposes to which it is not best suited.

9. What are the major advantages and disadvantages of using commercial inorganic fertilizers to help restore and maintain soil fertility? Why should organic fertilizers also be used on land treated with inorganic fertilizers?

10. Some cornucopians contend that average soil erosion rates in the United States and the world are low and that this problem has been overblown by environmentalists and can easily be solved by improved agricultural technology such as no-till cultivation and increased fertilizer use. Do you agree or disagree with this position? Explain.

soil, or other crops such as oats, barley, and rye. This method improves soil fertility, reduces erosion by covering land, and reduces pest infestation and plant diseases.

10

Water Resources

GENERAL OBJECTIVES

1. How is the existence of life on earth related to the unique physical properties of water?

2. How much usable fresh water is available for human use and how much of this supply are we using?

3. What are the major water resource problems in the world and in the United States?

4. How can the supply of usable fresh water in water-short areas of the world be increased?

5. How can water waste be reduced?

If there is magic on this planet, it is in water.

Loren Eisley

Water is our most abundant resource, covering about 71% of the earth's surface. This precious film of water—about 97% salt water and the remainder fresh—helps maintain the earth's climate and dilutes environmental pollutants. Essential to all life, water constitutes from 50% to 97% of the weight of all plants and animals and about 70% of your body. Water is also essential to agriculture, manufacturing, transportation, and countless other human activities.

Because of differences in average annual precipitation, some areas of the world have too little fresh water and others too much. With varying degrees of success, human beings have corrected these imbalances by capturing fresh water in reservoirs behind dams, transferring fresh water in rivers and streams from one area to another, tapping underground supplies, and attempting to reduce water use, waste, and contamination.

10-1 WATER'S UNIQUE PHYSICAL PROPERTIES

Most of water's usefulness results from its unique physical properties compared to those of other molecules of similar weight.

1. *Liquid water has a high boiling point of 100°C (212°F) and solid water has a high melting point of 0°C (32°F).* Otherwise, water at normal temperatures would be a gas rather than a liquid and the earth would have no oceans, lakes, rivers, plants, and animals.

2. *Liquid water has a very high heat of vaporization.* This means that water molecules absorb large quan-

tities of heat when they are evaporated by solar energy from bodies of water and release large amounts of heat when atmospheric water vapor condenses and falls back to the earth as precipitation. This ability to store and release large amounts of heat during physical changes is a major factor in distributing heat throughout the world. This property also means that evaporation of water is an effective cooling process for plants and animals—explaining why you feel cooler when perspiration evaporates from your skin.

3. *Liquid water has an extremely high heat capacity*—its ability to store large amounts of heat without a large temperature change. This property prevents large bodies of water from warming or cooling rapidly, helps protect living things from the shock of abrupt temperature changes, helps keep the earth's climate moderate, and makes water an effective coolant for car engines, power plants, and other heat-producing industrial processes.

4. *Liquid water is a superior solvent*, able to dissolve large amounts of a variety of compounds. This enables water to carry dissolved nutrients throughout the tissues of plants and animals, to flush waste products out of the tissues, to be a good all-purpose cleanser, and to remove and dilute the water-soluble wastes of civilization, if aquatic systems are not overloaded. But this ability of water to act as a solvent also means that it is easily polluted.

5. *Liquid water has an extremely high surface tension (the force that causes the surface of a liquid to contract) and an even higher wetting ability (the capability to coat a solid).* Together, these properties are responsible for liquid water's capillarity—the ability to rise from tiny pores in the soil into thin, hollow tubes, called capillaries, in the stems of plants. These properties along with water's solvent ability allow plants to receive nutrients from the soil, thus supporting the growth of plants and the animals that feed on them.

6. *Liquid water is the only common substance that expands rather than contracts when it freezes.* Consequently, ice has a lower density (mass per unit of volume) than liquid water. Thus ice floats on water, and bodies of water freeze from the top down instead of from the bottom up. Without this property, lakes and rivers in cold climates would freeze solid and most known forms of aquatic life would not exist. Because water expands on freezing, it can also break pipes, crack engine blocks (this is why we use antifreeze), and fracture streets and rocks.

10-2 SUPPLY, RENEWAL, AND USE OF WATER RESOURCES

Worldwide Supply and Renewal The world's fixed supply of water in all forms (vapor, liquid, and solid) is enormous. If we could distribute it equally, there would be enough to provide every person on earth with 292 trillion liters (77 trillion gallons). However, only about 0.003% of the world's water supply is available as fresh water for human use, and this supply is unevenly distributed.

About 97% of the earth's total supply of water is found in the oceans and is too salty for drinking, growing crops, and most industrial purposes except cooling. The remaining 3% is fresh water, but over three-fourths of it is unavailable for use by plants, human beings, and other animals, because it lies too far under the earth's surface or is locked up in glaciers, polar ice caps, atmosphere, and soil. This leaves 0.5% of the earth's water available as fresh water in rivers, lakes, and economically recoverable underground deposits (groundwater) to a depth of 1,000 meters (1.6 miles). However, when we subtract the portion of this water that is highly polluted or too difficult and expensive to tap, the remaining supply amounts to about 0.003% of the world's water. To put this in measurements that we can comprehend, if the world's water supply were only 100 liters (26 gallons), our usable supply of fresh water would be only about 0.003 liter (one-half teaspoon), as illustrated in Figure 10-1.

That tiny fraction of usable fresh water still amounts to an average of 879,000 liters (232,000 gallons) for each person on earth. The supply is continually collected, purified, and distributed in the natural *hydrologic (water) cycle* (Figure 4-11, p. 61). This natural purification process works as long as we don't pollute water faster than it is replenished or add chemicals that cannot be broken down by bacterial action.

Surface-Water Runoff The fresh water we use comes from two sources: groundwater and surface-water runoff (Figure 10-2). Precipitation that does not infiltrate into the ground or return to the atmosphere is known as **surface water** and becomes **runoff**—water that flows into nearby streams, rivers, lakes, wetlands, and reservoirs. This flow of water is renewed fairly rapidly (12 to 20 days) in areas with average precipitation. The land area that delivers runoff, sediment, and water-soluble substances to a major river and its tributaries is called a **watershed** or **drainage basin.**

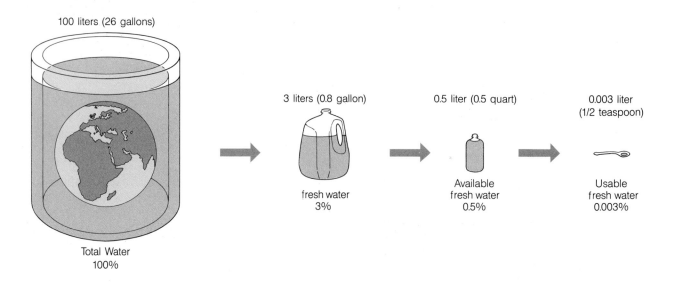

Figure 10-1 Only a tiny fraction of the world's water supply is available as fresh water for human use.

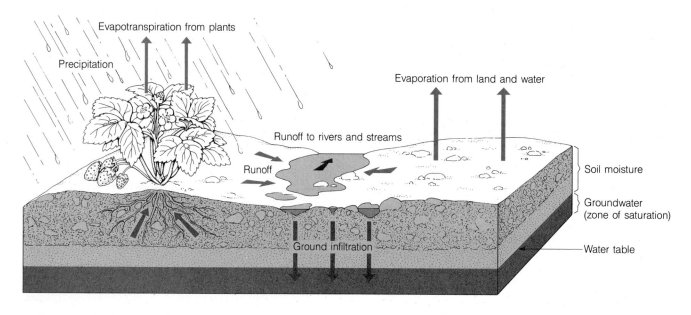

Figure 10-2 Major routes of local precipitation.

Surface water can be withdrawn from streams, rivers, lakes, and reservoirs for human activities, but only part of the total annual runoff is available for use. Some flows in rivers to the sea too rapidly to be captured, and some must be left in streams for wildlife and to supply downstream areas. In some years the amount of runoff is reduced by drought.

Groundwater Some precipitation seeps into the ground. Some of this infiltrating water accumulates as soil moisture and partially fills pores between soil particles and rocks within the upper soil and rock layers

of the earth's crust (Figure 10-2). Most of this water is eventually lost to the atmosphere by evaporation from the upper layers of soil and by evapotranspiration from leaves.

Under the influence of gravity, some infiltrating water slowly percolates through porous materials deeper into the earth and completely saturates pores and fractures in spongelike or permeable layers of sand, gravel, and porous rock such as sandstone. These water-bearing layers of the earth's crust are called **aquifers**, and the water in them is known as **groundwater** (Figure 10-3).

Aquifers are recharged or replenished naturally

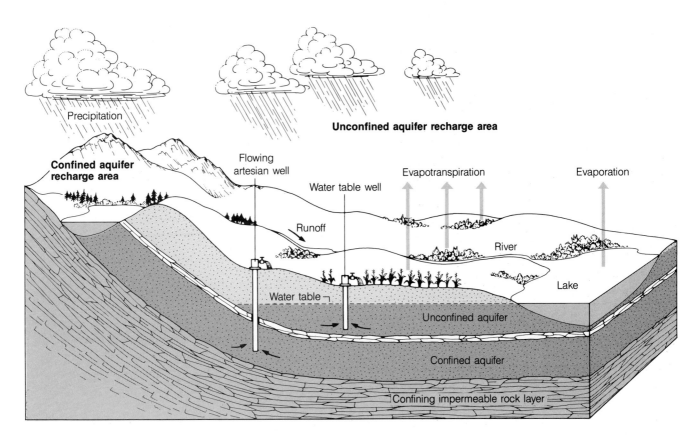

Figure 10-3 The groundwater system.

by precipitation, which percolates downward through soil and rock in what is called a **recharge area.** The recharge process is usually quite slow (decades to hundreds of years) compared to the rapid replenishment of surface water supplies. If the withdrawal rate of an aquifer exceeds its recharge rate, the aquifer is converted from a slowly renewable resource to a nonrenewable resource on a human time scale.

There are two types of aquifers: confined and unconfined. An **unconfined,** or **water-table, aquifer** forms when groundwater collects above a layer of relatively impermeable rock or compacted clay. The top of the water-saturated portion of an unconfined aquifer is called the **water table** (Figures 10-2 and 10-3). Thus groundwater is that part of underground water below the water table, and soil moisture is that part of underground water above the water table. Shallow, unconfined aquifers are recharged by water percolating downward from soils and materials directly above the aquifer.

To obtain water from an unconfined aquifer, a water table well must be drilled below the water table and into the unconfined aquifer. Because this water is under atmospheric pressure, a pump must be used to bring it to the surface. The elevation of the water table in a

particular area rises during prolonged wet periods and falls during prolonged drought. The water table can also fall when water is pumped out by wells faster than the natural rate of recharge, creating a vacated volume known as a cone of depression (Figure 10-4).

A **confined,** or **artesian, aquifer** forms when groundwater is sandwiched between two layers of relatively impermeable rock, such as clay or shale (Figure 10-3). This type of aquifer is completely saturated with water under a pressure greater than that of the atmosphere. In some cases the pressure is so great that when a well is drilled into the confined aquifer, water is pushed to the surface without the use of a pump. Such a well is called a flowing artesian well. With other confined-aquifer wells, known as nonflowing artesian wells, pumps must be used, because pressure is insufficient to force the water to the surface. Confined aquifers cannot be recharged from directly above them; they receive water from areas without overlying impermeable rock layers. Thus recharge areas for confined aquifers can be hundreds of kilometers away from wells where water is withdrawn, and the rate of natural recharge is not governed by local precipitation at the point of withdrawal as it is for unconfined aquifers.

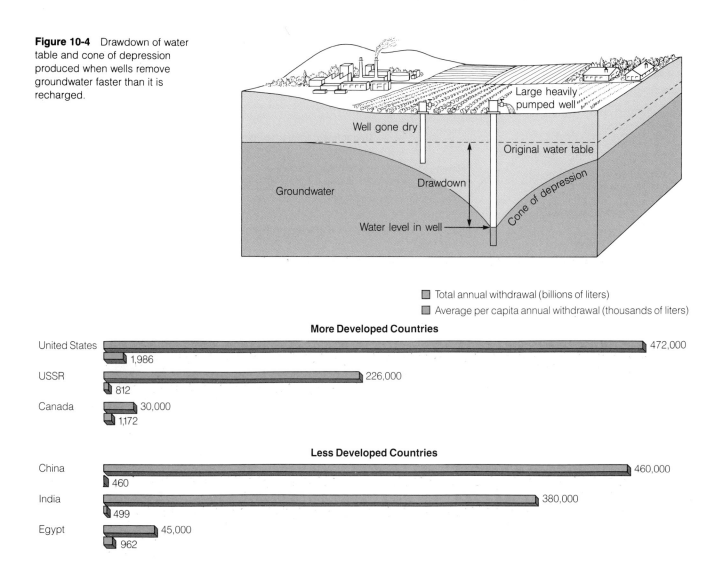

Figure 10-4 Drawdown of water table and cone of depression produced when wells remove groundwater faster than it is recharged.

Well gone dry

Large heavily pumped well

Original water table

Groundwater

Drawdown

Cone of depression

Water level in well

■ Total annual withdrawal (billions of liters)
■ Average per capita annual withdrawal (thousands of liters)

More Developed Countries

United States — 472,000
1,986

USSR — 226,000
812

Canada — 30,000
1,172

Less Developed Countries

China — 460,000
460

India — 380,000
499

Egypt — 45,000
962

Figure 10-5 Average total and per capita water withdrawal in selected countries in 1984. (Data from Worldwatch Institute and World Resources Institute)

World and U.S. Water Use The two common measures of water use are **withdrawal water use** and **consumption water use.** Water is withdrawn when it is taken from a surface or ground source and conveyed to the place of use. Water is consumed when, after withdrawal, it is no longer available for reuse in the local area because of evaporation, storage in the living matter of plants and animals, contamination, or seepage into the ground.

Total and average annual withdrawals per person vary considerably among various MDCs and LDCs (Figure 10-5). Almost three-fourths of the water withdrawn each year throughout the world is used for irrigation. The remainder is used for industrial processing, cooling electric power plants, and in homes and businesses (public use).

However, uses of withdrawn water vary widely from one country to another, depending on the relative amounts of agricultural and industrial produc-

tion. For example, in the United States about 41% of the water withdrawn is used for agriculture, 38% for cooling of electric power plants, 11% by industry, and 10% in homes and businesses. By contrast, in China 87% of the water withdrawn is used for agriculture, 7% by industry, 6% in homes and businesses, and a negligible amount for cooling of electric power plants.

Worldwide, up to 90% of all water withdrawn is returned to rivers and lakes for reuse and is not consumed. However, about 75% of the water supplied for irrigation is consumed and lost for reuse as surface water in the area from which it is withdrawn. Thus irrigation, which is projected to double between 1975 and 2000, places the greatest demand on the world's water supplies in terms of both withdrawal and consumption.

Total withdrawal and average withdrawal per person from U.S. surface and groundwater supplies have been increasing rapidly since 1900 and are projected

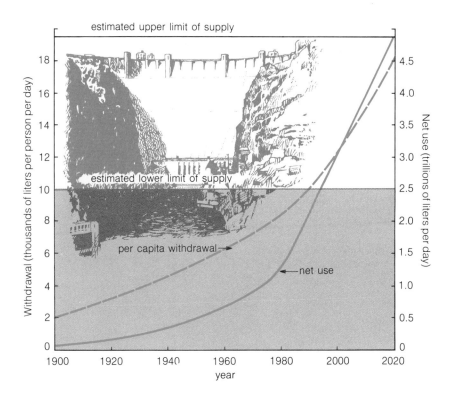

estimated upper limit of supply

estimated lower limit of supply

per capita withdrawal→

←net use

Withdrawal (thousands of liters per person per day)

Net use (trillions of liters per day)

year

Figure 10-6 Total and average per capita water withdrawal and consumption in the United States, 1900–2020 (projected). (Data from U.S. Water Resources Council and U.S. Geological Survey)

to increase more between 1985 and 2020 (Figure 10-6). Almost 80% of the water withdrawn in the United States is used for cooling electric power plants and for irrigation. About 54% of the water used for irrigation and 2% of that used for electric power plant cooling is consumed.

In 1985 average withdrawal for each American was 7,400 liters (1,950 gallons) a day, 1,700 liters (450 gallons) of which was consumed. In the West, average withdrawals per person are almost twice the level in the East, and average consumption per person is 10 times higher than in the East. About 340 liters (90 gallons) of the average daily withdrawal is used for domestic purposes (cooking, drinking, washing, watering, and flushing). This is about 3 times the average domestic use per person worldwide and 15 to 20 times that of people in LDCs.

10-3 WATER RESOURCE PROBLEMS

Too Little Water A number of experts consider *the availability of adequate supplies of fresh water to be the most serious long-range problem confronting the world and the United States.* At least 80 arid and semiarid countries, accounting for nearly 40% of the world's population, now experience serious periodic droughts and have considerable difficulty growing enough food to sup-

port their populations. Most of these countries are in Asia and Africa. During the 1970s, major drought disasters affected an average of 24.4 million people and killed over 23,000 a year—a trend continuing in the 1980s. By 1985 more than 154 million people in 21 tropical and subtropical countries in Africa were on the brink of starvation because of the combined effects of rapid population growth, prolonged drought, land misuse, war, and ineffective government policies for water and soil resource management and agricultural development (Figure 10-7).

In many LDCs, poor people must spend a good part of their waking hours fetching water, often from polluted streams and rivers. To get water, many women and children in LDCs walk 16 to 25 kilometers (10 to 15 miles) a day, carrying heavy water-filled jars on their return trip.

Although reduced average annual precipitation usually triggers a drought, rapid population growth and poor land use intensify its effects. In many LDCs, large numbers of poor people have no choice but to try to survive on drought-prone land by cutting trees, growing crops at higher, more erosion-prone elevations, cultivating poor soils, and allowing their livestock to overgraze grasslands. The resulting land degradation contributes to the severity of long-term drought by reducing the amount of rainfall absorbed and slowly released by vegetation and soils. In many cases, the result is desertification (see Spotlight on p. 163).

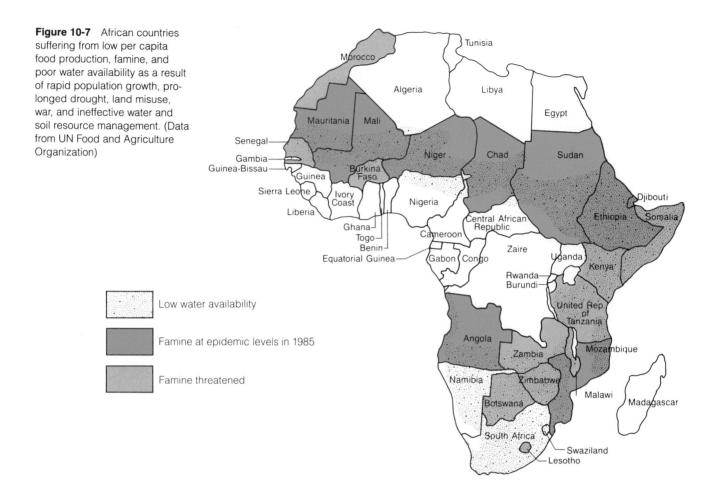

Figure 10-7 African countries suffering from low per capita food production, famine, and poor water availability as a result of rapid population growth, prolonged drought, land misuse, war, and ineffective water and soil resource management. (Data from UN Food and Agriculture Organization)

Low water availability

Famine at epidemic levels in 1985

Famine threatened

Too Much Water Other countries get enough precipitation on an annual basis but receive most of it at one time of the year. In India, for example, 90% percent of the annual precipitation falls between June and September—the monsoon season. This downpour runs off so rapidly that most of it cannot be captured and used. The massive runoff also leads to periodic flooding.

During the 1970s, major flood disasters affected 15.4 million people, killed an average of 4,700 people a year, and caused tens of billions of dollars in property damages—a trend that continued in the 1980s. Although floods are classified as natural disasters, human beings have contributed to the sharp rise in flood deaths and damages since the 1960s by removing water-absorbing vegetation and soil through cultivation of marginal lands, deforestation, overgrazing, and mining. Urbanization also increases flooding, even with moderate rainfall, by replacing vegetation with highways, parking lots, shopping centers, office buildings, homes, and numerous other structures.

Death tolls and damages from flooding have also increased because many poor people in LDCs have little choice but to live on land subject to severe peri-

odic flooding and because many people in LDCs believe that the benefits of living in flood-prone areas outweigh the risks. Many urban areas and croplands in LDCs and MDCs are situated on **floodplains**—flat areas along rivers subject to periodic flooding—and coastlands because these sites are level, have highly fertile topsoil deposited by rivers, are close to supplies of surface water and water transportation routes, and provide recreational opportunities.

Since 1925 the U.S. Army Corps of Engineers, the Soil Conservation Service, and the Bureau of Reclamation have spent more than $8 billion on flood-control projects such as straightening stream channels (channelization), dredging streams, and building dams, reservoirs, levees, and seawalls. Despite these efforts—and because these projects stimulate development on flood-prone land—property damage from floods in the United States has increased from about half a billion dollars a year in the 1960s to an average of about $3 billion a year in the 1980s.

A number of effective methods exist for preventing or reducing flood damage: replanting vegetation in disturbed areas to reduce runoff, building ponds in urban areas to retain rainwater and release it slowly

The conversion of rangeland (uncultivated land used for animal grazing), rain-fed cropland, or irrigated cropland to desertlike land with a drop in agricultural productivity of 10% or more is called **desertification.** Moderate desertification causes a 10% to 25% drop in productivity; severe desertification causes a 25% to 50% drop; and very severe desertification causes a drop of 50% or more and usually the formation of massive gullies and sand dunes.

Prolonged drought and hot temperatures may accelerate the desertification process. But its basic causes are overgrazing of rangeland by concentrating too many livestock on too little land area; improper soil and water resource management that leads to increased erosion, salinization, waterlogging, cultivation of marginal land with unsuitable terrain or soils, and deforestation and strip mining without adequate replanting.

It is estimated that about 900 million hectares (2 billion acres)—equivalent to an area ten times the size of Texas—have become desertified during the past 50 years (Figure 10-8). At least 50 million people, half of them in Africa, have experienced a major loss in their ability to feed themselves because of desertification. Another 400 million in moderately desertified areas have a reduced capacity to support themselves.

Each year, the amount of desertified land grows by at least 20 million hectares (49 million acres)—an area equal to that of South Dakota. According to the UN Environmental Programme, one-fifth of the world's people now live in areas that may become desertified over the next 20 years.

The spread of desertification can be halted or sharply reduced by improved management of rangeland, forest, soil, and water resources, and much currently desertified land can be reclaimed. The total cost of such prevention and rehabilitation would be about $141 billion. Although this amount may seem staggering, it is only five and one-half times the estimated $26 billion annual loss in agricultural productivity from desertified land. Thus once this potential productivity is restored, the costs of the program could be recouped in five to ten years. However, funds now devoted to preventing desertification and restoring desertified lands fall far short of the need.

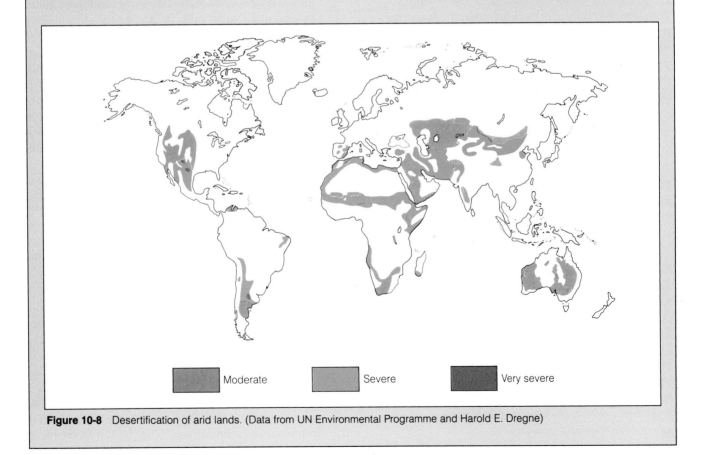

Moderate   Severe   Very severe

Figure 10-8   Desertification of arid lands. (Data from UN Environmental Programme and Harold E. Dregne)

Figure 10-9 About 70% of rural people and 25% of city dwellers in LDCs lack ready access to uncontaminated water. These children in Lima, Peru, are scooping up drinking water from a puddle.

to rivers, and diverting rainwater through storm sewers to holding tanks and ponds for use by industry. Floodplains should also be clearly identified, and laws or zoning regulations should be used to discourage their use for certain types of development. Sellers of property in these areas should be required to provide prospective buyers with information about average flood frequency.

Water in the Wrong Place In some countries with sufficient annual precipitation, the largest rivers, carrying much of the runoff, are far from agricultural and population centers where the water is needed. For instance, although South America has the largest average annual runoff of any continent, 60% of it flows through the Amazon in areas remote from most people.

Contaminated Drinking Water Although water scarcity, drought, and flooding are serious in some regions, drinking contaminated water is the most common hazard to people in much of the world. In 1980 the World Health Organization estimated that in LDCs, 70% of the people living in rural areas and 25% of the urban dwellers did not have enough safe drinkable water (Figure 10-9). WHO estimated that 25 million people die every year from cholera, dysentery, and other preventable waterborne diseases—an average of 68,500 deaths each day.

Irrigation Problems: Salinization and Waterlogging As irrigation water flows over and through the ground, it dissolves salts, increasing the salinity of the water. Much of the water in this saline solution is lost to the atmosphere by evaporation, leaving behind high concentrations of salts such as sodium chloride in the topsoil. The accumulation of these salts in soils is called **salinization** (Figure 10-10). Unless the salts are flushed or drained from the soil, their buildup promotes excessive water use, increases capital and operating costs, stunts crop growth, decreases yields, eventually kills crop plants, and makes the land unproductive.

An estimated one-third of the world's irrigated land is now affected by salt buildup. Salinization has already reduced agricultural productivity on 25% to 35% of all irrigated land in 17 U.S. western states (Figure 10-11) and may soon affect half of this land. Worldwide, it is projected that at least 50%, and probably close to 65%, of all currently irrigated land will suffer reduced productivity from excess soil salinity by 2000.

One way to reduce salinization is to flush salts out by applying much more irrigation water than is needed for crop growth. But this increases pumping and crop production costs and wastes enormous amounts of precious water in arid and semiarid regions. Pumping groundwater from a central well and applying it by a sprinkler system that pivots around the well maintains downward drainage and is especially effective in preventing salinization (Figure 10-12). However, at

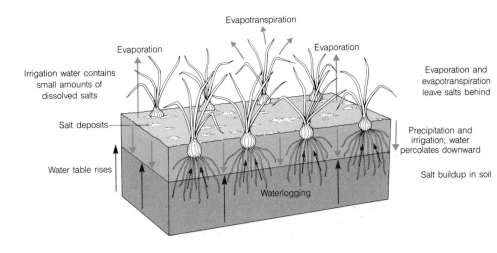

Figure 10-10 Salinization and waterlogging of soil on irrigated land without adequate drainage leads to decreased crop yields.

Evapotranspiration

Evaporation

Evaporation

Irrigation water contains small amounts of dissolved salts

Salt deposits

Water table rises

Waterlogging

Evaporation and evapotranspiration leave salts behind

Precipitation and irrigation; water percolates downward

Salt buildup in soil

Figure 10-11 Because of poor drainage and severe salinization, white alkaline salts have replaced crops that once grew in heavily irrigated Paradise Valley, Wyoming.

USDA/Bureau of Reclamation/Lyle C. Axhelm

least 30% of the water is consumed by evaporation, and eventually groundwater in unconfined aquifers can become too saline for irrigation and other human uses unless expensive drainage systems are installed.

In theory, once topsoil has become heavily salinized, it can be renewed by taking it out of production for two to five years, installing an underground network of perforated drainage pipe, and flushing the soil with large quantities of low-salt water. This scheme, however, is usually prohibitively expensive and only slows the buildup of soil salinity—it does not stop the process. Flushing salts from the soil also increases the salinity of irrigation water delivered to farmers further

downstream unless the saline water can be drained into evaporation ponds rather than returned to the river or canal.

A problem often accompanying soil salinity is **waterlogging** (Figure 10-10). To keep salts from accumulating and destroying fragile root systems, farmers often apply heavy amounts of irrigation water to wash or leach salts deeper into the soil profile. If drainage isn't provided, water accumulating underground can gradually raise the water table close to the surface, enveloping the roots of plants in saline water. This is a particularly serious problem in areas such as the heavily irrigated San Joaquin Valley in California, where

Figure 10-12 Center-pivot irrigation can reduce salinization but wastes 30% of irrigation water.

soils contain a clay layer impermeable to water. World-wide, at least one-tenth of all irrigated land suffers from waterlogging.

The U.S. Situation Overall, the United States has plenty of fresh water, but much of its annual runoff is not in the desired place, occurs at the wrong time, or is contaminated from agricultural and industrial activities. Most of the eastern half of the country usually has ample average annual precipitation, while much of the western half has too little. Many major urban centers in the United States are located in areas that already have inadequate water or are projected to have water shortages by 2000 (Figure 10-13). Because water is such a vital resource, you might find Figure 10-13 useful in evaluating where to live in coming decades.

In the eastern half of the United States, where there is usually no shortage of water, the major problems are flooding, inability to supply enough water to some large urban areas, and increasing pollution of rivers, lakes, and groundwater. For example, 3 million residents of Long Island, New York, must draw all their water from an underground aquifer that is becoming severely contaminated by industrial wastes, leaking septic tanks and landfills, and salt water from the ocean, which is drawn into the aquifer when fresh water is withdrawn.

The major water problem in arid and semiarid areas in the western half of the country is a shortage of runoff due to low average precipitation, high rates of evaporation, prolonged periodic drought, and rapidly declining water tables as farmers and cities deplete groundwater aquifers faster than they are recharged. Present water shortages and conflicts over water supplies will get much worse if more industries and people migrate west as projected and compete with farmers for scarce water.

10-4 WATER RESOURCE MANAGEMENT: INCREASING THE USABLE SUPPLY

Methods for Managing Water Resources Although we can't increase the earth's supply of water, we can manage what we have more effectively to reduce the impact and spread of water resource problems. There are two major approaches to water resource management: Increase the usable supply; decrease unnecessary loss and waste (Table 10-1). Most water resource experts believe that any effective plan for water management should rely on a combination of these approaches.

Water problems and available solutions often differ between MDCs and LDCs. LDCs may or may not have enough water, but they rarely have the money needed to develop water storage and distribution systems. Their people must settle where the water is. In MDCs, people tend to live where the climate is favorable and then bring in water through sophisticated systems. Some settle in a desert area such as Palm Springs and expect water to be brought to them. Others settle on a floodplain and expect the government to keep flood waters away.

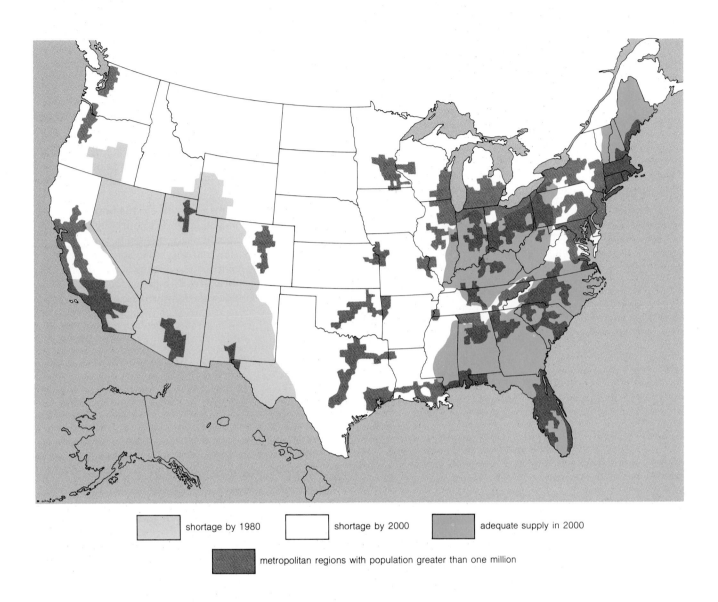

shortage by 1980   shortage by 2000   adequate supply in 2000

metropolitan regions with population greater than one million

Figure 10-13 Present and projected water-deficit regions in the United States compared with present metropolitan regions with populations greater than 1 million. (Data from U.S. Water Resources Council and U.S. Geological Survey)

Table 10-1 Major Methods for Managing Water Resources	
Increase the Supply	**Reduce Unnecessary Loss and Waste**
Build dams and reservoirs	Decrease evaporation of irrigation water
Divert water from one region to another	Redesign mining and industrial processes to use less water
Tap more groundwater	
Convert salt water to fresh water (desalinization)	Encourage the public to reduce unnecessary water waste and use
Tow freshwater icebergs from the Antarctic to water-short coastal regions	Increase the price of water to encourage water conservation
Seed clouds to increase precipitation	Purify polluted water for reuse (Chapter 19)

Schistosomiasis is caused by the trematode worm *Schistosoma,* which is transmitted between human and animal hosts by tiny snails found in freshwater streams, rivers, lakes, and irrigation canals (Figure 10-14). The adult worms lodge in the human host's veins and deposit eggs in surrounding organs and tissues, causing chronic inflammation, swelling, and pain. The urine and feces of newly infected people can generate the entire cycle again.

Victims of this disease, which affects about 200 million people throughout the world, suffer from cough, fever, enlargement of the spleen and liver, a general wasting away of the body, filling of the abdomen with fluid (which produces the characteristic bloated belly), and constant pain; they are more susceptible to other diseases and are often too weak to work. Although the disease itself is rarely fatal, people who are severely malnourished or severely infected may die.

In rural areas of Africa and Asia, avoiding contact with infested water is difficult for villagers, because they collect it for drinking, cooking, and washing clothes; they also bathe and swim in it and work in irrigation ditches. The building of dams for hydroelectric power and irrigation tends to intensify the spread of the disease because the slow-moving, often stagnant water in irrigation ditches makes them ideal breeding places for the snails that transmit the disease.

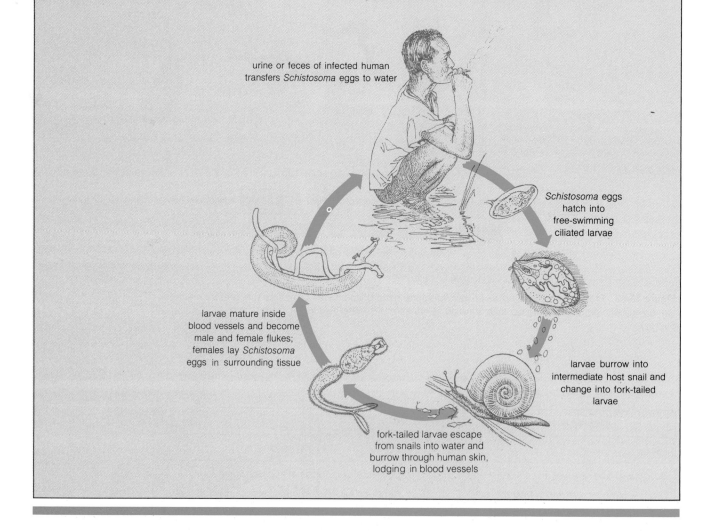

urine or feces of infected human transfers *Schistosoma* eggs to water

Schistosoma eggs hatch into free-swimming ciliated larvae

larvae burrow into intermediate host snail and change into fork-tailed larvae

fork-tailed larvae escape from snails into water and burrow through human skin, lodging in blood vessels

larvae mature inside blood vessels and become male and female flukes; females lay *Schistosoma* eggs in surrounding tissue

Dams and Reservoirs Some rainwater and water from melting snow that would otherwise be lost can be captured by dams on rivers and stored in large reservoirs behind the dams (Figure 5-7, p. 78). This increases the annual supply by collecting fresh surface water during wet periods and storing it for use during dry periods. In addition, dams control the flow of rivers and can reduce the danger of flooding in areas

Figure 10-15 California Water Plan and Central Arizona Project for large-scale transfer of water from one area to another.

below the dam, provide a controllable supply of water for irrigating land below the dam, generate relatively cheap electricity for local and regional residents, and allow people to live on fertile floodplain areas of major rivers below the dam. Large-scale dams also create reservoirs that can be used for swimming, boating, and fishing and thus aid the local economy.

However, the benefits of dams and reservoirs must be weighed against their costs. They are expensive to build, and reservoirs fill up with silt, becoming useless after 20 to 200 years, depending on local climate and land-use practices. The permanent flooding of land behind dams to form reservoirs displaces people and destroys vast areas of valuable agricultural land, wildlife habitat, white-water rapids, and scenic natural beauty.

Storage of water behind the dam also raises the water table, often waterlogging soil on nearby land and thereby decreasing crop or forestry productivity. A dam can also decrease rather than increase the available supply of fresh water, especially in semiarid areas, because water that would normally flow in an undammed river evaporates from the reservoir's surface or seeps into the ground below the reservoir. This evaporation also increases the salinity of water by leaving salts behind, decreasing its usefulness for irrigation and intensifying soil salinization.

By interrupting the natural flow of a river, a dam disrupts the migration and spawning of some fish, such as salmon, unless fish ladders are provided. Dams also reduce the flow of nutrients and fresh water into estuaries, decreasing their productivity. In the opinion of some outdoor sports enthusiasts, a dam replaces more desirable forms of water recreation (white-water canoeing, kayaking, rafting, stream fishing) with less desirable, more "artificial" forms (motorboating and sailboating, lake fishing). In rural areas of some LDCs, dams also reduce the natural flows of water that sweep away snails that can infect people with *schistosomiasis*, a debilitating, painful, incurable, and often fatal disease (see Spotlight on p. 168).

Faulty construction, earthquakes, sabotage, or war can cause dams to fail, taking a terrible toll in lives and property. In 1972 a dam failure in Buffalo Creek, West Virginia, killed 125 people, and another in Rapid City, South Dakota, killed 237 and caused more than $1 billion in damages. According to a 1986 study by the Federal Emergency Management Agency, the United States has 1,900 unsafe dams in populated areas. The agency reported that the dam safety programs of most states are inadequate because of weak laws and budget cuts.

Water Diversion Projects One of the most common ways to increase a limited supply of fresh water is to transfer water from water-rich areas to water-poor areas. Two interrelated massive water transfer projects in the United States are the California Water Plan, which transports water from water-rich northern California to arid, heavily populated southern California, and the federally financed $3.9 billion Central Arizona Project, which began pumping water from the Colorado River uphill to Phoenix in 1985 and is projected to deliver water to Tucson by 1991 (Figure 10-15).

Figure 10-16 Major areas of aquifer depletion, subsidence, saltwater intrusion, and groundwater contamination in the United States. (Data from U.S. Water Resources Council and U.S. Geological Survey)

Groundwater Overdrafts

High

Moderate

Minor or none

▲ Subsidence

● Saltwater intrusion

Significant groundwater pollution

For decades, northern and southern Californians have been feuding over this plan. In 1982 voters rejected a proposal to expand it by building a $1 billion canal to divert to southern California much of the water that now flows into San Francisco Bay. Opponents of the plan contended that it was a costly and unnecessary boondoggle that would degrade the Sacramento River, threaten fishing, and inhibit the flushing action that helps clean San Francisco Bay of pollutants. They argued that much of the water already sent south is wasted and that if irrigation efficiency was improved by only 10%, abundant water would be available for domestic and industrial uses in southern California. Proponents of the expansion contend that without more

water, a prolonged drought could bring economic ruin to much of southern California. The issue is far from dead.

Arizona is further complicating California's water problem. When the first portion of the Central Arizona Project was completed in 1985, southern California, especially the arid and booming San Diego region, began losing up to one-fifth of its water, which until then had been diverted from the Colorado River by the Colorado River Aqueduct (Figure 10-15). Although Arizona has been legally entitled since 1922 to one-fifth of the Colorado River's annual flow, without the new diversion system it lacked the ability to use more than half of its share. The surface water

Water drawn from the vast Ogallala Aquifer (Figure 10-17) is used to irrigate one-fifth of all U.S. cropland in an area too dry for rainfall farming. It supports $32 billion of agricultural production a year, mostly wheat, sorghum, cotton, corn, and 40% of the country's grain-fed beef.

Although the aquifer contains a large amount of water, it has an extremely low natural recharge rate, because it underlies a region with relatively low average annual precipitation. Today the amount of water being withdrawn is so enormous that overall the aquifer is being depleted eight times faster than its natural recharge rate. Even higher depletion rates, sometimes 100 times the recharge rate, are taking place in parts of the aquifer that lie in Texas, New Mexico, Oklahoma, and Colorado.

Water resource experts project that at the present rate of depletion, much of this aquifer could be dry by 2020, and much sooner in areas where it is only a few meters deep. Long before this happens, however, the high costs of pumping water from rapidly declining water tables will force many farmers to switch from irrigated farming to dryland farming (planting crops such as winter wheat and cotton) and to give up the cultivation of profitable but thirsty crops such as corn.

The amount of irrigated land already is declining in five of the seven states using this aquifer because of the high and rising cost of pumping water from depths as great as 1,825 meters (6,000 feet). If all farmers in the Ogallala region began using water conservation measures, depletion of the aquifer would be delayed but not prevented in the long run. However, the tragedy of the commons (Section 1-2) shows us that most farmers are likely to continue withdrawing as much water as possible from this commonly shared resource to increase short-term profits.

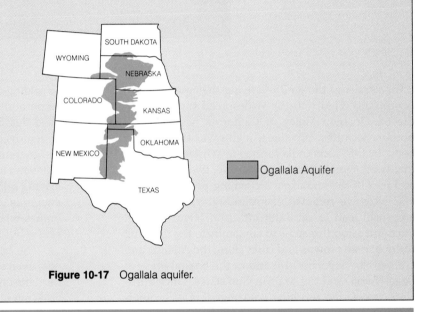

Figure 10-17   Ogallala aquifer.

---

diverted from the Colorado will partially replace groundwater overdrafts that have led to falling water tables in many parts of the state during the past 50 years.

Tapping Groundwater   One solution to water supply problems in some areas is heavier reliance on groundwater, which makes up about 95% of the world's supply of fresh water. About half of U.S. drinking water (96% in rural areas and 20% in urban areas), 40% of irrigation water, and 23% of all fresh water used is withdrawn from underground aquifers.

This increased use of groundwater gives rise to several problems: (1) **aquifer depletion** or **overdraft** when groundwater is withdrawn faster than it is recharged by precipitation; (2) **subsidence,** or sinking of the ground as groundwater is withdrawn; (3) **saltwater intrusion** into freshwater aquifers in coastal areas as groundwater is withdrawn faster than it is recharged; and (4) **groundwater contamination** from human activities (Figure 10-16).

In the United States, the major groundwater overdraft problem is in parts of the huge California-size Ogallala Aquifer extending across the farming belt from northern Nebraska to northwestern Texas (see Spotlight and Figure 10-17 above). Aquifer depletion is also a serious problem in northern China, Mexico City, and parts of India. The most effective solution to this growing problem is to reduce the amount of groundwater withdrawn by wasting less irrigation water and by abandoning irrigation in arid and semiarid areas.

When groundwater in an unconfined aquifer is withdrawn faster than it is replenished, the soil becomes

Figure 10-18 Large sinkhole formed in rural Alabama from withdrawal of groundwater from an unconfined aquifer.

U.S. Geological Survey

compacted and the land overlying the aquifer sinks, or subsides (Figure 10-18). Widespread subsidence in the San Joaquin Valley of California has damaged houses, factories, pipelines, highways, and railroad beds. In 1981 a sinkhole formed in Winter Park, Florida, swallowing several cars, a house, two businesses, and part of the municipal swimming pool.

Excessive removal of groundwater near coastal areas can lead to saltwater intrusion (Figure 10-19). Such intrusion threatens to contaminate the drinking water of many towns and cities along the Atlantic and Gulf coasts (Figure 10-16) and in the coastal areas of Israel, Syria, and the Arabian Gulf states. Once intrusion occurs, the resulting contamination of groundwater is difficult, if not impossible, to reverse.

Another growing problem in many MDCs such as the United States is groundwater contamination from agricultural and industrial activities, septic tanks, underground injection wells, and other sources. Because groundwater flow in aquifers is slow and not turbulent, contaminants that reach this water are diluted very little. In addition, organic waste contaminants are not broken down as readily as in rapidly flowing surface waters exposed to the atmosphere, because groundwater lacks decomposing bacteria and dissolved oxygen. As a result, it can take hundreds to thousands of years for contaminated groundwater to cleanse itself. Because of its location deep underground, pumping polluted groundwater to the surface, cleaning it up, and returning it to the aquifer is usually prohibitively expensive.

Desalinization Removing dissolved salts from ocean water or brackish (slightly salty) groundwater is an appealing way to increase freshwater supplies. Distillation and reverse osmosis are the two most widely used desalinization methods, although salt can also be removed by freezing salt water or by passing electric current through it. Distillation involves heating salt water to evaporate and then condense fresh water, leaving salts behind in solid form. In reverse osmosis, energy is used to force salt water through thin membranes whose pores allow the passage of water molecules but not of the dissolved salts.

The basic problem with all desalinization methods, even reverse osmosis, which requires about one-third the energy input of distillation, is that they require large amounts of energy and therefore are expensive. As a result, most experts project that desalted water will never be cheap enough for widespread use in irrigation, the main use of water throughout the world. Even more energy and money are required to pump desalted water uphill and inland from coastal desalinization plants. In addition, building and operating a vast network of expensive desalinization plants would produce mountains of salt to be disposed of. If the salt were returned to the ocean, it would increase the salt concentration near the coasts and threaten food resources in estuarine and continental shelf waters. Desalinization, however, can provide fresh water in selected coastal cities in arid regions such as Saudi Arabia, where the cost of obtaining fresh water by any method is high.

Towing Icebergs Some scientists believe that it may be economically feasible to use a fleet of tugboats to tow huge, flat, floating Antarctic icebergs to southern California, Australia, Saudi Arabia, and other dry

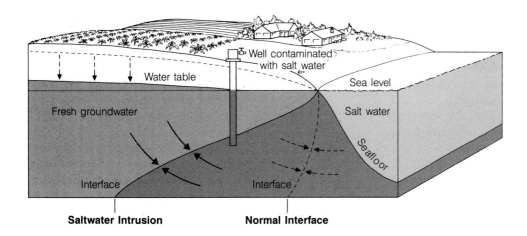

Figure 10-19 Saltwater intrusion along a coastal region. As the water table is lowered, the normal interface between fresh and saline groundwater moves inland.

Labels in figure: Well contaminated with salt water · Water table · Sea level · Fresh groundwater · Salt water · Seafloor · Interface · Interface · Saltwater Intrusion · Normal Interface

coastal places. But a number of unanswered questions and problems exist: How much would the scheme cost? How can such a massive object be "lassoed," wrapped, and towed? How can most of the iceberg be prevented from melting on its long journey through warm waters? If the towing project is successful, how would the fresh water from the slowly melting iceberg be collected and transmitted to shore? What effects might introducing such a cold mass of water into tropical water have on local weather and marine life? Who owns the icebergs in the Antarctic, and how could international conflicts over ownership be resolved?

Cloud Seeding Several countries, particularly the United States, have been experimenting for years with seeding clouds with chemicals to produce rain over dry regions and snow over mountains to increase runoff in such areas. In principle, cloud seeding involves finding a large, suitable cloud and injecting it with a powdered chemical such as silver iodide from a plane or from ground-mounted burners. The chemical particles serve as nuclei of condensation, causing small water droplets in the cloud to coalesce and form droplets or ice particles large enough to fall to the earth as precipitation.

Since 1977, clouds have been successfully seeded in 23 states, bringing rain to 7% of the U.S. land area; in some places average annual precipitation has increased 10%. But whether this represents an increase in total precipitation or merely a shift of precipitation from one area to another is unknown.

As with all methods of increasing the usable supply of water, cloud seeding has its drawbacks: It cannot be used effectively in very dry areas, where it is most needed, because rain clouds are rarely available. Large-scale use could change snowfall and rainfall patterns and alter regional or even global climate patterns in unknown and perhaps undesirable ways. Introduction of large quantities of silver iodide into soil and water systems could have harmful effects on people, wildlife, and agricultural productivity. There are also legal disputes over the ownership of water in clouds. For example, during the 1977 drought in the western United States, the attorney general of Idaho accused officials in neighboring Washington of "cloud rustling" and threatened to file suit in federal court.

10-5 WATER RESOURCE MANAGEMENT: WATER CONSERVATION

Importance of Water Conservation An estimated *30% to 50% of the water used in the United States is unnecessarily wasted.* This explains why many water resource experts consider water conservation the quickest and cheapest way to provide much of the additional water needed in dry areas.

The major reason for the large amount of water wasted in the United States is that the government, hoping to stimulate economic growth, keeps water prices artificially low by using taxes to build dams and water transfer projects, thus subsidizing the use of water by farmers, industries, and homeowners. Because the costs of these subsidies are borne by all taxpayers in the form of higher taxes, water users do not realize the full cost of the water they are using and have little incentive to conserve.

Another reason that water waste and pollution in the United States are higher than necessary is that the responsibility for water resource management in a particular water basin is divided among many state and local governments rather than being handled in terms of the entire watershed. For example, the Chicago metropolitan area has 349 separate water supply systems and 135 waste treatment plants divided among about 2,000 local units of government over a 6-county area.

In sharp contrast is the regionalized approach to water management used in England and Wales. The British Water Act of 1973 replaced more than 1,600 separate agencies with 10 regional water authorities based not on political boundaries but on natural watershed boundaries. In this successful ecological approach, each water authority owns, finances, and manages all water supply and waste treatment facilities in its region, including water pollution control, water-based recreation, land drainage and flood control, inland navigation, and inland fisheries. Each water authority is managed by a group of elected local officials and a smaller number of officials appointed by the national government.

Reducing Irrigation Losses   Since irrigation accounts for the largest fraction of water withdrawal, consumption, and waste, more efficient use of even a small amount frees water for other uses. Most irrigation systems distribute water from a groundwater well or surface canal by gravity flow through unlined field ditches. Although this method is cheap, it provides far more water than needed for crop growth, and at least 50% of the water is lost by evaporation and seepage.

As available water supplies dwindle and pumping prices rise, farmers find it more profitable to use a number of available techniques for reducing evaporation and using irrigation water more efficiently. For example, many farmers served by the Ogallala Aquifer have switched from gravity-flow canal systems to center-pivot sprinkler systems (Figure 10-12), which reduce water waste from 50% or more to 30%. Some farmers are switching to new, low-energy precision-application sprinkler systems, which spray water downward, closer to crops, rather than high into the air, cutting water waste to between 2% and 5% and energy requirements by 20% to 30%.

Highly water-efficient trickle or drip irrigation systems, developed in Israel in the 1960s, are economically feasible for high-profit fruit, vegetable, and orchard crops. In this approach, an extensive network of perforated piping, installed at or below the ground surface, releases a small volume of water and fertilizer close to the roots of plants, minimizing evaporation and seepage. Although drip irrigation accounts for less than 1% of total irrigated area worldwide, it is used on half of the irrigated land in Israel. Its use in the United States is still negligible but is increasing, especially in California and Florida.

Irrigation efficiency can also be improved by using computer-controlled systems to set water flow rates, detect leaks, and adjust the amount of water to soil moisture and weather conditions. Irrigation ditches can be lined with plastic to prevent seepage and waterlogging, and ponds can be constructed to store runoff for later use. Evaporation losses can be reduced by using conservation tillage (Section 9-5), covering ponds with floating alcohol-based liquids, and covering the soil with a mulch. Farmers can switch to new hybrid crop varieties that require less water or that tolerate irrigation with saline water. Between 1950 and 1985, Israel used many of these techniques to decrease waste of irrigation water from 83% to 5%, while allowing the country's irrigated land to expand by 44%.

Wasting Less Water in Industry   More than 80% of all water used in U.S. manufacturing is used in four industries: paper, chemicals, petroleum, and primary metals. These and most other manufacturing processes can use recycled water or be redesigned to use and waste less water. For example, depending on the process used, manufacturing a ton of steel can require as much as 200,000 liters (52,800 gallons) or as little as 5,000 liters (1,320 gallons) of water. To produce a ton of paper, a paper mill in Hadera, Israel, uses about one-tenth the amount of water as most paper mills. Manufacturing a ton of aluminum from recycled scrap rather than virgin resources can reduce water needs by 97% percent.

The potential for water recycling in U.S. manufacturing industries, however, has hardly been tapped, because much of the cost of water to industry is subsidized by taxpayers through federally financed water projects. Thus industries have little incentive to recycle water, which typically accounts for only about 3% of total manufacturing costs, even in industries that use large amounts.

Wasting Less Water in Homes   In the United States, leaks in pipes, water mains, toilets, bathtubs, and faucets alone waste an estimated 20% to 35% of water withdrawn from public supplies. There is little incentive to reduce leaks and waste in many cities, like New York, where there are no residential water meters and users are charged flat rates. In Boulder, Colorado, the introduction of water meters reduced water use by more than one-third. Individuals can develop their own plan for saving water and money (see Appendix 4).

Commercially available systems can also be used to purify and completely recycle wastewater from houses, apartments, and office buildings. Such a system, which can be installed in a small shed outside a residence, is serviced for a monthly fee about equal to that charged by most city water and sewer systems.

Born in a water-rich environment, we have never really learned how important water is to us. . . . Where it has been cheap and plentiful, we have ignored it; where it has been rare and precious, we have spent it with shameful and unbecoming haste. . . . Everywhere we have poured filth into it.

William Ashworth

DISCUSSION TOPICS

1. Which physical property or properties of water
   a. account for the fact that you exist?
   b. allow lakes to freeze from the top down?
   c. help protect you from the shock of sudden temperature changes?
   d. help regulate the climate?

2. If groundwater is a renewable resource, how can it be "mined" and depleted like a nonrenewable resource?

3. How do human activities contribute to drought? How could these effects be reduced?

4. How do human activities contribute to flooding? How could these effects be reduced?

5. In your community:
   a. What are the major sources of the water supply?
   b. How is water use divided among agricultural, industrial, power plant cooling, and public uses? Who are the biggest consumers of water?
   c. What has happened to water prices during the past 20 years?
   d. What water problems are projected?
   e. How is water being wasted?

6. Explain why dams and reservoirs may lead to more flood damage than might occur if they had not been built. Should all proposed large dam and reservoir projects be scrapped? What criteria would you use in determining desirable dam and reservoir projects?

7. How can the following problems be minimized or prevented: (a) soil salinity from irrigation? (b) saltwater intrusion in coastal areas?

8. Use the first and second laws of energy (Sections 3-4 and 3-5) to explain why desalinized seawater will probably never be an important source of fresh water for irrigation.

9. Should the price of water for all uses in the United States be increased sharply to encourage water conservation? Explain. What effects might this have on the economy, on you, on the poor, on the environment?

10. List ten major ways to conserve water on a personal level. Which, if any, of these practices do you now use or intend to use? (See Appendix 4.)

11

Food Resources and World Hunger

1. What major types of agricultural systems are used to provide food from domesticated crops and livestock throughout the world?
2. What are the world's major food problems?
3. How can the world's food problems be solved?

Hunger is a curious thing: At first it is with you all the time, working and sleeping and in your dreams, and your belly cries out insistently, and there is a gnawing and a pain as if your very vitals were being devoured, and you must stop it at any cost. . . . Then the pain is no longer sharp, but dull, and this too is with you always.

Kamala Markandaya

Each day, natural population growth produces 238,000 more people to feed, clothe, and house, and the world's population is projected to grow from 5 billion to 8 billion between 1987 and 2020. This means that during this 33-year period, we must produce as much food as humankind has produced since the dawn of agriculture about 10,000 years ago. Even if enough food is grown, how can it be made available to those who can't afford to buy it, and what are the environmental consequences of growing this much food?

11-1 WORLD AGRICULTURAL SYSTEMS: HOW IS FOOD PRODUCED?

Plants and Animals That Feed the World Although an estimated 80,000 species of plants are edible, only about 30 types of crops feed the world. Four crops—wheat, rice, corn, and potato—make up more of the world's total food production than all others combined.

Most of the remainder of the world's food is fish, meat, and meat products such as milk, eggs, and cheese. Almost all meat comes from just nine groups of livestock: cattle, sheep, swine, chickens, turkeys, geese, ducks, goats, and water buffalo. Meat and meat products are too expensive for most people, primarily because of the loss of usable energy resulting from

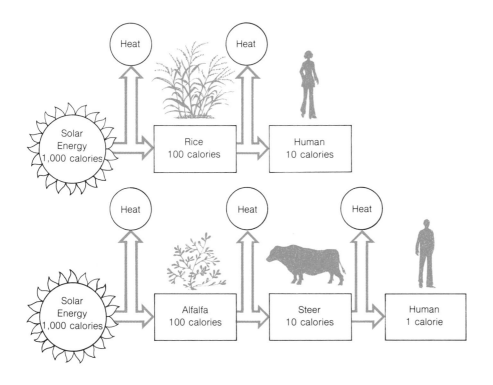

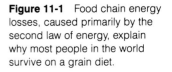

Figure 11-1 Food chain energy losses, caused primarily by the second law of energy, explain why most people in the world survive on a grain diet.

adding the animal link to food chains (Section 4-3 and Figure 11-1). That is, poor people can get more nourishment—more energy—per unit of money or labor from grain than from meat.

However, as their incomes rise, people begin to consume more grain *indirectly,* in the form of meat and meat products from domesticated animals. Because of the greater emphasis on meat-based diets in MDCs, almost half of the world's annual grain production and one-third of the fish catch are fed to livestock.

Major Types of Agriculture Two major types of agricultural systems are used to grow crops and raise livestock throughout the world: subsistence agriculture, practiced by two-thirds of the almost 4 billion people who live in small rural villages in today's LDCs, and industrialized agriculture, widely used in MDCs and spreading slowly to parts of some LDCs.

The goal of **subsistence agriculture** is to supplement solar energy with energy from human labor and draft animals to produce at least enough crops or livestock for survival and at most to have some left over to sell or put aside for hard times. About 60% of the world's cultivated land is still farmed by subsistence agriculture.

The goal of **industrialized agriculture** is to supplement solar energy with large amounts of energy derived from fossil fuels (especially oil and natural gas) to produce large quantities of crops and livestock for sale within the country where it is grown and to other countries. Yields per unit of land area are also increased by large inputs of commercial inorganic fertilizers, pesticides, and irrigation water. Figure 11-2 shows the relative inputs of land, human and animal labor, fossil fuel energy, and capital needed to produce one unit of food by modern industrialized agriculture and by the three major forms of subsistence agriculture.

U.S. Energy Use and Industrialized Agriculture The success of industrialized agriculture coupled with a favorable climate and fertile soils has been demonstrated by the dramatic increase in food production in the United States (Figure 11-3). Between 1820 and 1987, the percentage of the total U.S. population working on farms declined from about 72% to 2%, but during the same period, total food production doubled, and the output per farmer increased eightfold.

Although the number of farmers has declined drastically, to about 2 million (with only 650,000 full-time farmers), about 23 million people are involved in the U.S. agricultural system in activities ranging from growing and processing food to selling it at the supermarket. In terms of total annual sales, the agricultural system is the biggest industry in the United States—bigger than the automotive, steel, and housing industries combined.

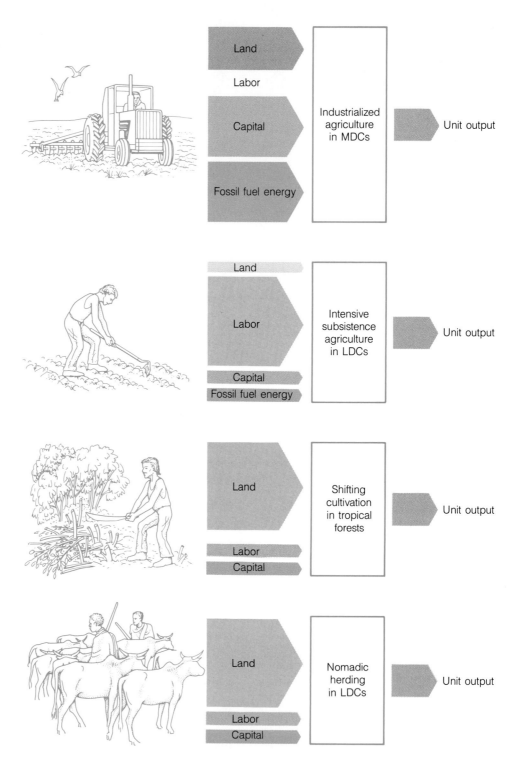

Figure 11-2 Relative inputs of major types of agricultural systems.

The agricultural system consumes about 17% of all commercial energy used in the United States each year (Figure 11-4). Most plant crops in the United States still provide more calories in food energy than the calories of energy (mostly from fossil fuels) used to grow them. But raising animals for food requires much more fossil fuel energy than the animals provide as food calories and protein.

The energy efficiency situation is much worse if we look at the entire food system in the United States. Counting fossil fuel energy inputs used to grow, store, process, package, transport, refrigerate, and cook all plant and animal food, *an average of about 10 calories of nonrenewable fossil fuel energy are needed to put 1 calorie of food energy on the table—an energy loss of 9 calories per calorie of food energy produced.*

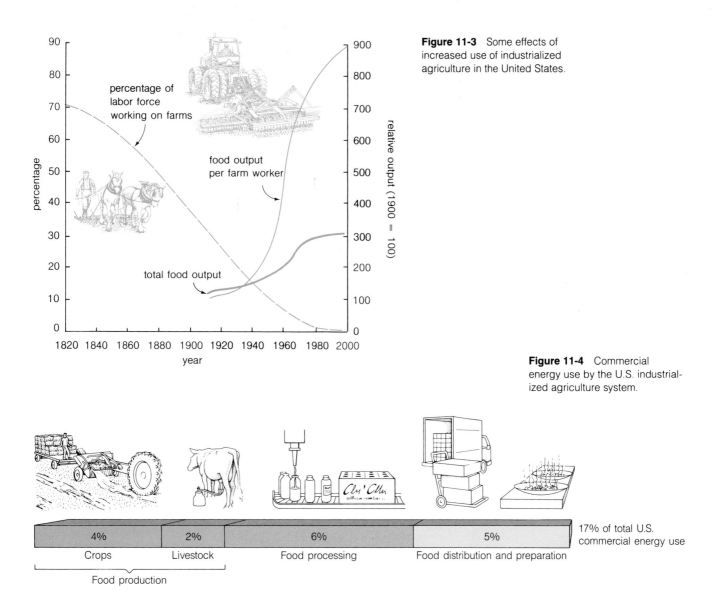

Figure 11-3 Some effects of increased use of industrialized agriculture in the United States.

Figure 11-4 Commercial energy use by the U.S. industrialized agriculture system.

| 4% | 2% | 6% | 5% | 17% of total U.S. commercial energy use |
| Crops | Livestock | Food processing | Food distribution and preparation | |

Food production

11-2 MAJOR WORLD FOOD PROBLEMS

Producing a large enough quantity of food to feed the world's population is only one of a number of complex and interrelated agricultural, economic, and environmental food problems. Other major problems are food quality, storage, and distribution, poverty (inability to grow or buy sufficient food regardless of availability), economic incentives for growing food, and the harmful environmental effects of agriculture.

Food Quantity: Population Growth and Food Production The good news is that thanks to improved agricultural technologies, practices, policies, and trade, world food production increased by 140% between 1950 and 1987 and kept ahead of the rate of population

on all continents except Africa. As a result, average food production per person increased by more than 25%, even though world population increased by nearly 2 billion. During the same period, food prices adjusted for inflation declined by 25%, and the amount of food traded on the world market quadrupled.

This combination of increased production and trade and declining real prices for food helped support a dramatic improvement in the average standard of living for many of the world's people. Perhaps the most spectacular success in food production has taken place in China, which manages to feed its people primarily by labor-intensive subsistence agriculture.

However, the worldwide increase in average food production per person since 1950 disguises the fact that average food production per person declined between 1950 and 1987 in 43 LDCs containing 1 out of every 7 people on earth. Furthermore, the rate of increase in world average food production per person

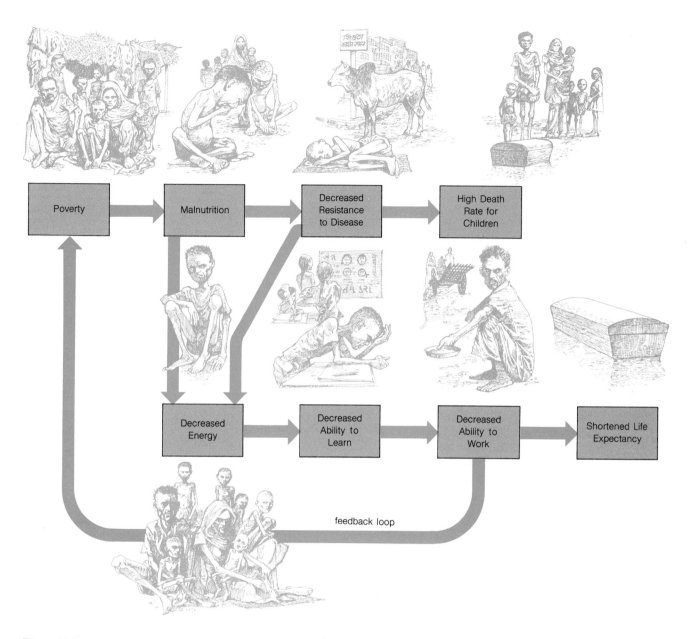

Figure 11-5 Interactions among poverty, malnutrition, and disease form a tragic cycle that perpetuates such conditions in succeeding generations of families.

has been steadily declining each decade, rising 15% between 1950 and 1960, 7% between 1960 and 1970, and only 4% between 1970 and 1980, a slowing trend that persists in the 1980s.

The largest declines have occurred in Africa, where average food production per person dropped 21% between 1960 and 1987 and is projected to drop another 30% during the next 25 years. Of Africa's 46 countries, 22 were facing catastrophic food shortages by 1985 (Figure 10-7, p. 162), and 1 out of 4 Africans were fed with grain imported from abroad—a dependence that is likely to increase.

This worsening situation in Africa is the result of a number of interacting factors: the fastest population growth rate of any continent, with 1 million more mouths to feed every 3 weeks; a 17-year drought; poor natural endowment of productive soils in many areas; overgrazing, deforestation, and extensive soil erosion and desertification (Figure 10-8, p. 163); and poor food distribution systems. Moreover, much of the best land is used for commercial production of plantation crops such as coffee and cacao, which are exported to well-fed countries. To prevent urban unrest, governments often keep food prices so low that farmers have little

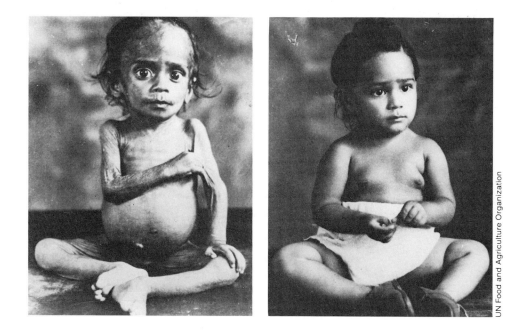

UN Food and Agriculture Organization

Figure 11-6 Most effects of severe marasmus can be corrected. This 2-year-old Venezuelan girl suffered from marasmus but recovered after 10 months of treatment and proper nutrition.

economic incentive to grow more crops. Frequent wars and growing dependence on food imports, both incurring rapidly rising foreign debts, result in severe underinvestment in agriculture and population control.

Undernutrition, Malnutrition, and Overnutrition Poor people living mostly on one or more plants such as wheat, rice, or corn often suffer from **undernutrition,** or insufficient caloric intake or food quantity. Survival and good health require that people must consume not only enough food but food containing the proper amounts of protein, carbohydrates, fats, vitamins, and minerals. People, such as those living mostly on one or more plants, whose diets are insufficient in these nutrients suffer from **malnutrition.**

Severe undernutrition and malnutrition lead to premature death, especially for children under age 5. Most severely undernourished and malnourished children, however, do not starve to death. Instead, about three-fourths of them die because their weakened condition makes them vulnerable to normally minor, nonfatal infections and diseases such as diarrhea, measles, and flu. The World Health Organization estimates that diarrhea kills at least 5 million children under age 5 a year.

Although the world currently produces more than enough food to feed everyone, each year 12 million to 20 million people die prematurely from undernutrition, malnutrition, or normally nonfatal diseases worsened by these conditions. This preventable death toll from hunger and hunger-related diseases is equivalent to the number who would die if between 94 and 157 fully loaded jumbo jets, half of the passengers children, crashed *each day* with no survivors. Because this condition of chronic undernutrition and malnutrition among the poor in LDCs is "normal" and spread out over much of the world, it often goes unnoticed and is not widely reported by the media.

Adults suffering from chronic undernutrition and malnutrition are vulnerable to infection and other diseases and are too weak to work productively or think clearly. As a result, their children also tend to be underfed and malnourished. If these children survive to adulthood, they are locked in a tragic malnutrition-poverty cycle that perpetuates these conditions in each succeeding generation (Figure 11-5).

Most world hunger is protein hunger, because poor people are forced to live on a low-protein, high-starch diet of grain. The two most widespread nutritional-deficiency diseases are marasmus and kwashiorkor. **Marasmus** (from the Greek "to waste away") occurs when a diet is low in both total energy (calories) and protein. Most victims of marasmus are infants in poor families where children are not breast-fed or where there is insufficient food after the children are weaned. A child suffering from marasmus typically has a bloated belly, thin body, shriveled skin, wide eyes, and an old-looking face (Figure 11-6). If treated in time with a balanced diet, however, most of these effects can be reversed.

Kwashiorkor (meaning "displaced child" in a West African dialect) occurs in infants and in children 1 to

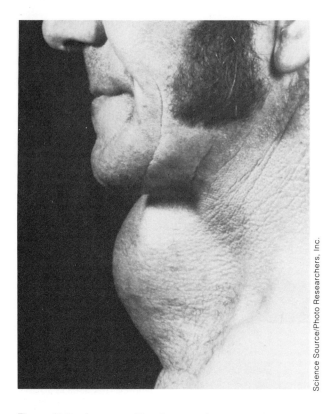

Figure 11-7 A person with goiter, an enlargement of the thyroid gland, caused by insufficient dietary iodine.

3 years old when, often because their mothers have a younger child to nurse, their diet changes from highly nutritious breast milk to one of grain or sweet potatoes, which is sufficient in calories but deficient in protein. If malnutrition is not prolonged, most of the effects can be cured with a balanced diet.

Without a daily intake of small amounts of vitamins that cannot be synthesized in the human body, various vitamin deficiency effects occur. Although a combination of balanced diets, vitamin-fortified foods, and vitamin supplements have greatly reduced the incidence of vitamin-deficiency diseases in MDCs, millions of cases occur each year in LDCs.

In many parts of Asia, people survive primarily on a diet of polished rice, made by removing the outer hulls. These individuals lack sufficient vitamin B_1, or thiamine (found in rice hulls), and often develop beriberi, which leads to stiffness of the limbs, enlargement of the heart, paralysis, pain, loss of appetite, and eventual deterioration of the nervous system. Each year at least 250,000 children are partially or totally blinded as a result of severe vitamin A deficiency.

Other nutritional-deficiency diseases are caused by the lack of certain minerals, such as iron and iodine. Too little iron can cause anemia, which saps one's energy, makes infection more likely, and increases a woman's chance of dying in childbirth. Iron-deficiency anemia affects about 10% of all adult men, a third of all adult women, and more than one-half of the children in tropical regions of Asia, Africa, and Latin America. Too little iodine can cause goiter, an abnormal enlargement of the thyroid gland in the neck (Figure 11-7). It affects as much as 80% of the population in the mountainous areas of Latin America, Asia, and Africa, where soils are deficient in iodine. Every year, iodine insufficiency also causes deafness or muteness in people in these areas.

This tragic loss of human life and life quality, especially in the world's children, could be prevented at relatively little cost. UNICEF officials estimate that between half and two-thirds of the annual childhood deaths from undernutrition, malnutrition, and associated diseases could be prevented at an average overall cost of only $5 to $10 per child. This program would involve a combination of the following simple measures:

- immunization against childhood diseases such as measles

- encouraging breast-feeding

- counteracting diarrhea with low-cost rehydration therapy in which infants drink a solution of a fistful of sugar and a pinch of salt in water

- preventing blindness by administering large doses of vitamin A twice a year (35 cents per dose)

- providing family planning services to help mothers space births at least two years apart

- increasing female education with emphasis on nutrition, sterilization of drinking water, and improved child care

While 15% of the people in LDCs suffer from undernutrition and malnutrition, about 15% of the people in MDCs suffer from **overnutrition,** which leads to obesity, or excess body fat. Although the causes of obesity are complex and not well understood, experts agree that a major cause is overeating—taking in food containing more energy than the body consumes. These overnourished people exist on diets high in calories, cholesterol-containing saturated fats, salt, sugar, and processed foods, and low in unprocessed fresh vegetables, fruits, and fiber. Partly as a result of these dietary choices, overweight people are at significantly higher than normal risk of diabetes, high blood pressure, stroke, and heart disease. Some elements of this diet are also associated with intestinal cancer, tooth decay, and other health problems.

Food Storage and Distribution Regardless of how much food is produced, much of it will rot or be consumed by pests unless a sophisticated system for storing, processing, transporting, and marketing it is employed. Because of inadequate food storage and distribution systems, food production in most LDCs is below its potential, and much of the food produced never reaches consumers.

Poverty: The Geography of Hunger If all the food currently produced in the world were divided equally among the earth's people, each person would receive more than three times the minimum amount needed to stay alive. Actually, the total amount of food produced today is more than enough to feed the 6.1 billion people anticipated by 2000, assuming that this food were distributed equally.

The world's supply of food, however, is not now distributed equally among the world's people, nor will it be. *Poverty—not lack of food production—is the chief cause of hunger and malnutrition throughout the world.* The world's desperately poor people do not have access to land where they can grow enough food of the right kind, and they do not have the money to buy enough food of the right kind, no matter how much is available.

The United Nations Food and Agriculture Organization and the World Bank estimate that 450 million to 800 million people are chronically undernourished and malnourished, most of them in LDCs. About one-third of the world's hungry live in India, even though it is self-sufficient in food production.

Increased worldwide total food production and average food production per person often mask widespread differences in food supply and quality between and within countries, and even within a particular family. For instance, although total and per person food supplies have increased in Latin America, much of this gain has been confined to Argentina and Brazil. In more fertile and urbanized southern Brazil, the average daily food supply per person is high, but in the semiarid, less fertile northeastern interior, many people are grossly underfed.

In MDCs, too, pockets of hunger exist. A 1985 report by a task force of doctors estimated that at least 20 million Americans—1 out of every 11—were hungry, mostly because of cuts in food stamps and other forms of government aid since 1980. Food is also inequitably distributed within families. Among the poor, children (ages 1 to 5), pregnant women, and nursing mothers are most likely to be underfed and malnourished, because the largest portion of the family food supply goes to working men.

Without a widespread increase in income and access to land, the number of chronically hungry and

Table 11-1 Environmental Effects of Food Production	
Effect	Text Discussion
Overfishing	Sections 11-5 and 13-5
Overgrazing	Section 12-6
Soil erosion and loss of soil fertility	Sections 9-3 and 9-4
Salinization and waterlogging of irrigated soils	Sections 10-3 and 11-4
Waterborne diseases from irrigation	Section 10-4
Loss of forests (deforestation)	Section 12-5
Endangered wildlife from loss of habitat	Sections 13-2 and 13-3
Loss of genetic diversity	Sections 11-3 and 13-1
Pollution from pesticides	Section 21-3
Water pollution from runoff of fertilizer and animal wastes	Chapter 19
Climate change from land clearing	Section 18-5
Air pollution from use of fossil fuels	Chapter 18
Health dangers from food additives	Section 11-7

malnourished people in the world could double to as many as 1.5 billion by 2000. Increasing the world's overall food production does little to solve this fundamental and preventable food problem for the world's poor.

Environmental Effects of Producing More Food Agricultural expert Lester R. Brown says, "The central question is no longer 'Can we produce enough food?' but 'What are the environmental consequences of attempting to do so?'" Industrialized agriculture can feed large numbers of people by using relatively little human labor to produce high yields on a relatively small percentage of the world's potential cropland. However, this form of agriculture probably has a greater overall environmental impact on the air, soil, and water than any other system in modern industrialized societies (Table 11-1). Severe environmental degradation also occurs when poor people, struggling for survival, farm highly erodible land on steep mountain slopes, do not allow cleared patches of tropical forests to lie fallow long enough to restore soil fertility, and allow their livestock to overgraze grasslands.

Figure 11-8 Scientists and two Indian farmers compare an older, full-size variety of rice (left) and a new, high-yield dwarf variety, grown in the second green revolution.

11-3 INCREASING CROP YIELDS AND USING NEW TYPES OF FOODS

Green Revolutions Most experts agree that the quickest and usually the cheapest way to grow more food is to raise the yield per unit of area of existing cropland. This is done by crossbreeding closely related wild and existing strains of plants to develop new, hybrid varieties that are better adapted to regional climate and soil conditions and can produce higher yields because they are able to make increased use of fertilizer, water, and pesticides. In countries where this method has succeeded in significantly increasing crop yields, the result has been called a **green revolution.**

Between 1950 and 1970, this approach led to dramatic increases in yields for most major crops in the United States and most other industrialized countries. In 1967, after 30 years of painstaking genetic research and trials, a modified version of the green revolution began spreading to many LDCs. New, high-yield, fast-growing dwarf varieties of rice and wheat, specially bred for tropical and subtropical climates, were introduced into LDCs such as Mexico, India, Pakistan, the Philippines, and Turkey.

The shorter, stronger, and stiffer stalks of the new varieties allow them to support larger heads of grain without toppling over (Figure 11-8). With high, properly timed inputs of fertilizer, water to keep the increased levels of fertilizer from killing the crops, and pesticides, wheat and rice yields from these new varieties can be two or three times higher than those from traditional varieties. With favorable growing conditions, overall yields per unit of land can be increased three to five times, because these fast-growing varieties allow farmers to grow two and even three crops a year (multiple cropping) on the same parcel of land.

Nearly 90% of the increase in world grain output in the 1960s and about 70% in the 1970s came from increased yields, mostly as a result of the second green revolution. It has been the major factor in allowing average food production per person to remain ahead of population growth in Asia.

In the 1980s and 1990s, at least 80% of the additional production of grains is expected to result from improved productivity of current cropland through the increased use of high-yield varieties, irrigation, fertilizer, pesticides, and multiple cropping.

Limitations of Green Revolutions Several factors can limit the spread and long-term success of present and future green revolutions. Without massive doses of fertilizer and water, the new crop varieties produce yields no higher and often lower than those from traditional grains. Thus areas without sufficient rainfall or irrigation potential cannot benefit from the new varieties; that is why the second green revolution has not spread to many areas.

A second problem is that in some countries, such as India, new varieties have increased overall food production at the expense of protein production. This occurred because the high-yield varieties of rice and wheat displaced legumes, an important source of protein in the Indian diet. In addition, because legumes naturally add nitrogen to the soil, their displacement impoverishes the soil, which requires additional inputs of expensive commercial fertilizer.

A third limitation is based on the concept of **diminishing returns,** in which a J-shaped curve of exponentially increasing crop productivity levels off and is converted to an S-shaped curve. Increased inputs of fertilizer, water, and pesticides cause yields to increase dramatically at first. But increasingly higher inputs eventually produce little or no increase in productivity, as has happened to yields of sorghum and corn in the United States. Once this point is reached, additional inputs merely add to the cost of crop production without increasing profits.

The diminishing-returns effect, however, typically takes 20 to 30 years to develop; thus yields in LDCs using existing green revolution varieties are projected to increase for some time. Geneticists also attempt to overcome this built-in biological limitation by developing even newer and improved varieties through crossbreeding.

Efforts are also being made to create new green revolutions much more rapidly by using genetic engineering and other biotechnology. Goals over the next 20 to 40 years include breeding new high-yield plant strains that have greater resistance to insects and disease, thrive on less fertilizer, make their own nitrogen fertilizer (for example, wheat with the ability of soybean to extract nitrogen from the air and convert it to nitrate fertilizer in its roots), do well in slightly salty soils, withstand periods of drought, and make more efficient use of solar energy during photosynthesis. If even a tiny fraction of this research is successful, the world could experience a new type of agricultural revolution in the early part of the next century based on rapid and enormous increases in crop and animal productivity. Some analysts, however, fear that such breakthroughs will be used primarily to make the rich richer instead of reducing poverty, the leading cause of world hunger, and will result in unexpected and harmful ecological effects, as previous green revolutions have.

Loss of Genetic Diversity   Another serious potential limitation of the two major green revolutions taking place in MDCs and some LDCs since 1950 is a loss of genetic diversity needed to develop future green revolutions. When fields of natural varieties are cleared and replaced with monocultures of crossbred varieties, much of the natural genetic diversity essential for developing new hybrids is lost forever. For example, a perennial variety of wild corn that, unlike cultivated corn, replants itself each year was recently dis-

covered. This variety is also resistant to a range of viruses and grows well on wet soils. However, the few thousand plants known to exist were found on a Mexican hillside that was in the process of being plowed up.

Genetic vulnerability was dramatically demonstrated in 1970, when most of the single-variety U.S. corn crop was wiped out by a blight-causing fungus. Seed companies quickly introduced a resistant seed variety and recovery was rapid, but the episode made farmers aware that genetic diversity is an essential form of insurance against disaster. Although monocultures predominate in MDCs, many of the world's 800 million farmers in LDCs plant several varieties of the same crops in their fields to prevent the total loss of a year's work.

To help preserve genetic variety, naturally growing native plants and strains of food crops throughout the world are being collected and maintained in 13 genetic storage banks and agricultural research centers located around the world.

Do the Poor Benefit?   Whether present and future green revolutions help reduce hunger among the world's poor depends on how they are used. In LDCs, the major resource available to agriculture is human labor. When green revolution techniques are applied to increase yields of labor-intensive subsistence agriculture in countries with equitable land distribution, the poor benefit, as has occurred in China.

However, when farmers in LDCs are encouraged to couple green revolution techniques with a shift from small-scale, labor-intensive subsistence cultivation to larger-scale, fossil-fuel-mechanized agriculture, the rural poor's ability to grow or buy sufficient food is reduced. The problem is that most of the benefits of the green revolution go to large landowners who have money or can obtain credit to buy the seed, fertilizer, irrigation water, pesticides, equipment, and fuel that the new techniques require. Meanwhile, tenant farmers suffer from rising land rents because of increased land value and are often forced off the land; farm laborers are displaced by the increased use of machinery. The net result is that many small landowners, farm laborers, and tenant farmers have been forced to migrate to cities in a desperate attempt to survive.

Coupling green revolution techniques with mechanized, industrialized agriculture also makes LDCs without their own supplies heavily dependent on large, MDC-based multinational companies for expensive supplies of seeds, fertilizer, farm machinery, and oil. This increases the LDCs' national debts, makes their agricultural and economic systems vulnerable to collapse from increases in oil and fertilizer prices, and

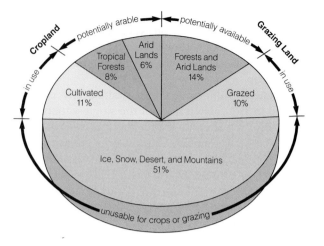

Figure 11-9 Classification of the earth's land. Theoretically, the world's cropland could be doubled in size by clearing tropical forests and irrigating arid lands. But converting this marginal land to cropland would destroy valuable forest resources, cause serious environmental problems, and usually not be cost-effective.

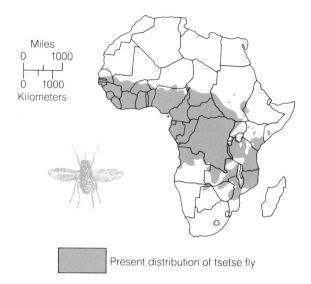

Present distribution of tsetse fly

Figure 11-10 Region of Africa infested by the tsetse fly.

reduces their rates of economic growth because their capital is diverted to pay for imported oil and other agricultural inputs.

Unconventional Foods Some analysts recommend greatly increased cultivation of various nontraditional plants in LDCs to supplement or replace traditional foods such as wheat, rice, and corn. Among dozens of other plants that could serve as important food sources is the winged bean, a protein-rich legume currently used extensively only in New Guinea and Southeast Asia. Its edible winged pods, leaves, tendrils, and seeds contain as much protein as soybeans, and its edible roots contain more than four times the protein of potato. Indeed, this plant yields so many different edible parts that it has been called "a supermarket on a stalk." The problem is getting farmers to cultivate such crops and convincing consumers to try new foods.

11-4 CULTIVATING MORE LAND

Availability of Arable Land Some agronomists have suggested that the world's cropland could be more than doubled by clearing tropical forests and irrigating arid lands mostly in Africa, South America, and Australia (Figure 11-9). Others believe such an expansion will not be achieved, because most of these arable, or potentially cultivatable, lands are too dry or remote or lack productive soils.

Some of the shortcomings of marginal agricultural land can be overcome by massive inputs of water and fertilizers, but doing this is expensive compared to increasing productivity on prime land already in use. Even if more cropland is developed, much of the increase will be used to offset the projected loss of almost one-third of today's cultivated cropland and rangeland from erosion, overgrazing, waterlogging, salinization, mining, and urbanization.

Location, Soil, and Insects as Limiting Factors About 83% of the world's potential new cropland is in the remote and lightly populated rain forests of the Amazon and Orinoco river basins in South America and in Africa's rain forests (see map inside front cover). Cultivating this land would require massive capital and energy investments to clear the land and to transport the harvested crops to populated areas. The resulting deforestation would greatly increase soil erosion and reduce the world's genetic diversity by eliminating vast numbers of the plant and animal species found in these incredibly diverse biomes.

Although the rain forests are blessed with plentiful rainfall and long or continuous growing seasons, the soils often are not suitable for intensive cultivation, because much of the plant nutrient supply is tied up in ground litter and vegetation rather than stored in the soil. Nearly 75% of the Amazon basin, roughly one-third of the world's potential cropland, consists of highly acidic and infertile soils. In addition, an estimated 5% to 15% of tropical soils (4% of those in the Amazon basin), if cleared, would bake under the trop-

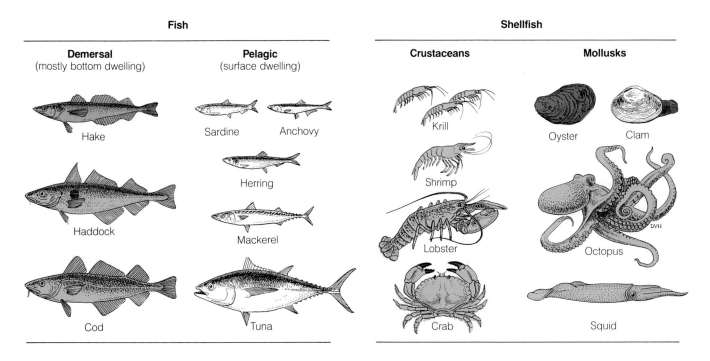

Fish		Shellfish	
Demersal (mostly bottom dwelling)	**Pelagic** (surface dwelling)	**Crustaceans**	**Mollusks**

Hake

Haddock

Cod

Sardine   Anchovy

Herring

Mackerel

Tuna

Krill

Shrimp

Lobster

Crab

Oyster   Clam

Octopus

Squid

Figure 11-11 Major types of commercially harvested fish and shellfish.

ical sun into brick-hard surfaces called laterites, useless for farming. Some scientists argue that agriculture in the tropics should be limited to plantation cultivation of trees adapted to the existing climates and soils, such as rubber trees, oil palms, and banana trees.

In Africa, potential cropland larger in area than the United States cannot be used for grazing or farming because it is infested by 22 species of the tsetse fly, whose bite can give both people and livestock incurable sleeping sickness (Figure 11-10). A $120 million eradication program has been proposed, but many scientists doubt it can succeed.

Water as a Limiting Factor More than half the remaining arable land lies in dry areas, where water shortages limit crop growth. This can be overcome with irrigation, but large-scale irrigation in these areas would be very expensive (typically $1,000 to $2,000 per hectare), require large inputs of fossil fuel to pump water long distances, deplete many groundwater supplies, and require constant and expensive maintenance to prevent seepage, salinization, and waterlogging. Unfortunately, Africa, the continent that needs irrigation the most, has the lowest potential for it because of the remote location of its major rivers and unfavorable topography and rainfall patterns.

Money as a Limiting Factor According to agriculture expert Lester Brown, "The people who are talking about cultivating more land are not considering the cost. If you are willing to pay the cost, you can farm the slope of Mount Everest." Thus the real questions are How much will it cost to increase the total amount of cropland in the world, and how will this cost affect the ability of poor people to pay for the additional food grown on this land?

11-5 CATCHING MORE FISH AND FISH FARMING

The World's Fisheries Fish and shellfish supply about 6% of all protein consumed by human beings and 24% of the animal protein worldwide—considerably more than beef, twice as much as eggs, and three times as much as poultry. Fish and shellfish are the major source of animal protein, iron, and iodine for more than half the world's people, especially in Asia and Africa.

About 91% of the annual commercial catch of fish and shellfish comes from the ocean and 9% from fresh water. About 99% of the world marine catch is taken within 370 kilometers (200 nautical miles) of the coast. Four main groups of marine species make up the bulk of the annual commercial catch (Figure 11-11). About 70% of the annual marine catch is eaten by people. The remainder, consisting largely of herringlike fish unacceptable for human consumption, is processed

Figure 11-12 Energy input needed to produce one unit of food energy for commercially desirable types of fish and shell-fish.

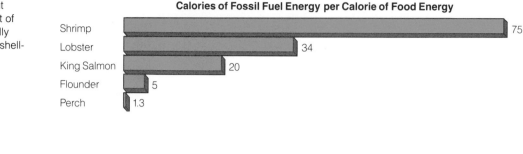

Calories of Fossil Fuel Energy per Calorie of Food Energy

Shrimp 75
Lobster 34
King Salmon 20
Flounder 5
Perch 1.3

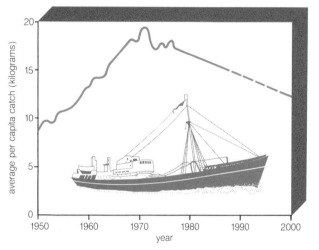

Figure 11-13 Average world fish catch per person declined in most years since 1970 and is projected to decline further between 1985 and 2000. (Data from United Nations and World-watch Institute)

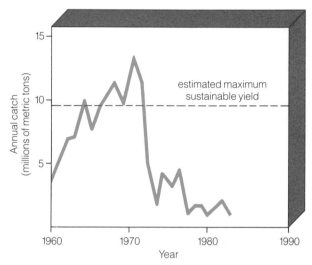

Figure 11-14 Collapse of the Peruvian anchovy catch as a result of a shift in an ocean current and gross overfishing. (Data from UN Food and Agriculture Organization)

into fish meal, most of which is fed to poultry, pigs, and cattle.

To achieve large catches, modern "distant-water" fishing fleets use sonar (bouncing high-frequency sound waves off solid objects), helicopters, aerial photography, and temperature measurement to locate schools of fish. They use lights and electrodes to attract the fish, and fine nets to "vacuum" the sea. Enormous, floating-fish-factory trawlers follow the fleets to process and freeze the catch. About 75% of the world's fish catch is taken by such large-scale operations, but small-scale fishing boats account for 40% of the fish consumed by people.

Using small and large boats to catch fish and shell-fish is essentially a hunting-and-gathering procedure taking place over a large area. Because fuel makes up 30% to 40% of the operating costs of fishing boats, energy inputs for each unit of food energy caught for most species are enormous (Figure 11-12).

Trends in the World Fish Catch Between 1950 and 1970, large-scale distant-water fishing fleets were a major factor in more than tripling the annual catch

from 21 million to 70 million metric tons (23 million to 77 million tons)—an increase greater than that of any other human food source during the same period. Between 1971 and 1976, however, the annual catch leveled off and rose only slightly by 1987. Meanwhile, world population continued to grow, so between 1970 and 1987 the average fish catch per person declined, and is projected to decline even further—back to the 1960 level—by 2000 (Figure 11-13).

This leveling off of the annual catch has resulted from a combination of natural oceanographic changes, overfishing, and pollution and destruction of estuaries and offshore areas where most commercial fish and shellfish are caught. **Overfishing** occurs when so many fish, especially immature ones, are taken that not enough breeding stock is left for adequate annual renewal. Surveys by the UN Food and Agriculture Organization indicated that by 1985 overfishing was the major factor in depleting stocks of 12 major species of commercially valuable fish to the point where hunting and gathering them is no longer profitable.

Between 1971 and 1978, for example, the Peruvian anchovy became commercially extinct from a combination of gross overfishing and the shifting away from

Figure 11-15 Harvesting of silver carp at an aquaculture farm near Chang-Chow, China.

UN Food and Agriculture Organization/F. Mattioli

shore of a cool current that brings nutrients up from the bottom (Figure 5-11, p. 82). Since then, the annual yield from this fishery, which previously accounted for 20% of the annual global harvest, has remained at a low level (Figure 11-14). Peruvian fishery officials had not heeded warnings by UN Food and Agriculture Organization biologists that during seven of the eight years between 1964 and 1971, the anchovy harvest had exceeded the estimated sustainable yield.

Fish Farming: Aquaculture If we have trouble catching more fish and shellfish at affordable prices using our present hunting-and-gathering approach, why not raise and harvest fish crops in land-based ponds or fenced-in coastal lagoons and estuaries? This approach, known as **aquaculture,** or fish farming, can produce much higher yields per unit of land area than conventional fishing. It also uses a relatively low amount of fuel and thus is not dependent on the price of oil, which could make most conventional commercial fishing prohibitively expensive if oil prices rise sharply as projected some time after 1995.

Aquaculture, which may have originated in China more than 4,000 years ago, supplies about 12% of the total world catch and is a major source of protein for the poor in many LDCs, especially in Asia. In LDCs, aquaculture operations usually involve fertilizing small ponds (usually less than 1 hectare) with animal wastes, fish wastes, or commercial fertilizer to produce phytoplankton. These are eaten by zooplankton and bottom animals, which in turn are eaten by fish such as carp, which grow rapidly and can be easily harvested with nets when they reach a desired size (Figure 11-15). One problem, however, is that the fish can be killed by pesticide runoff from nearby croplands, as has happened in aquaculture ponds in the Philippines, Indonesia, and Malaysia.

Very high yields can be obtained by feeding fish directly with grain or grain by-products supplemented with high-quality protein such as soy meal or fish meal from less valuable fish. Nutrient-rich estuaries can also be farmed to produce large yields of desirable marine species such as shrimp, lobster, oysters, and salmon in fenced-off bays, large tanks, or floating cages. In the United States, most of the catfish and crawfish, nearly all rainbow trout, and 40% of the oysters are harvested from fish farms. Aquaculture in MDCs, however, is designed to produce expensive fish and shellfish species for consumption by the affluent. This may be highly profitable, but contributes little to increasing food and protein supplies for the poor.

Can the Annual Catch Be Increased Significantly? Some scientists believe that between 1987 and 2000, the world's annual sustainable fish and shellfish catch could be increased by 15% and perhaps by 100% by a combination of methods. First, it is hoped that the signing of the 1982 United Nations Convention on the Law of the Sea by 159 countries will be effective in regulating overfishing in 99% of the world's prime fishing grounds. This treaty recognizes that all coastal countries have the right to control the amount of fishing allowed by their own fishing fleets and by foreign ships within 364 kilometers (200 nautical miles) of their coasts. However, the Law of the Sea will benefit only a small number of LDCs (mostly on the western coasts of Africa and Latin America), because two-thirds of the total value of the ocean's fish catch has been taken off the coasts of MDCs.

Additional increases could also be brought about by a sharp decrease in the one-fifth of the annual catch now wasted, mainly from throwing back potentially useful fish taken along with desired species, lack of adequate refrigerated storage to prevent spoilage, and inefficient processing. The world fish catch can also be expanded by increasing the harvest of unconventional species such as Antarctic krill and several species of squid. Some marine biologists, however, believe that the krill catch will not increase significantly because of bad weather in Antarctic waters, the necessity to process the catch within three hours if it is to be consumed by people, and the high energy and economic costs of harvesting in distant waters. Some scientists also warn that extensive annual harvesting of krill could lead to declines in the populations of krill eaters, including several species of whale, such as the already threatened blue whale, and several species of fish, seabirds, seals, and penguins. Perhaps the greatest problem is making krill palatable to human consumers.

It is also projected that annual freshwater and saltwater aquaculture production could be increased more than threefold between 1986 and 2000. The Food and Agriculture Organization estimates that only one-tenth of the land suitable for aquaculture, including low-lying, flood-prone land not suitable for crop production, is currently being used. Other fishery experts believe that the world's annual marine fish catch may already be at or near its maximum sustainable yield because of overfishing and pollution of estuaries.

11-6 MAKING FOOD PRODUCTION PROFITABLE, PROVIDING FOOD AID, AND DISTRIBUTING LAND TO THE POOR

Government Agricultural Policies Governments can influence crop and livestock prices, and thus the supply of food, by (1) keeping food prices artificially low, making consumers happy but decreasing profits for farmers; (2) providing farmers with subsidies to keep them in business; or (3) providing no price controls or subsidies and allowing free market competition to determine food prices.

Many governments in LDCs keep food prices in cities deliberately low to prevent political unrest. But this draws more poor people from rural areas to cities, further aggravating urban problems and unemploy-

ment. Because prices are too low for them to make a decent living, rural farmers do not produce enough food to feed the country's population. High food export taxes can also discourage farmers from growing food for export because of lack of sufficient profit.

Government price supports, cash subsidies, and import restrictions can be used to stimulate crop and livestock production by guaranteeing farmers a certain minimum yearly return on their investment. For example, despite population problems, the average food supply per person in India has been rising for more than two decades. One reason is the green revolution, but the main reason is removal of government price controls, which were keeping wheat prices at artificially low levels, and their replacement with price supports, which made the price of domestic wheat the same as the import price. With this economic incentive, Indian farmers began producing much more wheat.

Government price supports and other subsidies make farmers happy. They are also popular with most consumers, because subsidies make food prices seem low. What most consumers don't realize is that they are really paying higher prices for their food indirectly in the form of higher taxes to provide the subsidies. There is no free lunch.

The U.S. Farm Situation If government price supports are too generous and the weather is good, farmers may produce more food than can be sold. Food prices and profits then drop because of the oversupply. Unless even higher government subsidies are provided to prop up farm income by buying unsold crops or paying farmers to idle some of their land, a number of debt-ridden farmers go bankrupt, as has occurred in the United States in the 1980s.

Between 1980 and 1987, several hundred thousand part-time and full-time U.S. farmers quit farming or went bankrupt. Many farm-oriented businesses, including those providing farmers with machinery and other supplies, and general businesses in farming towns also suffered severe economic losses. By 1986 about one-fourth of the country's 650,000 full-time farmers, who produce about 90% of the nation's food, were having trouble paying off their loans.

The farm-debt crisis has its roots in the 1960s and 1970s, when the government encouraged farmers to grow food to help fight world hunger and to increase food exports to help offset the mounting bill for imported oil after the 1973 OPEC oil embargo. Most farmers responded by expanding their operations, and many borrowed heavily to buy additional land and equipment. Their heavy borrowing was backed by the rapidly escalating value of their cropland.

During the first half of the 1980s, however, the bottom dropped out for farmers who had borrowed heavily and gambled on increased food demand to keep cropland values and crop prices high. Between 1981 and 1987, the average value per acre of farmland dropped by 33% (by 40% to 55% in some midwestern states), net farm income and exports fluctuated, and tax-supported federal subsidies to farmers increased to buy up surpluses and keep production down to prevent even more farmers from going bankrupt. Two major factors contributed to the decrease in land value. First, overproduction reduced crop prices. Second, U.S. food exports declined in most years because of increases in production in many countries as a result of the green revolution, inability of many debt-ridden LDCs to buy the food they needed, and the ability of other countries to sell surplus crops for less because a stronger dollar and federal price supports made U.S. crops more expensive in the world market.

The overriding problem is that the U.S. agricultural system is too successful for its own good. A number of analysts believe that the only way out of this dilemma is to gradually wean U.S. farmers from all federal subsidies and let them respond to market demand—a politically difficult process.

The first step would be to ensure that federal subsidies go only to farmers in economic trouble. At present, nearly half of all subsidies go to the country's largest and most successful farms. In 1986 the crown prince of Lichtenstein, a Texas landowner, received more than $2 million in federal aid intended for struggling farmers. By law no farmer is supposed to get more than $50,000 a year in federal subsidies. But many big farmers have gotten around this by dividing the ownership of their acreage among several people, usually relatives, each eligible for the maximum subsidy.

Next the government would phase out farm subsidies over several years. Some of the money saved would be used to help needy farmers pay off part of their debts, provided they practice an approved soil and water conservation program. Once all federal subsidies were eliminated, all farmers would respond to the demands of the market so that only those who were good farmers and financial managers would remain in business. Reducing the debt of currently overextended farmers between 1986 and 1991 would cost taxpayers about $75 billion, less than the $80 billion to $100 billion in farm subsidies projected for the same period.

Phasing out farm subsidies, however, is difficult to accomplish because of the political influence of large corporate farmers, whose profits are higher and risks are lower because of such subsidies. In addition, many small- and medium-size farmers have become dependent on federal subsidies that now provide one-third to one-half of their income regardless of how much they grow. These farmers and people dependent on them—small-town bankers and businesses, food processors, farm machinery dealers, fertilizer makers, and the like—dominate the economies of their states, which would elect members of Congress opposed to such a program.

International Relief Aid Since 1945, in terms of total dollars expended per year, the United States has been—and continues to be—the world's largest donor of nonmilitary foreign aid. Aid donated directly to countries (bilateral aid) or to international institutions such as the World Bank for distribution to other countries (multilateral aid) is used primarily for agriculture and rural development, food relief, population planning, health, and economic development projects. In addition to helping other countries, foreign aid stimulates U.S. economic growth and provides Americans with jobs. Seventy cents of every dollar the United States gives directly to other countries is used to purchase American goods and services. Despite the humanitarian benefits and economic returns of such aid, the percentage of the U.S. gross national product allocated for nonmilitary foreign aid has declined from a high of 1.6% in the 1950s to only 0.25% in 1986—an average of only $34 per American.

Private charity organizations such as CARE and Catholic Relief Services provide at least $2 billion a year of additional foreign aid. In 1985 many of the world's popular musical performers used benefit concerts and record sales to provide aid to nearly bankrupt farmers in the United States and starving people in Ethiopia and other parts of Africa. Some people call for greatly increased food relief aid from government and private sources, while others question the value of such aid (see Spotlight on p. 192).

Distributing Land to the Poor An important step in reducing world hunger and malnutrition is land reform, by which the landless rural poor in LDCs are given access to land to produce enough food for their survival. Such reform would increase agricultural productivity, because small, labor-intensive farms produce more per unit of land area than larger fossil-fuel-intensive farms. Moreover, land reform would reduce the flow of landless poor to overcrowded urban areas and create employment in rural areas. China and Taiwan have had the most successful land reforms. Unfortunately, land reform is difficult to institute in countries where government leaders are unduly influenced by wealthy and powerful landowners.

Most people view food relief as a humanitarian effort to prevent people from dying prematurely from lack of sufficient food. However, some analysts contend that providing food to starving people in countries where population growth rates are high does more harm than good in the long run, condemning even greater numbers to premature death in the future. Massive food aid can also depress local food prices, decrease food production, stimulate mass migration from farms to already overcrowded cities, and decrease a country's long-term ability to provide food for its people.

Another problem is that much food aid does not reach hunger victims: Transportation networks and storage facilities are inadequate, so that some of the food rots or is devoured by pests before it can reach the hungry, and theft by officials who sell the food for personal profit or use it for political gain is rampant.

Critics of food relief are not against foreign aid. Instead, they believe that such aid should be concentrated on efforts to control population growth and on helping LDCs become self-sufficient in growing their own food and in developing resource-efficient forms of economic growth that will help them compete in the world marketplace without excessive dependence on MDCs. They believe that such aid should be provided to countries committed to effective plans for controlling population growth and to equitable distribution of land. What do you think?

11-7 FOOD ADDITIVES

Use and Types of Food Additives   In LDCs, many rural and urban dwellers consume harvested crops directly. In MDCs and in a growing number of cities in LDCs, harvested crops are used to produce processed foods for sale in grocery stores and restaurants. A large and increasing number of natural and synthetic chemicals, called **food additives,** are deliberately added to such processed foods to retard spoilage, to enhance flavor, color, and texture, and to provide missing amino acids and vitamins. Although some food additives are useful in extending shelf life and preventing food poisoning, most are added to improve appearance and sales. For example, the following letter to the editor of the Albany *Times-Union* lists only a few of the 93 chemicals that may be added to the "enriched" bread you buy in a grocery store:

Give us this day our daily calcium proprianate (spoilage retarder), sodium diacetate (mold inhibitor), monoglyceride (emulsifier), potassium bromate (maturing agent), calcium phosphate monobasic (dough conditioner), chloramine T (flour bleach), aluminum potassium sulfate acid (baking powder ingredient), sodium benzoate (preservative), butylated hydroxyanisole (antioxidant), mono-isopropyl citrate (sequestrant); plus synthetic vitamins A and D.

Forgive us, O Lord, for calling this stuff BREAD.

J. H. Read, Averill Park*

_____
*Used by permission of the *Times-Union*, Albany, New York.

All food, of course, is a mixture of chemicals, but today at least 2,800 chemicals are deliberately added to processed foods in the United States. Each year, the average American consumes about 55 kilograms (120 pounds) of sugar, 7 kilograms (15 pounds) of salt, and about 4.5 kilograms (10 pounds) of other food additives. Each day, the average American eats 1 teaspoon of artificial colors, flavors, and preservatives. Table 11-2 summarizes the major classes of food additives. The most widely used groups of additives—coloring agents, natural and synthetic flavoring agents, and sweeteners—are used solely to make food look and taste better.

The extremes of the controversy over food additives range from "Essentially all food additives are bad" and "We should eat only natural foods" to "There's nothing to worry about, because there is no absolute proof that chemical X has ever harmed a human being." As usual, the truth probably lies somewhere in between.

Natural versus Synthetic Foods   The presence of synthetic chemical additives does not necessarily mean that a food is harmful, and the fact that a food is completely natural is no guarantee that it is safe. A number of natural or totally unprocessed foods contain potentially harmful and toxic substances.

Polar bear or halibut liver can cause vitamin A poisoning. Lima beans, sweet potatoes, cassava, sugarcane, cherries, plums, and apricots contain chemicals (glucosides) that our intestines convert to small

Table 11-2 Commonly Used Food Additives and Food Processes

Class	Function	Examples	Foods Typically Treated
Preservatives	To retard spoilage caused by bacterial action and molds (fungi)	Processes: drying, smoking, curing, canning (heating and sealing), freezing, pasteurization, refrigeration Chemicals: salt, sugar, sodium nitrate, sodium nitrite, calcium and sodium propionate, sorbic acid, potassium sorbate, benzoic acid, sodium benzoate, citric acid, sulfur dioxide	Bread, cheese, cake, jelly, chocolate syrup, fruit, vegetables, meat
Antioxidants (oxygen interceptors, or freshness stabilizers)	To retard spoilage of fats (excludes oxygen or slows down the chemical breakdown of fats)	Processes: sealing cans, wrapping, refrigeration Chemicals: lecithin, butylated hydroxyanisole (BHA), butylated hydroxytoluene (BHT), propyl gallate	Cooking oil, shortening, cereal, potato chips, crackers, salted nuts, soup, toaster tarts, artificial whipped topping, artificial orange juice, many other foods
Nutritional supplements	To increase nutritive value of natural food or to replace nutrients lost in food processing*	Vitamins, essential amino acids	Bread and flour (vitamins and amino acids), milk (vitamin D), rice (vitamin B_1), corn meal, cereal
Flavoring agents	To add or enhance flavor	Over 1,700 substances, including saccharin, aspartame (NutraSweet®), monosodium glutamate (MSG), essential oils (such as cinnamon, banana, vanilla)	Ice cream, artificial fruit juice, toppings, soft drinks, candy, pickles, salad dressing, spicy meats, low-calorie foods and drinks, most processed heat-and-serve foods
Coloring agents	To add aesthetic or sales appeal, to hide colors that are unappealing or that show lack of freshness	Natural color dyes, synthetic coal tar dyes	Soft drinks, butter, cheese, ice cream, cereal, candy, cake mix, sausage, pudding, many other foods
Acidulants	To provide a tart taste or to mask undesirable aftertastes	Phosphoric acid, citric acid, fumaric acid	Cola and fruit soft drinks, desserts, fruit juice, cheese, salad dressing, gravy, soup
Alkalis	To reduce natural acidity	Sodium carbonate, sodium bicarbonate	Canned peas, wine, olives, coconut cream pie, chocolate eclairs
Emulsifiers	To disperse droplets of one liquid (such as oil) in another liquid (such as water)	Lecithin, propylene glycol, mono- and diglycerides, polysorbates	Ice cream, candy, margarine, icing, nondairy creamer, dessert topping, mayonnaise, salad dressing, shortening
Stabilizers and thickeners	To provide smooth texture and consistency; to prevent separation of components; to provide body	Vegetable gum (gum arabic), sodium carboxymethyl cellulose, seaweed extracts (agar and algin), dextrin, gelatin	Cheese spread, ice cream, sherbet, pie filling, salad dressing, icing, dietetic canned fruit, cake and dessert mixes, syrup, pressurized whipped cream, instant breakfasts, beer, soft drinks, diet drinks
Sequestrants (chelating agents, or metal scavengers)	To tie up traces of metal ions that catalyze oxidation and other spoilage reactions in food; to prevent clouding in soft drinks; to add color, flavor, and texture	EDTA (ethylenediaminetetraacetic acid), citric acid, sodium phosphate, chlorophyll	Soup, desserts, artificial fruit drinks, salad dressing, canned corn and shrimp, soft drinks, beer, cheese, frozen foods

*Adding small amounts of vitamins to breakfast cereals and other "fortified" and "enriched" foods in America is basically a gimmick used to raise the price. The manufacturer may put vitamins worth about 5¢ into 340 grams (12 ounces) of cereal and then add 45% to the retail price. Vitamin pills are normally a far less expensive source of vitamins than fortified foods. The best way to get vitamins, however, is through a balanced diet.

In 1958 an amendment, known as the Delaney clause, was added to U.S. food and drug laws. It was named after Representative James J. Delaney of New York, who fought to have this amendment passed despite great political pressure and heavy lobbying by the food industry. This amendment prohibits the deliberate use of any food additive shown to cause cancer in test animals or people. The amendment is absolute, allowing for no extenuating circumstances or consideration of benefits versus risks. Between 1958 and 1987, the FDA used this amendment to ban only nine chemicals.

Critics say the Delaney clause is too rigid and not needed because the FDA already has the power to ban any chemical it deems unsafe. In general, the food industry would like to see the amendment removed, and some scientists and politicians would like it to be modified to allow a consideration of benefits versus risks. Others point out that it requires chemicals to be banned even when the dosage causing cancer in test animals is 10 to 10,000 times greater than the amount a person might be expected to consume. These critics also argue that cancer tests in animals don't necessarily apply to human beings.

Supporters of the Delaney clause say that because people can't serve as guinea pigs, animal tests are the next best thing. Such tests don't prove that a chemical will cause cancer in human beings, but they do strongly suggest that a risk is present. Moreover, all but two substances known to cause cancer in people also cause cancer in laboratory animals. Supporters also argue that the high doses of chemicals are necessary to compensate for the test animals' relatively short life spans and relatively fast metabolism and excretion rates. Tests using low doses not only would be inaccurate but would require thousands of test animals to establish that an effect was not the result of chance. Such tests would be prohibitively expensive.

It is also argued that because the clause is absolute, it protects FDA officials from undue pressure from the food industry and politicians. If the FDA had to weigh benefits versus risks, political influence and lobbying by the food industry could delay banning a dangerous chemical while it underwent years of study.

Indeed, instead of revoking the Delaney clause, some scientists feel it should be strengthened and expanded. Some even argue that the clause gives the FDA too much discretion, including the right to reject the validity of well-conducted animal experiments that show carcinogenicity. These critics cite the FDA's infrequent use of the clause as evidence that the law is too weak. What do you think?

amounts of deadly hydrogen cyanide. Eating cabbage, cauliflower, turnips, rutabagas, mustard greens, collard greens, or brussels sprouts can cause goiter in susceptible individuals. Blood pressure can be raised by certain chemicals (amines) found in bananas, pineapples, various acid cheeses (such as Camembert), and some beers and wines. Chemicals (aflatoxins) produced by fungi sometimes found on corn and peanuts are extremely toxic to human beings and are not legal in U.S. foods at levels above 20 ppb. Clams, oysters, cockles, and mussels can concentrate natural and artificial toxins in their flesh.

In addition, natural foods can be contaminated with food-poisoning bacteria, such as *Salmonella* and the deadly *Clostridium botulinum*, through improper processing, food storage, or personal hygiene. Each year, more than 20 million Americans come down with food poisoning and 9,000 of these people die. The botulism toxin from *Clostridium botulinum* is one of the most toxic chemicals known. As little as one ten-millionth of a gram (0.0000001 gram) can kill an adult, and it is estimated that 227 grams (half a pound) would be enough to kill every human being on earth. However, because of modern food processing methods, only about 10 to 20 cases of botulism occur annually in the United States. An increasing number of food-poisoning cases—more than a quarter-million each year—are caused by germs that have become genetically resistant to antibiotics and are found in meat from cows, pigs, and chickens raised on antibiotic-laden feed.

Consumer Protection: FDA and the GRAS List   In the United States, the safety of foods and drugs has been monitored by the Food and Drug Administration (FDA) since its establishment by the Pure Food and Drug Act of 1906. However, it was not until 1958 that federal laws required that the safety of any new food additive be established by the manufacturer and approved by the FDA before the additive was put into common use. Today the manufacturer of a new addi-

Table 11-3 Suggested Food Additives to Avoid*

Additive	Major Uses	Possible Problems
Coal tar dyes (reds no. 3, 8, 9, 19, and 37, and orange no. 17)	Cherries in fruit cocktail, candy, beverages	May cause cancer; poorly tested
Citrus red no. 2	Skin of some Florida oranges	May cause cancer
Yellow no. 5	Gelatin dessert, candy, baked goods	May cause cancer; allergic reaction in some people; widely used
BHA	Antioxidant in chewing gum, potato chips, oils	May cause cancer; stored in body fat; can cause allergic reaction; safer alternatives available
BHT	Same as BHA	Appears safer than BHA, but needs better testing; safer alternatives available
Propyl gallate	Antioxidant in oils, meat products, potato stock, chicken soup base, chewing gum	Not adequately tested; use is frequently unnecessary
Quinine	Flavoring in tonic water, quinine water, bitter lemon	Poorly tested; may cause birth defects
Saccharin	Noncaloric sweetener in diet foods	Causes cancer in animals
Sodium nitrite, sodium nitrate	Preservative and flavoring agent in bacon, ham, hot dogs, luncheon meats, corned beef, smoked fish	Prevents formation of botulism bacteria but can lead to formation of cancer-causing nitrosamines in stomach
Sodium bisulfite, sulfur dioxide	Preservative and bleach in wine, beer, grape juice, dehydrated potatoes, imported shrimp, dried fruit, cake and cookie mixes, canned and frozen vegetables, breads, salad dressings, fruit juices, soft drinks, some baked goods and snacks, some drugs	Causes severe allergic reaction in about 500,000 Americans; implicated in 12 deaths between 1982 and 1986

*Data from Center for Science in the Public Interest.

tive must carry out extensive toxicity testing, costing up to a million dollars per item, and the results must be submitted to the FDA. The FDA itself does no testing but merely evaluates data submitted by manufacturers. The FDA can also use the Delaney clause to ban any food additives shown to cause cancer in test animals (see Spotlight on p. 194).

However, these federal laws did not apply to the hundreds of additives in use before 1958. Instead of making expensive, time-consuming tests, the FDA drew up a list of the food additives in use in 1958 and asked several hundred experts for their professional opinions on the safety of these substances. A few substances were deleted, and in 1959 a list of the remaining 415 substances was published as the "generally recognized as safe," or **GRAS** (pronounced "grass"), **list.** Since 1959, further testing has led the FDA to ban several substances on the original GRAS list.

As a regulatory agency, the FDA is caught in the crossfire between consumer groups and the food industry. It is criticized by consumer groups as being overly friendly to industry and for hiring many of its executives from the food industry—a practice the FDA contends is the only way it can recruit the most experienced food scientists. At the same time, the food industry complains that the FDA sometimes gives in too easily to demands from consumer groups. Both industry and consumer groups have criticized the agency for bureaucratic inefficiency.

What Can the Consumer Do? Avoiding all food additives is almost impossible for a consumer in an affluent country. Indeed, as mentioned, many additives perform important functions, and there is no guarantee that natural foods will always be better and safer. However, to minimize risk, individuals can follow a prudent diet and avoid or use with caution those additives that have come under suspicion, according to the Center for Science in the Public Interest (Table 11-3).

1. *Heavy reliance on human labor,* one of the most abundant resources in LDCs.

2. *Heavy reliance on organic fertilizers from crop and animal wastes* rather than expensive, fossil-fuel-based, commercial inorganic fertilizers.

3. *Minimal use of nonrenewable fossil fuel energy* to reduce pollution and dependence on expensive oil imports.

4. *Emphasis on conservation of soil, soil nutrients, and water* to enhance short- and long-term crop productivity and reduce water pollution.

5. *Emphasis on cultivation of perennial crops that replant themselves* rather than annual crops whose seeds must be purchased each year.

6. *Emphasis on polyculture,* in which several varieties of the same crop or several different crops are planted in the same field, to reduce vulnerability to crop losses from pests and diseases.

7. *Emphasis on biological rather than chemical control of pests and weeds* to reduce water pollution, dependence on expensive fossil-fuel-based chemicals, and genetic resistance of pest species (Section 21-5).

8. *Land reform* giving the rural poor access to enough land to grow their own food and produce a surplus for sale.

11-8 SUSTAINABLE-EARTH AGRICULTURE

Sustainable-Earth Agriculture in LDCs Cornucopians believe that the key to reducing world hunger is to transfer highly mechanized industrialized agriculture to LDCs. Neo-Malthusians, however, say that this approach will not end hunger among the poor, because it will make LDCs increasingly dependent on MDCs, allow the rich in MDCs and LDCs to get richer at the expense of the poor, and greatly increase pollution and environmental degradation. They believe that the key to reducing world hunger and the harmful environmental impacts of agriculture is for LDCs to become self-sufficient in food production by developing a **sustainable-earth agricultural system** (see Spotlight above). China, the world's leader in sustainable-earth agriculture, has shown that this approach can essentially eliminate hunger.

Sustainable-Earth Agriculture in MDCs Some elements of a sustainable-earth agricultural system for LDCs can be readily applied to MDCs. The increased use of no-till and low-till cultivation (Section 9-4) will reduce soil erosion and water pollution. Likewise, increased use of water-conserving forms of irrigation (Section 10-5) will reduce groundwater depletion, river pollution, and erosion, salinization, and waterlogging of soils. Developing government agricultural policies that encourage and reward the use of such strategies should be a major priority.

Rising costs of pesticides, inorganic fertilizers, and fuel, along with decreasing yields have motivated about 35,000, or 5%, of the full-time farmers in the United States to shift to industrialized **organic farming**—growing crops and livestock with organic fertilizers and biological pest control. Studies have shown that U.S. organic farms use an average of 40% less energy per unit of food produced, and many have 5% to 15% higher yields than conventional industrialized farms. A Department of Agriculture study indicated that a widespread shift to organic farming would increase net farm income, lower farm debts, reduce soil erosion and nutrient depletion, meet domestic food needs, reduce oil imports, and lower the environmental impact of agriculture. However, such a switch would also raise consumer food costs and reduce the amount of food available for export. It would also lead to a sharp reduction in the sales of fertilizers and pesticides; thus it would be strongly opposed by the politically powerful agricultural chemical industry.

Reducing Food Waste If the peoples of affluent countries reduced their enormous waste of food, it is often said, more food would be available for export to feed the poor. An estimated 25% of all food produced in the United States is wasted; it rots in the supermarket or refrigerator or is thrown away off the plate. This wasted food, theoretically, could feed more than 60 million people a U.S. meat-based diet and 150 million people a grain-based diet.

It is argued that if cattle in the United States were not fattened in feedlots, enough grain would be available to feed about 400 million people—equal to 69% of Africa's population. Even if feedlots were not eliminated, merely decreasing annual meat consumption in the United States by 10% could release enough grain to feed 60 million people. Similarly, commercial fertilizers spread on U.S. lawns, golf courses, and cemeteries could, in principle, be used to produce grain for 65 million people each year. Pet foods consumed in the United States, which has the world's highest ratio of pets to people, contain enough protein to feed 21 million people each year.

Many analysts point out, however, that food made available by the reduction of waste would not necessarily go to the poor, because they cannot afford to buy it. It is also argued that conserving food in affluent countries would result in less food for the poor by causing temporary food surpluses, followed by price declines, and eventually cuts in food production so that less food would be available for export.

What Can You Do?   In addition to becoming more knowledgeable about world food problems and possible solutions to these problems, you can examine your own lifestyle to find ways to reduce the unnecessary waste of food, energy, and matter resources along with the pollution and environmental degradation resulting from the use of these resources. For example, you can use organic cultivation techniques to grow some of your own food in backyard plots, window planters, rooftop gardens, or cooperative community gardens in unused urban spaces. In addition to saving money, you will have a more nutritious diet based on fresh rather than processed convenience foods. Today almost half of all U.S. households grow some of their own food—an amount worth about $14 billion. Unfortunately, most of it is grown with larger amounts of commercial fertilizers and pesticides per unit of land than are used on most commercial cropland.

Other ways you can reduce food waste: Put no more food on your plate than you intend to eat, ask for smaller portions in restaurants, recycle garbage in compost piles to produce organic fertilizer for growing your own food, and feed food waste to pets.

To fight hunger on a larger scale, you can become politically involved, supporting state and national leaders whose policies will reduce hunger in the United States and the world and will reduce the harmful environmental effects of agriculture.

The most important fact of all is not that people are dying from hunger, but that people are dying unnecessarily. . . . We have the resources to end it; we have proven solutions for ending it; . . . What is missing is the commitment.

The Hunger Project

DISCUSSION TOPICS

1. What are the major advantages and disadvantages of (a) labor-intensive subsistence agriculture; (b) energy-intensive industrialized agriculture; (c) organic farming?

2. Explain why most people who die from lack of a sufficient quantity or quality of food do not starve to death.

3. Explain why you agree or disagree with the following statement: There really isn't a severe world food problem, because we already produce enough food to provide everyone on earth more than three times the minimum amount needed to stay alive.

4. Summarize the advantages and limitations of each of the following proposals for increasing world food supplies and reducing hunger over the next 30 years: (a) cultivating more land by clearing tropical jungles and irrigating arid lands; (b) catching more fish in the open sea; (c) harvesting krill from the ocean; (d) producing fish and shellfish with aquaculture; (e) increasing the yield per area of cropland.

5. Should price supports and other subsidies paid to U.S. farmers out of tax revenues be eliminated? Explain. Try to have a farmer discuss this problem with your class.

6. Is food relief helpful or harmful? Explain.

7. Should tax breaks and subsidies be used to encourage more U.S. farmers to switch to organic farming? Explain.

8. Give your reasons for agreeing or disagreeing with the following suggestions made by some environmentalists and health scientists:

    a. All new and presently used food additives should be reviewed and tested not only for toxicity and carcinogenicity but also for their ability to induce birth defects and long-term genetic effects.

    b. All testing of food additives should be performed by a third party, independent of the food industry.

    c. All food additives, including specific flavors, colors, and sodium content (for people on salt-free diets), should be listed on the label or container of all foods and drugs.

    d. All food additives should be banned unless extensive testing establishes that they are safe and that they enhance the nutritive content of foods or prevent food spoilage or contamination by harmful bacteria and molds.

9. Do you believe that the Delaney clause should be revoked, left as is, altered to allow an evaluation of risks and benefits, or strengthened and broadened? Explain.

12

Land Resources: Wilderness, Parks, Forests, and Rangelands

GENERAL OBJECTIVES

1. How is land used in the world and the United States and how should publicly owned lands be managed?

2. Why is wilderness important and how much should be be set aside and protected in the United States?

3. Why are U.S. national and state parks important and how can they be protected from overuse and pollution?

4. Why are forests important and how can their renewability be maintained?

5. Why are rangelands important and how can their renewability be maintained?

We abuse land because we regard it as a commodity belonging to us. When we see land as a community to which we belong, we may begin to use it with love and respect.

Aldo Leopold

Many of the three out of four people in MDCs who live in urban areas are unaware or forget that their well-being is linked to the soil, plants, and animals found in the earth's croplands, forests, rangelands, parks, and wilderness areas. Protecting these nonurban land resources from overexploitation and environmental degradation is one of our most important challenges.

12-1 LAND USE IN THE WORLD AND THE UNITED STATES

Why Are Nonurban Land Resources Important? Land in the world and in the United States is used for several major purposes (Figure 12-1). Urban areas containing 43% of the world's population occupy less than 1% of the world's total land area. Urban areas containing 76% of the U.S. population occupy only 2% of the country's land area. Thus at first glance it would appear that the world has more than enough nonurban land resources to support today's 5 billion people and perhaps the 10.4 billion projected by the end of the next century.

However, such an assumption does not take into account the vast areas of cropland, forest, rangeland, watersheds, estuaries, and other types of nonurban land and water resources needed to sustain urban areas. Nonurban air, water, and land also serve as receptacles for urban-generated air and water pollutants and solid and toxic wastes.

Private and Public U.S. Land Ownership In the United States, 55% of all land is privately owned by

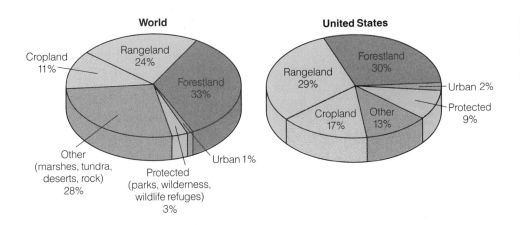

Figure 12-1 Land use in the world and the United States. (Data from U.S. Bureau of Commerce, Conservation Foundation, and UN Food and Agriculture Organization)

individuals and corporations. About 42% of all U.S. land consists of public lands owned jointly by all citizens but managed for them by federal, state, and local governments. About 35% of the country's land is owned jointly by its citizens and managed for them by the federal government; about 73% percent of this land is in Alaska and 22% is in western states (Figure 12-2).

About 40% of these public lands are *national resource lands*, administered by the Bureau of Land Management in the Department of Interior; 24% are *national forests*, administered by the Forest Service in the Department of Agriculture; 11% consist of 434 *national wildlife refuges*, administered by the Fish and Wildlife Service in the Department of Interior; and 3.5% consist of 388 *national park* units administered by the National Park Service in the Department of Interior.

Federally administered public lands contain a significant portion of the country's timber, grazing, energy, and mineral resources (Figure 12-3). As a result, there has been a long history of conflict between various groups over whether federal public lands should be transferred to private ownership, who should have access to the resources on and under these lands, and at what price (see Section 2-4 and Spotlight on p. 201).

12-2 WILDERNESS

U.S. National Wilderness Preservation System This system includes 275 roadless areas found within the national parks, national wildlife refuges, and national forests, which are managed, respectively, by the National Park Service, Fish and Wildlife Service, and Forest Service. According to Congress, **wilderness** consists of those areas "where the earth and its community of life are untrammeled by man, where man himself is a visitor who does not remain." Such areas are to be managed and preserved in their essen-

tially untouched condition "for the use and enjoyment of the American people in such a manner as will leave them unimpaired for future use and enjoyment as wilderness."

Wilderness areas are open only for such recreational activities as hiking, sport fishing, camping, nonmotorized boating, and in some areas sport hunting and horseback riding. Roads, timber harvesting, grazing, mining, commercial activities, and human-made structures are prohibited, except where such activities occurred before an area's designation as wilderness. Motorized vehicles, boats, and equipment are banned except for emergency uses such as fire control and rescue operations. Exploration and identification of mineral, energy, and other resources is allowed as long as such activities do not involve the use of motorized vehicles or equipment.

U.S. Wild and Scenic Rivers In 1968 Congress passed the National Wild and Scenic Rivers Act to prevent further development of rivers and river segments with outstanding scenic, recreational, geologic, wildlife, historic, or cultural values. The only activities allowed on these protected areas are camping, swimming, nonmotorized boating, sport hunting, and sport and commercial fishing, with the exception that new mining claims are permitted in scenic and recreational river areas. The system is administered by the Interior and Agriculture departments in cooperation with state agencies.

Use and Abuse of Wilderness Areas Popular wilderness areas, especially some in California, North Carolina, and Minnesota, are visited by so many people that fragile vegetation is damaged, soil is eroded from trails and campsites, water is polluted from bathing and dishwashing, litter is scattered along trails, and instead of quiet and solitude, users face the

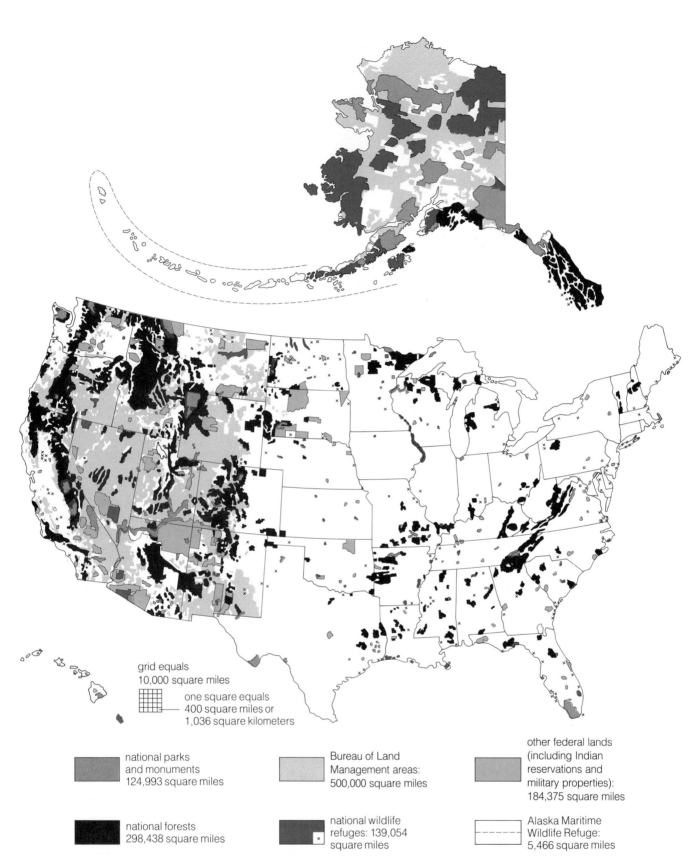

grid equals
10,000 square miles

one square equals
400 square miles or
1,036 square kilometers

national parks
and monuments
124,993 square miles

Bureau of Land
Management areas:
500,000 square miles

other federal lands
(including Indian
reservations and
military properties):
184,375 square miles

national forests
298,438 square miles

national wildlife
refuges: 139,054
square miles

Alaska Maritime
Wildlife Refuge:
5,466 square miles

Figure 12-2 Landholdings of the federal government. (Data from U.S. Department of the Interior, Geological Survey)

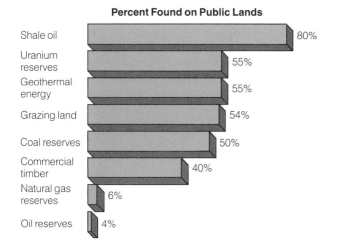

Percent Found on Public Lands

- Shale oil — 80%
- Uranium reserves — 55%
- Geothermal energy — 55%
- Grazing land — 54%
- Coal reserves — 50%
- Commercial timber — 40%
- Natural gas reserves — 6%
- Oil reserves — 4%

Figure 12-3 Percentages of key resources on U.S. public lands.

congestion they are trying to escape. To protect the most popular areas from damage, government agencies have had to limit the number of people hiking or camping at any one time and designate campsites.

To prevent overuse, historian and wilderness expert Roderick Nash advocates dividing wilderness areas and areas within them into different categories. The accessible and most popular areas would be inten-

sively managed and have trails, bridges, hikers' huts, outhouses, assigned campsites, and extensive ranger patrols. Large, remote wilderness areas would be relatively unmanaged and accessible only to people who qualify for special licenses by demonstrating their wilderness skills. A third category would consist of large, unique areas to be left undisturbed as genetic pools of plant and animal resources, with no human entry allowed.

Why Preserve Wilderness? Since 1964 there has been a continuing battle over how much land should be added to the U.S. wilderness system. Officials of timber, mining, and energy industries and ranchers operating on public lands have opposed expansion of the wilderness system. They accuse preservationists of wanting to lock up the woods for use as parks by the affluent and physically fit and argue that resources on and under public lands should be used now and in the future to promote economic growth and provide mineral and energy resources to enhance national security. They believe most public lands should continue to be managed under the principle of multiple use, which permits both conservation and development uses.

Since wilderness is irreplaceable, wilderness enthusiasts have urged Congress to add more area to

---

Spotlight   How Should Public Land Resources Be Used?

Ecosystem-centered, or Biocentric, Approach

Preservationist View: "Preserve It." Portions of public land resources should be preserved from development by timber, mining, and other economic interests. Emphasis should be on preserving the biological diversity and sustainability of natural ecosystems. Such areas should be used only for nondestructive forms of outdoor recreation and left as ecological preserves. The National Wilderness Preservation System, which makes up 11% of all public lands, represents this view.

Human-centered, or Anthropocentric, Approaches

Short-Term Economic View: "Use It." Public lands should be transferred to private interests and used to provide the highest short-term economic gain for its owners and to promote national economic growth.

Multiple-Use View. Instead of being left untouched or unexploited, public land resources should be used for a variety of purposes, including timbering, mining, grazing, recreation, aesthetics, and wildlife and water conservation, and managed in ways that do not damage or deplete them

for future generations. About 60% of all public land, found in national forests (excluding wilderness) and national resource lands, is officially managed on this basis. However, conservationists contend that far too often multiple-use management ends up being single-use management because of political influence by timber, mining, and ranching interests. The management of 13% of public lands as national wildlife refuges (8%) and the national parks (5%) represents a compromise between multiple-use and preservationist views.

the wilderness system, especially in the lower 48 states, where wilderness use is the most intensive and where only 1.7% of the total land area is at present designated as wilderness. They contend that the 60% of public lands managed officially or unofficially for multiple use are the country's most resource-rich public lands and should provide ample resources to enhance national security and promote economic growth.

Wilderness enthusiasts argue that by preserving wilderness, we (1) provide wild places where we can experience majestic beauty and natural biological diversity and where we can renew the spirit and enhance mental health by getting away from noise and stress; (2) set aside a repository of resources that could be used later in a true emergency situation, rather than now for short-term economic gain; (3) protect diverse ecosystems as an ecological insurance policy against human abuse of the land and elimination of too much of the earth's natural biological diversity; and (4) provide an ecological laboratory in which we can discover how nature works and measure and observe how much human beings have altered the earth. On ethical grounds it is also argued that wilderness should be preserved because the plant and animal species it contains have an inherent right to exist without human interference.

12-3 PARKS

U.S. National Park System   Since 1872 the National Park System has grown to 338 units, including 49 major parks (mostly in the West), and 289 national recreation areas, monuments, memorials, battlefields, historic sites, parkways, trails, rivers, seashores, and lakeshores. The major national parks provide spectacular scenery on a scale not usually found in state and local parks, preserve wildlife that can't coexist with human beings, and surround and protect wilderness areas within them.

During the 1970s, several urban parks (known as national recreation areas), national seashores, lakeshores, and other units usually located close to heavily populated urban areas were added to bring parks closer to the people. Since they opened in the 1970s, the Golden Gate National Recreation Area near San Francisco and the Gateway National Recreation Area in Greater New York City have been two of the most widely used units in the National Park System.

National parks can be used only for camping, hiking, sport and commercial fishing, and motorized and nonmotorized boating. Motor vehicles are permitted only on roads, and off-road vehicles are not allowed. In addition to the activities permitted in the parks,

national recreation areas can be used for sport hunting, new mining claims, and new oil and gas leasing. About 49% of the land in the National Park System is protected as wilderness areas, and its use is restricted even more. The National Park System is supplemented by a number of state and local parks.

Internal Stresses on Parks   Between 1950 and 1986, recreational visits to National Park System units increased twelvefold and visits to state parks sevenfold. Because they are more numerous and located closer to most urban areas (especially in the East, where there are few national parks), state parks are used more intensively than national parks. Recreational use of state and national parks and other public lands is projected to increase even more in the future, putting additional stress on many already overburdened parks.

Under the onslaught of people during the peak summer season, the most popular national parks are often overcrowded with cars and trailers and plagued by noise, traffic jams, litter, vandalism, deteriorating trails, polluted water, drugs, and crime. Theft of timber, petrified wood chips, and cacti from national parks is a growing problem. Park Service rangers, now trained more in law enforcement than resource conservation and management, must wear guns and spend an increasing amount of their time acting as park police officers. In addition to the stresses brought by human visitors, some parks face degradation by growing populations of certain native animal species whose natural predators have been sharply reduced or eliminated.

External Threats to Parks   The greatest danger to many parks today is from external threats, such as mining, timber harvesting, grazing, coal-burning power plants, water diversion, and urban development in nearby areas. A 1980 survey by the National Park Service revealed that scenic resources were threatened in more than 60% of the national parks, while visibility, air and water quality, and wildlife were endangered in about 40% of the parks.

At Grand Canyon National Park, visitors often cannot see the canyon's opposite rim because of smog, and the noise from helicopters carrying sightseers into the canyon ruins the wilderness experience for others backpacking in the canyon. Underground rivers in Kentucky's Mammoth Cave National Park carry sewage from nearby communities. The amount and diversity of plant and animal life found in the Everglades National Park in the southern tip of Florida have dropped sharply because much of the water that once flowed southward into this swampland wildlife preserve has been diverted and used for irrigation and

Reagan Administration	Conservationists
Increase funding for maintenance and repair of existing parks.	Agree.
Eliminate funding for purchase of additional parkland.	More land should be added to accommodate increasing use, prevent choice parcels from being developed. Additions near existing parks can also be used to reduce external environmental threats.
Transfer ownership of some national park units such as the Gateway and Golden Gate National Recreation Areas and other recently created urban parks to states and localities.	Disagree. With declining federal and state funding, states and localities are in no position to keep them up.
Increase mining, energy development, and other income-producing resource development on public land near national parks.	Strictly regulate such activities to reduce pollution threats to parks.
Allow private enterprise to develop and run luxury hotels, restaurants, and other commercial facilities in parks to bring in more revenue.	Reduce congestion, crime, and pollution in national parks by prohibiting addition of any new commercial facilities and removing most existing commercial facilities to private or federally owned areas outside parks, as has been done for Acadia National Park in Maine.
Turn more of the management of camping, recreational, and educational activities in parks to private concessionaires.	Sharply increase usage fees paid by park concessionaires to return more of the high profits they make to taxpayers; place concessionaires under much stricter control to ensure they provide adequate services at fair prices.
Sharply increase automobile entrance fees to the point where they provide at least 10% of park operating costs.	Agree, but would go further and ban motor vehicles from heavily used parks to reduce congestion, noise, crime, and pollution. Visitors would be shuttled by bus to and from satellite parking lots, as is done in parts of Yosemite. Inside in the park, visitors could travel by foot, bicycle, or shuttle bus.
Encourage private donations for park restoration (such as the highly successful privately financed restoration of the Statue of Liberty) and increased use of volunteer workers in park system units.	Agree, but this should not be used as an excuse to reduce the Park Service budget.

urban development. Planned geothermal energy development may take the steam out of geysers in Yellowstone and Lassen (northeastern California) national parks. Similar harmful environmental impacts affect many state and local parks.

What Should Be Done?   Since 1980 there has been considerable disagreement between the Reagan administration and conservationists over how to reduce stresses on the National Park System (see Spotlight above). To maintain, improve, and expand the country's priceless state and national parks, Americans must be willing to pay higher entry and use fees, higher taxes, or both. The alternative is to allow these vital resources to become increasingly overcrowded and degraded and pass them on to future generations in an impaired state.

Importance of Forests Potentially renewable forest resources cover about one-third of the earth's surface. These forests provide fuelwood, sawlogs for construction and wood products, and wood pulp primarily for paper products valued at $150 billion a year.

In addition to their commercial value, forests have vital ecological functions that are often unrecognized and unappreciated. They help control climate by influencing wind, temperature, humidity, and rainfall. They add oxygen to the atmosphere and assist in the global recycling of water, carbon, and nitrogen. Forested watersheds act like giant sponges that absorb, hold, and gradually release water, thus recharging springs, streams, and groundwater aquifers. By regulating the downstream flow of water, forests help control soil erosion, the amount of sediment washing into rivers and reservoirs, and the severity of flooding. Forests provide habitats for organisms that make up much of the earth's genetic diversity. They also help absorb noise and some air pollutants, cool and humidify the air, and nourish the human spirit by providing solitude and beauty.

Too often, economists evaluate forests only on the short-term market value of their products, without considering the value of their long-term ecological benefits. It is estimated, for example, that a typical tree provides $196,250 worth of ecological benefits in a 50-year life span; sold as timber, it provides only about $590. In its 50 years the tree produces $31,250 worth of oxygen, $62,500 in air pollution reduction, $31,250 in soil fertility and erosion control, $37,500 in water recycling and humidity control, $31,250 in wildlife habitat, and $2,500 worth of protein. While such a calculation is a general estimate, not reflecting actual market values that can be redeemed, it illustrates dramatically how vital forests are to human beings and other forms of life.

Management of Commercial Forests The cultivation and management of forests to produce a renewable supply of timber is called **silviculture.** Experience has shown that by careful application of the principles of multiple use and sustained yield, forest resources can be used for a variety of purposes and harvested in a manner and at a rate that preserves these renewable resources for present and future generations.

There is increasing emphasis on the use of **intensive forest management,** which typically involves clearing an area of all vegetation, planting it with even-aged stands (Figure 12-4), and then fertilizing and spraying the resulting tree plantation with pesticides.

Figure 12-4 Monoculture tree plantation of white pine near Asheville, North Carolina.

Once the trees reach maturity, the entire stand is harvested and the area is replanted. Genetic crossbreeding and genetic engineering techniques can be used to improve both the quality and quantity of wood produced from such plantations.

Most foresters and resource managers favor intensive forest management. They see it not only as a way to increase short-term profits but as the best way to increase the amount of timber produced per unit of area to meet the increasing worldwide demand for wood. They believe that careful use of the principle of sustained yield allows intensively managed forest resources to be harvested and regenerated in a manner and at a rate that conserve these potentially renewable resources for future generations.

Other foresters and a number of ecologists, however, are concerned that overemphasis on intensive forest management will lead to a severe reduction in the diversity of plant and animal life in the world's commercial forestlands. Replacing diverse tree species and vegetation with single species of commercial trees could leave large areas of forest highly vulnerable to destruction by pests and diseases. These analysts are not against intensive management altogether but believe that no more than one-fourth of the world's commercially exploitable forestland should be managed in this way.

Figure 12-5 Selective cutting of Douglas fir in Mount Baker National Forest, Washington.

Figure 12-6 Shelterwood cutting of hemlock. Residual stand after first cut in an experimental forest in Grays Harbor County, Washington.

Tree Harvesting and Regeneration Methods
Trees are harvested and regenerated by five main methods: selective cutting, shelterwood cutting, seed-tree cutting, clearcutting, and whole-tree harvesting. In **selective cutting,** intermediate-aged or mature trees in an uneven-aged forest stand are cut either singly or in small groups at intervals (Figure 12-5). This encourages the growth of younger trees and produces a stand with trees of different species, ages, and sizes; over time, the stand will regenerate itself.

This harvesting method is favored by those wishing to use forests for both timber production and recreation; if the harvest is too limited, however, not enough timber may be produced to make the process economically feasible. In addition, the need to reopen roads and trails periodically for selective harvests can cause erosion of certain soils. A similar method of cutting, not considered a sound forestry practice, is called *high grading.* Here the most valuable commercial tree species are cut without regard for the quality or distribution of remaining trees.

Some tree species do best when grown in full sunlight in forest openings or in large, cleared and seeded areas. Even-aged stands of such species (Figure 12-4) are usually harvested by shelterwood cutting, seed-tree cutting, or clearcutting. **Shelterwood cutting** involves removing all mature trees in an area in a series

of cuts over one or more decades. In the first harvest, unwanted tree species and dying, defective, and diseased trees are removed, leaving properly spaced, healthy, well-formed trees as seed stock (Figure 12-6). In the next stage, ten or more years later, the stand is cut further so that seedlings can receive adequate sunlight and heat and can become established under the shelter of a partial canopy of remaining trees. Later, a third harvest removes the remaining mature canopy trees, allowing the new stand to develop in the open as an even-aged forest. This method leads to very little erosion. Without careful planning and supervision, however, loggers may take too many trees in the initial cut, especially the most commercially valuable trees.

Seed-tree cutting harvests nearly all trees on a site in one cut, with a few of the better commercially valuable trees left uniformly distributed as a source of seed to regenerate the forest. Allowing a variety of species to grow at one time, seed-tree cutting is a form of multiple-use management. However, it is not used for many species, because if the remaining seed trees are lost to wind and ice, the site will be left without a sufficient seed source for reforestation.

In **clearcutting,** all trees are removed from a given area in a single cutting to establish a new, even-aged stand, usually of fast-growing, shade-intolerant species (Figure 12-7). The clearcut area may consist of a

Figure 12-7 Clearcutting of white pine in St. Joe National Forest, Montana.

Figure 12-8 Patch clearcutting in Kootenai National Forest, Montana.

Figure 12-9 Whole-tree harvesting. Tree is fed into chipper, which deposits chips into truck at right.

whole stand, a group, a strip, or a series of patches (Figure 12-8). After clearing, the site is reforested naturally from seed released by the harvest or, increasingly, by planting genetically superior seedlings raised in a nursery. Timber companies prefer clearcutting, because it increases the volume of timber harvested per acre, reduces road building, and shortens the time needed to establish a new stand of trees. Clearcut openings and the fringes along uncut areas also improve the forage and habitat for some herbivores, such as deer and elk, and some shrubland birds.

Conservationists and ecologically oriented foresters recognize that clearcutting can be useful for some species if properly done. Their concern is that the size of the clearcut areas is too often determined by the economics of logging rather than by consideration of forest regeneration, and the method is sometimes used on species that could be harvested by less ecologically destructive methods. Excessive use of clearcutting on steeply sloped land can lead to severe erosion and sediment water pollution. In addition, it creates ugly scars (Figures 12-7 and 12-8) that take years to heal, reduces the recreational value of the forest for years, destroys habitats for many wildlife species, and replaces a genetically diverse stand of trees with a vulnerable monoculture.

Figure 12-10 Highly destructive effects of a crown fire in Hunts Gulch, Idaho.

Figure 12-11 Ground fire in a California forest.

Whole-tree harvesting is a variation of clearcutting in which a machine cuts each tree at ground level and transports it to a chipping machine, where massive blades reduce the entire tree to small chips in about one minute (Figure 12-9). Some whole-tree harvesting machines pull up the entire tree so that roots are also utilized. This approach, which is used primarily to harvest stands for use as pulpwood or fuelwood, can increase the yield of a temperate forest by 300% by using all wood materials in a stand, including defective trees and dead standing timber. Many foresters and ecologists, however, oppose this method, because the periodic removal of all tree materials eventually removes most soil nutrients. Research is under way to determine the rates of nutrient depletion in various regions and to determine how cutting methods could be modified to reduce the ecological impact of whole-tree harvesting.

Protecting Forests from Fires About 85% of all forest fires in the United States are started by people, either accidentally or deliberately. It is important to distinguish between two types of forest fires: ground fires and crown fires. **Crown fires** are extremely hot fires that can destroy all vegetation, kill wildlife, and accelerate erosion (Figure 12-10). They tend to occur in forests where all fire has been prevented for several decades, allowing accumulations of dead wood and

ground litter that burn intensely enough to ignite tree tops.

Ground fires are low-level fires that burn only undergrowth (Figure 12-11). They usually do not harm mature trees, and most wildlife can escape them. In areas where excessive ground litter has not accumulated, periodic ground fires reduce the buildup of undergrowth and ground litter and help prevent the more destructive crown fires. They also help release and recycle valuable plant nutrients tied up in litter and undergrowth, increase the activity of nitrogen-fixing bacteria, help control diseases and insects, and provide the intense heat needed for the germination of seeds of some conifers, such as giant sequoia and jack pine. Some wildlife species, such as deer, moose, elk, muskrat, woodcock, and quail, depend on periodic ground fires to maintain their habitats and to provide food from vegetation that sprouts after such fires.

Because of these benefits, ecologists and foresters have increasingly prescribed the use of carefully controlled ground fires as an important tool in the management of some forests (especially those dominated by conifers such as giant sequoia and Douglas fir) and rangelands.

Protecting Forests from Diseases and Insects
Diseases and insects destroy more commercial timber than fires do, both in the United States and through-

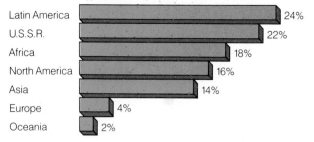

Percent of World's Total Forest Area

Latin America — 24%
U.S.S.R. — 22%
Africa — 18%
North America — 16%
Asia — 14%
Europe — 4%
Oceania — 2%

Figure 12-12 Percentage of the world's forests in various groups of countries. (Data from UN Food and Agriculture Organization)

out the world. Parasitic fungi cause most tree-damaging diseases, such as chestnut blight, white pine blister rust, and Dutch elm disease. One way to control tree diseases is to ban imported timber that might bring in new types of parasites. Other methods include developing disease-resistant species, reducing air pollution, which makes trees more susceptible to disease, and identifying and removing dead, infected, and susceptible trees.

Insect pests that cause considerable damage to trees include the gypsy moth, spruce budworm, pine weevil, larch sawfly, and several species of pine beetles. Pest-control methods include isolating and removing infested trees, encouraging natural insect control by preserving forest diversity, introducing other insects that prey on pest species (biological control), using sexual attractants (pheromones) to lure insects to traps, releasing sterilized male insects to reduce the population growth of pest species, and pesticide spraying (Section 21-5).

Protecting Forests from Air Pollution Air pollution is a rapidly growing new threat to many of the world's forests. One of the most serious air pollution problems affecting forests is **acid deposition,** commonly called acid rain. It occurs when sulfur dioxide and nitrogen oxide air pollutants, released by the burning of fossil fuels in power plants and cars, are transformed chemically in the atmosphere to sulfuric and nitric acids and dry acidic particulate matter that fall to the earth in rain, snow, or fog (Section 18-2). Another air pollutant that causes severe tree damage is ozone. Trees downwind of major coal-burning power plants and industrial areas and trees at high altitudes are exposed to the largest concentrations of damaging air pollutants.

The main solution to this problem is to use air pollution control devices on coal-burning power and industrial plants and on cars to reduce emissions of sulfur and nitrogen oxides (Section 18-7).

12-5 STATUS OF WORLD AND U.S. FORESTS

World Forests The world's remaining forests are divided unevenly among regions (Figure 12-12 and map inside the front cover) because of variations in climate, land use, total land area, and past forest exploitation. Since the Agricultural Revolution began, human activities have reduced the earth's original forested area by an estimated one-third, and by one-half in LDCs. In most MDCs, the amount of forested area has remained the same and in some cases increased since 1900, primarily because urbanization and industrialized agriculture reduced the need to convert forestland to farmland. Overall, however, the world is losing about 1% of its forested land each year as forests—especially tropical forests—in LDCs are cleared for farming and grazing and cut for fuelwood and lumber without adequate reforestation.

The Fuelwood Crisis in LDCs About 70% of the people in LDCs, most of whom live in rural areas, depend on free or cheap wood as their principal fuel for heating and cooking. By 1985 about 1.5 billion people—almost 1 out of every 3 on earth—in 63 LDCs were unable to obtain enough fuelwood to meet their minimum needs or were forced to meet their needs by consuming wood faster than it was being replenished. The UN Food and Agriculture Organization projects that by the end of this century, 3 billion people in these plus 14 other LDCs either face acute fuelwood scarcity (500 million people) or will be depleting remaining supplies (2.5 billion people).

Fuelwood scarcity has several harmful consequences. It places an additional burden on the poor, especially women, who must walk long distances, carry heavy loads, and spend a large amount of their potentially productive time collecting fuelwood. Deforestation is also accelerated, especially in areas near villages and cities where commercial markets for fuelwood and charcoal (produced as a fuel from fuelwood) exist. Deforestation in turn increases soil erosion, flooding, and desertification, and reduces agricultural production. Food production is also decreased when families who cannot obtain enough fuelwood burn dried animal dung and crop residues, thus preventing vital natural fertilizers from reaching the soil.

LDCs can reduce the severity of the fuelwood crisis by planting more trees to increase the supply and by burning wood more efficiently or switching to another fuel such as kerosene to reduce consumption. Governments of countries such as China, South Korea, and Nepal have been successful in instituting massive tree-planting programs at the village level. Villagers are provided seed or seedlings by government foresters and encouraged to plant fast-growing fuelwood trees and shrubs in fields along with crops (agroforestry), in plantations, and in unused patches of land around homes and along roads and waterways.

However, most LDCs suffering from fuelwood shortages have inadequate forestry policies and budgets and lack trained foresters. As a result, they are planting 10 to 20 times less than what is needed to offset forest losses and meet increased demands for fuelwood and other forest products.

Tropical Deforestation   Loss of the world's tropical forests is considered to be one of the world's most serious environmental problems. Occupying an area roughly equal to that of the continental United States, these forests are home to at least half of the earth's species of plants and animals. They are by far the earth's most diverse biome—and they are being depleted and degraded faster than any other biome. About one-third of the world's original expanse of these forests has already been cleared and used to grow crops, graze livestock, and provide timber and fuelwood. Africa has lost 52% of its tropical forests, Asia 42%, Central America 37%, and South America 36%.

There is controversy over the rate at which these forests are currently being cleared and degraded because of insufficient data and the tendency of some governments to understate the problem for political reasons. However, on the evidence of surveys made by remote-sensing satellites and other available data, each year an estimated 1% of the world's tropical forests are being completely cleared, and another 1% are being degraded. Some experts estimate that the rate of removal and degradation is almost two times higher. If present rates continue, all remaining tropical forests will be gone or seriously disturbed by 2035.

Four direct causes of tropical deforestation and degradation have been identified (Figure 12-13): **(1)** poor people clearing land to grow food; **(2)** poor people gathering fuelwood faster than it is regenerated (especially in Africa and Asia); **(3)** commercial logging by international companies (especially in the Pacific Islands, West and Central Africa, and parts of Latin America); and **(4)** ranchers clearing land (primarily in Central America and Brazil's portion of the Amazon basin) to graze cattle and produce low-cost beef, mostly for export to MDCs, where it is used for hamburger in many fast-food chains and in pet food, luncheon meats, chili, stews, and frozen dinners. Indirect causes of deforestation are rapid population growth, poverty (which forces landless people to settle and cultivate unowned forestland), and failure of governments to require international timber companies to regenerate cleared areas.

Ecologists warn that loss and degradation of these incredibly diverse biomes could cause the premature

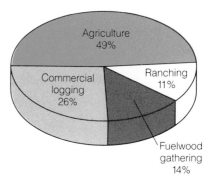

Figure 12-13   Major direct causes of deforestation and degradation of tropical forests. (Data from Norman Myers)

extinction of 1 million species—10% to 20% of the earth's total—by the beginning of the next century. Many of these species may be crucial in the development of hybrid and genetically engineered food plants needed to support future green revolutions, new medicines to fight disease, and a host of other products important to both MDCs and LDCs. In addition, tropical forests are home for 150 million to 200 million people, who survive by shifting cultivation, and they protect watershed and regulate water flow for other farmers, who grow food for over 1 billion people.

In 1985 the World Resources Institute and the World Bank proposed that MDCs fund a five-year $5.3 billion plan to help tropical LDCs protect and renew tropical forests. This plan called for setting aside 14% of the world's tropical forests as reserves and parks to protect them from development; massive planting of fuelwood and multipurpose trees carried out primarily by villagers; establishing industrial tree plantations in areas where soils are best suited for this type of forest management; rehabilitating tropical watersheds to reduce flooding and sedimentation; and strengthening of forestry research and training in tropical LDCs. In addition, conservationists believe that commercial timber companies should be required to reforest lands they harvest in tropical LDCs. If such a program is carried out, it could represent a turning point in preventing the irreversible loss of most of the world's tropical forests.

Forests in the United States   Since the first colonists arrived at Jamestown in 1607, the United States has lost about 45% of its original forested area. Since 1920, however, the country's total forested area has remained about the same, covering about one-third of all U.S. land area (Figure 12-14).

About two-thirds of the country's forests are classified as commercial forestland, suitable for growing potentially renewable crops of economically valuable tree species and not protected from commercial log-

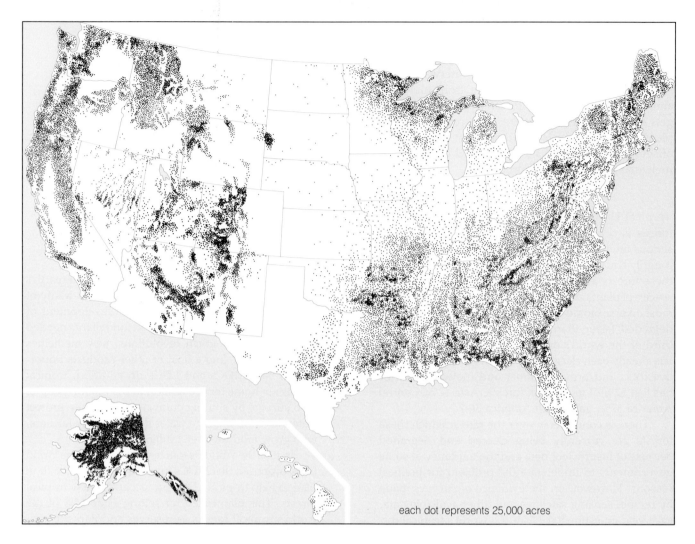

each dot represents 25,000 acres

Figure 12-14 Forestlands in the United States. (Data from Council on Environmental Quality)

ging as part of parks, wilderness, or other restricted areas. The annual harvest of commercial timber comes mostly from the 14% of the country's commercial forestland owned by timber companies, the 18% found in national forests, and another 10% on other federal and state public lands. Most of the 58% of U.S. commercial forestland owned by individuals and families consists of fragmented and small sites managed poorly in terms of timber production.

Importance and Use of National Forests The National Forest System includes 155 individual national forests (Figure 12-2) and 19 national grasslands. Excluding the 15% of this land protected as wilderness areas, this system is managed according to the principles of sustained yield and multiple use. These lands are used for timbering, grazing, agriculture, mining, oil and gas leasing, recreation (national forests receive more recreational visits than any other federal public

lands), sport hunting, sport and commercial fishing, and conservation of watershed, soil, and wildlife resources. Emphasis, however, is on maintaining a forest reserve to furnish renewable timber supplies for the country.

Almost half of national forestlands are open to commercial logging, providing about 15% of the country's total annual timber harvest—enough wood to build about 1 million houses a year. Each year, private timber companies bid for rights to cut a certain amount of timber from areas designated by the Forest Service. Forest Service funds provided by taxpayers are used to build and maintain roads to harvest areas and to reforest harvested areas.

Conflicting Demands on National Forests Since 1950, greatly increased demands have been placed on the forests for timber sales, development of domestic mineral and energy resources, recreational use, wilderness protection, and wildlife conservation. Environmentalists have accused the Forest Service of emphasizing commercial logging at the expense of other uses, pointing to the doubling of timber sales and harvesting between the early 1950s and 1960 and the continuation of high levels of harvesting since then.

In 1974 conservationists' outrage over extensive and often improperly managed clearcutting in national forests during the 1960s and early 1970s led to passage of the National Forest Management Act. This act set certain limitations and restrictions on timber harvesting methods and required the Forest Service to prepare comprehensive forest management plans for each national forest region, with public participation in this process.

Despite some improvement in management of the National Forest System for recreational use and wildlife, water, and soil resource conservation, conservationists charge that commercial logging is still by far the dominant use in terms of management decisions and budget allocation. In 1985, for example, 30% of the Forest Service budget was used for timber sales (including reforestation and road building and maintenance), compared to only 9% for recreation and 6% for soil, water, and wildlife conservation.

Under intense pressure from the timber industry and the Reagan administration, the Forest Service has developed long-term management plans calling for (1) timber sale volumes to double between 1986 and 2030; (2) building 580,000 miles of new roads between 1986 and 2030 (amounting to over 14 times the mileage in the entire interstate highway system); (3) departing from sustained-yield management to cut towering stands of 200-to-400-year-old Douglas fir and other commercially important trees in the country's remaining virgin forests of the Pacific Northwest, where timber companies have already cut most of the mature timber on their own lands; and (4) replacing these old-growth stands with more productive young stands of one or two faster-growing species to prevent timber shortages and higher prices for houses and wood products.

Conservationists challenge the need to increase timber harvests in these and other national forests to avoid timber shortages and higher prices, pointing to large quantities of uncut timber already sold to timber companies at bargain-basement prices. By losing money on 22% to 42% of its annual timber sales between 1980 and 1985, the Forest Service has provided timber companies with tax-supported subsidies. They contend that as long as timber companies can increase profits by cutting timber from public lands at low cost, they will continue to pressure elected officials to increase the amount of timber cut from national forests regardless of need.

Forestry industry representatives argue that such subsidies help taxpayers by keeping lumber prices down. But conservationists note that each year, taxpayers already provide the lumber industry with tax breaks almost equal to the cost of managing the entire National Forest System.

Conservationists propose four ways to reduce exploitation of publicly owned timber resources and provide true multiple use of national forests as required by law: (1) Reduce annual timber harvest levels by half rather than increasing them. (2) Leave between 15% and 25% of remaining old-growth timber in each national forest uncut. (3) Have Congress require that no timber from national forests be sold at a loss. (4) Devote a much larger portion of the Forest Service budget to conservation uses in national forests and to improving the management and increasing the timber output of commercial forestland owned by individuals and families.

12-6 RANGELANDS

Nature of Rangelands Land on which the vegetation is predominantly grasses, grasslike plants, or shrubs such as sagebrush is called **rangeland.** Rangelands include most of the world's tropical grasslands (savanna) and uncultivated temperate grasslands (see map inside the front cover). Most animals, including people, cannot digest the cellulose that makes up most rangeland vegetation, or *forage*. But ruminant animals like cattle, sheep, and goats can digest cellulose and produce protein-rich meat, milk, butter, and cheese for human consumption. These livestock animals also provide manure for fertilizing soil and useful nonedible goods such as wool, leather, and tallow (used to make soap and candles).

About 29% of the total land area of the United States consists of rangelands, located mostly in the arid and semiarid western half of the country (Figure 12-15). About 34% of these rangelands is publicly owned land, most of it managed by the Bureau of Land Management and the Forest Service. About three-fourths of public and privately owned rangeland in the United States is actively grazed by domestic animals at some time during each year. The public rangelands are managed under the principle of multiple use and are used for other purposes, such as mining, energy resource development, recreation, and conservation of soil, water, and wildlife.

Characteristics of Rangeland Vegetation Many rangeland weeds and bushes have a single main taproot and can thus be easily uprooted. By contrast, rangeland grass plants have a fibrous taproot system with multiple branches that make the plants very dif-

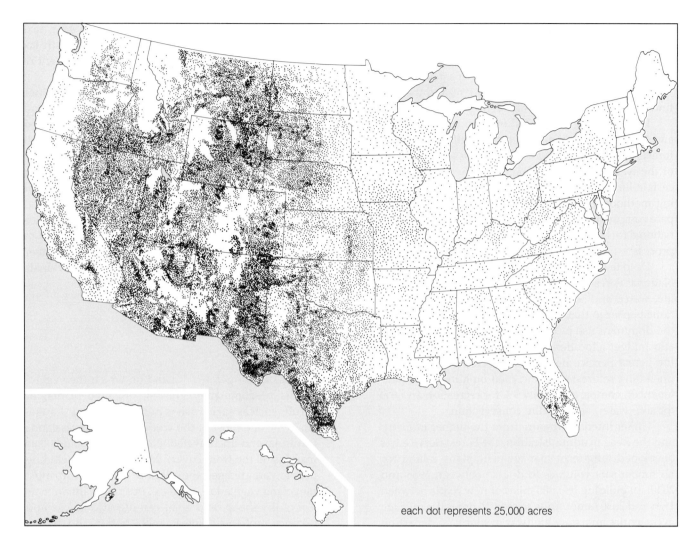

each dot represents 25,000 acres

Figure 12-15 Rangelands in the United States. (Data from Council on Environmental Quality)

ficult to uproot. This explains why these grasses help prevent soil erosion.

When the leaf tip of most plants has been eaten, no further leaf growth occurs. In contrast, each leaf of rangeland grass grows from its base, not its tip. Thus even when the upper half of the stem and leaves of rangeland grass have been eaten by livestock or wild herbivores such as deer, antelope, and elk, it can grow back to its original length in a short time. However, the lower half of the plant, known as the **metabolic reserve,** must remain if the plant is to survive and grow new leaves. As long as only the upper half is eaten, rangeland grass is a renewable resource that can be grazed again and again.

Rangeland Carrying Capacity and Overgrazing

Each type of rangeland has a limited carrying capacity—the maximum number of wild and domesticated herbivores a given area can support without risk of degradation from consumption of the metabolic reserve. **Overgrazing** can kill the root system of grass so that little if any grass is left (Figure 12-16). Then invader species of weeds and shrubs unpalatable to livestock take over, or all vegetation disappears, leaving the land barren and vulnerable to erosion. Severe overgrazing coupled with drought can convert potentially productive rangeland to desert.

Restoring overgrazed lands is difficult and expensive. During the 5-to-12-year period required for restoration, the number of animals grazing per unit of land area must be decreased or even reduced to zero.

Throughout the world, most rangelands used extensively for livestock show many signs of overgrazing. In the United States, 61% of private rangeland is considered to be in fair condition (45%) or poor condition (16%). Despite more than 50 years of management under the Taylor Grazing Act of 1934, a 1984 survey revealed that 64% of public rangelands were in fair or poor condition.

USDA, Soil Conservation Service

Overgrazing continues on a large portion of public rangeland primarily because grazing fees on this land are low (about one-fifth those for grazing on comparable private land) and ranchers pressure the federal government to keep grazing allotments as high as possible. In 1986 U.S. taxpayers provided subsidies of at least $30 million to ranchers with federal grazing permits. Conservationists propose that the government raise grazing fees to reduce overgrazing and give taxpayers a better return on these publicly owned resources, or replace the grazing permit system with the sale of grazing rights by competitive bids, as is done for timber-cutting contracts in national forests. Ranchers who are not able to get permits also favor open bidding for public grazing rights. They argue that the present permit system gives politically well-connected ranchers an unfair economic advantage.

Rangeland Management The most important method of managing rangeland is to control the number and kinds—the stocking rate—of animals grazing on a given area to ensure that its carrying capacity is not exceeded. This is often difficult in practice, because carrying capacity varies with climate, seasons of the year, slope, soil type, types of forage, and past land use.

Controlling the distribution of grazing animals over rangeland is another important way to prevent overgrazing. Ranchers can control distribution by building fences to prevent animals from grazing on degraded rangeland, rotating livestock from one area to another, providing supplemental feeding at selected sites, and strategically locating water holes (Figure 12-17) and supplies of salt. Livestock need both salt and water, but not together. Placing salt blocks in ungrazed areas away from water sources prevents livestock from congregating and overgrazing near water sources.

A more expensive and less widely used method of rangeland management is to suppress growth of undesirable vegetation with herbicides, mechanical removal, or controlled burning. Growth of desirable vegetation can be encouraged by artificial seeding and fertilization, but this is expensive relative to the economic return on livestock. Reseeding is worth the cost, however, to prevent desertification of badly degraded public rangeland.

Many ranchers still promote the use of poisons, trapping, and shooting to control jackrabbits and prairie dogs, which compete with livestock for forage, and to kill off predatory species such as coyotes, which sometimes kill livestock, especially sheep. However, experience has shown that this approach usually provides only temporary relief, is rarely worth the cost, and sometimes can make matters worse.

For example, since 1940 a highly controversial poisoning program has been waged by western ranchers and the U.S. Department of the Interior against the coyote. However, as such predator populations are reduced, populations of small herbivores, especially

Figure 12-17 Livestock tend to concentrate and overgraze around water sources unless techniques are used to distribute the animals to other areas.

USDA, Soil Conservation Service/B. C. McLean

rabbits, that coyotes feed on grow unchecked and compete with livestock for rangeland vegetation. This can reduce rangeland productivity and cause much larger economic losses than the coyotes.

Between 1972 and 1985, the EPA outlawed the use of poisoned bait for predator control on public and private land because the poisons could accidentally kill nontarget animals, including dogs, endangered species, and human beings. In 1985, however, under pressure from ranchers, western congressional representatives, and the Reagan administration, the EPA decided to again allow the use of poison in collars placed on livestock.

Conservation groups opposed any use of poison, because it can kill nontarget species including dogs, golden eagles, and human beings. They suggest that fences, guard dogs, predator repellants, and penning sheep and cattle together and then allowing them to graze together (the cattle butt and kick at predators and thus protect the sheep) be used to keep coyotes and other predators away.

Forests precede civilizations, deserts follow them.
François-René de Chateaubriand

DISCUSSION TOPICS

1. Should more wilderness areas and wild and scenic rivers be preserved in the United States? Explain.

2. Discuss the pros and cons of each of the following suggestions for the use of national parks:

   a. Entrance, activity, and private concessionaire fees should be increased to the point where they pay for 25% of the costs of operating the National Park System.

   b. All private vehicles should be kept out.

   c. Campgrounds, lodges, and other commercial facilities in parks should be moved to nearby areas outside the parks.

3. Should the annual budget for both restoration of existing national parks and purchase of additional parkland be increased? Explain.

4. Should a large fraction of the world's existing uneven-aged, mixed forests be converted into even-aged tree plantations? Explain.

5. What difference, if any, could the loss of most of the world's tropical forests have on your life?

6. Should tax dollars continue to be used for building and maintaining roads into areas of national forests harvested by private companies and for reforesting such areas, or should some or all of these expenses be borne by the companies who profit from such harvests? Explain.

7. Should many of the old-growth forests on public lands, primarily in the western United States, be logged? Explain.

8. Should fees for grazing on federally owned lands be eliminated and replaced with a competitive bidding system? Why or why not? Why would such a change be politically difficult?

9. Should the poisoning and hunting of livestock predators be allowed on federal rangelands? Why or why not? Try to have both a rancher and a wildlife scientist present to your class their viewpoints on this controversial issue.

10. Should trail bikes, dune buggies, snowmobiles, and other off-road vehicles be banned from all national forests, parks, and wilderness areas? Why or why not?

13

Wild Plant and Animal Resources

GENERAL OBJECTIVES

1. Why are wild species of plants and animals important?

2. What natural and cultural factors cause wild species to become endangered and extinct?

3. How can wild species be protected from premature extinction as a result of human activities?

4. How can populations of desirable animal game species be managed to ensure their availability for sport hunting without endangering their long-term survival?

5. How can populations of desirable species of freshwater and marine fish be managed to ensure their availability for sport and commercial fishing without endangering their long-term survival?

Love the animals, love the plants, love everything. If you love everything, you will perceive the divine mystery in things. Once you perceive it, you will begin to comprehend it better every day. And you will come at last to love the whole world with an all-embracing love.

Fyodor Dostoyevski, The Brothers Karamazov

In the 1850s Alexander Wilson, a prominent ornithologist, watched a single migrating flock of passenger pigeons darken the sky for more than 4 hours. He estimated that this flock of more than 2 billion birds was 384 kilometers (240 miles) long and 1.6 kilometers (1 mile) wide.

By 1914 the passenger pigeon (Figure 13-1) had disappeared forever. How could the species that was once the most numerous bird in North America become extinct in only a few decades? The major reasons for the extinction of this species were commercial hunting and loss of habitat and food supplies as forests were cleared for farms and cities. Passenger pigeons made excellent eating and good fertilizer. They were easy to kill because they flew in gigantic flocks and nested in long, narrow colonies. People used to capture one pigeon alive and tie it to a perch called a stool; soon a curious flock alighted beside this "stool pigeon" and were shot or trapped by nets that might contain more than 1,000 birds. Beginning around 1858, massive killing of passenger pigeons became a big business. Shotguns, fire, traps, artillery, and even dynamite were used. Some live birds served as targets in shooting galleries. In 1878 one professional pigeon trapper made $60,000 by killing 3 million birds at their nesting grounds near Petoskey, Michigan.

By the early 1880s, intensive commercial hunting ceased, because the species had been reduced to only several thousand. At this point, recovery was essentially impossible, and the population continued to decline because passenger pigeons laid only one egg per nest and were susceptible to death from infectious disease and from severe storms during their annual fall migration to Central and South America. By 1896 the last massive breeding colony had vanished. In 1914 the last known passenger pigeon on earth died in the Cincinnati Zoo.

Does it really matter that a wild species such as the passenger pigeon became extinct and that the existence of numerous wild plant species, such as an orchid known as the small whorled pegonia, and animal species, such as the whooping crane, is threatened—primarily by human activities?

13-1 WHY PRESERVE WILD PLANT AND ANIMAL SPECIES?

Economic and Health Significance Certain wild species, known as **wildlife resources,** are important because of their actual or potential economic value to people. Wildlife resources that provide sport in the form of hunting or fishing are known as **game species.**

Wildlife resources provide people with a wide variety of direct economic benefits as sources of food, spices, flavoring agents, scents, soap, cooking oils, lubricating oils, waxes, dyes, natural insecticides, paper, fuel, fibers, leathers, furs, natural rubber, medicines, and other important materials. Most of the plants that supply 90% of the world's food today were domesticated from wild plants in the tropics. Other wild species not currently classified as wildlife resources may be needed by agricultural scientists to develop new crop strains that have higher yields and increased resistance to diseases, pests, heat, and drought.

Pollination by insects is essential for many food and nonfood plant species. Predatory insects, parasites, and disease-causing bacteria and viruses are increasingly used for the biological control of various weeds and insect pests, thus helping reduce losses of crops and trees.

About 40% of the prescription and nonprescription drugs used throughout the world have active ingredients extracted from plants and animals. Annual sales of drugs based on naturally derived chemicals amount to at least $40 billion worldwide and $20 billion in the United States. Aspirin, probably the world's most widely used drug, was developed according to a chemical "blueprint" supplied by a compound extracted from the leaves of tropical willow trees. Penicillin is produced by a fungus, and certain species of bacteria produce other lifesaving antibiotics such as tetracycline and streptomycin. A chemical that causes leaves to change color in the fall is being studied as a possible cure for colon cancer.

Many animal species are used to test drugs and vaccines and to increase our understanding of human health and disease. The nine-banded armadillo (Figure 13-2), for example, is being used to study leprosy and prepare a vaccine for this disease. The Florida

Figure 13-1 The extinct passenger pigeon. The last known passenger pigeon died in the Cincinnati Zoo in 1914.

manatee, an endangered mammal, is being used to help understand hemophilia. Many new drugs will come from currently unclassified plant and animal species, most located in tropical forests and the ocean. For example, an estimated 10% of the world's marine species contain anticancer chemicals.

Despite their present and future economic and health importance to human beings, very little is known about most of the earth's 1.7 million identified species and nothing about the estimated 5 million to 30 million undiscovered species. Less than 1% of the earth's identified plant species have been thoroughly studied to determine their possible usefulness. Loss of this biological and genetic diversity reduces our ability to respond to new problems and opportunities—as though we have thrown away millions of gifts without unwrapping them.

Aesthetic and Recreational Significance Many wild species are a source of beauty, wonder, joy, and recreational pleasure for large numbers of people. Observing leaves change color in autumn, smelling the aroma of wildflowers, watching an eagle soar overhead or a porpoise glide through the water are pleasurable experiences that cannot be measured in dollars.

Ecological Significance The most important contributions of wild species may be their roles in main-

Figure 13-2 The nine-banded armadillo is used in research to find a cure for leprosy.

Jen and Des Bartlett/Photo Researchers, Inc.

taining the health and integrity of the world's ecosystems (see Spotlight on p. 218). Ecosystem services of wild plant and animal species include provision of food from the soil and the sea, production and maintenance of oxygen and other gases in the atmosphere, filtration and detoxification of poisonous substances, moderation of the earth's climate, regulation of freshwater supplies, decomposition of wastes, recycling of nutrients essential to agriculture, production and maintenance of fertile soil, control of the majority of potential crop pests and disease carriers, maintenance of a vast storehouse of genetic material that is the source of all future adaptations to environmental change, and storage of solar energy as chemical energy in food, wood, and fossil fuels.

Because we know little about the workings of even the simplest ecosystems, we cannot be sure which species play crucial roles today, which ones have genes crucial for our survival and the survival of other species, and how many species can be removed before an ecosystem will collapse or suffer serious damage. Conservationist Aldo Leopold suggested that "the first rule of intelligent tinkering is to keep all the parts."

Ethical Significance   Some believe that for humans to hasten the extinction of any species is ethically and morally wrong. Some ethical theorists go further, arguing that each individual wild creature has an inherent right to survive without human interference, just as each human being has the inherent right to survive. In practice, most advocates of an ethical position argue that only a species—not each individual organism—has an inherent right to survive, regardless of whether it has any present or potential future use for people.

13-2   HOW SPECIES BECOME ENDANGERED AND EXTINCT

Extinction of Species   At least 90% of the half a billion or so different species estimated to have lived on earth have either become extinct or have evolved into a form sufficiently different to be identified as a new species. Over the 3.6 billion years since life is believed to have begun, new species have formed at a higher rate than the extinction rate; hence the present accumulation of at least 5 billion—perhaps 30 billion—different species.

Since agriculture began about 10,000 years ago, species extinction has increased at an alarming rate, especially since 1900 as human settlements have expanded worldwide. The rate is now accelerating rapidly. Rough estimates indicate that between 8000 B.C. and A.D. 1975, the average extinction rate of mammal and bird species increased about a thousandfold (Figure 13-4).

If the extinction of species of plants and insects is included, the estimated extinction rate in 1975 was 100

People tend to divide plants and animals into "good" and "bad" species and to assume that we have a duty to wipe out the villains. Consider the American alligator, which is hunted for its hide in its marsh and swamp habitat (Figure 13-3). Between 1950 and 1960, Louisiana lost 90% of its alligators; the alligator population in the Florida Everglades also was threatened.

Many people might say, "So what?" But they are overlooking the key role the alligator plays in subtropical wetland ecosystems such as the Everglades. Alligators dig deep depressions, or "gator holes," which collect fresh water during dry spells and provide a sanctuary for aquatic life as well as fresh water and food for birds and other animals. Large alligator nesting mounds also serve as nest sites for birds such as herons and egrets. As alligators move from gator holes to nesting mounds, they help keep waterways open. In addition, by eating large numbers of gar, a fish that preys on other fish, alligators help maintain populations of game fish such as bass and bream.

In 1968 the U.S. government placed the American alligator on the endangered species list. Protected from hunters, by 1975 the alligator population had made a comeback in many areas. Indeed, it had reestablished its population too successfully in some places. People began finding alligators in their backyards and swimming pools. Although the American alligator still remains on the endangered species list in some areas, it has been downgraded from endangered to threatened in Louisiana, Texas, and Florida and in some areas of Georgia and South Carolina. Limited hunting is now allowed in some areas to keep its population from growing too large.

Figure 13-3 In 1968 the American alligator was classified as an endangered species in the United States. After being protected, its populations in various areas have increased to the point where it has been reclassified as a threatened species.

U.S. Fish and Wildlife Service

species a year—*an average of 1 species every 3 days* (Figure 13-5). Biologist Edward O. Wilson estimated the extinction rate by 1985 had increased tenfold to 1,000 species a year—*an average of 3 species a day.* Wilson and several other biologists and conservation experts warn that if deforestation (especially of tropical moist forests), desertification, and destruction of wetlands and coral reefs continue at their present rates, at least 500,000 and perhaps 1 million species will become extinct as a result of human activities between 1975 and 2000. Using the lower estimate, this amounts to an average extinction rate by 2000 of 20,000 species a year, or *1 species every 30 minutes*—a 200-fold increase in the extinction rate in only 25 years. Most of these species will be plants and insects that have yet to be classified as species, much less evaluated for their use to people and for their roles in ecosystems.

Although animal extinctions receive the most publicity, plant extinctions are more important ecologically, because most animal species depend directly or indirectly on plants for food. An estimated 10% of the world's plant species are already threatened with extinction and an estimated 15% to 25% of all plant species face extinction by 2000.

Some analysts contend that the projected extinction rate for 2000 is only a wild guess and that it may greatly overstate the situation. But even if the average extinction rate is only 1,000 a year by the end of this century, this will rank as one of the greatest mass extinctions since the dawn of life on this planet.

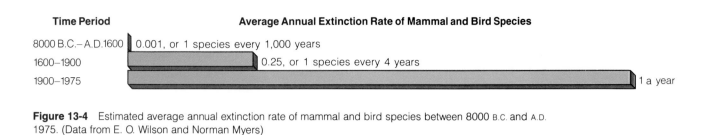

Time Period	Average Annual Extinction Rate of Mammal and Bird Species
8000 B.C.– A.D.1600	0.001, or 1 species every 1,000 years
1600–1900	0.25, or 1 species every 4 years
1900–1975	1 a year

Figure 13-4 Estimated average annual extinction rate of mammal and bird species between 8000 B.C. and A.D. 1975. (Data from E. O. Wilson and Norman Myers)

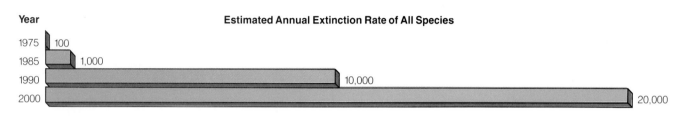

Year	Estimated Annual Extinction Rate of All Species
1975	100
1985	1,000
1990	10,000
2000	20,000

Figure 13-5 Estimated annual extinction rate of all species between 1975 and 2000. (Data from E. O. Wilson and Norman Myers)

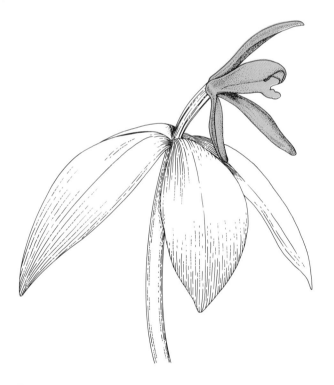

Figure 13-6 The endangered small whorled pegonia. By 1985 only about 3,200 plants of this orchid species existed in 13 states from Maine to northern Georgia.

Threatened and Endangered Species An **endangered species** is one having so few individual survivors that the species could soon become extinct in all or most of its natural range. Examples in the United States are the whooping crane, the California condor, and an orchid called the small whorled pegonia (Fig-ure 13-6). **Threatened species,** such as the grizzly bear and the bald eagle in the United States, are still abundant in their range but are declining in numbers and likely to become endangered in the foreseeable future.

Habitat Disturbance and Loss The greatest threat to wild plant and animal species is destruction or alteration of habitat: the area where species seek food, find shelter, and breed. As the human population grows, urban settlements, draining and filling of wetlands, clearing forests, and production of food, minerals, energy, and other resources destroy or disrupt habitats for many wild species. Disturbance and loss of habitat has been a major factor in the extinction of such magnificent U.S. bird species as the ivory-billed woodpecker, the near extinction of the whooping crane, and the critical endangerment of the California condor (Figures 13-7 and 13-8).

Commercial Hunting There are three major types of hunting: **commercial hunting,** in which animals are killed for profit from sale of their furs or other parts; **subsistence hunting,** the killing of animals to provide enough food for survival; and **sport hunting,** the killing of animals for recreation. Although subsistence hunting was once a major cause of extinction of some species, it has now declined sharply in most areas. Sport hunting is now closely regulated in most countries; game species are endangered only when protective regulations don't exist or are not enforced.

On a worldwide basis, commercial hunting threatens a number of large animal species. The jag-

Figure 13-7 The whooping crane is an endangered species in the United States. Since being protected, its population in the wild increased from about 29 to 138 individuals by 1986.

Figure 13-8 The critically endangered California condor is no longer found in the wild. By 1987 only 27 individuals remained in captivity in zoos.

uar, tiger, snow leopard, and cheetah are hunted for their furs. Alligators are hunted commercially for their skins, elephants for their ivory tusks (accounting for the slaughter of about 90,000 elephants a year), and rhinoceroses for their horns.

For example, a single rhino horn—a mass of compact hair—is worth as much as $24,000 on the black market. It is used to make handles for ornamental knives in North Yemen and ground into a powder and used for medicinal purposes, especially reducing fever, in parts of Asia. It is also thought by many Asians to be an aphrodisiac, or sexual stimulant, even though it consists of a substance (keratin) that can be obtained by eating hair trimmings and fingernails. Although 60 countries have agreed not to import or export rhino horns, illegal traffic goes on because of its high market value. Between 1970 and 1986, the number of black rhinos in Africa dropped from 65,000 to about 5,000, and only about 100 white rhinos were left by 1986 (Figure 13-9). If poaching continues at present rates, all species of rhino will be extinct within a decade.

Predator and Pest Control Extinction or near extinction can also occur because of attempts to exterminate pest and predator species that compete with

people and livestock for food and with people for game species. The Carolina parakeet was exterminated in the United States around 1914 because it fed on fruit crops. Its disappearance was hastened by the fact that when one member of a flock was shot, the rest of the birds hovered over its body, making themselves easy targets.

Ranchers, hunters, and government employees involved in predator control programs have sharply reduced populations of large predators such as the timber wolf, the mountain lion, and the grizzly bear over most of the continental United States. Campaigns to protect rangeland for grazing livestock by poisoning prairie dogs and pocket gophers have eliminated these rodents' natural predator, the black-footed ferret (Figure 13-10) from the wild.

Pets, Medical Research, and Zoos Worldwide, more than 6 million live wild birds are sold each year, most of them ending up as pets in countries such as the United States, Great Britain, and West Germany. Large numbers of these animals die during shipment, and after purchase many others are killed or abandoned by their owners. As a direct result of this trade, at least nine bird species are now listed as threatened or

Figure 13-9 Only about 100 critically endangered African white rhinos remain.

Figure 13-10 The highly endangered black-footed ferret no longer is found in the wild in the United States. Only 18 remain in captivity.

endangered but continue to be smuggled illegally into the United States and Europe. For example, bird collectors may pay as much as $10,000 for a threatened hyacinth macaw illegally smuggled out of Brazil.

Some species of exotic plants, especially orchids and cacti, are also endangered because they are gathered, often illegally, and sold to collectors and used to decorate houses, offices, and landscapes. A single rare orchid may be sold for $5,000 to a collector. Nearly one-third of the cactus species native to the United States, especially in Texas and Arizona, are thought to be endangered, because they are collected and sold for use as potted plants.

About 71 million animals—mice, rats, dogs, cats, primates, birds, frogs, guinea pigs, rabbits, and hamsters—are used each year in the United States for toxicity testing, biomedical and behavioral research, and drug development. Although most test animal species are not endangered, medical research coupled with habitat loss is a serious threat to endangered wild primates such as the chimpanzee and the orangutan.

Under pressure from animal-rights groups, scientists are trying to find alternative testing methods that do not subject animals to suffering or—better yet—do not use animals at all. Promising alternatives include the use of cell and tissue cultures, simulated tissues

and body fluids, bacteria, and computer-generated mathematical models that enable scientists to estimate the toxicity of new compounds from knowledge of chemical structure and properties.

Public zoos, botanical gardens, and aquariums are under constant pressure to exhibit rare and unusual animals such as the orangutan. For each exotic animal or plant that reaches a zoo or botanical garden alive, many others die during capture or shipment. Since 1967, reputable zoos and aquariums have agreed not to purchase endangered species, although some abuses still occur.

Pollution   Chemical pollution is a relatively new but growing threat to wildlife. Industrial wastes, mine acids, acid deposition, and excess heat from electric power plants have wiped out some species of fish, such as the humpbacked chub, in local areas. Slowly biodegradable pesticides, especially DDT and dieldrin, have been magnified in food chains and have caused reproductive failure and eggshell thinning of important birds of prey, such as the peregrine falcon, eastern and California brown pelicans, osprey, and bald eagle. Banning persistent pesticides in North America and Europe has allowed most species to recover. Yet these chem-

Figure 13-11 European wild rabbits around a water hole in Australia before a virus was introduced to control this species.

Australian Information Service

icals are still exported in large quantities by U.S. companies for use in LDCs.

Introduction of Alien Species When an alien species is introduced into a new geographical area, it may be able to establish itself without seriously affecting the population size of native species. In other cases, however, an alien species can cause a population decrease or even extinction of one or more existing species by preying on them, beating them in the competition for food, or destroying their habitat. It can also cause a population explosion of existing species by killing off their natural predators. Island species are particularly vulnerable, because many have evolved in ecosystems with few if any natural herbivores or carnivore predators.

In 1859 a farmer in southern Australia imported a dozen pairs of wild European rabbits as a game animal. Within six years these 24 rabbits had mushroomed to 22 million that by 1907 had reached every corner of the country. By the 1930s their population had reached an estimated 750 million. They competed with sheep for grass and cut the sheep population in half. They also devoured food crops, gnawed young trees, fouled water holes (Figure 13-11), and accelerated soil erosion in many places. In the early 1950s, about 90% of the rabbit population was killed by the deliberate human introduction of a virus disease. There

is concern, however, that members of the remaining population may eventually develop immunity to this viral disease through natural selection and again become the scourge of Australian farmers and ranchers.

The kudzu vine was deliberately imported into the southeastern United States from Africa to help control soil erosion. It does control erosion, but its growth is so prolific that it has also covered hills, trees, houses, roads, stream banks (Figure 13-12), and even entire patches of forest. People have dug it up, cut it up, burned it, and tried to kill it with herbicides, all without success. Table 13-1 lists examples of other accidental and deliberate introductions of alien species into the United States.

Characteristics of Extinction-prone Species Some species have certain natural characteristics that make them more susceptible to extinction by human activities and natural disasters than other species (Table 13-2). Each species has a critical population density and size, below which survival may be impossible because males and females have a hard time finding one another. Once this point is reached, population size continues to decline even if the species is protected, because its death rate exceeds its birth rate. The remaining small population is also highly vulnerable to extinction from fires, floods, and other catastrophic events.

Figure 13-12 Kudzu covering stream banks and a patch of forest in Georgia. So far no one has devised a way to stop the spread of this prolific and hardy alien plant throughout the southeastern United States.

USDA, Soil Conservation Service/Paul Tabor

13-3 PROTECTING WILD SPECIES FROM EXTINCTION

The Species Approach: Treaties and Laws Organizations such as the International Union for the Conservation of Nature and Natural Resources (IUCN), the International Council for Bird Preservation, and the World Wildlife Fund have identified threatened and endangered species and led efforts to protect them. The IUCN, for example, regularly compiles lists of threatened and endangered species and publishes them in *The Red Data Book.*

Several international treaties and conventions now protect wild species, but most involve only two or a small number of countries. One of the most far-reaching treaties is the 1975 Convention on International Trade in Endangered Species (CITES), developed after 10 years of work by the IUCN and administered by the UN Environmental Programme. This treaty, now signed by 93 countries, bans hunting or capturing of 700 endangered and threatened species.

Although enforcement varies, the treaty has reduced illegal trade of such species. However, Singapore, a major international center for distributing wildlife and wildlife products, and several other countries involved in wildlife trade have not signed this agreement. Even when illegal-wildlife smugglers are caught, the penalties are usually too low to hurt over-

all profits. In 1979 a Hong Kong fur dealer apprehended for illegally importing 319 Ethiopian cheetah skins valued at $160,000 was fined only $1,540.

Although a number of countries officially offer protection to endangered or threatened species, the most strictly enforced protection is provided by the United States, the Soviet Union, and Canada. Since 1903 several pieces of legislation have been passed in the United States to protect endangered species. One of these, the Endangered Species Act of 1973 (including amendments in 1982 and 1987), is one of the toughest environmental laws enacted by any country. It authorizes the National Marine Fisheries Service of the Department of Commerce to identify and list endangered and threatened marine species, and the Fish and Wildlife Service (FWS) to identify all other plant and animal species endangered or threatened in the United States and abroad. Any decision by either agency to add or remove a species from the list must be based solely on biological grounds, without economic considerations. By 1987 the federal list of endangered and threatened animals and plants protected under this act contained 928 species, including 385 found in the United States and its territories.

The Endangered Species Act also prohibits interstate and international commercial trade of endangered or threatened plant or animal species (with certain exceptions) or products made from such species, and it prohibits killing, hunting, collecting, or injuring

Table 13-1   Damage Caused by Plants and Animals Imported into the United States

Name	Origin	Mode of Transport	Type of Damage
Mammals			
European wild boar	Russia	Intentionally imported (1912), escaped captivity	Destruction of habitat by rooting; crop damage
Nutria (cat-sized rodent)	Argentina	Intentionally imported, escaped captivity (1940)	Alteration of marsh ecology; damage to levees and earth dams; crop destruction
Birds			
European starling	Europe	Intentionally released (1890)	Competition with native songbirds; crop damage; transmission of swine diseases; airport interference
House sparrow	England	Intentionally released by Brooklyn Institute (1853)	Crop damage; displacement of native songbirds
Fish			
Carp	Germany	Intentionally released (1877)	Displacement of native fish; uprooting of water plants with loss of waterfowl populations
Sea lamprey	North Atlantic Ocean	Entered via Welland Canal (1829)	Destruction of lake trout, lake whitefish, and sturgeon in Great Lakes
Walking catfish	Thailand	Imported into Florida	Destruction of bass, bluegill, and other fish
Insects			
Argentine fire ant	Argentina	Probably entered via coffee shipments from Brazil (1918)	Crop damage; destruction of native ant species
Camphor scale insect	Japan	Accidentally imported on nursery stock (1920s)	Damage to nearly 200 species of plants in Louisiana, Texas, and Alabama
Japanese beetle	Japan	Accidentally imported on irises or azaleas (1911)	Defoliation of more than 250 species of trees and other plants, including many of commercial importance
Plants			
Water hyacinth	Central America	Intentionally introduced (1884)	Clogging waterways; shading out other aquatic vegetation
Chestnut blight (a fungus)	Asia	Accidentally imported on nursery plants (1900)	Destruction of nearly all eastern American chestnut trees; disturbance of forest ecology
Dutch elm disease, *Cerastomella ulmi* (a fungus, the disease agent)	Europe	Accidentally imported on infected elm timber used for veneers (1930)	Destruction of millions of elms; disturbance of forest ecology

From *Biological Conservation* by David W. Ehrenfeld. Copyright © 1970 by Holt, Rinehart and Winston, Inc. Modified and reprinted by permission.

Table 13-2	Characteristics of Extinction-prone Species	
Characteristic	Examples
Low reproductive rate	Blue whale, polar bear, California condor, Andean condor, passenger pigeon, giant panda, whooping crane
Specialized feeding habits	Everglades kite (apple snail of southern Florida), blue whale (krill in polar upwelling areas), black-footed ferret (prairie dogs and pocket gophers), giant panda (bamboo)
Feed at high trophic levels	Bengal tiger, bald eagle, Andean condor, timber wolf
Large size	Bengal tiger, African lion, elephant, Javan rhinoceros, blue whale, American bison, giant panda, grizzly bear
Limited or specialized nesting or breeding areas	Kirtland's warbler (nests only in 6- to 15-year-old jack pine trees), whooping crane (depends on marshes for food and nesting), orangutan (now found only on islands of Sumatra and Borneo), green sea turtle (lays eggs on only a few beaches), bald eagle (preferred habitat of forested shorelines), nightingale wren (nests and breeds only on Barro Colorado Island, Panama)
Found in only one place or region	Woodland caribou, elephant seal, Cooke's kokio, and many unique island species
Fixed migratory patterns	Blue whale, Kirtland's warbler, Bachman's warbler, whooping crane
Preys on livestock or people	Timber wolf, some crocodiles
Certain behavioral patterns	Passenger pigeon and white-crowned pigeon (nests in large colonies), redheaded woodpecker (flies in front of cars), Carolina parakeet (when one bird is shot, rest of flock hovers over body), Key deer (forages for cigarette butts along highways—it's a "nicotine addict")

any protected animal species. It also directs federal agencies not to carry out, fund, or authorize projects that would jeopardize endangered or threatened species or destroy or modify their habitats. This last provision has been highly controversial (see Spotlight on p. 226).

Conservationists complain that the Endangered Species Act is not being carried out as intended by Congress primarily because of budget cuts and lack of staff. At current rates it will take the Fish and Wildlife Service 29 years to evaluate the 4,000 other species on the waiting list for possible protection. A dozen or so species have already disappeared while awaiting classification, and many others are expected to become extinct before they can be protected. Conservationists also note that relatively few plants have been given protection, despite their ecological importance as the base of food webs. This has occurred because much less is known about plants than animals (thus lengthening the review process for plants), lack of funds, and traditional emphasis on animal species.

Listing a species is only the first step. Once a species is listed, the FWS is supposed to prepare a plan to enhance its recovery. Of the 385 listed U.S. species, only 58% had approved recovery plans by 1986, and only about half of these plans were being actively implemented. Since 1973 when the Endangered Spe-

cies Act was enacted, fewer than 25 endangered species with recovery plans have made progress toward recovery, and only 3 species (all birds native to small islands in the Pacific Ocean) have recovered enough to be taken off the the endangered and threatened species lists.

The Species Approach: Wildlife Refuges In 1903 President Theodore Roosevelt established the first federal wildlife refuge in the United States at Pelican Island on the east coast of Florida to protect the endangered brown pelican (Figure 13-13). By 1987 the National Wildlife Refuge System contained 437 refuges managed by the Fish and Wildlife Service. About 88% of the area encompassed by this system is in Alaska. This system is supplemented by a number of state and privately owned wildlife refuges managed primarily for the benefit of hunters.

Although more than three-fourths of the refuges are wetlands for protection of migratory waterfowl, many other species are also protected in such refuges. Some refuges have been set aside for specific endangered species and have helped species such as the Key deer and the brown pelican of southern Florida and the trumpeter swan to recover.

Congress has not established guidelines (such as multiple use or sustained yield) for management of

In 1975 conservationists filed suit against the Tennessee Valley Authority to stop construction of the $137 million Tellico Dam on the Little Tennessee River in Tennessee because the area to be flooded by the resulting reservoir was the only known breeding habitat of an endangered fish species, the snail darter—a three-inch-long minnow. Although the dam was 90% completed, construction was halted by the court action for several years.

In 1978 Congress amended the Endangered Species Act to permit a seven-member review committee to grant an exemption if it believed that the economic benefits of a project would outweigh the potential harmful ecological effects. At their first meeting, the review committee denied the request to exempt the Tellico Dam project on the grounds that it was an economically unsound, "pork barrel" project. Despite this decision, political influence by members of Congress from Tennessee and others afraid of losing present and future projects in their states prevailed, and in 1979 Congress passed special legislation exempting the Tellico Dam from the Endangered Species Act. The dam's reservoir is now full of water. The snail darters that once dwelled there were transplanted to nearby rivers.

In 1981 snail darter populations were found in several remote tributaries of the Little Tennessee River, and in 1983 their status was downgraded by the FWS from endangered to threatened. The important question raised by this incident is to what degree, if any, economic considerations should influence the protection of endangered and threatened species. What do you think?

Figure 13-13   The first federal wildlife refuge was established at Pelican Island in Florida in 1903 to protect the brown pelican from extinction. Although its numbers have increased, it is still an endangered species.

the head of the FWS to open as many refuges as possible to hunting.

Pollution is also a problem in a number of refuges. A 1986 study by the FWS revealed that perhaps one in five refuges is contaminated with toxic chemicals. A 1983 survey by the FWS showed that 86% of the federal refuges were experiencing water quality problems and 67% were facing air quality and visibility problems.

the National Wildlife Refuge System as it has for other public land systems. As a result, the FWS has allowed a number of refuges to be used for a variety of purposes, including hunting (260 of the 437 refuges in 1987), trapping, timber cutting, grazing, farming, oil and gas development, mining, and recreational activities. Since 1980 the Reagan administration has encouraged expansion of such commercial activities to offset the costs of operating refuges.

Conservationists charge that some of these commercial uses are getting out of hand, are not always controlled properly, and can interfere with wildlife protection. They are especially opposed to oil and gas development on refuges and to the 1984 directive by

The Species Approach: Gene Banks, Zoos, Botanical Gardens, and Aquariums   Plant gene banks (collections of varieties of plants) of most known and many potential varieties of agricultural crops and other plants now exist throughout the world, and scientists have urged that many more be established. Despite their importance, gene banks have significant disadvantages and need to be supplemented by preservation of a variety of representative ecosystems throughout the world. Storage is not possible for many species, such as potatoes, fruit trees, orchids, and many tropical plant species. Many seeds rot and must periodically be replaced. Accidents such as power failures, fires, and unintentional disposal of seeds can cause

© Zoological Society of San Diego, 1985

Figure 13-14 The Arabian oryx barely escaped extinction in 1969 after being overhunted in the deserts of the Middle East. Captive breeding programs in zoos in Arizona and California have saved this antelope species from extinction.

irrecoverable losses. Furthermore, stored species do not continue to evolve and thus become less fit for reintroduction into their native habitats, which may have undergone various environmental changes.

Zoos, botanical gardens, and aquariums are increasingly being used as a last-ditch effort to preserve a representative number of individuals of critically endangered species that would otherwise become extinct. When it is judged that an animal species will not survive on its own, eggs may be collected and hatched in captivity (known as egg pulling), or captive breeding programs may be established in zoos or private research centers. Other techniques include artificial insemination of species that don't breed well in captivity and using adults of one related species to serve as foster parents to hatch collected eggs and raise offspring of another species.

In some, but not all, cases, captive breeding and egg pulling programs allow a population to increase enough so that the species can be successfully reintroduced into the wild. For example, captive breeding programs at zoos in Phoenix, San Diego, and Los Angeles have been used to save the nearly extinct Arabian oryx antelope (Figure 13-14), which is now being returned in small numbers to the wild.

Captive breeding can help save some critically endangered species, but it has several disadvantages. It is expensive, and zoos and botanical gardens have room for relatively few of the critically endangered species because of the large number of individuals (a minimum of 100 and ideally 250 to 500 for an animal species) needed for each species to avoid extinction through accident, disease, or loss of genetic variability through inbreeding. Currently the world's zoos contain only 20 endangered species with populations of 100 or more individual animals. Thus just as doctors must make difficult decisions about which individuals receive transplants of scarce organs from donors, wildlife experts must decide which species should be saved.

The Ecosystem Approach Most wildlife biologists argue that the major threat to most wildlife species today—namely, the destruction of habitat—cannot be halted by concentrating on preserving individual species. Instead, they believe that the most effective way to prevent the loss of wild species is to establish and maintain a worldwide system of reserves, parks, and other protected areas that would represent at least 10% of the world's land area and include at least 5 examples of each of the world's 193 major ecosystem types or biogeographical realms. Ideally each reserve would be large enough to sustain viable populations of most of its existing species.

Such a global network of ecological reserves would help protect the earth's existing biological and genetic diversity, provide sufficient natural habitats for reintroducing endangered species now in zoos and other artificial habitats, provide scientists with opportunities for research, and cost less to run than management of endangered and threatened species one by one.

By 1985 more than 3,500 major protected areas had been established throughout the world, totaling 4.3 million square kilometers (1.6 million square miles). Although this is an important beginning, it represents less than 3% of the earth's land area, and many of the world's different ecosystem types have not been included or consist of too little protected area. Furthermore, many of the protected areas are too small to protect their populations of wild species.

In 1980 the International Union for the Conservation of Nature and Natural Resources, the UN Environmental Programme, and the World Wildlife Fund published a long-range plan for conserving the world's biological resources. The three major goals of this World Conservation Strategy are to (1) maintain essential ecological processess and life-support systems on which human survival and economic activities depend; (2) preserve genetic diversity; and (3) ensure that any use of species and ecosystems is sustainable.

By 1985 some 40 countries had started or completed national conservation strategies. If this program is supported adequately by MDCs, it offers a glimmer of hope for preserving the world's biological and genetic diversity and thus enhancing the long-range sustainability of the resource base on which human survival and economic activities depend.

13-4 WILDLIFE MANAGEMENT

Management Approaches The science and art of **wildlife management** involves manipulating populations of wild species and their habitats for human benefit, the welfare of other species, and the preservation of endangered and threatened species. The first step in the process of wildlife management is to decide which species or groups of species are to be managed in a particular area.

Once management goals have been decided, the wildlife manager must develop a management plan. Ideally such a plan is based on an understanding of the population dynamics and the cover, food, water, space, and other habitat requirements of each species to be managed. Often, however, such information is not available or reliable and is usually quite difficult, expensive, and time-consuming to obtain.

This helps explain why wildlife management is as much an art as a science. In practice, it involves considerable guesswork, trial and error, and adaptation of plans to political pressures from conflicting groups, unexpected short- and long-term consequences of interfering in nature's processes, and lack of sufficient

funds. The two major approaches used to manage desired species are population regulation and manipulation of habitat vegetation.

Population Regulation by Controlled Hunting Wildlife managers usually use population management to manipulate the numbers, sexes, and age distributions of populations of wild game species. In the United States, populations of game animals are managed by laws that (1) specify certain times of the year for the hunting of a particular species with various types of equipment, such as bows and arrows, muzzle-loading guns, shotguns, or rifles; (2) restrict the length of hunting seasons; (3) regulate the number of hunters permitted in an area; and (4) limit the size, number, and sex of animals allowed to be killed. Some individuals and conservation groups, however, are opposed to sport hunting (see Spotlight on p. 230).

Wildlife managers also attempt to control or eliminate wild species considered pests by farmers, ranchers, loggers, hunters, hikers, and other groups. Methods for controlling animal pests include (1) killing by hunting, trapping, and poisoning; (2) using fences or other devices to prevent them from reaching certain areas; (3) moving them to another location; and (4) using chemicals to control their fertility and thus reduce their population size.

Manipulation of Habitat Vegetation and Water Supplies Wildlife managers can encourage growth of plant species that are the preferred food and cover for an animal species whose population is to be managed. They do this primarily by controlling the stages of ecological succession of vegetation in various areas.

Grizzly bear, wolf, caribou, and bighorn sheep are examples of **wilderness species,** which flourish only in relatively undisturbed climax vegetational communities, such as large areas of mature forest, tundra, grassland, and desert. Their survival depends to a large degree on the establishment of large state and national wilderness areas and wildlife refuges. Wild turkey, marten, and gray squirrel are examples of **late-successional species,** whose habitats require establishment and protection of moderate-size mature forest refuges.

Mid-successional species such as elk, moose, deer, grouse, and snowshoe hare are found around abandoned croplands and partially open areas created in forests by logging of small stands of timber, controlled burning, or applying herbicides. The removal of old vegetation promotes the growth of vegetation favored as food by mid-successional mammal and bird species. It also increases the amount of edge habitat, where two communities such as a forest and a field come

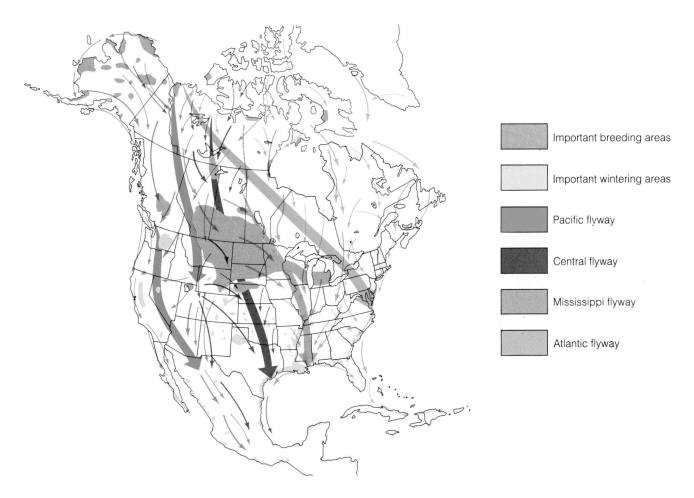

Figure 13-15 Major breeding and wintering areas and fall migration flyway routes used by migratory waterfowl in North America.

Legend:
- Important breeding areas
- Important wintering areas
- Pacific flyway
- Central flyway
- Mississippi flyway
- Atlantic flyway

together. This zone allows animals to feed on vegetation in the fields and quickly escape to cover in the nearby forest.

Early-successional species such as rabbit, quail, ring-necked pheasant, and dove find food and cover in the weedy pioneer plants that invade an area that has been cleared of vegetation for human activities and then abandoned.

Management of Migratory Waterfowl Migratory waterfowl such as ducks, geese, and swans require some special management approaches. Many of these species nest in Canada during the summer and migrate to the United States and Central America along generally fixed routes called *flyways* during the fall hunting season (Figure 13-15). Thus international agreements are needed to prevent destruction of their winter and summer habitats and to prevent overhunting. In addition, it is necessary to provide protected stopovers along flyways, where migrating birds can rest and feed. In 1934 Congress passed the Migratory Bird Hunting Stamp Act, which required waterfowl hunters to purchase a special hunting permit, called a duck stamp, each year. The sale of these permits has brought in $300 million, which has been used for waterfowl research and to acquire, maintain, and develop waterfowl refuges.

To develop new waterfowl habitats and improve existing ones, game managers can (1) periodically drain ponds to keep them from becoming clogged with vegetation; (2) create channels and openings in dense marsh vegetation to allow birds to feed and move about; and (3) construct artificial ponds, nesting islands, and nesting sites.

13-5 FISHERY MANAGEMENT

Freshwater Fishery Management The goals of **freshwater fish management** are to encourage the growth of populations of desirable commercial and sport fish species and to reduce or eliminate popula-

Sport hunters, hunting groups, and state game officials contend that Americans should be free to hunt as long as they obey state and local game regulations and don't damage wildlife resources. Wildlife managers point out that since human beings have eliminated most natural predators of deer and a number of other large game animals, carefully regulated sport hunting can keep the populations of game species within the carrying capacity of the available habitat and prevent excessive destruction of vegetation and loss of other species.

Conservation groups like the Sierra Club and Defenders of Wildlife also consider hunting an acceptable management tool to keep num-bers in line with habitat capacity. Defenders of sport hunting point out that fees from hunting licenses and taxes on sport firearms and ammunition have provided more than $1.5 billion since 1931 for the acquisition, restoration, and mainte-nance of wildlife habitat areas and for wildlife research.

On the other hand, some indi-viduals and conservation groups such as the Humane Society oppose sport hunting on the grounds that it inflicts needless cruelty on animals. The Humane Society has filed suit to block all hunting in national wild-life refuges on the grounds that state game commissions often set hunting limits to cater to hunters' demands, not to keep wildlife in balance. The Humane Society also points out that sport hunting tends to reduce the genetic quality of remaining wildlife populations, because hunters usually go after the strongest and healthiest individuals. In contrast, natural predators tend to improve population quality by eliminating weak and sick individ-uals. Antihunting groups contend that populations of wild animal spe-cies, including deer, are eventually controlled by lack of food and other natural factors. These groups recom-mend that in areas where vegetation is being destroyed, wildlife officials should reintroduce natural preda-tors, not hunters. What do you think?

---

tions of less desirable species. Several techniques are used. Laws and regulations govern the timing and length of the fishing season for various species, estab-lish the minimum fish size that can be taken, set catch quotas, and require that commercial fish nets have a large enough mesh to ensure that young fish are not harvested. Natural and artificial habitats can be estab-lished, protected, and maintained by providing hid-ing places, removing debris, preventing excessive growth of aquatic plants to minimize oxygen deple-tion, and using small dams to control water flow. Hab-itats of desirable species can be protected from buildup of sediment and other forms of pollution. Predators, parasites, and diseases can be controlled by habitat improvement, breeding genetically resistant fish vari-eties, and using antibiotics and disinfectants. Hatch-eries are used to restock ponds, lakes, and streams with species such as trout and salmon.

Marine Fishery Management   The history of the world's commercial ocean fishing and whaling indus-try is an excellent example of the tragedy of the com-mons—the abuse and overuse of a potentially renew-able resource. Users of the marine fisheries tend to maximize their catch for short-term economic gain at the expense of long-term economic collapse from overfishing. As a result, many species of commer-cially valuable fish and whales found in international and coastal waters have been overfished to the point of **commercial extinction;** that is, they are so rare that it no longer pays to hunt them.

Several techniques can be used to manage marine fisheries and prevent commercial extinction. Annual quotas can be established for heavily used species, and regulations governing fishing gear and net size can be established and enforced. International and national laws have been used to extend the offshore fishing zone of coastal countries to 322 kilometers (200 nautical miles). Foreign fishing vessels are now allowed to take certain quotas of fish within such zones only with government permission. Food and game fish species, such as the striped bass along the Pacific and Atlantic coasts of the United States, can be introduced. Desirable species can be attracted by constructing arti-ficial reefs from boulders, building rubble, and tires.

The Whaling Industry   The pattern of the whaling industry has been to hunt the most commercially val-uable species until it becomes too scarce to be of com-mercial value and then turn to another species. In 1900 an estimated 4.4 million whales swam the ocean. Today only about 1 million remain, with only a few species such as the sperm whale and the minke whale making up most of the total.

Probably only a few hundred to a few thousand blue whales, the world's largest animals, remain—perhaps too few for the species to recover. The sharp decline in its population can be attributed to a greedy whaling industry, as well as to three natural characteristics of the blue whale. First, they are large and thus easy to spot. Second, they can be caught in large numbers because they congregate in their Antarctic feeding grounds for about 8 months of each year. Third, they multiply very slowly, taking up to 25 years to mature sexually and having one offspring every 2 to 5 years. Once the total population has been reduced below a certain level, mates may no longer be able to find each other, and natural death rates will exceed natural birth rates until extinction occurs. Within the next few decades, blue whales could become extinct, even though they are now protected by law.

Today 40 whaling and nonwhaling countries belong to the International Whaling Commission (IWC), established in 1946 out of concern over the sharp decline in stocks of many whale species from overfishing. The commission was empowered to regulate the annual harvest by setting hunting quotas to ensure a sustainable supply of all commercially important species. But most fishery experts agree that the annual quotas set by the IWC were so high that during its first 20 years of existence, it presided over serious depletion of nearly all the world's whale populations.

The discovery in the 1960s that the blue whale was in danger of extinction led to an international "Save the Whales" movement. In the late 1960s, conservationist and other groups began pressuring the IWC to ban all commercial whaling. Greenpeace, the largest worldwide environmentalist group (1.5 million members), has used bold, commandolike tactics to embarrass whaling countries, stop pirate whalers who ignore the quotas, and gain worldwide publicity and support for whale protection. In 1972 the IWC discussed a ban on commercial whaling, and since then the United States and several other former whaling countries have urged that such a ban be imposed.

It was not until 1982, however, that the commission voted to phase out all commercial whaling by the end of 1986. According to IWC rules, member countries that file formal protests to commission rulings do not have to abide by them. In 1982, Japan, the Soviet Union, and Norway, the three major whaling countries, filed formal protests against the ban and announced their intentions to continue commercial whaling operations. However, after the United States threatened to reduce or cut off fishing privileges in U.S. waters, Japan, Norway, and the Soviet Union agreed to stop commercial whaling by the end of 1988. Iceland has announced its intention to increase the number of whales taken for research purposes—still allowed under IWC rules—and to use the meat for domestic use, much of it as feed for mink farms. Although at least 6,000 whales were killed in 1986—when such killing was supposed to stop—the end of commercial whaling may be in sight.

It is the responsibility of all who are alive today to accept the trusteeship of wildlife and to hand on to posterity, as a source of wonder and interest, knowledge, and enjoyment, the entire wealth of diverse animals and plants. This generation has no right by selfishness, wanton or intentional destruction, or neglect to rob future generations of this rich heritage. Extermination of other creatures is a disgrace to humankind.

World Wildlife Charter

DISCUSSION TOPICS

1. Discuss your reaction to the statement "Who cares that the passenger pigeon is extinct and the blue whale, whooping crane, small whorled pegonia, and other plant and animal species face extinction because of human activities? They are important only to a bunch of bird-watchers, Sierra Clubbers, and other ecofreaks." Be honest about your reaction and present arguments for your position.

2. Do you agree or disagree with the idea that since most of the species that have existed on earth have become extinct by natural processes, we should not be concerned about extinction of wild species because of human activities? Explain.

3. Some argue that all species have an inherent right to exist regardless of whether they are useful to human beings and that they should at least be preserved in a natural habitat somewhere on earth. Do you agree or disagree with this position? Explain. Would you apply this idea to (a) anopheles mosquitoes, which transmit malaria; (b) tigers that roam the jungle along the Indian-Nepalese border and killed at least 105 people between 1978 and 1983; (c) bacteria that cause smallpox or other infectious diseases; (d) rats that compete with people for many food sources; and (e) rattlesnakes?

4. Use Table 13-2 to predict a species that may soon be endangered. What, if anything, is being done for this species? What pressures is it being subjected to? Try to work up a plan for protecting it.

5. Are you for or against sport hunting? Explain.

14

Nonrenewable Mineral Resources

GENERAL OBJECTIVES

1. What methods are used to locate and extract nonrenewable minerals from the earth's crust?

2. What are some of the harmful environmental consequences of mining, processing, and using nonrenewable minerals?

3. Are we likely to run out of affordable supplies of any essential nonrenewable nonfuel minerals in the world and in the United States in the foreseeable future?

4. How can present known supplies of key mineral resources be increased?

5. How can present and potential supplies of key mineral resources be extended through resource conservation?

We seem to believe we can get everything we need from the supermarket and corner drugstore. We don't understand that everything has a source in the land or sea, and that we must respect these sources.

Thor Heyerdahl

What do cars, spoons, beverage cans, coins, electrical wiring, bricks, and sidewalks have in common? Few people stop to think that these products and thousands of others they use every day are made from nonrenewable raw materials extracted from the earth's solid crust—the upper layer of the lithosphere (Figure 4-2, p. 52). Nonrenewable nonfuel resources are discussed in this chapter, and nonrenewable fuel resources such as coal, oil, natural gas, and uranium are discussed in Chapters 15 and 16.

14-1 LOCATING AND EXTRACTING MINERAL RESOURCES

Mineral Resource Abundance and Distribution Human ingenuity has found ways to locate and extract more than 100 nonrenewable nonfuel minerals and to transform them into most of the everyday items we use and then discard, reuse, or recycle (Figure 14-1). Any naturally occurring concentration of a free element or compound of two or more elements in solid form is called a **mineral deposit**. Although a few minerals, such as gold and silver, occur as free elements, most are found as various compounds of only eight elements that make up 99.3% of the earth's crust (Figure 14-2). The earth's remaining elements are found only in trace amounts.

Geochemical processes occurring over hundreds of millions of years during the earth's early history selectively dissolved, transported, and deposited elements and their compounds unevenly. As a result, there is a tremendous difference between the average crustal abundances of elements shown in Figure 14-2 and the amount of an element actually found at a single location. Although rich deposits of a particular

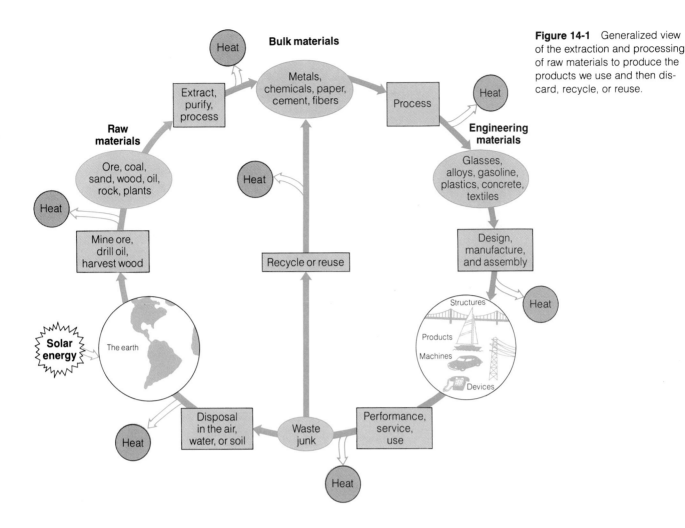

Figure 14-1 Generalized view of the extraction and processing of raw materials to produce the products we use and then discard, recycle, or reuse.

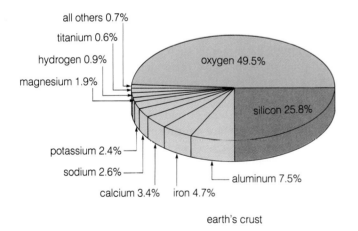

Figure 14-2 Percentage by weight of elements in the earth's crust.

mineral such as iron or copper may exist at some locations, little or none is found in most places.

A mineral deposit with a high enough concentration of at least one metallic element to permit it to be extracted and sold at a profit is called an **ore** or **ore deposit.** An ore that contains a relatively large concentration of a desired metallic element is called a **high-grade ore;** one with a relatively low concentration is known as a **low-grade ore.**

Making Mineral Resources Available Use of a mineral resource involves several major steps. First, a deposit containing enough of the desired mineral to make removing it profitable must be located. Second, some form of mining is used to extract the mineral from the deposit. Third, the mineral is processed to remove impurities and in some cases (especially metallic ores) converted to a different chemical form by smelting or other chemical processes. For example, aluminum is found in the earth's crust in ore form as aluminum oxide (Al_2O_3). After the ore is purified and melted, electrical current is passed through the molten oxide to convert it to aluminum metal (Al) and oxygen gas (O_2). Most smelters and other mineral-processing plants are located near mines, because transporting the huge volume of ore-bearing rock is costly. Finally, the desired form of the mineral is used directly (for example, crushed stone may be applied

Figure 14-3 A giant shovel used for strip-mining coal. The cars behind the shovel look like toys.

National Coal Association/Bucyrus Erie Company

to roads) or manufactured into various products (aluminum metal is converted to aluminum foil, cans, or cookware).

Locating Deposits Finding concentrated deposits of useful minerals is difficult and expensive, because they are very unevenly distributed on earth. Typically the task involves a combination of geological knowledge about crustal movements and mineral formation, use of various instruments and measurements, and luck. Photos taken from airplanes or images relayed by satellites can sometimes reveal geological features such as mounds or rock formations usually associated with deposits of certain minerals. Instruments can also be mounted on aircraft and satellites to detect concentrated deposits of minerals that affect the earth's magnetic field or modify the earth's gravity field. Then at promising sites, samples are extracted from test holes, tunnels, or trenches and analyzed for their mineral content.

Extraction Once an economically acceptable mineral deposit has been located, it is removed by surface or subsurface mining, depending on the location of the deposit in the earth's crust. Mineral deposits located near the earth's surface are removed by **surface mining.** Mechanized equipment removes the overlying layer of soil and rock—known as **overburden**—and vegetation so that the underlying mineral deposit can be extracted with large power shovels (Figure 14-3). You can get some idea of such equipment by imagining a shovel 32 stories high, with a metal boom and

attached bucket as long as a football field, capable of gouging out 152 cubic meters (5,366 cubic feet) of land every 55 seconds and dropping this 295,000-kilogram (325-ton) load a distance of a city block away. This is an accurate description of Big Muskie, a $25 million power shovel used for surface mining of coal in the United States.

Surface mining extracts about 90% of the metallic and nonmetallic minerals and almost two-thirds of the coal used in the United States. Almost half of the land disturbed by surface mining in the United States has been mined for coal; the remainder has been mined for nonmetallic minerals such as sand, gravel, stone, and phosphate rock (38%) and metallic minerals such as iron, copper, and aluminum (14%).

Several different types of surface mining are used, depending on the type of mineral and local topography. **Open-pit surface mining** involves using large machinery to dig a hole in the earth's surface and remove a mineral deposit, primarily stone, sand, gravel, iron, and copper. Sand and gravel are removed from thousands of small pits in many parts of the country. Building rocks such as limestone, granite, and marble are taken from larger pits called quarries. In Minnesota's Mesabi Range and in some western states, copper (see photo on p. 139) and iron ores are removed from huge open-pit mines.

Area strip-mining is carried out on flat or rolling terrain. Bulldozers and power shovels strip away the overburden and dig a trench to remove the mineral deposit (Figure 14-4). Then another, parallel trench is dug and its overburden is placed in the adjacent trench from which the mineral deposit has already been removed. When no attempt is made to restore the

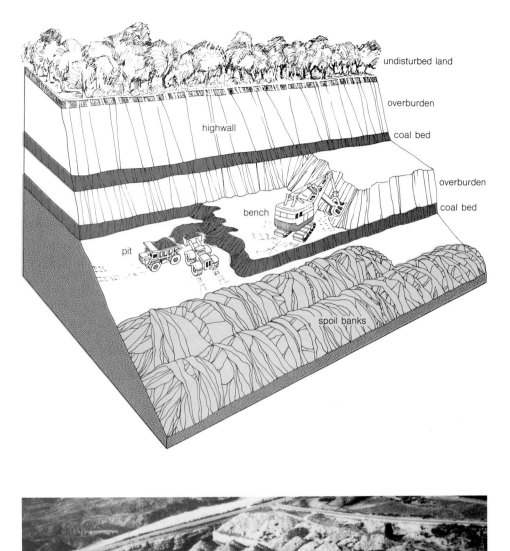

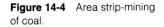

Figure 14-4 Area strip-mining of coal.

undisturbed land

overburden

highwall

coal bed

overburden

coal bed

bench

pit

spoil banks

Figure 14-5 Effects of area strip-mining of coal. Although restoration of newly strip-mined areas is now required, many previously mined areas have not been restored.

National Archives

area, the result is a wavelike series of highly erodible hills of rubble known as spoil banks (Figure 14-5). This technique is used primarily for mining coal in many western and midwestern states, and phosphate rock, especially in Florida, North Carolina, and Idaho.

Contour strip-mining is used in hilly or mountainous terrain. A power shovel cuts a series of shelves or terraces into the side of a hill or mountain, dumping the overburden from each new terrace onto the one below. If the land is not restored, the result is an ugly

Figure 14-6 Severely eroded hillsides on Bolt Mountain, West Virginia, as a result of contour strip mining of coal without proper restoration.

wall of dirt in front of a highly erodible, steep bank of soil and rock (Figure 14-6). In the United States, contour strip-mining is used primarily for extracting coal in the mountainous Appalachian region. In areas where the overburden is too thick to be removed economically, coal and some other minerals may be removed by huge drills, called augers, which burrow horizontally into a mountain deposit.

When a mineral deposit lies so deep in the ground that removing it by surface mining is too expensive, it is extracted by **subsurface mining.** For solid materials such as coal and some metallic ores, this is usually done by digging a deep vertical shaft, blasting subsurface tunnels and rooms to get to the deposit, and hauling the coal or ore to the surface. In the **room-and-pillar method,** as much as half the coal is left in place as pillars to prevent the mine from collapsing. Underground deposits of coal and some ores are also removed by the **longwall method;** a narrow tunnel is created and then supported by movable metal pillars. After a cutting machine has removed the coal or ore from a portion of a mineral seam, the roof supports are moved forward, allowing the earth behind them

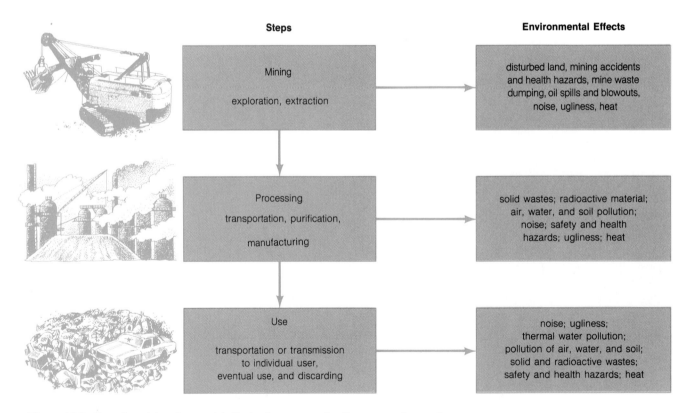

Figure 14-7 Some harmful environmental effects of resource extraction, processing, and use.

to collapse. As a result, no tunnels are left after the mining operation has been completed.

Subsurface techniques are also necessary for extracting crude oil and natural gas. Wells are drilled into underground rock reservoirs, allowing the pressurized oil or gas to rise to the surface, or a gas or hot water is injected into the well to force remaining oil to the surface (Sections 15-3 and 15-4).

14-2 ENVIRONMENTAL IMPACT OF MINING AND PROCESSING MINERAL RESOURCES

Overall Impact The mining, processing, and use of any nonfuel or fuel mineral resource cause some form of land disturbance along with air and water pollution (Figure 14-7). Most land disturbed by mining can be restored to some degree, and some forms of air and water pollution can be controlled (Chapters 18 and 19). But these efforts are expensive and also require

U.S. Department of Interior, Office of Surface Mining

Figure 14-8 With the land regraded to its original contour and grass planted to hold the soil in place, there is little evidence that this was once a surface coal-mining site.

energy, which, in being produced and used, can again pollute the environment.

Mining Impacts The harmful environmental effects of mining depend on the specific type of mineral extracted, the size of the deposit, the method used (surface or subsurface), and the local topography and climate. For each unit of mineral produced, subsurface mining disturbs less than one-tenth as much land as surface mining and generally produces less waste material. But subsurface mining is more dangerous and expensive than surface mining. Roofs and walls of underground mines occasionally collapse, sometimes trapping and killing miners. Explosions of dust and natural gas in mines can also kill and injure miners. Miners can also contract lung diseases as a result of prolonged inhalation of coal and other types of dust. Sometimes the surfaces above extensively mined areas cave in or subside, causing roads and houses to crack and buckle, railroad tracks to bend, sewer lines to crack, and gas mains to break and possibly explode.

Compared to other uses of land such as agriculture and urbanization, surface mining uses a relatively small amount of the earth's surface. For example, between 1930 and 1986, only about 0.26% of the total land area of the United States was surface-mined or used to dispose of wastes (spoils) from such mining. Nevertheless, surface mining has a severe environmental impact, because the land is stripped bare of vegetation and is not always restored. Only about 64% of the land area disturbed by coal mining, 26% by nonmetals, and 8% by metals has been restored. The exposed soil and mining wastes are subject to erosion by wind and water and can pollute the atmosphere and nearby aquatic ecosystems.

Ideally restoration of surface-mined land includes filling in and regrading to restore the original contour, replacing topsoil (which can be saved and placed on top of regraded mining rubble), and reestablishing vegetation to anchor the soil. Restoration costs typically range from $2,500 to $12,500 a hectare ($1,000 to $5,000 an acre).

The success of restoration efforts is highly dependent on average precipitation, slope of the land, and how well federal and state surface mining regulations are enforced. Restoration is simpler and more effective in areas with more than ten inches of rainfall a year and with flat or slightly rolling topography (Figure 14-8). In the arid and semiarid regions of the western United States, which contain about three-fourths of the country's surface-minable coal, full restoration is usually not possible. For many large open-pit mines

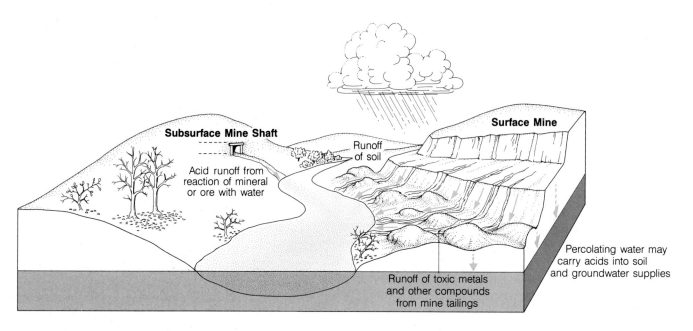

Figure 14-9 Degradation and pollution of a stream and groundwater by runoff of acids and toxic chemicals from surface and subsurface mining operations.

and quarries, especially in arid and semiarid regions, there is little hope of restoration. It has been suggested, however, that large and medium-size pit mines could be lined with concrete and used as sites for underground parking garages or storage of petroleum, chemicals, or other supplies.

Another environmental problem of subsurface and surface mining, but more frequent with subsurface mining, is runoff of acids, eroded soil (silt), and toxic substances into nearby surface and ground waters (Figure 14-9). Rainwater seeping through surface-mine spoils and through abandoned subsurface mines, especially coal mines rich in sulfur compounds, causes chemical reactions that produce sulfuric acid. This acid and the toxic compounds leached from mine spoils can run off into nearby rivers and streams, contaminating water supplies and killing aquatic life. These pollutants can also percolate downward and contaminate groundwater.

Processing Impacts Processing extracted mineral deposits to remove impurities produces large quantities of rock and other waste materials called **tailings,** which are piled on the ground or dumped into ponds. Wind blows particles of dust and toxic metals from piles of tailings into the atmosphere; water leaches toxic substances into nearby surface or ground water supplies. Although some mining companies have made considerable investments to reduce contamination from tailings, there is still a long way to go in requiring and monitoring such efforts.

Without adequate pollution control equipment, mineral-smelting plants emit massive quantities of air pollutants, especially sulfur dioxide, soot, and tiny particles of toxic elements and compounds (such as arsenic, cadmium, and lead) found as impurities in many ores. For example, decades of uncontrolled sulfur dioxide emissions from copper-smelting operations near Copperhill and Ducktown, Tennessee, killed all vegetation for miles around and acidified the soil to such an extent that it has not yet recovered (Figure 14-10).

Smelting plants also cause water pollution and produce liquid and solid hazardous wastes that must be disposed of safely or converted into less harmful substances. In addition, workers in some smelting industries have an increased risk of cancer. The lung cancer death rate for arsenic smelter workers is almost three times the expected rate, that for cadmium smelter workers is more than twice the expected rate, and lead smelter workers have higher than normal incidences of lung and stomach cancer.

14-3 WILL THERE BE ENOUGH MINERAL RESOURCES?

How Much Is There? Estimating how much of a particular nonrenewable mineral resource exists on earth and how much of it may be located and extracted at an affordable price is a complex and controversial process. The term **total resources** refers to the total

Figure 14-10 Sulfur dioxide and other fumes from a copper smelter in Tennessee killed the luxurious forest that once flourished on this land.

A. Keith/U.S. Geological Survey

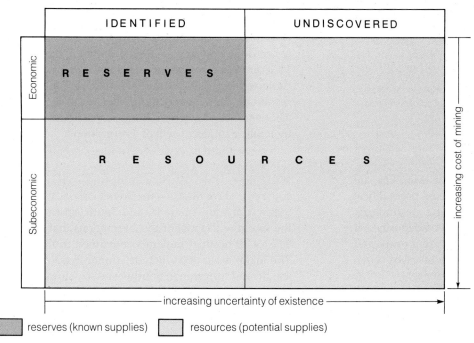

Figure 14-11 General classification of mineral resources by the U.S. Geological Survey.

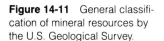

reserves (known supplies)   resources (potential supplies)

amount of a particular material that exists on earth. Reliable estimates of the total available amount of a particular resource are difficult to make, however, because the entire world has not been explored for each resource.

The U.S. Geological Survey estimates actual and potential supplies of a mineral resource by dividing the estimated total resources into two broad categories: identified and undiscovered (Figure 14-11). **Identified resources** are specific bodies of a particular

mineral-bearing material whose location, quantity, and quality are known or have been estimated from geological evidence and measurements. **Undiscovered resources** are potential supplies of a particular mineral resource believed to exist on the basis of broad geologic knowledge and theory, although specific locations, quality, and amounts are unknown.

These two categories are then subdivided into reserves and resources, depending on the estimated costs of mining them and the degree of certainty of

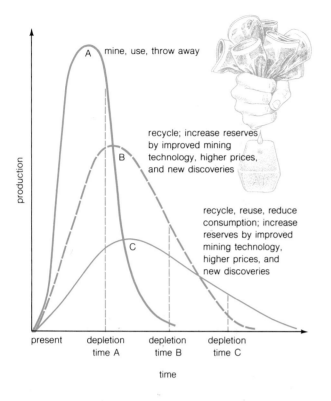

Figure 14-12 (labels within figure)

production

A mine, use, throw away

B recycle; increase reserves by improved mining technology, higher prices, and new discoveries

recycle, reuse, reduce consumption; increase reserves by improved mining technology, higher prices, and new discoveries

C

present   depletion time A   depletion time B   depletion time C

time

Figure 14-12 Depletion curves for a nonrenewable resource, based on three different sets of assumptions. Dashed vertical lines show when 80% depletion occurs.

Depletion time is the period required to use a certain fraction—usually 80%—of the known reserves or estimated resources of a mineral, at an assumed rate of use.

We can plot a series of projections of different depletion times in the form of depletion curves by making different assumptions about the resource supply and its rate of use. For example, one estimate of depletion time might be based on the assumption that the resource is not recycled or reused and that its estimated reserves will not increase (curve A, Figure 14-12). A longer depletion time estimate can be obtained if we assume that recycling will extend the life of existing reserves and that improved mining technology, price rises, and new discoveries will expand present reserves by some factor, say 2 (curve B).

We can project a still longer depletion time by assuming that present reserves will be expanded even more, perhaps five or ten times, by new discoveries and through recycling, reuse, and reduced consumption (curve C). Of course, finding a substitute for a resource cancels all these curves and requires a new set of depletion curves for the new resource. Figure 14-12 illustrates why there is so much controversy over projected supplies of nonrenewable nonfuel and fuel resources. One can get optimistic or pessimistic projections of the depletion time of a particular resource by making different sets of assumptions.

their existence. **Reserves, or economic resources,** are identified resources that can be extracted profitably at present prices with current mining technology. **Resources** are identified and unidentified deposits that cannot be recovered profitably with present prices and technology but may be converted to reserves when prices rise or mining technology improves.

Most published estimates of the available supply of a particular nonrenewable mineral refer to reserves, not resources or total resources. The actual supply that becomes available is normally higher than estimated, because some of the resources will be converted to reserves by discoveries of new supplies, improved mining technology, and shortage-caused price increases, which permit profitable mining of low-grade deposits. However, a large portion of potentially recoverable resources will not become available, because finding and extracting them will take more money and energy than they are worth.

How Fast Are Supplies Being Depleted? The future availability of a mineral resource depends not only on its actual or potential supply but also on how rapidly this supply is being depleted to the point where extracting and processing what remains is too costly.

Who Has the World's Nonfuel Mineral Resources? Five MDCs—the Soviet Union, the United States, Canada, Australia, and South Africa—supply the world with most of the 20 minerals that make up 98% of all nonfuel minerals consumed in the world. The major exceptions include copper in South America, tin and tungsten in Southeast Asia, aluminum ore in the Caribbean, and cobalt in Zaire.

The United States and the Soviet Union are the world's two largest producers and consumers of raw and processed nonfuel mineral resources. Unlike other countries, the Soviet Union has long been essentially self-sufficient for almost all of its mineral needs. Despite its rich resource base, the United States imports all or a large percentage of many key nonfuel minerals (Table 14-1), in some cases because the U.S. rate of consumption exceeds domestic production and in other cases because higher-grade ores found in other countries are cheaper to extract than more plentiful lower-grade U.S. reserves. Japan and many western European countries are even more dependent on imports of vital nonfuel minerals.

Currently the United States stockpiles 93 **strategic materials** vital to industry and defense, 80 of them nonfuel minerals, to cushion against supply interruptions and sharp price rises. The stockpiles are sup-

Table 14-1 U.S. Import Dependence for Selected Key Nonfuel Minerals in 1985

Mineral	Percentage Imported	Major Suppliers	Key Uses
Columbium	100	Brazil, Canada, Thailand	High-strength alloys for construction, jet engines, machine tools
Industrial diamonds	100	South Africa, United Kingdom, Soviet Union	Machinery, mineral services, abrasives, stone and ceramic products
Manganese	99	South Africa, Gabon, France	Alloys for impact-resistant steel, dry-cell batteries, chemicals
Aluminum ore	96	Jamaica, Guinea, Suriname	Aluminum production, building materials, abrasives
Cobalt	95	Zaire, Zambia, Canada	Alloys for tool bits, aircraft engines, high-strength steel
Tantalum	94	Thailand, Malaysia, Brazil	Nuclear reactors, aircraft parts, surgical instruments
Platinum	91	South Africa, United Kingdom, Soviet Union	Oil refining, chemical processing, telecommunications, medical and dental equipment
Graphite	90+	Mexico, China, Brazil	Foundry operations, lubricants, brake linings
Chromium	82	South Africa, Zimbabwe, Soviet Union	Alloys for springs, tools, engines, bearings
Tin	79	Thailand, Malaysia, Indonesia	Cans and containers, electrical products, construction, transportation
Rutile	61	Australia, Sierra Leone, South Africa	Paint, plastics, paper, welding-rod coatings
Vanadium	41	South Africa, Canada, Finland	Iron and steel alloys, titanium alloys, sulfuric acid production

posed to be large enough to last through a three-year war, but most supplies are well below this level. Although OPEC has been able to raise oil prices, other **cartels,** or groups of resource-rich countries banding together to control supplies and raise prices, are not considered a serious threat to supplies of minerals.

Unlike oil, which is consumed directly, mineral raw materials contribute such a small percentage to the cost of finished products that increases in their prices have relatively small effects. So far no cartel, single country, or company has been successful in controlling the supply and price of any nonfuel mineral except industrial diamonds, which have been rigidly controlled for 50 years by De Beers Consolidated Mines of South Africa.

According to the U.S. Bureau of Mines and the U.S. Geological Survey, the United States has adequate domestic reserves of most key minerals—except

chromium, cobalt, platinum, tin, gold, and palladium—for at least the next several decades. However, the Geological Survey estimates that present reserves of most key minerals will not satisfy U.S. needs for more than 100 years without increased recycling, conservation, and substitutes.

14-4 INCREASING MINERAL RESOURCE SUPPLIES: THE SUPPLY-SIDE APPROACH

Economics and Resource Supply Although geologic processes determine how much of a mineral resource is available, economics determines what portion of the supply will actually be used. According to standard economic theory, a competitive free market

National Oceanic and Atmospheric Administration

Figure 14-13 Manganese nodules found on the ocean floor.

should control the supply and demand of goods and services: If a resource becomes scarce, prices rise; if there is an oversupply, they fall. Cornucopians contend that rising prices based on increased demand will stimulate new discoveries and development of more efficient mining technology, making it profitable to mine ores of increasingly lower grades.

However, a number of economists argue that this idea often does not apply to the supply and demand of nonfuel mineral resources in most MDCs, which use the bulk of these materials. For example, in the United States (and in many other industrial countries), both industry and government have gained so much control over supply, demand, and prices of mineral raw materials and products that a competitive free market doesn't exist. In addition, costs of nonfuel mineral resources account for only a small percentage of the total costs of most goods. As a result, resource scarcities do not cause a large enough rise in the final price of products to encourage consumers to reduce demand soon enough to avoid depletion. Indeed, artificially low prices increase demand and encourage faster resource depletion.

Finding New Land-based Mineral Deposits

There is little doubt that geologic exploration—guided by better geologic knowledge, satellite surveys, and other techniques—will extend present reserves of most minerals. According to geologists, rich deposits will probably be found in unexplored areas in LDCs; but in MDCs and many LDCs, the most easily accessible high-grade deposits have already been discovered. Remaining deposits are more difficult and more expensive to find and mine and usually are less concentrated.

Exploration for new resources, therefore, requires large capital investment and is a risky financial venture. Typically, if geologic theory identifies 10,000 sites where a deposit of a particular resource might be found, only 1,000 sites are worth costly exploration; only 100 warrant even more costly drilling, trenching, or tunneling; and only 1 out of the 10,000 is likely to be a producing mine. Even if large new supplies are found, no mineral supply can stand up to continued exponential growth in its use. For example, a 1-billion-year supply of a resource would be exhausted in only 584 years if the level at which it was used increased at 3% a year.

Obtaining More Minerals from Seawater and the Ocean Floor

Potential ocean resources are found in three areas: seawater, sediments and deposits on the shallow continental shelf and slope, and sediments and nodules on the deep ocean floor. The huge quantity of seawater appears to be an inexhaustible source of minerals, but most of the 90 chemical elements found there occur in such low concentrations that recovering them takes more energy and money than they are currently worth. For example, to get a mere 0.003% of the annual U.S. consumption of zinc from the ocean would require processing a volume of seawater equivalent to the combined annual flows of the Delaware and Hudson rivers. Only magnesium, bromine, and common table salt (sodium chloride) are abundant enough in seawater to be extracted profitably at present prices with current technology.

Offshore deposits and sediments in shallow waters are already important sources of crude oil, natural gas, sand, gravel, and ten other minerals. Extraction of these resources is limited less by supply or mining technology than by the increasing costs of the energy needed to find and remove them and the harmful effects of oil leaks and spills and extensive mining on marine food resources and wild species.

There is considerable interest in locating and removing manganese-rich rocks, or nodules, found in large quantities on the deep ocean floor at a few sites. These potato-size nodules (Figure 14-13) contain 30% to 40% manganese (used in certain steel alloys) and small amounts of other strategically important metals such as nickel and cobalt. It is proposed that a device much like a giant vacuum cleaner be developed to suck these nodules up from muds of the deep ocean floor and deliver them through a three-mile-long pipe to a ship above the mining site. Environmentalists recognize that such seabed mining would probably cause less harm than mining on land. They are concerned, however, that vacuuming nodules off the seabed and stirring up deep ocean sediments could destroy seafloor organisms and have unknown effects on poorly understood deep-sea food webs; surface waters might also be polluted by the discharge of sediments from mining ships and rigs.

Economic uncertainties make it unclear whether these nodules will be extracted in the near future. There is no guarantee that metal prices will be high enough to permit a reasonable profit on such a large and risky investment, especially since ample and much cheaper supplies of these metals are expected to be available for many decades. An even greater threat is posed by international legal and political squabbles over ownership of the nodules, because most large deposits are located in international waters.

Improved Mining Technology and Mining Low-Grade Deposits Cornucopians talk of improved mining technology that will allow us to drill deeper into the earth to obtain more minerals. However, the likelihood of obtaining materials from greater depths is slim, because of the extreme heat and pressure at such depths and the enormous costs of locating and extracting such resources.

Cornucopians also assume that we can increase supplies of any mineral by mining increasingly lower grades of ore. They point to the fact that advances in technology during the past few decades have allowed the mining of low-grade deposits of some copper, for example, without significant cost increases. Neo-Malthusians point out, however, that as increasingly lower-grade deposits are mined, we eventually run into geologic, energy, water, and environmental factors that place limits on the amounts of minerals that can be extracted and processed—long before actual supplies of these minerals are exhausted.

Another problem is that the ability to mine and process increasingly poorer grades of ore depends on an inexhaustible source of cheap energy. Indeed, the ability to locate and process lower-grade ores between 1935 and 1975 was based primarily on the availability of abundant supplies of cheap oil and natural gas. But most energy experts agree that in the future, energy will neither be unlimited nor cheap.

Available supplies of fresh water may also limit the supply of some mineral resources, because large amounts of water are needed to extract and process most minerals. Many areas with major mineral deposits are poorly supplied with water. Finally, exploitation of increasingly lower grades of ore may be limited by the environmental impact of the catastrophic increase in waste material produced during mining and processing. Mining and processing low-grade ores may lead to such an increase in disturbed land and pollution of air and water that the cost of land reclamation and pollution control will eventually exceed the value of the minerals produced.

Substitution Cornucopians believe that even if supplies of key minerals become very expensive or scarce, human ingenuity will find substitutes. They argue that plastics, high-strength glass fibers and ceramic materials made mostly from silicon (the second most abundant element in the earth's crust), or four of the most abundant metals (aluminum, iron, magnesium, and titanium) can be substituted for most scarce metals.

Although substitutes can probably be found for many scarce resources, there are problems. Finding substitutes and phasing them into complex manufacturing processes is costly and requires long lead times. During the transition period, serious economic hardships could occur as prices of the increasingly scarce resource rise catastrophically. And finding suitable substitutes for some key materials may be extremely difficult, if not impossible. Some substitutes may be inferior to the minerals they replace, and some may become scarce and prohibitively expensive themselves because of greatly increased demand.

14-5 EXTENDING MINERAL RESOURCE SUPPLIES: THE CONSERVATION APPROACH

Recycling Recycling items containing iron and aluminum, which account for 94% of all metals used, as well as other nonrenewable mineral resources has a number of advantages. It extends the supply of the mineral by reducing the amount of virgin materials that must be extracted from the earth's crust to meet demand. It also usually saves energy, causes less pollution and land disruption than use of virgin resources, cuts waste disposal costs, and prolongs the life of landfills by reducing the volume of solid wastes.

For example, using scrap iron (Figure 14-14) instead of iron ore to produce steel conserves virgin iron ore and coal, requires 65% less energy and 40% less water, and produces 85% less air pollution and 76% less water pollution. Recycling aluminum produces 85% less air pollution and 97% less water pollution and requires 92% less energy than mining and processing virgin aluminum ore.

Despite these advantages, only about one-fourth of the world's iron and aluminum is recovered for recycling. Recycling rates vary in different countries, with about one-half of these two materials recycled in the Netherlands and about one-third in Japan and the United States. It is encouraging that 52% of new aluminum beverage cans used in the United States were recycled in 1986 at more than 5,000 recycling centers set up by the aluminum industry, other private inter-

Figure 14-14 Each junk car will yield about 909 kilograms (1 ton) of iron and steel scrap, which can be recycled into new products.

ests, and local governments. People who returned the cans received slightly more than a penny a can for their efforts. But almost half of the aluminum cans produced each year are still thrown away, each representing a waste of energy equivalent to half a beverage can of gasoline.

Beverage container deposit laws can be used to decrease litter and encourage recycling of nonrefillable glass and metal containers and in some cases plastic containers, which now make up 25% of the total. Consumers pay a deposit (usually a nickel) on each beverage container they buy; deposits are refunded when containers are turned in to retailers, redemption centers, or reverse vending machines, which return cash when consumers put in empty beverage cans and bottles. By 1986 container deposit laws had been adopted in Sweden, Norway, the Netherlands, the Soviet Union, parts of Canada and Japan, and in ten states (about one-fourth of the U.S. population).

Such laws work. Experience in the United States has shown that a whopping 90% of cans and bottles are turned in for refund, litter is decreased by 70% or more, expensive landfills don't fill up as quickly, energy and mineral resources are saved, and jobs are created. Environmentalists, noting that 61% of Americans live in states with no laws to encourage recycling of beverage containers, believe a nationwide deposit law should be passed by Congress (see Spotlight on p. 245).

In recent years, Americans have heard many slogans: "Waste is a resource out of place," "Urban waste is urban ore," "Trash is cash," "Landfills are urban mines," and "Trash cans are really resource con-

tainers." The state of New Jersey and cities such as Chicago and New York have set recycling goals of 25% by 1991. Other cities, such as Berkeley, California, and Portland, Oregon, aim at recycling 50% of their waste materials and now recycle about 22%.

Nevertheless, only about 10% of all potentially recoverable waste material in the United States is now recycled, compared to 40% to 60% in densely populated countries such as Japan and the Netherlands, which import most of their fuel and nonfuel minerals. Recycling rates are also high in some LDCs such as Mexico, India, and China, where small armies of poor people go through urban garbage disposal sites by hand. They remove paper and sell it to paper mills, metal scraps to metal-processing factories, bones to glue factories, and rags to furniture factories for use in upholstery.

Obstacles to Recycling in the United States Several factors have hindered recycling efforts in the United States. One is the failure of many U.S. metals industries to modernize. Since 1950, countries such as Japan and West Germany have built new steel plants based on modern processes that use large amounts of scrap steel, much of it bought from the United States. During this same period, the U.S. steel industry did not reinvest much of its profits in modernizing and replacing older plants and continued to rely heavily on older processes that require virgin iron ore. As a result, the industry has now lost much of its business to foreign competitors and no longer has the capital to modernize—illustrating how overemphasis on short-

A well-funded lobby of steel, aluminum, and glass companies, metal-workers' unions, supermarket chains, and most major brewers and soft drink bottlers has vigorously opposed passage of a national beverage container deposit law as well as such laws in individual states. Merchants don't like to have returned bottles and cans piling up in their stores—although reverse vending machines could solve this problem. Labor unions are afraid that some workers in bottle and can manufacturing industries will lose their jobs. Beverage makers fear the extra nickel deposit per container will hurt sales. Some consumers think returning containers is too much trouble. They would rather toss their cans and bottles away and let someone else worry about them—out of sight, out of mind.

"Keep America Beautiful" and other expensive ad campaigns financed by these groups have helped prevent passage of container deposit laws in a number of states. These industries favor litter-recycling laws, which levy a tax on industries whose products pose a potential threat as litter or landfill clutter. Revenues from the tax are used to establish and maintain state-wide recycling centers. By 1987 seven states, containing about 14% of the U.S. population, had this type of law.

Environmentalists point out that surveys indicate litter taxes are not nearly as effective as beverage container deposit laws. EPA and General Accounting Office studies estimate that a national container deposit law would have a number of desirable effects, including saving consumers at least $1 billion annually. Roadside beverage container litter would be reduced by 60% to 70%, saving taxpayers money now used for cleanup. Urban solid waste would be reduced by at least 1%, saving taxpayers $25 million to $50 million a year in waste disposal costs. Mining and processing of virgin aluminum ore would be decreased by 53% to 74% and the use of iron ore by 45% to 83%. Air, water, and solid waste pollution from the beverage industry would drop by 44% to 86%, and the energy saved would be equivalent to that needed to provide the annual electrical needs for 2 million to 7.7 million homes. There would also be a net increase of 80,000 to 100,000 jobs: Collecting and refilling beverage containers is more labor-intensive than producing new ones. Surveys have shown that such a law is supported by 73% of the Americans polled. What do you think?

term economic gain can lead to long-term economic pain, decline, and loss of jobs for thousands of Americans.

Despite increased awareness of the need for recycling, most Americans have been conditioned by advertising and example to a throwaway lifestyle designed to increase short-term economic growth regardless of the long-term environmental and economic costs (Figure 14-15). Waste collection and disposal costs account for a major portion of local tax revenue expenditures. Consumers pay for these costs indirectly in the form of higher taxes rather than directly by a waste disposal tax placed on all recoverable items. As a result, consumers have no easily identifiable economic incentive to recycle and conserve recoverable resources.

Growth of the recycling, or secondary-materials, industry in the United States is hindered by tax breaks, depletion allowances, and other tax-supported federal subsidies for primary mining and energy industries designed to encourage them to get virgin resources out of the ground as fast as possible. In contrast, recycling industries receive relatively few tax breaks and other subsidies. The lack of large, steady markets for recycled materials makes recycling industries risky, "boom-and-bust" financial ventures that don't attract large amounts of investment capital.

With economic and political incentives to encourage recycling, the United States could easily recycle half and perhaps two-thirds of the matter resources it uses each year. Such a shift could be accomplished over a ten-year period through the following measures: (1) including waste disposal costs in the price of all items; (2) providing favorable federal and state subsidies for secondary-materials industries; (3) decreasing subsidies for primary-materials industries; (4) encouraging federal, state, and local governments to require the highest feasible percentage of recycled materials in all products they purchase, thus guaranteeing a sufficient market to encourage investment; and (5) using advertising and education to discourage the throwaway mentality.

Figure 14-15 Evidence of the throwaway mentality at the site of an outdoor rock concert.

UPI/Bettmann Newsphotos

Reusable Containers Reuse involves employing the same product again and again in its original form, as in the case of beverage bottles that can be collected, rewashed, and refilled by bottling companies. Reuse extends resource supplies and reduces energy use and pollution even more than recycling. For example, three times more energy is needed to crush and remelt a glass bottle to make a new one than to clean and refill it. Cleaning and refilling a bottle also takes far less energy than melting a recycled aluminum can and making a new one. If reusable bottles replaced the 80 billion throwaway beverage cans produced annually in the United States, enough energy would be saved to provide electricity for 13 million people.

Instead, about 85% of all U.S. beverage containers are nonreusable bottles and cans, which are either thrown away or recycled. Environmentalists strongly support a national beverage container deposit bill as a step in the right direction, but would like to go further and ban all nonreusable beverage containers, as has been done in Denmark.

The Low-Waste Society: Beyond Recycling and Reuse In addition to increased recycling and reuse of nonfuel mineral resources, environmentalists also call for increased resource conservation, especially in MDCs, to reduce unnecessary waste of matter and energy resources. They call for the United States and other MDCs to shift from a high-waste society (Figure 3-16, p. 47) to a sustainable-earth society—a low-waste society based on recycling, reuse, and resource conservation (Figure 3-17, p. 48).

Reducing unnecessary waste of nonrenewable mineral resources can extend supplies even more dramatically than recycling and reuse. Furthermore, recycling and reuse require energy. Thus at some point the supply of energy resources and the environmental impacts associated with them can limit resource recycling and reuse. Table 14-2 compares the present throwaway resource system used in the United States, a resource recovery and recycling system, and a sustainable-earth, or low-waste, resource system.

Manufacturers can also conserve resources by using less resource per product. For example, the trend toward smaller and lighter cars saves nonfuel mineral resources and also saves energy by increasing gas mileage. Solid-state electronic devices and microwave transmissions have significantly reduced materials requirements. The transistor, for example, requires about one-millionth the material needed to make the vacuum tube it replaces. Optical fibers drastically reduce the demand for copper and aluminum wire.

Another approach is to make products that last longer. The economies of the United States and most industrial countries are built on the principle of planned obsolescence. Product lives are designed to be much shorter than they could be so that people will buy more things to stimulate the economy and raise short-

Table 14-2 Three Systems for Handling Discarded Materials

Item	For a High-Waste Throwaway System	For a Moderate-Waste Resource Recovery and Recycling System	For a Low-Waste Sustainable-Earth System
Glass bottles	Dump or bury	Grind and remelt; remanufacture; convert to building materials	Ban all nonreturnable bottles and reuse (not remelt and recycle) bottles
Bimetallic "tin" cans	Dump or bury	Sort, remelt	Limit or ban production; use returnable bottles
Aluminum cans	Dump or bury	Sort, remelt	Limit or ban production; use returnable bottles
Cars	Dump	Sort, remelt	Sort, remelt; tax car lasting less than 15 years, weighing more than 818 kilograms (1,800 pounds), and getting less than 13 kilometers per liter (30 miles per gallon)
Metal objects	Dump or bury	Sort, remelt	Sort, remelt; tax items lasting less than 10 years
Tires	Dump, burn, or bury	Grind and revulcanize or use in road construction; incinerate to generate heat and electricity	Recap usable tires; tax all tires not usable for at least 64,400 kilometers (40,000 miles)
Paper	Dump, burn, or bury	Incinerate to generate heat	Compost or recycle; tax all throwaway items; eliminate overpackaging
Plastics	Dump, burn, or bury	Incinerate to generate heat or electricity	Limit production; use returnable glass bottles instead of plastic containers; tax throwaway items and packaging
Garden wastes	Dump, burn, or bury	Incinerate to generate heat or electricity	Compost; return to soil as fertilizer; use as animal feed

term profits. Many consumer societies can empathize with Willy Loman, the main character in Arthur Miller's play *Death of a Salesman:* "Once in my life I would like to own something outright before it's broken! I'm always in a race with the junkyard."

In addition to lasting longer, products could be designed so that they could be repaired easily. Currently many items are intentionally designed to make repair impossible or prohibitively expensive, even though the broken or worn part may represent only a small percentage of the value of the product. The modular design of computers and other electronic devices allows certain circuits to be easily and quickly replaced without replacing the entire item. This is a step in the right direction. Another step is to develop remanufacturing industries, which would disassem-

ble, repair or improve, and reassemble used and broken items. New products could be designed to facilitate remanufacturing.

Some argue that a shift away from a production-disposal, high-waste society to a service-repair, low-waste society will result in a loss of jobs and economic decline. Actually, the reverse is true. The loss of profits and jobs in increasingly automated, machine-intensive production will be more than offset by the increase in profits and jobs in labor-intensive service, repair, and recycling businesses.

Solid wastes are only raw materials we're too stupid to use.

Arthur C. Clarke

DISCUSSION TOPICS

1. Why should an urban dweller be concerned about the environmental impact from increasing surface mining of land for mineral resources?

2. Debate the following resolution: The United States is an overdeveloped country that uses and unnecessarily wastes too many of the world's resources relative to its population size.

3. Debate each of the following propositions:

   a. The competitive free market will control the supply and demand of mineral resources.

   b. New discoveries will provide all the raw materials we need.

   c. The ocean will provide all the mineral resources we need.

   d. We will not run out of key mineral resources because we can always mine lower-grade deposits.

   e. When a mineral resource becomes scarce, we can always find a substitute.

   f. When a nonrenewable resource becomes scarce, all we have to do is recycle it.

4. Use the second law of energy (thermodynamics) to show why the following options are normally not profitable:

   a. extracting most minerals dissolved in seawater

   b. recycling minerals that are widely dispersed

   c. mining increasingly low-grade deposits of minerals

   d. using inexhaustible solar energy to mine minerals

   e. continuing to mine, use, and recycle minerals at increasing rates

5. Explain why you support or oppose the following:

   a. eliminating all tax breaks and depletion allowances for extraction of virgin resources by mining industries

   b. passing a national beverage container deposit bill

   c. requiring that all beverage containers be reusable

6. Why is it difficult to get accurate estimates of mineral resource supplies?

15

Nonrenewable Energy Resources: Fossil Fuels

GENERAL OBJECTIVES

1. How have people used various sources of energy throughout history?

2. What criteria can we use to evaluate present and future energy resource alternatives?

3. What are the major uses, advantages, and disadvantages of oil as an energy resource?

4. What are the major uses, advantages, and disadvantages of natural gas as an energy resource?

5. What are the major uses, advantages, and disadvantages of coal as an energy resource?

We are an interdependent world and if we ever needed a lesson in that we got it in the oil crisis of the 1970s.
Robert S. McNamara

Useful high-quality energy is the lifeblood of human societies—driving virtually all activities that shape individual lifestyles and national and world economic systems. This chapter examines the history of energy use in the world and the United States, criteria for evaluating present and future energy alternatives, and advantages and disadvantages of using nonrenewable fossil fuels and geothermal energy. The two chapters that follow evaluate nonrenewable nuclear and geothermal energy (Chapter 16) and perpetual and renewable energy resources (Chapter 17).

15-1 BRIEF HISTORY OF ENERGY USE

Primitive to Modern Times In each phase of cultural history human ingenuity has increased the average amount of energy used per person to supplement the direct input of solar energy (Figure 15-1). In the 1700s, when the Industrial Revolution began, most of the energy used by the United States and other industrializing countries came from perpetual and renewable sources (domesticated animal labor, wood, flowing water, and wind), mostly from locally available supplies.

By 1850 wood provided about 91% of the commercial energy used in industrializing European countries and the United States (Figure 15-2). By 1900 coal replaced wood as the major energy resource in such countries, primarily because of new coal-mining technology and because coal was a more concentrated (higher-quality) source of energy. Since 1910 oil and natural gas have largely replaced coal because they burn cleaner and are easier and cheaper to transport than coal, and because oil, unlike coal, can be refined to produce liquid fuels for vehicles.

By 1984 about 82% of the commercial energy used throughout the world and 91% of that used in the United States was provided primarily by the burning of three nonrenewable fossil fuels—oil, coal, and natural gas—and a small amount by the nuclear fission of nonrenewable uranium atoms to produce electricity (Figures 15-2 and 15-3). The remaining 18% of the world's energy and 9% of that in the United States was provided by *perpetual* and *renewable* direct and indirect solar energy resources—mostly biomass (wood, dung, and crop residues) and hydropower.

Thus between 1700 and 1984, today's MDCs shifted from a decentralized energy system based on locally available renewable and perpetual resources to a centralized energy system based on use of nonrenewable fossil fuels (especially oil and natural gas) increasingly produced in one part of the world and transported to and used in another part. This shift fueled the rapid economic development of the MDCs, especially since 1950. At the same time, this fossil-fuel age has made most MDCs dependent on a finite resource base that is being rapidly exhausted. In addition, countries, communities, and individuals who once obtained most of the energy they needed from local resources are now dependent on large national and multinational energy companies, government policies, and other countries for most of their energy and the prices they must pay.

Energy Use and Problems in Less Developed Countries Most increases in energy consumption per person since 1900 have taken place in MDCs, and the gap in average energy use per person between the MDCs and LDCs has widened. At one extreme, the United States, with about 5% of the world's population, accounts for 25% of the world's commercial energy consumption. At the other extreme, India, with about 15% of the world's people, uses only about 1.5% of the world's commercial energy. In 1987, 244 million Americans used more energy for air conditioning alone than 1.06 billion Chinese used for all purposes.

The most important source of energy for LDCs is potentially renewable biomass—especially fuelwood—which serves as the main source of energy for roughly half the world's population. This information does not appear in summaries of world commercial energy use (Figure 15-3), which are distorted by the massive energy consumption in MDCs and the fact that much fuelwood is gathered by rural people and thus not sold commercially.

While one-fourth of the world's population in MDCs worries about future shortages of oil, half the world's population already faces a fuelwood shortage energy crisis because of widespread deforestation by logging companies, farmers, and poor people stripping land of firewood for short-term survival. With-

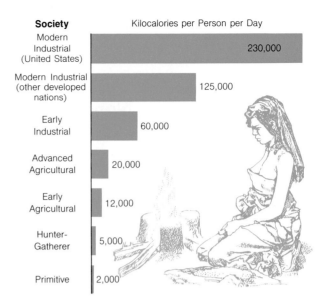

Figure 15-1 Average direct and indirect daily energy use per person at various stages of human cultural development.

out adequate replanting, deforestation gradually converts this vital renewable energy resource to a nonrenewable one. About one of every six people on earth also faces a food shortage energy crisis on a daily basis. This shortage results not so much from insufficient food production worldwide as from poverty, which prevents these people from growing or buying sufficient food for good health.

The Oil Crisis of the 1970s By 1940 the depletion of many low-cost domestic oil deposits made it cheaper for the United States to import much of its oil. During 1973 the United States imported about 30% of its oil (Figure 15-4); almost half of this came from the 13 countries in the Organization of Petroleum Exporting Countries (OPEC).* Other MDCs, such as Japan and most western European countries, are even more dependent on imported oil than the United States because their domestic supplies are scant or nonexistent.

By 1973 the OPEC countries, which possess 57% of the world's proven oil reserves (the United States has only 4%), accounted for 56% of the world's oil production and about 84% of all oil exports. Saudi Arabia, with the largest and most accessible oil reserves, can produce oil cheaper than any other country at costs ranging from 20 cents a barrel from old fields to no more than $3 from newer fields. By comparison, it

---

*OPEC was formed in 1960 at the urging of Venezuela. Today its 13 member countries are Algeria, Ecuador, Gabon, Indonesia, Iran, Iraq, Kuwait, Libya, Nigeria, Qatar, Saudi Arabia, the United Arab Emirates, and Venezuela.

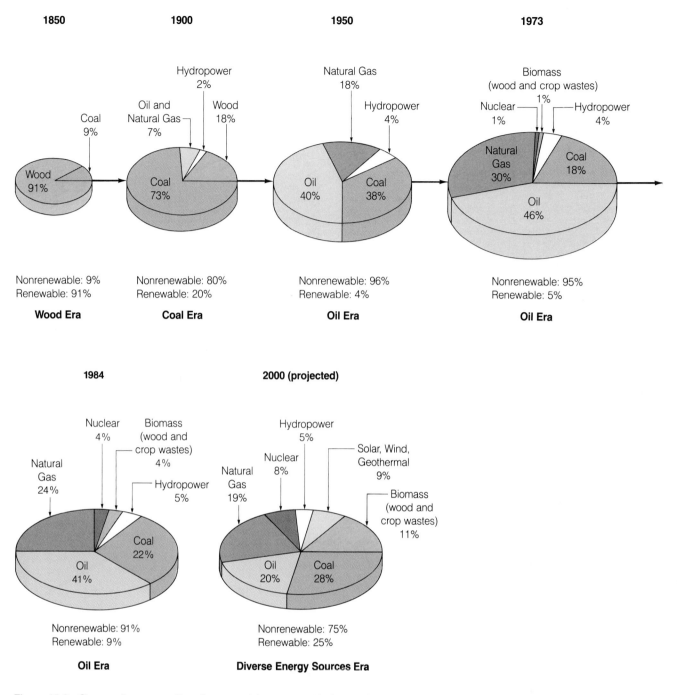

Figure 15-2 Changes in consumption of commercial nonrenewable (shaded) and renewable (unshaded) energy resources in the United States between 1850 and 1984 with a projection for the year 2000. Relative circle size indicates the total amount of energy used. (Sources: U.S. Department of Energy and Center for Renewable Resources; National Audubon Society for 2000 A.D. projection)

costs Norway and Great Britain from $5 to $9 a barrel to produce oil from the North Sea. Mexico's production cost is $5 to $7 a barrel, and costs range in the United States from $5 a barrel in old fields to $12 a barrel from offshore wells.

This dependence of most MDCs on OPEC countries for imported oil set the stage for the first phase of the oil crisis of the 1970s. On October 18, 1973, during the 18-day Yom Kippur War between Israel and the alliance of Egypt and Syria, the Arab members of OPEC reduced oil exports to Western industrial countries and prohibited all shipments of their oil to the United States because of its support of Israel. The embargo lasted until March 1974 and caused a fivefold increase in the average world price of crude oil (Figure 15-5), contributing to double-digit inflation in the United States and many other countries, high interest rates, soaring international debt, and a global economic

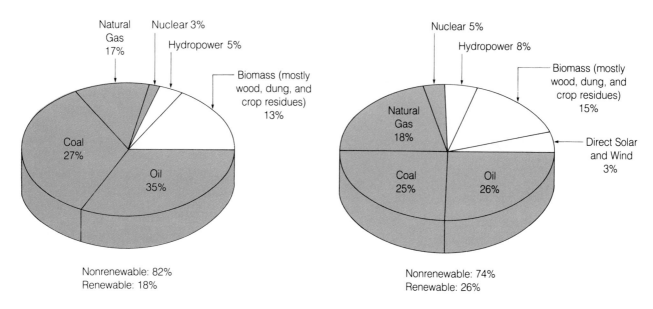

1984

Natural Gas 17% Nuclear 3%
Hydropower 5%
Coal 27%
Oil 35%
Biomass (mostly wood, dung, and crop residues) 13%

Nonrenewable: 82%
Renewable: 18%

2000 (projected)

Nuclear 5%
Hydropower 8%
Natural Gas 18%
Coal 25%
Oil 26%
Biomass (mostly wood, dung, and crop residues) 15%
Direct Solar and Wind 3%

Nonrenewable: 74%
Renewable: 26%

Figure 15-3 World consumption of commercial nonrenewable (shaded) and renewable (nonshaded) energy by source in 1984 with projections for 2000. (Sources: U.S. Department of Energy and Worldwatch Institute)

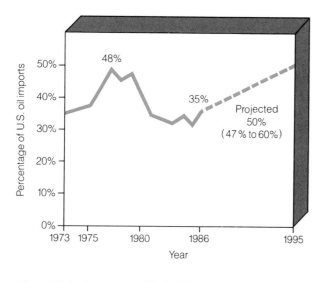

Figure 15-4 Percentage of U.S. oil imported between 1973 and 1986 with projections to 1995. (Data from U.S. Department of Energy and Spears and Associates, Tulsa, Okla.)

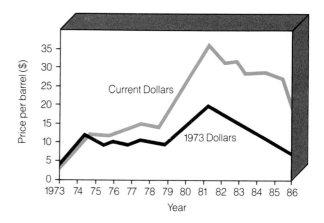

Figure 15-5 Average world crude oil prices between 1973 and 1986. (Data from Department of Energy and Department of Commerce)

recession. Americans accustomed to cheap and plentiful fuel waited for hours to buy gasoline and turned down the thermostats in homes and offices.

Despite the sharp price increase, U.S. dependence on imported oil increased from 30% to 48% between 1973 and 1977 (Figure 15-4). OPEC imports increased from 48% to 67% during the same period. The increased dependence resulted primarily from the government's failure to lift oil price controls that kept prices artificially low and discouraged energy conser-

vation. This false message to consumers set the stage for the second phase of the oil crisis, when available world oil supplies again decreased in 1979 after the revolution in Iran shut down most of that country's production. Gasoline waiting lines became even longer, and by 1981 the average world price of crude oil rose to about $35 a barrel.

The oil crisis of the 1970s was not due to an actual physical shortage of oil. Instead, it resulted from OPEC's significant control of the world's supply, distribution,

The sharp drop in oil prices between 1982 and 1986 (Figure 15-5) had a number of benefits for MDCs such as the United States and for LDCs heavily dependent on imported oil: stimulation of economic growth (except in the oil industry), creation of new jobs (except in the oil industry), and reduction of the rate of inflation.

At the same time, the price drop had a number of undesirable effects: **(1)** a sharp decrease in the search for new oil in the United States and most other countries; **(2)** economic chaos in many oil-producing countries, especially those with large international debts (such as Mexico) and in major oil-producing states (such as Texas, Oklahoma, and Louisiana); **(3)** loss of many jobs in the oil and related industries; **(4)** failure or near-failure of many U.S. banks with massive outstanding loans to oil companies and oil-producing LDCs such as Mexico, which had a $97 billion debt by 1986; and **(5)** reduction in the rate of improvements in energy efficiency and decreased development of energy alternatives to replace oil within 50 to 60 years, when the world is projected to begin facing true physical shortages of affordable oil.

When world oil prices drop below $15 a barrel, countries with relatively high oil-producing costs, such as Norway, the United King-dom, Mexico, Canada, and the United States, find little profit in producing and selling oil from newer wells, located mostly in inaccessible and hostile areas such as the North Sea, the Gulf Coast, and northern Alaska. When prices reach $10 a barrel or lower, oil companies lose money from such wells and oil exploration and production drops sharply. This slowdown, combined with decreasing efforts to reduce unnecessary energy waste, can make the United States—the world's largest importer of oil—and other oil-importing MDCs vulnerable to sharp rises in oil prices in the future. What do you think should be done?

and price of oil. This situation arose from several factors: (1) the rapid and significant economic growth of MDCs during the 1960s, stimulated primarily by low oil prices that were often held down by government price controls; (2) the greatly increased use of oil worldwide with little concern for reducing unnecessary use and waste until 1979; and (3) the heavy dependence of most MDCs on oil imports from OPEC nations, which have most of the world's proven and unproven oil reserves and the capability to produce oil at much lower prices than most MDCs.

The Oil Glut of the 1980s   Between 1979 and 1982 a combination of energy conservation (the biggest factor), substitution of other energy sources for oil, increased oil production by non-OPEC countries (such as Mexico and Great Britain), and several years of worldwide economic recession led to a 10% drop in world oil consumption. By 1982 these factors, coupled with disagreements among OPEC countries that prevented them from reducing production enough to sustain high prices, led to an oversupply of oil. As a result, the average price of crude oil dropped from $35 a barrel in 1981 to slightly below $15 a barrel in 1986; it then rose to around $20 a barrel by mid-1987. This meant that inflation-adjusted crude oil prices in 1986 were about the same as those in 1974 (Figure 15-5). As a

result, OPEC's share of the world oil market dropped from 48% to about 27% between 1978 and 1987. This oil glut has had good and bad effects (see Spotlight above).

The Next Oil Crisis   Although the oil glut of the 1980s has had some good short-term effects, it could lead to long-term economic disaster for the United States and other oil-importing countries within the next 7 to 20 years. Most energy analysts believe that the oil glut of the 1980s is only temporary and project significant increases in the price of oil—rising to at least $32 a barrel and perhaps as high as $98 a barrel—sometime between 1990 and 2015, when world oil use is projected to increase to the point where demand exceeds supply. When this happens, the OPEC countries, with 57% of the world's proven oil reserves and 23% of the estimated undiscovered supplies, are projected to increase their share of the world's oil market from 27% in 1987 to 60% in the 1990s, again dominating world oil markets and prices.

Because of decreasing domestic production since 1970, the Department of Energy and most major oil companies project that by 1995 the United States could be dependent on imported oil for 60% of its oil consumption—much higher than in 1977 (Figure 15-4). This situation would drain the already debt-ridden

Much of the argument over U.S. energy policy is whether we should provide more electricity by building more coal or nuclear power plants, large or small-scale hydroelectric power plants, solar power plants, or vast farms of wind turbines. However, according to physicist and energy expert Amory Lovins, *"supplying more electricity is irrelevant to the energy problem that we have.* Even though electricity accounts for two-thirds of the federal energy research and development budget and for about half of national energy investment, it is the wrong kind of energy to meet our needs economically. Arguing about what kind of new power station to build—coal, nuclear, solar—is like shopping for the best buy in Chippendales to burn in your stove. *It is the wrong question."*

Electricity is a high-quality, expensive form of energy that is ideally suited only for certain tasks such as running lights, motors, electronic devices, and metal smelters. But these special uses require only 8% of all delivered U.S. energy

needs. Today's electric power plants already produce more than two times the amount of electricity needed for these uses. As a result, about 40% of the electricity produced in the United States is now used for uneconomic and highly wasteful uses, such as space heating, water heating, and air conditioning. Using electricity to heat and cool air and to heat water is equivalent to buying the heat content of oil costing $128 per barrel—over 8 times the average world price during 1986.

Thus the real question facing the United States is, How should we provide the energy needed for space heating and cooling, water heating, and nonrail transportation at the lowest cost? Producing more electricity by any means is essentially irrelevant to this key question.

According to Lovins, eliminating unnecessary waste of electricity, increased use of passive and some active solar techniques for water heating and space heating, and making present lights, motors,

appliances, and metal smelters more cost-effective and efficient would make it possible "to run the U.S. economy, with no changes in lifestyles, using no thermal power plants, whether old or new, and whether fueled with oil, gas, coal, or uranium. We would need only the present hydroelectric capacity, readily available small-scale hydroelectric projects, and a modest amount of wind power."

Even if we still wanted more electricity, the next cheapest sources would include industrial cogeneration, solar ponds, filling empty turbine bays in existing large hydroelectric power plants, farms of modern wind machines, additional small-scale hydroelectric turbines, and perhaps solar photovoltaic cells (if prices come down as anticipated in the 1990s). Lovins argues that "only after all these cheaper alternatives have been exhausted should we even consider building a new central power station of any kind in the United States." What do you think?

United States of vast amounts of money, leading to severe inflation and widespread economic recession, perhaps even a major depression. The Congressional Research Service warns that an increase in the price of oil to between $50 and $98 a barrel could reduce the U.S. Gross National Product by as much as 29%, cut jobs by as much as 28%, and increase the likelihood of war as the world's MDCs compete for greater control over oil supplies to avoid total economic collapse.

Thus without greatly increased efforts to reduce energy waste and develop alternatives to oil, the short-term economic bonanza of cheap oil in the 1980s could turn into long-term economic disaster in the 1990s or the first decade of the next century. Unfortunately, since 1981 the United States has done little to prepare for a future oil crisis. As low oil and gasoline prices lulled most consumers and elected officials into a false sense of security, polls showed that less than 5% of the American public listed energy as an important national problem.

15-2   EVALUATING ENERGY RESOURCES

Future Energy Resources   There is considerable controversy over which combination of energy alternatives should be developed to provide most of the energy we need in the future. As we near the end of the world's affordable supplies of oil and perhaps of natural gas, some energy experts call for increased use of nonrenewable coal—the world's most abundant fossil fuel—to produce electricity and high-temperature heat for industrial processes and as a source of synthetic gas and liquid fuels (synfuels) for use in home heating and motor vehicles. Others call for greatly increased dependence on nonrenewable nuclear fission to produce electricity, and eventually on almost inexhaustible nuclear fusion, when and if it can be developed (Section 3-2 and Chapter 16).

Conservationists believe that either alternative is

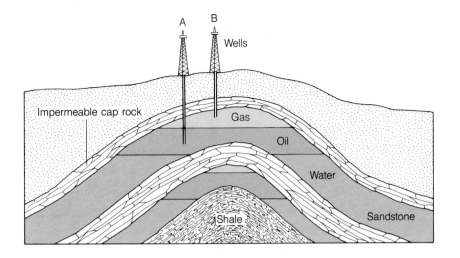

unwise: Both coal and nuclear power are centralized, nonrenewable energy resources that are undesirable from long-term economic, social, environmental, and national security viewpoints. Instead, conservationists call for an energy strategy based on greatly increased energy efficiency to extend remaining fossil fuel supplies long enough to phase in greatly increased dependence on a variety of renewable and perpetual energy sources based primarily on locally available supplies (Figures 15-2 and 15-3 and Chapter 17).

Questions to Ask   To determine which combination of energy alternatives might provide energy for the future, it is necessary to think and plan in three time frames that cover the 50-year period normally needed to develop and phase in new energy resources: the short term (1989 to 1999), the intermediate term (1999 to 2009), and the long term (2009 to 2039).

First, we must decide how much we need, or want, of various kinds of energy, such as low-temperature heat, high-temperature heat, electricity, and liquid fuels for transportation (see Figure 3-8, p. 38). The next step is to project the combination of energy alternatives— including improving energy efficiency—that provide the necessary energy services at the lowest cost and most acceptable environmental impacts (see Spotlight on p. 254). Thus, for each energy alternative, we need to know (1) the total estimated supply available in each time frame; (2) the estimated net useful energy yield (Section 3-6); (3) the projected costs of development and lifetime use; and (4) the potential environmental impacts.

Environmental Impact of Energy Alternatives Energy policy is also environmental policy because energy use is directly or indirectly responsible for most land disruption, air pollution, and water pollution (Figure 14-7, p. 236). For example, nearly 80% of all U.S. air pollution is caused by fuel combustion in cars, furnaces, industries, and power plants.

Most analysts believe that the two greatest global environmental threats from present energy use are (1) possible changes in global climate from carbon dioxide emitted when any fossil fuel—especially coal and synthetic gaseous and liquid fuels produced from coal— is burned (Section 18-5); and (2) potential large-scale release of long-lived radioactive materials into the environment from one or more nuclear power plant accidents or inadequate storage of radioactive wastes (Chapter 16).

Choosing any single energy option or combined options involves making trade-offs between several different potential environmental impacts based on risk-benefit analysis. The major environmental impacts for each energy alternative are discussed in the remainder of this chapter and in the two chapters that follow.

15-3   OIL

Conventional Crude Oil   A gooey liquid, **crude oil** or petroleum is made up mostly of hydrocarbon compounds (90% to 95% of its weight) and small quantities of compounds containing oxygen, sulfur, and nitrogen. Typically, deposits of crude oil and natural gas are trapped together deep underground, beneath a dome of impermeable cap rock and above a lower dome of sedimentary rock such as shale, with the natural gas lying above the crude oil (Figure 15-6). Crude oil is also found beneath the sea floor.

If there is enough pressure from water and natural gas under the dome of rock, some of the crude oil will be pushed to the surface when the well is drilled. However, such wells, called *gushers*, are relatively rare.

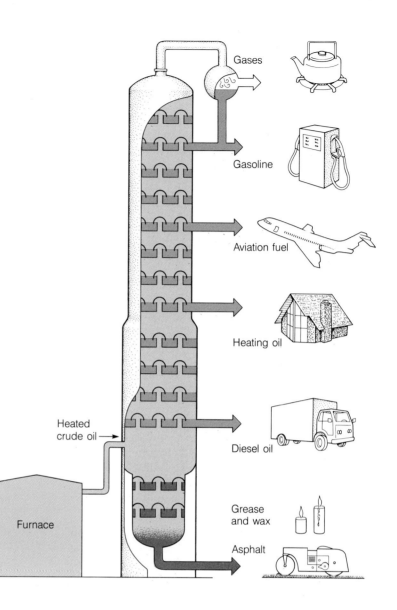

Figure 15-7 Refining of crude oil: Major components are removed at various levels, depending on their boiling points, in a giant distillation column.

Gases

Gasoline

Aviation fuel

Heating oil

Heated crude oil →

Diesel oil

Grease and wax

Furnace

Asphalt

Primary oil recovery involves pumping out all of the oil in gushers and other deposits that will flow by gravity into the hole. When this oil has been removed, water can be injected into the well to force out some of the remaining crude oil. This is known as **secondary oil recovery.** Usually primary and secondary recovery remove only about one-third of the crude oil in a well.

Two barrels of a thicker oil, called **heavy oil,** are left in a typical well for each barrel removed by primary and secondary recovery. As oil prices rise, it may become economical to remove about 10% of the heavy oil by **enhanced oil recovery.** For example, steam can be forced into the well to soften the heavy oil so that it can be pumped to the surface, or some of the heavy oil can be ignited to increase the flow rate of the surrounding oil so that it can be pumped to the surface. But enhanced oil recovery processes are expensive and require the energy equivalent of one-third of a barrel of oil to pump each barrel to the surface, thus reducing the net useful energy yield. Additional energy is needed

to increase the flow rate and remove impurities, especially sulfur and nitrogen compounds, before the heavy oil can be sent via pipeline to an oil refinery. Recoverable heavy oil from known U.S. crude oil reserves could supply U.S. oil needs for only about seven years at 1984 usage rates.

Once it is removed, most crude oil is sent by pipeline to a refinery. There it is distilled to separate it into component chemicals, which boil at different temperatures and are removed from various levels of giant distillation columns (Figure 15-7). Some of these chemicals, known as **petrochemicals,** are sent to petrochemical plants for use as raw materials in the manufacture of most industrial chemicals, fertilizers, pesticides, plastics, synthetic fibers, paints, medicines, and numerous other products. Production of these vital petrochemicals requires 3% of the world's fossil fuel production and 7% of U.S. production. This use of oil explains why the prices of a wide range of products based on petrochemicals correspondingly rise after crude oil prices rise.

Figure 15-8 Major deposits of natural gas and crude oil in the United States. (Source: Council on Environmental Quality)

natural gas and crude oil fields

How Long Will Supplies of Conventional Crude Oil Last? OPEC is expected to have long-term control over world oil supplies and prices because most of the world's oil reserves are in the Middle East, whereas most oil consumption takes place in North America, western Europe, and Japan. Saudi Arabia alone has 25% of the world's proven oil reserves (and the United States only 4%). Figure 15-8 shows the locations of the major crude oil and natural gas fields in the United States. Most are in Texas, Louisiana, and Oklahoma and in the continental shelf of the Pacific, Atlantic, Gulf, and Alaskan coasts.

Experts disagree over how long identified and unidentified crude oil resources will last. Cornucopians argue that higher prices will stimulate the location and extraction of unidentified crude oil resources, as well as the extraction and upgrading of heavy oils from oil shale and tar sands and from oil too thick to pump out of existing oil wells. They point out that global proven oil reserves increased almost ninefold between 1950 and 1973.

Neo-Malthusians, however, argue that cornucopians misunderstand the arithmetic and consequences of exponential growth in the use of any non-renewable resource. They point out that between 1973 and 1986, global proven oil reserves increased by only 5%, despite much higher oil prices and greatly increased exploration between 1973 and 1980. According to estimates by the U.S. Department of Energy and the American Petroleum Institute, 80% of the world's known reserves of crude oil will be depleted by 2013 if annual oil consumption remains at the 1984 rate and by 2006 if the annual depletion rate increases by a modest 2%.

Consider the following implications of the world's present exponential growth in oil use: Assuming the 1984 rate of crude oil consumption is maintained, (1) Saudi Arabia, with the world's largest known crude oil reserves, could supply the world's total needs for only ten years if it were the world's only source; (2) Mexico, with the world's sixth largest crude oil reserves, could supply the world's needs for only about three years; (3) each year the world consumes oil roughly equivalent to the entire proven reserves of Venezuela or Libya; (4) the estimated crude oil reserves under Alaska's North Slope—the largest deposit ever found in North America—would meet world demand for only six months or U.S. demand for two to four years;

Figure 15-9 Sample of oil shale rock and the shale oil extracted from it. Oil shale projects have now been canceled in the United States because of excessive cost.

and (5) if drilling off the east coast of the United States meets the most optimistic estimates, which is quite unlikely, the resulting additions to crude oil reserves would satisfy world oil needs for one week or U.S. needs for less than three months. Thus cornucopians who argue that new discoveries will solve world oil supply problems must somehow figure out how to discover the equivalent of a new Saudi Arabian deposit *every ten years* merely to maintain the world's 1984 level of oil use.

The ultimately recoverable supply of crude oil based on undiscovered deposits (most believed to be in the Middle East) is estimated at more than three times today's proven reserves. Even if *all* estimated supplies of crude oil are discovered and developed—which most oil experts consider unlikely—and sold at a price of $50 to $95 a barrel, 80% of this reserve would be depleted by 2076 at 1984 usage rates and by 2037 if oil usage increased by 2% a year. Thus relatively little of the world's crude oil is likely to remain by the 2059 bicentennial of the world's first oil well.

Although the Soviet Union is currently the world's largest producer, the United States uses more of the world's oil (25%) than any other country. U.S. oil exploration has been more intensive than in any other country, but domestic oil production peaked in 1970 and has declined since then despite increased exploration. In 1979 the United States produced 79% of the oil it consumed but is projected to produce only 40% by 1995. Any major new discoveries are likely to be in deep water or in remote areas with frigid climates, where production costs are very high and net useful energy yields are low. Because of unsuccessful exploratory drilling, in 1985 the U.S. Geological Survey reduced its estimate of undiscovered oil off the coasts of the United States by 55%. Officials also expect production from Alaska's North Slope fields, which pro-

vided almost one-fourth of U.S. oil production in recent years, to begin falling by 1990.

Oil production by the Soviet Union may also decline because it is heavily dependent on old fields with declining yields and is plagued by difficulties in opening up new fields in frigid areas of Siberia far from population centers. Since the Soviet Union now produces one-fifth of the world's oil, any significant decline in its oil production will have a major effect on world supply and price. Oil production in the North Sea is also projected to peak before 1990 and then to decline steadily during the 1990s. This analysis indicates why most energy experts expect oil prices to rise sharply sometime between 1995 and 2010.

Major Advantages and Disadvantages of Oil Oil has several important advantages that account for its widespread use. It has been and still is relatively cheap (Figure 15-5), can be transported easily within and between countries, and is a versatile fuel that can be burned to propel vehicles, provide low-temperature heating of water and buildings, and provide high-temperature heat for industrial processes and production of electricity. It also has a high net useful energy yield (Figure 3-15, p. 46).

However, oil also has some disadvantages. Affordable supplies may be depleted within 40 to 80 years. Its burning releases carbon dioxide gas, which could alter global climate, and air pollutants such as sulfur oxides and nitrogen oxides. Its use can also cause water pollution from oil spills and contamination of underground water by the brine solution injected into oil wells. Finally, with the increasing use of less accessible and more remote deposits, the net useful energy yield of oil will decrease.

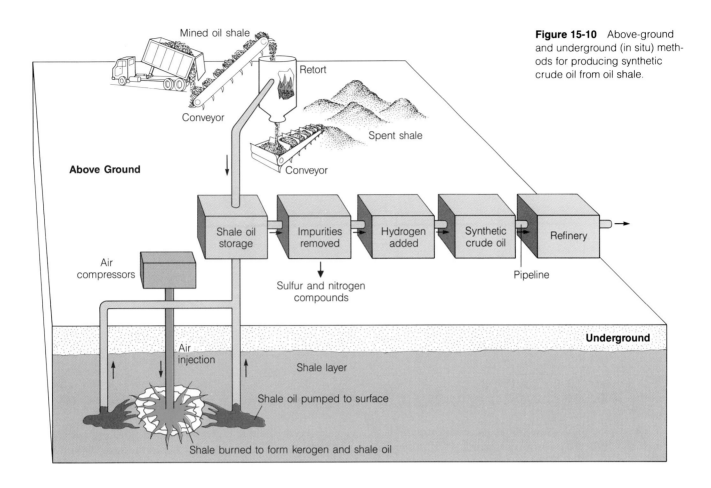

Figure 15-10 Above-ground and underground (in situ) methods for producing synthetic crude oil from oil shale.

Heavy Oils from Oil Shale and Tar Sands Some places on earth contain rich underground deposits containing heavy oil ranging from thick oil, which doesn't flow like conventional crude oil, to tar. The two major sources of such heavy oil are deposits of oil shale and tar sands.

Oil shale is a fine-grained rock (Figure 15-9) that contains varying amounts of a solid, waxy mixture of hydrocarbon compounds called **kerogen** disseminated throughout the rock. After being removed by surface or subsurface mining, shale rock is crushed and heated to a high temperature in a large retort to vaporize the solid kerogen (Figure 15-10). The kerogen vapor is condensed, forming a slow-flowing, dark brown, and heavy oil called **shale oil** (Figure 15-9). Before shale oil can be sent by pipeline to a refinery, it must be processed to increase its flow rate and heat content and to remove sulfur, nitrogen, and other impurities.

The world's largest known deposits of oil-bearing shale are in Colorado, Utah, and Wyoming. Because 80% of this shale is below public lands, oil companies must obtain leases from the federal government to exploit these resources. Significant oil shale deposits are also found in Canada, China, and the Soviet Union. It is estimated that the potentially recoverable heavy

oil from oil shale in the United States can supply the country with crude oil for 44 years if consumption remains at 1984 levels, and for 32 years if consumption rises by 2% a year.

It takes the energy equivalent of about one-third of a barrel of conventional crude oil to mine, retort, and purify one barrel of shale oil, thus sharply reducing its net useful energy yield compared to conventional oil (Figure 3-15, p. 46). As a result, the price of extracting and processing this oil is high. Until conventional oil prices rise sharply, shale oil is not an economically feasible source of oil without government subsidies or new extraction and processing technology.

Environmental problems may also limit shale oil production. Shale oil processing requires large amounts of water, which is scarce in the semiarid areas where the richest deposits are found. When the oil is processed and burned carbon dioxide is released into the atmosphere; without adequate air pollution controls nitrogen oxides and several possibly cancer-causing substances are also released. There would be significant land disruption from the mining and disposal of large volumes of shale rock, which breaks up and expands when heated. Various salts, cancer-causing substances, and toxic metal compounds could be leached

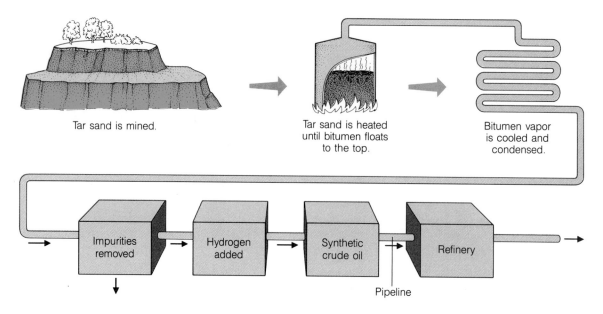

Figure 15-11 Generalized summary of production of synthetic crude oil from tar sands.

from the processed shale rock into nearby water supplies.

One way to avoid some of these environmental problems is to extract oil from oil shale rock underground—known as *in situ processing* (Figure 15-10). After the shale rock is broken up with explosives or water under high pressure, it is retorted and distilled underground by pumping in air and setting the deposit on fire. As the oil shale vapor is driven through the mine, it comes in contact with cooler shale rock, condenses, and drips to the bottom of the fractured area, where it can be pumped to the surface. However, experimental projects have shown that with present technology it is difficult to fracture the shale evenly enough to allow uniform combustion. In addition, groundwater often seeps into the retort area and extinguishes the fire. Sulfur dioxide emissions per barrel of shale oil are also higher than those from surface retorting.

By 1987, no oil shale extraction and processing technology had reached the commercial stage in the United States. Moreover, most pilot projects were abandoned because both surface mining and in situ methods proved too expensive even with large government subsidies, which were canceled in 1986.

Tar sands (or oil sands) are deposits of a mixture of fine clay, sand, water, and variable amounts of **bitumen,** a black, high-sulfur, tarlike heavy oil. Typically, tar sand is removed by surface mining and heated with steam at high pressure to make the bitumen fluid enough to float to the top. The bitumen is removed and then purified and upgraded to synthetic crude oil before being refined (Figure 15-11). So far it is not technically or economically feasible to remove deeper

deposits by underground mining or by in situ extraction.

The world's largest known deposits of tar sands lie in a cold, desolate area in northern Alberta, Canada. Other fairly large deposits are in Venezuela, Colombia, and the Soviet Union. There are smaller deposits in the United States, about 90% of which are located in Utah.

For several years two plants have been supplying about 15% of Canada's oil demand and most of the demand in its northwestern provinces of Alberta and Saskatchewan by extracting and processing heavy oil from tar sands at a cost of between $15 and $25 a barrel. Economically recoverable deposits of heavy oil from tar sands could supply Canada's projected oil needs for about 36 years at the 1984 consumption rate but would supply total world oil needs at the same rate for only about 2 years. The U.S. Office of Technology Assessment estimated that U.S. deposits of tar sands would become economically feasible sources of synthetic crude oil only with an average world crude oil price of $48 to $62 a barrel. If developed, these deposits would supply all U.S. oil needs at 1984 usage rates for only about three months.

Although large-scale production of synthetic crude oil from tar sands may meet a significant fraction of Canada's oil needs, producing this oil is beset with problems. The net useful energy yield is low because it takes the energy equivalent of one-third of a barrel of conventional oil to produce the steam and electricity needed to extract and process one barrel of bitumen. More energy is needed to remove sulfur impurities and to upgrade the bitumen to synthetic crude oil before it can be sent to an oil refinery. Other problems include the need for large quantities of water for processing and the release of air and water pollutants similar to those produced when oil shale is processed and burned.

15-4 NATURAL GAS

Conventional Supplies of Natural Gas In its underground gaseous state, **natural gas** consists of 50% to 90% methane (CH_4) gas and smaller amounts of heavier gaseous hydrocarbon compounds such as propane (C_3H_8) and butane (C_4H_{10}). Although most natural gas used so far lies above deposits of crude oil (Figure 15-6), it is also found by itself in other underground deposits.

When a natural gas deposit is tapped, propane and butane gases are liquefied and removed as **liquefied petroleum gas (LP gas).** The remaining gas (mostly methane) is then dried, cleaned of hydrogen sulfide and other impurities, and pumped into pressurized pipelines for distribution over land. LP gas is stored in pressurized tanks for use mostly in rural areas not served by natural gas pipelines.

Very low temperature can be used to convert natural gas to **liquefied natural gas (LNG),** which can be transported by sea in specially designed, refrigerated tanker ships. However, LNG is extremely volatile, unstable, and flammable; an explosion in a tanker would create a massive fireball. Conversion of natural gas to LNG, however, reduces the net useful energy yield by about one-fourth.

Most of the present U.S. reserves of natural gas are located with the country's deposits of crude oil (Figure 15-8). America's largest known deposits of natural gas lie in Alaska's Prudhoe Bay, thousands of kilometers from natural gas consumers in the lower 48 states. Geologists estimate that up to eight times as much natural gas awaits discovery in Alaska's North Slope area.

How Long Will Natural Gas Supplies Last?
Conventional supplies of natural gas are projected to last somewhat longer than those of crude oil. Between 1973 and 1984, proven reserves of natural gas doubled, whereas those of petroleum and coal have not risen. Much of this increase came from large discoveries in the Soviet Union, which now has 40% of the world's proven reserves and is the world's largest natural gas producer. Other countries with large proven natural gas reserves include Iran (14%), the United States (6%), Qatar (4%), Algeria (4%), Saudi Arabia (3%), and Nigeria (3%).

Additional discoveries of natural gas are expected, especially in LDCs, as a result of intensified exploration and improved methods of locating deposits. Although known reserves in the United States are projected to last only until 1993 at 1984 usage rates, additional U.S. supplies are expected to be found. However, in 1985 the U.S. Geological Survey reduced its estimate of offshore deposits of undiscovered gas likely to be found by 48% as a result of unsuccessful exploratory drilling.

The world's identified reserves of natural gas are projected to last until 2033 at 1984 usage rates and to 2018 if usage increases by 2% a year. The supply that would be found and recovered at much higher prices would last about 200 years at 1984 usage rates and 80 years if usage increased by 2% a year.

Unconventional Sources of Natural Gas As the price of natural gas from conventional sources rises, some analysts believe it may become economical to drill deeper into the earth and extract and process natural gas from unconventional sources. These include concrete-hard, deep geologic formations of tight sands, deep geopressurized zones containing deposits of hot water under such high pressure that large quantities of natural gas are dissolved in the water, coal seams, and deposits of Devonian shale rock.

Energy experts agree that there are large deposits of natural gas in such sources but disagree over whether the gas can be recovered at affordable prices. If a reasonable amount of this natural gas could be recovered, world supplies would be extended for several hundred to a thousand years, allowing natural gas to become the most widely used fuel for space heating, industrial processes, producing electricity, and transportation.

Advantages and Disadvantages of Natural Gas Two major advantages of conventional natural gas are its ability to burn hotter and cleaner than any other fossil fuel and its relatively low price. It is also a versatile fuel, is transported easily over land by pipeline, and has a high net useful energy yield (Figure 3-15, p. 46). Its burning, however, produces carbon dioxide, although the amounts per unit of energy produced are lower than those from other fossil fuels (Figure 15-16, p. 265). Natural gas is also difficult, expensive, and dangerous to transport by tanker as volatile, unstable LNG.

Natural gas from unconventional sources has the same advantages and disadvantages as conventional natural gas except that it is more difficult and expensive to recover and process—thus its net useful energy yield is lower. Furthermore, the deep-drilling technology needed to remove this resource is not fully developed.

Figure 15-12 Stages in the formation of different types of coal over millions of years.

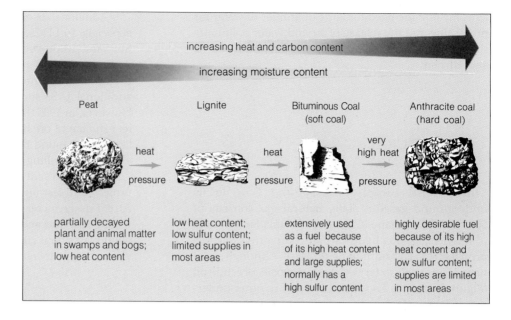

increasing heat and carbon content

increasing moisture content

| Peat | Lignite | Bituminous Coal (soft coal) | Anthracite coal (hard coal) |

heat → pressure → heat → pressure → very high heat → pressure

partially decayed plant and animal matter in swamps and bogs; low heat content

low heat content; low sulfur content; limited supplies in most areas

extensively used as a fuel because of its high heat content and large supplies; normally has a high sulfur content

highly desirable fuel because of its high heat content and low sulfur content; supplies are limited in most areas

15-5 COAL

Conventional Types of Coal   Natural **coal** is a solid formed in several stages as the remains of plants are subjected to intense heat and pressure over millions of years. It is mostly carbon with varying amounts of water and small amounts of nitrogen and sulfur. Coal with high sulfur and nitrogen content produces high emissions of sulfur dioxide and nitric oxide (also produced by reaction of the N_2 and O_2 in air at high temperatures) into the atmosphere when it is burned without adequate air pollution control devices. About 60% of the coal extracted in the world and 70% in the United States is burned in boilers to produce steam to generate electrical power.

There are several major kinds of coal, each with a different carbon content, moisture content, sulfur content, and fuel value (heat content), depending on how much the original plant material was modified by heat and the weight of overlying materials (Figure 15-12). Anthracite, or hard coal, burns cleaner with less smoke than other types, but it is not as common and is usually more expensive. Bituminous coal, or soft coal, is much more abundant but usually has a high sulfur content. Subbituminous coal and lignite are generally low in sulfur but also have a low fuel value. Peat, which is not a true coal, has a low heat content and a high moisture content (70% to 95%) and is burned for fuel in areas such as Ireland where supplies are plentiful.

Distribution of Coal   Coal is the world's most abundant fossil fuel. About 68% of the world's proven coal reserves and 85% of the estimated undiscovered coal deposits are located in three countries: the United States (29% of proven reserves), the USSR (28%), and China (11%). These countries also account for about 60% of present total world coal production.

Major U.S. coalfields are located primarily in 17 states (Figure 15-13). Anthracite, the most desirable form of coal, makes up only about 2% of U.S. coal reserves. About 45% of U.S. coal reserves—consisting mostly of high-sulfur, bituminous coal with a relatively high heat content—is found east of the Mississippi River in the Appalachian region, particularly in Kentucky, West Virginia, Pennsylvania, Ohio, and Illinois. Because most of this coal has such a high sulfur content, the percentage of all U.S. coal extracted from fields east of the Mississippi River fell from about 93% to 64% between 1970 and 1985.

About 55% of U.S. coal reserves are found west of the Mississippi River. Most of these deposits consist of low-sulfur (typically 0.6%), subbituminous and lignite coals, which can be surface-mined more safely and cheaply than the underground deposits of bituminous coal found east of the Mississippi. Because about 70% of western coal reserves are under federally owned lands, major increases in the development of these resources will depend primarily on the actions of government agencies and to a lesser degree of private groups, including Native American tribes, that hold the remaining western coal reserves and water rights.

Health and Environmental Hazards of Coal Mining   Underground mining of coal is the second most hazardous occupation in the world (after commercial fishing) because of injuries and deaths from cave-ins and from explosions caused by a single spark

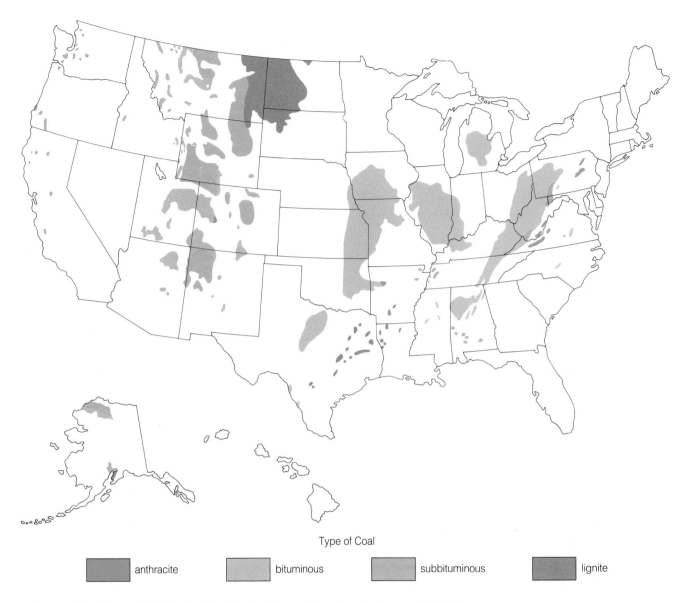

Type of Coal

| | anthracite | | bituminous | | subbituminous | | lignite |

Figure 15-13 Major coalfields in the United States. (Source: Council on Environmental Quality)

igniting underground air laden with coal dust or methane gas. Between 1900 and 1985 underground mining in the United States killed more than 100,000 miners, permanently disabled at least 1 million, and caused at least 250,000 retired miners to suffer from **black lung disease** caused by breathing coal dust and other particulate matter over a prolonged period (Figure 15-14). For the victims of this form of emphysema, even the slightest movement is difficult and death is slow and painful. Past failures to enact and enforce stricter mine safety laws now cost U.S. taxpayers over $1 billion a year in federal disability benefits paid to coal miners with black lung disease. Although U.S. coal mining is safer than in most other countries, worker safety could be improved significantly by stricter enforcement of existing laws and by enactment of tougher new laws.

Two major environmental problems resulting from underground mining of coal are acid mine drainage and subsidence. **Acid mine drainage** occurs when surface water enters an abandoned underground mine and dissolves and carries sulfuric acid and toxic metal compounds into nearby streams and rivers (Figure 14-9, p. 238). These chemicals, formed when air and water react with sulfur compounds (such as iron sulfide or pyrite) in rocks or soil in the mine, kill and damage aquatic plant and animal life, make the water rust-colored and unfit for drinking and swimming, and cause millions of dollars of corrosion damage to metal bridges, canal locks, barges, and ships. In the United States over 11,000 kilometers (7,000 miles) of streams (90% of which are in Appalachia) are severely affected by drainage of acids from underground coal mines.

Figure 15-14 Comparison of the lungs of a nonsmoking non-miner (left) and a nonsmoking coal miner, who died from black lung disease, a severe form of emphysema.

Figure 15-15 Subsidence from an abandoned and inadequately supported underground coal mine caused this home to collapse.

Acid mine drainage can be controlled by filling sinkholes and rerouting gulleys to prevent surface water from entering abandoned mines. It can also be reduced by treating the water draining from mines with crushed limestone to neutralize the acidity, but in practice this does not always work well.

Subsidence occurs when a mine shaft partially collapses during or after mining, creating a depression in the surface of the earth above the mine (Figure 15-15). In the United States over 800,000 hectares (2 million acres) of land, much of it in central Appalachia, has subsided as a result of underground coal mining.

Virtually all coal mined west of the Mississippi River and half that produced in Appalachia is removed by some form of *surface mining* (Section 14-2). Surface mining now accounts for about two-thirds of the coal extracted in the United States, because it is cheaper per ton of coal removed, more efficient (removing 80% to 90% of the coal in a deposit compared to 40% to 50% with underground mining), more profitable, less labor-intensive, and safer than underground mining.

Without adequate land restoration, surface mining can have a devastating impact on land (Section 14-2). It can destroy the natural vegetation and habitats for many types of wildlife. Soil erosion from unrestored surface-mined land is up to 1,000 times more severe than that from the same area under natural conditions. Minerals and salts, which leach out of unrestored land when rainwater and melting snow percolate through overburden deposits, can also kill aquatic life, pollute streams, and contaminate groundwater because many aquifers are located near or in coal seams.

More than 1 million acres of American land dis-turbed by surface mining have long ago been abandoned by coal companies and not restored. To help control land disturbance from surface mining of coal, the Surface Mining Control and Reclamation Act of 1977 was enacted. According to this act, (1) surface-mined land must be restored to its approximate original contour and, if requested by the owner, also revegetated so that it can be used for its original purposes; (2) surface mining is banned on some prime agricultural lands in the West, and farmers and ranchers can veto mining under their lands even though they do not own the mineral rights; (3) mining companies must minimize the effects of their activities on local watersheds and water quality by using the best available technology, and they must prevent acid from entering local streams and groundwater; (4) money from a $4.1 billion fund, financed by a fee on each ton of coal mined, is to be used to restore surface-mined land not reclaimed before 1977; and (5) responsibility for its enforcement is delegated to the states, but the Department of the Interior has enforcement power when states fail to act and on federally owned lands.

If strictly interpreted, enforced, and adequately funded, this law could help protect valuable ecosystems. Since the law was passed, however, there has been growing pressure from the coal industry (much of which is owned by major oil companies) to weaken or declare it unconstitutional. In the early 1980s the Reagan administration cut the federal inspection and enforcement staff by 70%, thus reducing the effectiveness of the federal government in enforcing the act. To cite only one of numerous examples of enforcement laxity, in 1983 Utah state officials were carrying out fewer than half the inspections required by law.

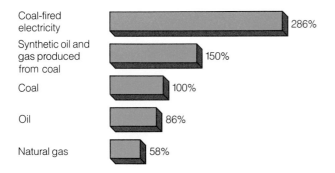

Coal-fired electricity	286%
Synthetic oil and gas produced from coal	150%
Coal	100%
Oil	86%
Natural gas	58%

Figure 15-16 Emissions of carbon dioxide per unit of energy produced by burning fossil fuels as percentages of those from burning coal.

How Long Will Supplies Last?

Based on known reserves, coal is the most abundant conventional fossil fuel in the world and in the United States. Identified world coal reserves should last for about 276 years at 1984 usage rates and 86 years if usage increases by 2% a year. The world's estimated ultimately recoverable supply of coal would last about 900 years at 1984 usage rates and 149 years if usage increased by 2% a year.

Identified coal reserves in the United States are projected to last about 300 years at 1984 usage rates. Unidentified U.S. coal resources could extend these supplies at such high rates for perhaps 100 years, at a much higher average cost.

Air Pollution from Burning Coal

Without effective air pollution control devices, burning coal emits sulfur dioxide, particulate matter (mostly fly ash), cancer-causing substances, and small amounts of radioactive materials naturally found in coal deposits into the atmosphere. Air pollution from coal-burning power and industrial plants can be sharply reduced by the pollution control devices (Chapter 18) required by federal law on all U.S. coal-fired plants built since 1978. Although removal of particulate matter and sulfur dioxide reduces air pollution, it produces large quantities of solid waste that must be handled in a way that does not pollute water supplies and soil (Chapter 20).

Existing coal-fired boilers in the United States are responsible for 70% of the country's sulfur dioxide emissions and 20% to 25% of nitrogen oxide emissions. Both of these pollutants contribute to acid deposition that damages many forests and aquatic ecosystems (Section 18-2). They also cause an estimated 5,000 premature deaths, 50,000 cases of respiratory disease, and several billion dollars in property damage each year.

Requiring all older coal-burning plants to have air pollution control devices that remove 95% of sulfur dioxide as well as other harmful emissions would reduce deaths from burning coal to about 500 per year. This improvement in consumer health would increase the cost of electricity, but coal-burning plants would still produce electricity more cheaply than nuclear plants.

To many analysts the most serious disadvantage of coal is that it produces more carbon dioxide per weight when burned than oil or natural gas—especially when it is burned to produce electricity or burned in the form of synthetic natural gas or oil (Figure 15-16). At present there is no technologically and economically feasible method for preventing this carbon dioxide from reaching the atmosphere.

Burning Coal More Cleanly and Efficiently

One method for burning coal more cleanly and efficiently is to burn a mixture of finely powdered coal and water. The finely crushed coal particles can be mixed with water to produce a molasses-like mixture that can be burned in existing oil burners without too many modifications. By 1986 coal-water mixtures were being used in a few small-scale, experimental electricity plants in the United States and Canada.

Another promising method for burning coal more efficiently, cleanly, and cheaply than in conventional coal boilers is **fluidized-bed combustion (FBC)** (Figure 15-17). In FBC a stream of hot air is blown through a series of small jets into a boiler to suspend a mixture of powdered coal and crushed limestone. This process removes 90% to 98% of the sulfur dioxide gas produced during combustion and emits considerably less nitrogen oxides than present federal air pollution standards permit. FBC boilers also can burn a variety of fuels and can be retrofitted to conventional boilers.

Successful small-scale FBC plants have been built in Great Britain, Sweden, Finland, the Soviet Union, West Germany, and China. In the United States, 20 small-scale FBC plants burning coal or wood were in operation by 1986 and are projected to begin replacing conventional coal boilers by the 1990s.

Synfuels: Gaseous and Liquid Fuels from Coal

Techniques for producing gaseous and liquid fuels, called **synfuels,** from coal have been available for many years. **Coal gasification** involves converting coal to a gas that can be burned more cleanly as a fuel and, unlike coal, can be transported through a pipeline. In one process, coal is converted to coke, which is heated with steam to produce a gaseous mixture of carbon monoxide and hydrogen, known as *producer gas*. However, because producer gas has a low heating value, it is uneconomical to transport by pipeline.

A second and more useful process converts coal to **synthetic natural gas (SNG),** which, like natural gas, has a high heating value and can be transported by pipeline. One way of producing SNG is to burn

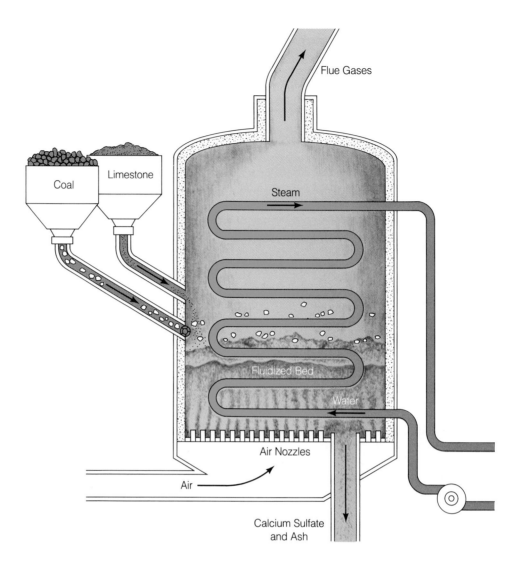

Figure 15-17 Fluidized-bed combustion of coal.

Flue Gases

Coal

Limestone

Steam

Fluidized Bed

Water

Air Nozzles

Air

Calcium Sulfate and Ash

powdered coal in air to produce carbon monoxide, which is then combined with hydrogen gas in the presence of a catalyst (to speed up the rate of the reaction) to form methane and water vapor (Figure 15-18).

Coal liquefaction involves converting coal to a liquid hydrocarbon fuel such as methanol or synthetic gasoline. A number of different liquefaction processes have been developed. In South Africa a commercial plant has been converting synthetic natural gas made from coal to gasoline and other motor fuels for more than 25 years and now supplies 10% of the country's liquid fuel needs. When two new plants are completed the three plants are expected to meet 50% of South Africa's current oil needs.

Advantages and Disadvantages of Solid Coal and Synfuels Coal is the most abundant conventional fossil fuel in the world and in the United States and has a high net useful energy yield for producing high-temperature heat for industrial processes and for the generation of electricity (Figure 3-15, p. 46). In countries with adequate coal supplies, burning solid coal is the cheapest way to produce high-temperature heat and electricity compared to oil, natural gas, and nuclear energy.

But solid coal has several major disadvantages that may limit its use. Burned coal releases more carbon dioxide than other fossil fuels. It is also dangerous to mine and expensive to move from one place to another, and is not useful in solid form as a transportation fuel.

Converting solid coal to gaseous or liquid synfuels allows it to become a more versatile fuel that can be burned more readily to heat homes and power vehicles. Synfuels are easier and often cheaper to transport than solid coal and when burned produce much less air pollution than solid coal.

It is much more expensive, however, to build a synfuel plant than an equivalent coal-fired power plant fully equipped with air pollution control devices. Other major problems of synfuels include their low net useful energy yield (Figure 3-15, p. 46), accelerated depletion of world coal supplies because 30% to 40% of the energy content of coal is lost in the conversion process; the massive amounts of water required for processing; release of large amounts of carbon dioxide per unit of

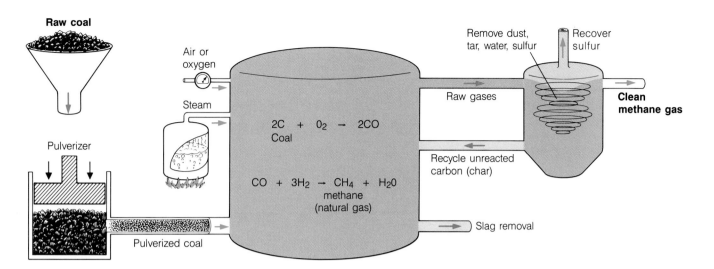

Figure 15-18 Coal gasification. Generalized view of one method for converting solid coal into synthetic natural gas.

weight when processed and burned; increased air pollution from polycyclic aromatic hydrocarbons (PAHs), many of which cause cancer; and greater land disruption from surface mining because of increased use of coal per unit of energy produced.

Some of these problems (except carbon dioxide emissions, high costs, and low net energy yields) could be avoided or reduced by using an in situ method to convert coal to SNG underground in the coal seam itself. But despite potential advantages and years of experimentation, in situ coal gasification is still not competitive with conventional coal mining and aboveground coal gasification.

The major factor holding back large-scale production of synfuels in the United States is their high cost compared to conventional oil and natural gas—equivalent to buying oil in 1985 at $90 a barrel. In 1980 Congress set up the U.S. Synthetic Fuels Corporation to provide up to $88 billion in grants to oil companies to promote the development of a synfuels industry. By 1984, however, oil companies found it too expensive to continue such experimental projects, even with large government subsidies, and most were abandoned. In 1986 Congress disbanded the U.S. Synthetic Fuels Corporation. Most analysts expect synfuels to play only a minor role as an energy resource until oil prices rise substantially sometime after 2000.

Oil and natural gas will play an important but diminishing role for some time, but how long is less clear. Coal will likely grow in importance, but how much we should burn considering the serious side effects of its use is a tough question.

Daniel Deudney and Christopher Flavin

DISCUSSION TOPICS

1. Try to trace your own direct and indirect energy consumption each day to see why it probably averages 230,000 kilocalories (Figure 15-1).

2. Explain why you agree or disagree with the following statements:
   a. We can get all the oil we need by extracting and processing heavy oil left in known oil wells.
   b. We can get all the oil we need by extracting and processing heavy oil from oil shale deposits.
   c. We can get all the oil we need by extracting heavy oil from tar sands.
   d. We can get all the natural gas we need from unconventional sources.

3. Should present U.S. mine safety and surface mining laws be strengthened or weakened, or should the enforcement of existing laws be greatly improved? Defend your choice.

4. Coal-fired power plants in the United States cause an estimated 10,000 deaths a year, primarily from atmospheric emissions of sulfur oxides, nitrogen oxides, and particulate matter. These plants also cause extensive damage to many buildings and some forests and aquatic systems. Should air pollution emission standards for *all* new and existing coal-burning plants be tightened significantly, even if this raises the price of electricity sharply and makes it cheaper to produce electricity by using conventional nuclear fission? Explain.

5. Should all coal-burning power and industrial plants in the United States be required to convert to fluidized-bed combustion? Explain. What are the alternatives?

6. Do you favor a U.S. energy strategy based on greatly increased use of coal-burning plants to produce electricity between 1988 and 2020? Explain. What are the alternatives?

7. List the energy services you would like to have, and note which of these *must* be furnished by electricity.

16

Nonrenewable and Perpetual Energy Resources: Geothermal and Nuclear Energy

GENERAL OBJECTIVES

1. What are the major types of geothermal energy and what are their uses, advantages, and disadvantages?

2. What are the advantages and disadvantages of using conventional nuclear fission to produce electricity?

3. What are the advantages and disadvantages of using breeder nuclear fission to produce electricity?

4. If and when it is developed, what are the advantages and disadvantages of using nuclear fusion to produce electricity?

We nuclear people have made a Faustian compact with society; we offer an inexhaustible energy source tainted with potential side effects that if not controlled could spell disaster.

Alvin M. Weinberg

Geothermal energy—heat from the earth's interior—and nuclear energy—heat from the fission of nuclei of certain isotopes (Section 3-2)—are used to produce limited amounts of electricity in various parts of the world. Some people urge that the world greatly increase its dependence on energy from nuclear fission to produce electricity because it is a safe, technically well-developed, economically acceptable energy resource that does not produce carbon dioxide and has less environmental impact than coal. Others consider expanded use of this energy resource, as well as its present use, to be unnecessary, uneconomic, unsafe, and unethical compared to other alternatives.

16-1 NONRENEWABLE AND PERPETUAL GEOTHERMAL ENERGY

Nonrenewable Geothermal Energy Heat from the earth's molten core, or **geothermal energy,** is transferred over thousands to millions of years to normally nonrenewable underground deposits of dry steam (steam with no water droplets), wet steam (a mixture of steam and water droplets), and hot water at various places. When these deposits are close enough to the earth's surface, wells can be drilled to bring this dry steam, wet steam, or hot water to the earth's surface to be used for space heating and to produce electricity or high-temperature heat for industrial processes.

Currently, about 20 countries are tapping geothermal deposits, providing space heating for over 2 million homes in cold climates and enough electricity for over 1.5 million homes. Figure 16-1 shows that most accessible high-temperature geothermal depos-

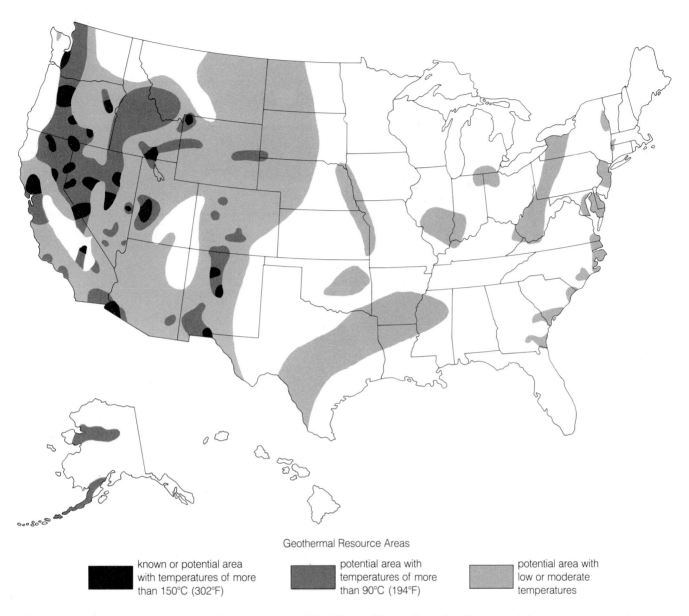

Geothermal Resource Areas

■	known or potential area with temperatures of more than 150°C (302°F)	▓	potential area with temperatures of more than 90°C (194°F)	░	potential area with low or moderate temperatures

Figure 16-1 Major deposits of geothermal resources in the United States. (Source: Council on Environmental Quality)

its in the United States lie in the western states, especially California and the Rocky Mountain states. Hawaii also has a number of possible sites.

Dry steam deposits are the preferred geothermal resource, but also the rarest. A large dry steam well near Larderello, Italy, has been producing electricity since 1904 and is a major source of power for Italy's electric railroads. Two other major dry steam sites are the Matsukawa field in Japan and the Geysers steam field about 145 kilometers (90 miles) north of San Francisco. The Geysers field has been producing electricity since 1960 and by 1987 was supplying more than 2% of California's electricity—enough to meet all electrical needs of a city the size of San Francisco—at less than half the cost of a new coal or nuclear plant. New

units can be added every 2 to 3 years (compared to 6 years for a coal plant and 12 years for a nuclear plant), and by 2000 this field may supply 25% of the state's electricity.

Underground *wet steam deposits* are more common but harder and more expensive to convert to electricity. These deposits contain water under such high pressure that its temperature is far above the boiling point of water at normal atmospheric pressure. When this superheated water is brought to the surface, some of it flashes into steam because of the sharp decrease in pressure. The mixture of steam and water is spun at a high speed (centrifuged) to separate out the steam, which is used to spin a turbine to produce electricity (Figure 16-2).

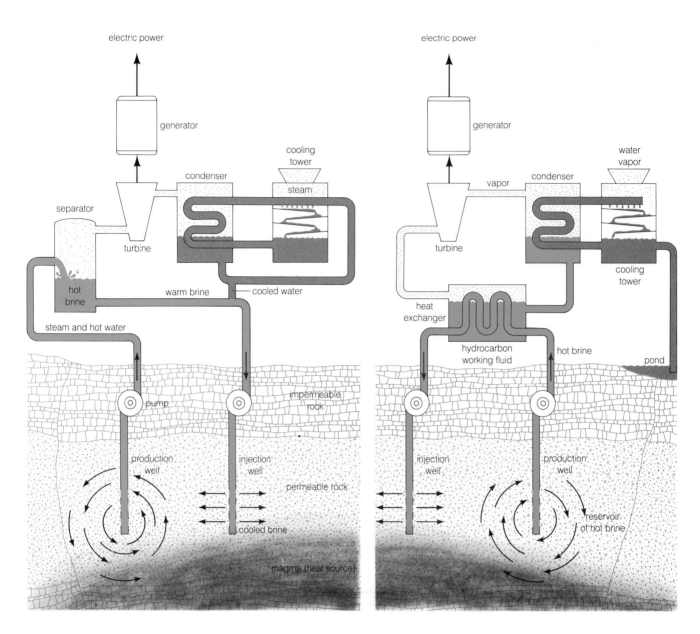

Figure 16-2 Direct-flash (left) and binary-cycle (right) methods for extracting and using geothermal energy to produce electricity.

The largest wet steam geothermal electric power plant in the world is in Wairaki, New Zealand. Other wet steam power plants operate in Mexico, Japan, and the Soviet Union. Four small-scale demonstration plants built in the United States since 1980 are producing electricity, but drilling problems and metal corrosion from the salty water have reduced yields so much that the cost of this energy is equivalent to paying $40 a barrel for oil.

Hot water deposits are more common than dry steam and wet steam deposits. Almost all the homes, buildings, and food-producing greenhouses in Reykjavik, Iceland, a city with a population of about 85,000, are heated by hot water drawn from deep hot water geothermal deposits under the city. At 180 locations in the United States, mostly in western states, hot water deposits have been used for years to heat homes and farm buildings and to dry crops.

The hot salty water (brine) pumped up from such wells can also be used to produce electricity in a binary cycle system (Figure 16-2). A demonstration binary-cycle system went into operation in 1984 in Herber in California's Imperial Valley. The main problem is that the brine corrodes metal parts and clogs pipes. An underground binary-cycle system is also being tested: Heat from the hot water vaporizes a liquid, which is brought to the surface to spin the turbine. Not only does this approach avoid corrosion and wastewater problems, it leaves water and steam in the well for continual reheating rather than depleting the resource.

A fourth potential source of nonrenewable geothermal energy is *geopressurized zones,* consisting of high-temperature, high-pressure reservoirs of water (often saturated with natural gas because of the high pressure), usually trapped deep under ocean beds of shale or clay. Still in the exploratory phase, geopressurized zones could be tapped by very deep drilling; but with present drilling technology they would provide energy at a cost equivalent to two to three times the price of conventional oil.

Major advantages of nonrenewable geothermal energy include a 100- to 200-year supply of energy for areas near deposits, moderate cost, moderate net useful energy yields for large and easily accessible deposits, and no emissions of carbon dioxide. However, the scarcity of easily accessible deposits and the fact that the energy cannot be used to power vehicles are two substantial disadvantages. Without pollution control its use results in moderate to high air pollution from hydrogen sulfide, ammonia, and radioactive materials, as well as moderate to high water pollution from dissolved solids (salinity) and runoff of various toxic compounds of elements such as boron and mercury. Noise, odor, and local climate changes can also be problems. Most experts, however, consider the environmental effects of geothermal energy to be less or no greater than those of fossil fuel and nuclear power plants.

Perpetual Geothermal Energy   There are also vast potentially perpetual sources of geothermal energy in the form of *molten rock (magma)* found near the earth's surface; *hot dry-rock zones,* where molten rock has penetrated the earth's crust and heats subsurface rock to high temperatures; and low- to moderate-temperature *warm rock deposits,* useful for preheating water and geothermal heat pumps for space heating and air conditioning.

The U.S. Geological Survey estimates that bodies of molten *rock* located no more than 9.6 kilometers (6 miles) below the earth's surface in the continental United States could supply 800 to 8,000 times all the commercial energy the country consumed in 1984. But extracting energy from magma is complicated and expensive. Currently scientists at the Sandia National Laboratory plan to test the technological feasibility of extracting heat from magma at an affordable price at two promising sites in California.

Warm and hot dry-rock deposits lying deep underground are potentially the largest and most widely distributed geothermal resource in the United States and most countries, but with present technology they are too difficult to locate, tap, and use on a large scale. By 1984, researchers in the United States and the United Kingdom had drilled several test wells and successfully extracted heat from dry-rock deposits by fracturing the hot rock with hydraulic pressure to create a reservoir, pumping water in, and bringing the resulting steam to the surface to run turbines for creating electricity. In 1986 a large demonstration hot dry-rock plant in New Mexico produced enough electricity and heat for a town of about 2,000 people.

16-2   CONVENTIONAL NONRENEWABLE NUCLEAR FISSION

A Controversial Fading Dream   Originally nuclear power was heralded as a clean, cheap, and safe source of energy that with 1,800 projected plants could provide as much as 21% of the world's commercial energy and 25% of U.S. commercial energy by the year 2000. However, by 1987, after 36 years of development, 397 commercial nuclear reactors in 26 countries were providing only 15% of the world's electricity—amounting to about 3% of the world's commercial energy. Five countries—the United States, France, West Germany, Japan, and the Soviet Union—had 72% of the world's nuclear power generating capacity by 1986 (Figure 16-3).

Industrialized countries like Japan and France, which have few fossil fuel resources, consider increased dependence on nuclear power a necessity to reduce their dependence on increasingly expensive imported oil. Although the Soviet Union is the world's largest producer of both oil and natural gas, it is committed to greatly increased use of nuclear power so that it can sell more of its oil and natural gas to other countries, especially in Europe.

Primarily because of concerns over safety, MDCs such as Denmark, Austria, Luxembourg, and Norway have chosen not to develop this energy source, and Sweden plans to close its ten operating reactors by 2010. Since the Chernobyl nuclear accident in 1986 most LDCs have scaled back or eliminated their plans to build nuclear power plants. As a result, nuclear power's share of the world's electricity will very likely be lower in the year 2000 than it was in 1986.

The United States, which has abundant coal and a variety of other energy alternatives, can decide to what degree it wants to use nuclear power with less urgency than MDCs with limited energy alternatives. Projected future use of nuclear fission for producing electricity in the United States has decreased sharply since 1975. No new plants have been ordered since 1978 and 114 previous orders have been canceled. By 1986, a total of 101 commercial nuclear reactors in 32 states (mostly in the eastern half of the United States) supplied 16% of the country's electricity and about 4% of its total commercial energy. If the 27 reactors still

Figure 16-3 Use of nuclear fission reactors to produce electricity in various countries in 1986. (Sources of data: Atomic Industrial Forum and International Atomic Energy Agency)

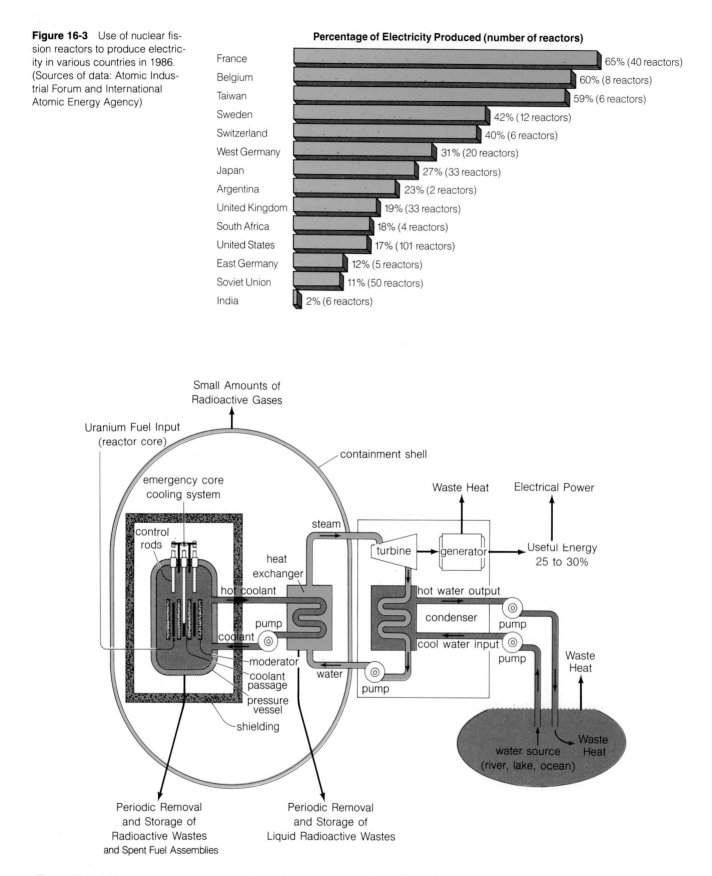

Percentage of Electricity Produced (number of reactors)

France — 65% (40 reactors)
Belgium — 60% (8 reactors)
Taiwan — 59% (6 reactors)
Sweden — 42% (12 reactors)
Switzerland — 40% (6 reactors)
West Germany — 31% (20 reactors)
Japan — 27% (33 reactors)
Argentina — 23% (2 reactors)
United Kingdom — 19% (33 reactors)
South Africa — 18% (4 reactors)
United States — 17% (101 reactors)
East Germany — 12% (5 reactors)
Soviet Union — 11% (50 reactors)
India — 2% (6 reactors)

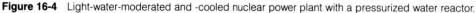

Figure 16-4 Light-water-moderated and -cooled nuclear power plant with a pressurized water reactor.

Figure 16-5 Bundles of fuel rods filled with pellets of enriched uranium-235 oxide serve as the fuel core for a conventional nuclear fission reactor.

under construction are completed, nuclear power could supply 20% of the country's electricity and 5% of its commercial energy by 1995—far below the 25% of commercial energy predicted in the 1960s.

How Does a Nuclear Fission Reactor Work? When the nuclei of certain atoms such as uranium-235 and plutonium-239 are split apart by neutrons, energy is released and converted mostly to high-temperature heat in a nuclear fission chain reaction (Figure 3-5, p. 34). The rate at which this process occurs can be controlled in the nuclear fission reactor in a nuclear power plant, where the high-temperature heat released is used to spin a turbine connected to a generator that produces electrical energy.

Almost three-fourths of these reactors (62% in the United States) are **light-water reactors (LWRs).** Key parts of an LWR are the core, fuel assemblies, fuel rods, control rods, moderator, and coolant (Figure 16-4). The core of an LWR typically contains about 180 fuel assemblies, each of which contains about 200 long, thin fuel rods made of nonradioactive zirconium alloy or stainless steel (Figure 16-5). Each fuel rod is packed with eraser-size pellets of uranium oxide (UO_2) fuel consisting of 3% fissionable uranium-235 and 97% nonfissionable uranium-238 (Figure 16-6). The ura-

nium-235 in each fuel rod can produce energy equal to that from about three railroad cars of coal.

Interspersed between the fuel assemblies are control rods made of materials that capture neutrons. These rods are moved in and out of the reactor to regulate the rate of fission and thus the amount of power the reactor produces.

All reactors circulate or place some type of material, known as a moderator, between the fuel rods and fuel assemblies to slow the neutrons emitted by the fission process and thus sustain the chain reaction. Three-fourths of the world's reactors use ordinary water, called *light water,* as a moderator. The moderator in about 20% of the world's reactors (50% of those in the Soviet Union, including the ill-fated Chernobyl reactor) is solid graphite, a form of carbon. In addition to producing electricity, graphite-moderated reactors are particularly useful for producing plutonium-239 for use in nuclear weapons.

All reactors also have a coolant circulating through the reactor core, removing generated heat to prevent fuel rods and other materials from melting and to produce electricity. Most water- and graphite-moderated reactors use water as a coolant; a few gas-cooled reactors use some unreactive gas such as helium or argon.

A typical LWR has an energy efficiency of only 25% to 30% compared to 40% for a coal-burning plant. Although more expensive to build and operate, graphite-moderated, gas-cooled reactors, widely used in the United Kingdom, are more energy-efficient (38%) because they operate at a higher temperature.

Nuclear power plants, each with one or more reactors, are only one part of the *nuclear fuel cycle* necessary for using nuclear energy to produce electricity (Figure 16-7). *In evaluating the safety and economics of nuclear power, it is necessary to look at the entire cycle— not just the nuclear plant itself.*

After about three years in a reactor, the concentration of fissionable uranium-235 in a fuel rod becomes too low for the chain reaction to proceed, or the rod becomes severely damaged from ionizing radiation. Thus each year about one-third of the spent fuel elements in a reactor are removed and stored in large, concrete-lined pools of water at the plant site for several years. After they have cooled and lost some of their radioactivity, they can be sealed in heavily shielded, crash-proof casks and transported to other storage pools away from the reactor or to a permanent nuclear waste repository or dump. Because neither of these options exist in the United States, spent fuel is stored at plant sites, where adequate storage space is rapidly running out.

A third option is to transport spent fuel to a fuel-reprocessing plant, where remaining fissionable ura-

Figure 16-6 Fuel pellets of enriched uranium-235 oxide.

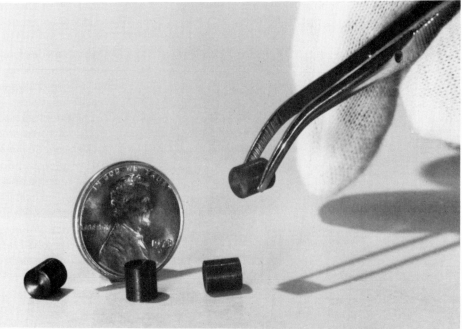

Westinghouse Hanford Company

nium-235 and plutonium-239 (produced as a by-product of the fission process) are removed and sent to a fuel fabrication plant for use in a conventional nuclear fission reactor or a breeder nuclear fission reactor (Figure 16-7). A large commercial fuel-reprocessing plant is in operation in Great Britain and others are under construction in Japan and West Germany. The United States has delayed development of commercial fuel-reprocessing plants because such facilities would handle and ship nuclear fuel in a form that could be used to make nuclear weapons, and because of technical difficulties and high construction and operating costs.

Nuclear Reactor Safety Fission converts some fuel to radioactive fragments, and under intense neutron bombardment the metals in the fuel rods and other metal parts in the core are converted to radioactive isotopes. Because these radioactive fission products produce much heat, they continue to heat the fuel even after a reactor has been shut down. Thus water or some other coolant must be continuously circulated through the core to prevent a **meltdown** of the fuel rods and the reactor core, which can result in the release of massive quantities of highly radioactive materials into the environment.

To greatly reduce the chances of a meltdown and other serious reactor accidents, commercial reactors in the United States (and most countries) have a number of safety features. In the United States these include (1) thick walls and concrete and steel shields surrounding the reactor vessel; (2) a system for auto-

matically inserting control rods into the core to stop fission under certain emergency conditions; (3) a steel-reinforced concrete containment building, designed to prevent radioactive gases and materials from reaching the atmosphere as a result of most conceivable accidents except a complete core meltdown or a massive chemical explosion; (4) large filter systems and chemical sprayers inside the containment building to remove radioactive dust from the air and further reduce the chances of radioactivity reaching the environment; (5) systems to condense steam released from a ruptured reactor vessel and thus prevent pressure from rising beyond the holding power of containment building walls; (6) an emergency core-cooling system to flood the core automatically with tons of water within one minute to prevent meltdown of the reactor core; (7) two separate power lines servicing the plant and several backup diesel generators to provide backup power for the massive pumps in the emergency core-cooling system; (8) X-ray inspection of key metal welds during construction and periodically after the plant goes into operation to detect possible sources of leaks from corrosion; and (9) an automatic backup system to replace each major component of the safety system in the event of a failure.

Although such elaborate safety systems make complete reactor core meltdown extremely unlikely, such an event is possible through a series of equipment failures, operator errors, or both. For example, the core might lose its cooling water through a break in one of the pipes that conduct cooling water and steam to and from the reactor core. If the emergency

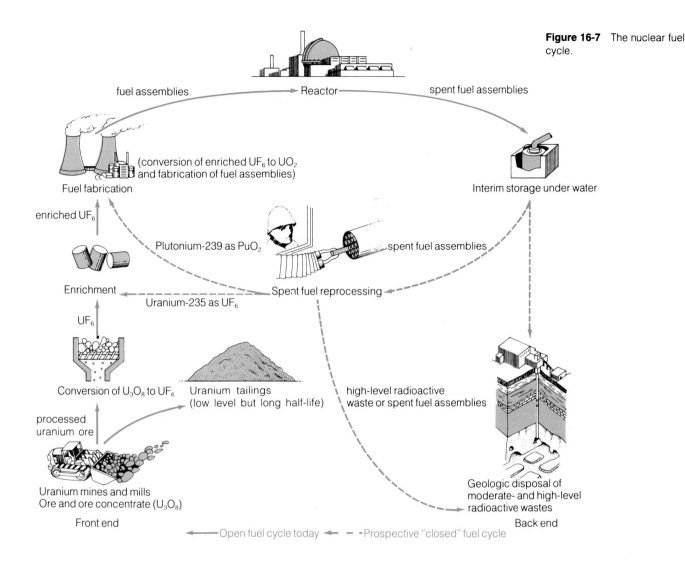

Figure 16-7 The nuclear fuel cycle.

fuel assemblies → Reactor → spent fuel assemblies

Fuel fabrication
(conversion of enriched UF_6 to UO_2 and fabrication of fuel assemblies)

enriched UF_6

Plutonium-239 as PuO_2

Enrichment

Uranium-235 as UF_6

Spent fuel reprocessing

spent fuel assemblies

Interim storage under water

UF_6

Conversion of U_3O_8 to UF_6

Uranium tailings (low level but long half-life)

high-level radioactive waste or spent fuel assemblies

processed uranium ore

Uranium mines and mills
Ore and ore concentrate (U_3O_8)

Geologic disposal of moderate- and high-level radioactive wastes

Front end

Back end

← Open fuel cycle today ← – · Prospective "closed" fuel cycle

core-cooling system also failed, the loss of coolant would allow the reactor core to overheat and eventually melt down through its thick concrete slab into the earth. Depending on the geological characteristics of the underlying soil and rock, the melted core might sink 6 to 30 meters (20 to 100 feet) and gradually dissipate its heat, or it might burn itself more deeply into the earth's crust and contaminate groundwater with radioactive materials. In 1979 one of the reactors at the Three Mile Island plant in Pennsylvania underwent a partial meltdown through a combination of equipment failures and operator errors (see Spotlight on p. 276).

Another possibility is that a powerful hydrogen gas or steam explosion inside the reactor containment vessel could split the containment building open and spew highly radioactive materials high into the atmosphere. The resulting cloud of radioactive materials could kill and injure hundreds to tens of thousands of people and contaminate large areas with radioactive isotopes for hundreds to thousands of years. Such an

explosion could result from a partial loss of cooling caused by a leak in the cooling system or a turbine failure. In a graphite-moderated reactor the explosion would also ignite the graphite, which would burn like a giant pile of coal, releasing more radioactive fission products into the lower atmosphere surrounding the site. A series of operator errors led to such a catastrophic explosion and graphite fire at the Soviet Union's Chernobyl nuclear reactor in 1986 (see Spotlight on p. 277).

An exhaustive study of the TMI accident concluded:

The principal deficiencies in commercial reactor safety today are not hardware problems, they are management problems . . . problems that cannot be solved by the addition of a few pipes or valves . . . or, for that matter, by a resident federal inspector.

Although operator training was improved after the 1979 TMI accident, in 1981 a federal inspector entered the control room at Commonwealth Edison's reactor

Winter 1957

Perhaps the worst nuclear disaster in history occurred in the Soviet Union in the southern Ural Mountains around the city of Kyshtym, believed then to be the center of plutonium production for Soviet nuclear weapons. Though the cause of the accident and the number of people killed and injured remain a secret, a massive amount of radiation was released—allegedly from an explosion of large quantities of radioactive wastes carelessly stored in shallow trenches. Today the area is deserted, hundreds of square miles have been sealed off, a river has been diverted around the area, and the names of about 30 towns and villages in the region have disappeared from Soviet maps.

October 7, 1957

The water-cooled, graphite-moderated Windscale facility for producing plutonium for nuclear weapons north of Liverpool, England, underwent an explosion and caught fire as the Chernobyl plant did 19 years later. By the time the fire was put out, 200 square miles of countryside had been contaminated with radioactive material. An estimated 33 people died prematurely from cancers traced to effects of the accident.

March 22, 1975

The flame from a candle used by a maintenance worker to test for air leaks at the Brown's Ferry commercial nuclear reactor near Decatur, Alabama, set off a fire that knocked out five emergency core-cooling systems. Although the reactor's cooling water dropped to a dangerous level, backup systems prevented any radioactive material from escaping into the environment. At the same plant, in 1978, a worker's rubber boot fell into a reactor and led to an unsuccessful search costing $2.8 million. Such incidents, based on unpredictable human errors, are fairly common in nuclear plants.

March 28, 1979

In what is considered the worst accident in the history of U.S. commercial nuclear power, one of the two reactors at the Three Mile Island (TMI) nuclear plant near Harrisburg, Pennsylvania, lost its coolant water because of a series of mechanical failures and human operator errors not foreseen in safety studies. The reactor's core became partially uncovered, underwent a partial core meltdown, and small but unknown amounts of ionizing radiation escaped into the atmosphere (Figure 16-8). Investigators found that had a stuck valve stayed open for just another 30 to 60 minutes, a complete meltdown would have resulted. Although no one is known

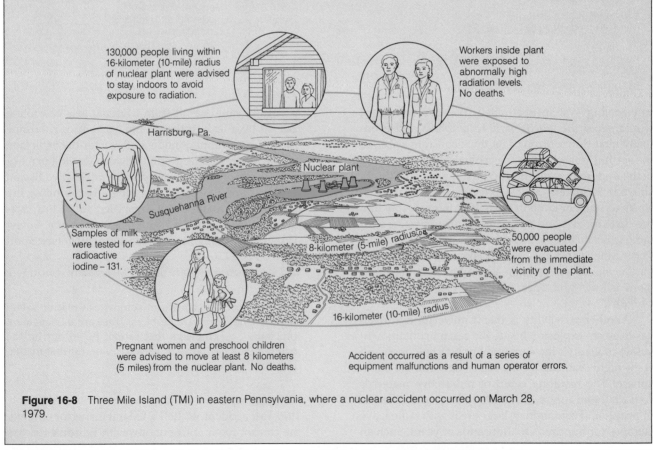

130,000 people living within 16-kilometer (10-mile) radius of nuclear plant were advised to stay indoors to avoid exposure to radiation.

Workers inside plant were exposed to abnormally high radiation levels. No deaths.

Harrisburg, Pa.

Nuclear plant

Susquehanna River

Samples of milk were tested for radioactive iodine – 131.

8-kilometer (5-mile) radius

16-kilometer (10-mile) radius

50,000 people were evacuated from the immediate vicinity of the plant.

Pregnant women and preschool children were advised to move at least 8 kilometers (5 miles) from the nuclear plant. No deaths.

Accident occurred as a result of a series of equipment malfunctions and human operator errors.

Figure 16-8   Three Mile Island (TMI) in eastern Pennsylvania, where a nuclear accident occurred on March 28, 1979.

to have died as a result of the accident, the long-term health effects on workers and nearby residents is still being debated because data published on the amount of radiation released during the accident are contradictory and incomplete. The cleanup of the damaged TMI reactor, which will probably cost $1 billion to $1.5 billion (compared to the $700 million construction cost of the reactor), threatens the utility with bankruptcy and may not be completed until 1990 or later. Confusing and misleading statements about the seriousness of the accident issued by Metropolitan Edison (which owns the plant) and by the Nuclear Regulatory Commission seriously eroded public confidence in the safety of nuclear power.

June 9, 1985

Despite new safety standards and improved operator training since the TMI accident, 16 equipment failures and a human operator who punched the wrong button led to a partial loss of cooling water in a reactor at the Davis-Besse nuclear plant near Toledo, Ohio. Fortunately, the problem, similar to that at TMI, was corrected in time by auxiliary cooling pumps, and no ionizing radiation was released to the environment. Nuclear power critics contend that TMI and hundreds of other serious incidents like this one have not led to a complete meltdown primarily because of luck. Nuclear industry officials claim that the fact that a meltdown has not occurred demonstrates that the multiple-backup safety systems work.

April 26, 1986

Two gas explosions inside one of the four graphite-moderated, water-cooled reactors at the Soviet Chernobyl nuclear power plant north of Kiev blew the roof off the reactor building and set the graphite core on fire. The accident occurred when engineers deliberately turned off most of the reactor's key automatic safety and warning systems to keep them from interfering with an unauthorized safety experiment they were conducting (Figure 16-9). The explosions and the resulting fire spewed highly radioactive materials into the atmosphere, where they were carried by winds over parts of the Soviet Union and much of eastern and western Europe as far as 2,000 kilometers (1,250 miles) from the plant. During the 10 days that it took firefighters to get the intensely hot graphite fire under control, large amounts of additional radioactive materials were released into the nearby area and 135,000 people living within 30 kilometers (18 miles) of the plant were evacuated by an armada of 1,100 buses. These peo-

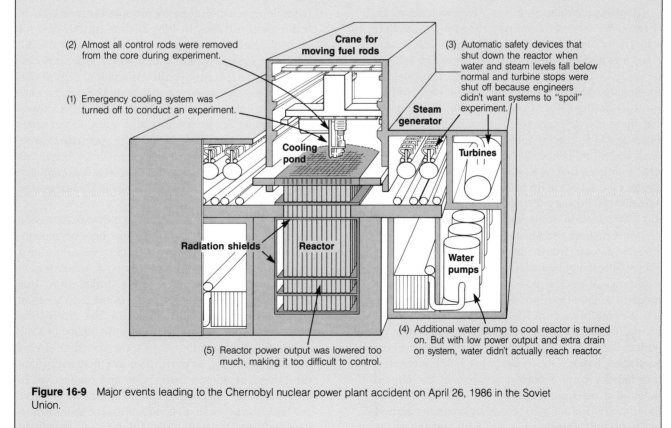

Figure 16-9 Major events leading to the Chernobyl nuclear power plant accident on April 26, 1986 in the Soviet Union.

ple have been resettled in other villages and probably will not be allowed to return to their homes for at least four years. Five months after the accident 31 plant workers and firefighters had died from exposure to high levels of ionizing radiation, 200 others were suffering from acute radiation sickness (many of whom will die prematurely from cancer in coming years), and a land area of 2,590 square kilometers (1,000 square miles) around the reactor was contaminated with radioactive fallout. All nearby forests will have to be cut down and the topsoil

removed and buried. Some farmland in the area may have to be abandoned for many decades. Soviet and western medical experts estimate that at least 5,000 and perhaps 100,000 additional people in the Soviet Union will die prematurely over the next 70 years from cancer caused by exposure to ionizing radiation. Thousands of others will be afflicted with thyroid tumors, cataracts, and sterility. One-half to three times as many additional premature deaths are likely outside the Soviet Union. Estimated damages run from $3 billion to $5

billion, but may be as high as $14 billion, taking into account long-term health effects. Some American nuclear experts contend that had the reactor been designed with the key safety features of reactors in the United States and most of the rest of the world, no radioactive materials would have been released. In 1987 the United States shut down a Chernobyl-type military reactor at Hanford, Washington, to make safety improvements after a study documented 54 serious safety violations at the facility during 1985 and 1986.

near Morris, Illinois, and found two operators asleep; an inspector for Florida Light and Power found a reactor operating at full power with no one in the control room. Since 1981 at least four other incidents of sleeping on duty have been reported.

Although numerous studies have been made during the 31 years after the first nuclear power plant began operating in the United States, there is still no officially accepted study of just how safe or unsafe these plants are and no study of the safety of the entire nuclear fuel cycle. Despite the uncertainty and controversy over reactor safety, in 1985 the Nuclear Regulatory Commission (NRC) estimated that there is a 45% chance of a complete core meltdown at a U.S. reactor sometime during the next 20 years and a 12% chance of two such meltdowns. The NRC estimates that in a worst-case scenario such an accident would cause up to 100,000 deaths in the first year, 600,000 injuries, 400,000 ultimate cancer deaths, and $300 billion in damages.

Most citizens and businesses suffering injuries or property damage from a major nuclear accident would receive little if any financial reimbursement. Since the beginnings of nuclear power in the 1950s, insurance companies have been unwilling to underwrite more than a small fraction of the estimated risks. Because having to bear the costs of liability would have prevented the U.S. nuclear industry from ever being developed, federal law in effect since 1957 limits insurance liability from a nuclear accident in the United States to a maximum of only $640 million. Of this amount, $160 million is to be paid by insurance carried by the utility company on each plant, and the rest is to be paid by the federal government—the taxpayers.

Repeated attempts to have this law repealed or to raise liability limits have failed.

Concern has also arisen over whether nuclear accident evacuation zones and plans in the United States are adequate. After the TMI accident a 16-kilometer (10-mile) evacuation zone was adopted around all U.S. commercial reactors. In 1986 nuclear critics called for an extension of evacuation zones to at least the 30-kilometer (18-mile) zone Soviet officials found necessary after the Chernobyl nuclear accident. These critics point out that areas around many U.S. reactors are up to ten times more densely populated than those around most Soviet reactors and many urban areas near U.S. reactors would be almost impossible to evacuate. In the case of a nuclear accident most Americans would get into their cars and clog up evacuation routes, a situation that didn't occur in the Soviet Union because few people had cars.

However, instead of expanding the evacuation zone the U.S. nuclear industry and utilities have been pushing the NRC to sharply reduce the evacuation area around U.S. reactors to as low as 1.6 kilometers (1 mile), contending that new safety studies show that less radiation would escape in the event of an accident than previously thought. Reducing the size of the evacuation zone would also limit the ability of state and local governments to block the licensing of new plants by refusing to participate in emergency planning they considered inadequate and ineffective. In 1987 the NRC adopted a regulation that would give the NRC the power to override refusal of state authorities to accept evacuation plans that states considered inadequate.

Figure 16-10 Construction of double-walled steel tanks for storing high-level liquid radioactive wastes at the federal government's Hanford nuclear weapon production facility near Richland, Washington. Each tank holds 4 million liters (1 million gallons) and will be covered with seven feet of earth.

Disposal and Storage of Radioactive Wastes Each part of the nuclear fuel cycle for military and commercial nuclear reactors produces a mixture of solid, liquid, and gaseous, low-level to high-level radioactive wastes that must be stored for about ten times their half-lives until their radioactivity has dropped to extremely low levels. From the 1940s to 1970 most low-level radioactive waste produced in the United States (and most other countries) was dumped into the ocean in steel drums. Since 1970 low-level wastes have been buried at 13 U.S. government facilities for defense-related wastes and 3 sites run by private firms under state and federal regulations. Materials are transported to the sites in steel drums. Several drums are then placed in a large container, which is buried in a trench and covered with several feet of dirt.

High-level radioactive wastes consist mostly of spent fuel rods from commercial nuclear power plants and an assortment of wastes from nuclear weapons facilities. By 1987 about 49,000 highly radioactive spent fuel assemblies from U.S. nuclear power plants were stored temporarily in deep pools of water at nuclear plant sites, pending the development of a method for long-term storage or disposal. But space at reactor sites is running out, and 4,000 more spent assemblies are added each year. According to the EPA, spent fuel rods must be stored safely for 10,000 years before they decay to acceptable levels of radioactivity.

High-level liquid wastes from U.S. nuclear weapons production, equal in volume to about 200 Olympic-size swimming pools, are also awaiting permanent storage. These wastes, which remain lethal for centuries, are currently stored in underground tanks in government facilities in Idaho Falls, Idaho, Barnwell, South Carolina, and Hanford, Washington. These tanks must be continuously monitored for corrosion and leaks.

Nearly 2 million liters (530,000 gallons) of highly radioactive wastes have already leaked from older, single-shell tanks built between 1943 and 1965 at the Hanford and Savannah River sites. However, a National Academy of Sciences study concluded that these leaks have not caused any significant radiation hazard to public health. Storage tanks constructed after 1968 have a double shell; if the inner wall corrodes, the liquid will spill into the space between the two walls, where it can be detected in time to be pumped into another tank (Figure 16-10).

In 1986 once-secret documents revealed that ionizing radiation had been deliberately and accidentally released from the Hanford nuclear waste dump over a 40-year period, and that local residents were not notified. A 1986 EPA report revealed 14,000 incidents of publicly unreported radioactive waste leaks at the Savannah River Weapons Plant, where nuclear waste storage tanks stand near shallow waters that drain into the Savannah River and above the Tuscaloosa aquifer, a major source of drinking water for parts of South Carolina, Georgia, Alabama, and northern Florida.

In 1987 several government studies also revealed that most of the aging U.S. nuclear weapons production facilities supervised by the Department of Energy were being operated with gross disregard for the safety of their workers and people in nearby areas. These miltary production facilities are not required to meet the more stringent safety requirements set by the NRC for commercial nuclear power plants and other related facilities.

After 31 years of research and debate there is still no widely agreed upon scientific solution to how high-level radioactive wastes can be stored safely for the 10,000 years currently required by EPA regulations (see Spotlight on p. 280). Regardless of the storage method, most U.S. citizens strongly oppose the location of a nuclear waste disposal facility anywhere near them. By 1985 at least 22 states had enacted laws banning radioactive waste disposal.

The long-term safe storage or disposal of high-level radioactive wastes is believed to be technically possible. However, it is essentially impossible to establish that any method will work over the thousands of years required before the wastes decay to safe levels. Some of the proposed methods and their possible drawbacks:

1. *Bury it deep underground.* The currently favored method is to concentrate the waste, convert it to a dry solid, fuse it with glass or a ceramic material, seal it in a metal canister, and bury it permanently in deep underground salt, granite, or other stable geologic formations that are earthquake-resistant and waterproof (Figure 16-11). Some geologists question this approach, arguing that extensive drilling and tun-

neling can destabilize such rock structures and that present geologic knowledge is not sufficient to predict the paths of groundwater flows that could contaminate groundwater drinking supplies with radioactive wastes.

2. *Shoot it into space or into the sun.* Even if technically feasible, costs would be very high and a launch accident of a rocket or space vehicle could disperse high-level radioactive wastes over a wide area of the earth's surface.

3. *Bury it under the Antarctic ice sheets or the Greenland ice caps.* The long-term stability of the ice sheets is unknown, and they could be destabilized by heat from the wastes; retrieval would be difficult or impossible if the method failed.

4. *Dump it into downward-descending, deep ocean bottom sediments.* The long-term stability and motion of these sediments are unknown and wastes could eventually be spewed out somewhere else by volcanic activity; waste containers might leak and contaminate the ocean before being carried downward; retrieval would probably be impossible if the method failed.

5. *Change it into harmless or less harmful iosotopes.* Currently there is no known way to do this; even if it should become technically feasible, costs would probably be extremely high and new toxic materials and lower-level radioactive wastes created would also require safe disposal.

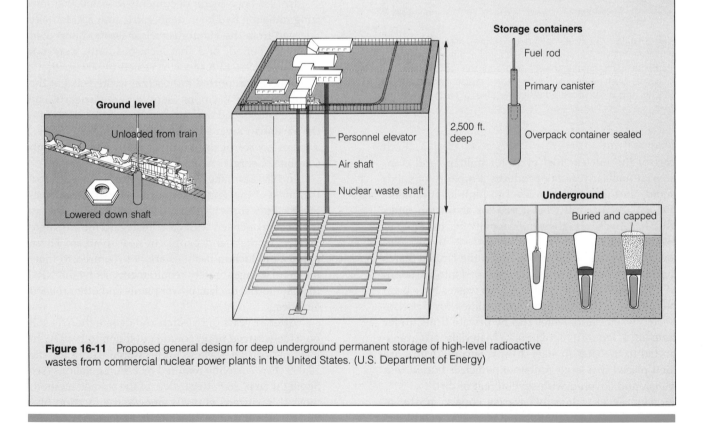

Figure 16-11 Proposed general design for deep underground permanent storage of high-level radioactive wastes from commercial nuclear power plants in the United States. (U.S. Department of Energy)

In 1983 the Department of Energy began building the first underground repository in the United States, to be used only for long-term storage of high-level radioactive wastes produced by the nuclear weapons program. This $1 billion structure is being built in a salt bed deep under federal land about 40 kilometers (25 miles) east of Carlsbad, New Mexico. Shortly after construction began, it was discovered that water had aleady entered the underground storage chambers through the ventilation shaft. If water ever penetrates the waste, some experts fear that some of the highly toxic radioactive plutonium-239 could be carried fairly rapidly to the nearby Pecos River. Other experts believe the site is safe.

In 1982 Congress passed the Nuclear Waste Policy Act, which established a timetable for the Department of Energy to choose a site and build the country's first deep underground repository for long-term storage of high-level radioactive wastes from *commercial* nuclear reactors. In 1985 the Department of Energy announced plans to build the first repository, at a cost of $6 billion to $10 billion, based on the design shown in Figure 16-11. The repository will be built at one of three sites: a compacted volcanic ash formation in Yucca Mountain, Nevada, northwest of Las Vegas; a volcanic basalt formation under the Hanford federal nuclear weapons facility in southeastern Washington; or a salt deposit in Deaf Smith County, Texas, west of Amarillo.

After each site is tested for five years, the president of the United States is to decide in 1991 which site will be used. Because this first repository will be filled quickly with wastes already awaiting storage, a second site is supposed to be selected in the eastern United States, where most reactors are located, after the first site has opened.

The Department of Energy estimates that the costs of building and running the two sites will be from $21 billion to $35 billion, which is being paid for by a 0.1 cent per kilowatt-hour charge added to all electricity bills since 1983. The U.S. General Accounting Office, however, estimates that the cost may be as high as $114 billion. The first site was supposed to be ready by 1998, but in 1987 the DOE announced it would be delayed until at least 2003; most observers believe it will be delayed to 2008 or later. By 1986 intense opposition from citizens and elected officials in states where the potential sites are to be located and from geologists identifying possible geologic and water-contamination problems with each of the three sites had given many members of Congress second thoughts about the entire project.

Citizens in cities and states along the proposed routes for transporting the highly radioactive wastes to each of the possible western repository sites are also concerned. If all waste is transported by truck there will be about 6,405 shipments every year passing through parts of 45 states—an average of 17 shipments a day for 30 years. If all waste is transported by rail, there will be about 830 shipments annually. A study by the NRC concluded that a serious accident involving a truck or train passing through a heavily urbanized area could result in damages of $2 billion to $3 billion, only part of which would be covered by the insurance liability limits set by federal law on any type of nuclear accident.

Decommissioning Nuclear Power Plants   The useful operating life for a nuclear power plant is between 30 and 40 years. Since the core and many other parts contain large quantities of radioactive materials, the plant cannot simply be abandoned or demolished by a wrecking ball like coal-fired and other power plants. The *decommissioning process* is the final step of the nuclear fuel cycle. Worldwide more than 20 commercial reactors (4 in the United States) had been shut down and were awaiting decommissioning by 1986. Another 225 large commercial units (67 in the United States) will be retired between 2000 and 2010.

Three ways to decommission a nuclear reactor have been proposed: (1) entombment by covering the reactor with reinforced concrete and erecting barriers to keep out intruders; (2) dismantlement by decontaminating and taking the reactor apart immediately after shutdown and shipping all radioactive debris to a radioactive waste-burial facility; and (3) mothballing for several decades to 100 years prior to dismantlement by erecting a barrier and setting up a 24-hour security guard system.

Each method would involve shutting down the plant, removing the spent fuel from the reactor core, draining all liquids, flushing all pipes, and sending all radioactive materials to appropriate waste storage sites. Although entombment was once viewed as the easiest and cheapest way out, research has indicated that some remaining radioactive materials would still be dangerous long after the concrete tomb had crumbled. Currently most U.S. utilities favor immediate dismantlement, whereas utilities in France, Canada, and West Germany are planning to mothball their reactors for several decades before dismantlement.

Dismantlement of the first commercial nuclear reactor in the United States near Shippingport, Pennsylvania, began in 1986. This five-year project, involving a small reactor, is expected to cost at least $100 million. Utility companies estimate that dismantlement of a typical reactor 10 to 20 times larger and containing several hundred times as much radioactivity as the Shippingport reactor should cost about $170 million, but most analysts consider this figure much too low and put the cost at $1 billion to $3 billion per reactor. Adding dismantlement costs to the already high price of producing electricity by nuclear fission has led some analysts to believe that nuclear power will never become economically feasible compared to other alternatives.

Proliferation of Nuclear Weapons Since the late 1950s the United States has been giving away and selling to other countries various forms of nuclear technology. By 1986 at least 14 other countries had entered the international market as sellers of nuclear technology. For decades the U.S. government has denied that the information, components, and materials used in the nuclear fuel cycle could be used to make nuclear weapons. In 1981, however, a Los Alamos National Laboratory report admitted: "There is no technical demarcation between the military and civilian reactor and there never was one."

Between 4 and 9 kilograms (9 to 20 pounds) of either plutonium-239 or uranium-233, a mass about the size of an orange, and 11 to 25 kilograms (24 to 55 pounds) of uranium-235 are needed to make a small atomic bomb capable of blowing up a large building or a city block and contaminating a much larger area with radioactive materials for centuries. Although difficult, it is believed that a handful of trained people could make such a "blockbuster" nuclear bomb if they could get enough fissionable bomb-grade material. A crude 10-kiloton nuclear weapon placed properly and detonated during working hours could topple the World Trade Center in New York City, easily killing more people than those killed by the 20,000-kiloton atomic bomb the United States dropped on Hiroshima, Japan, in 1945.

Spent reactor fuel is so highly radioactive that theft is unlikely, but separated plutonium-239 is easily handled. Although bomb-grade plutonium-239 is heavily guarded, it could be stolen from nuclear weapons facilities, especially by employees. Each year about 3% of the approximately 126,000 people working with U.S. nuclear weapons are relieved of duty because of drug use, mental instability, or other security risks. By 1978 at least 320 kilograms (700 pounds) of plutonium-239 was missing from commercial and government-operated reactors and storage sites in the United States—enough to make 32 to 70 atomic bombs, each capable of blowing up a city block. No one knows whether this missing plutonium was stolen or whether it represents careless measuring and bookkeeping techniques.

Concentrated bomb-grade plutonium fuel could also be stolen from a commercial fuel-reprocessing plant, stolen from more than 150 research and test reactors operating in 30 countries, or hijacked from shipments to six experimental breeder nuclear fission power plants already in operation in parts of Europe. Bomb-grade fuel might also be manufactured by using one of the simpler and cheaper technologies for isotope separation currently being developed to concentrate 3% uranium-235 to weapons-grade material.

Actually, those who would steal or produce plutonium-239 need not bother to make atomic bombs. They could simply use a conventional explosive charge to disperse the plutonium into the atmosphere from atop any tall building. Dispersed in this manner, 1 kilogram (2.2 pounds) of plutonium oxide powder could theoretically contaminate 7.7 square kilometers (3 square miles) with radioactivity. This radiation, which would remain at dangerous levels for at least 100,000 years, could cause lung cancers among those who inhaled contaminated air or dust in such areas.

One suggestion for reducing the possibility of diversion of plutonium fuel from the nuclear fuel cycle is to contaminate it with other substances that render it more dangerous to handle and unfit as weapons material. But so far no acceptable "spiking agent" has emerged that could not be removed by reprocessing or isotope separation.

Soaring Costs: The Achilles Heel of Nuclear Power The largest cutback in commercial nuclear power has taken place in the United States, primarily because of economics and reduced need for electricity because of energy conservation and use of other methods for producing electricity. After 35 years of development and a $154 billion investment, including $44 billion in government subsidies, the 101 nuclear power reactors in operation in the United States produced no more of the country's commercial energy than that provided by wood and crop wastes with hardly any government subsidies.

After the destruction of Hiroshima and Nagasaki, scientists who developed the atomic bomb and elected officials responsible for use of these weapons were determined to show the world that the peaceful uses of atomic energy would outweigh the immense harm it had done. Utility companies were skeptical but began ordering nuclear power plants in the late 1950s for four major reasons. First, the Atomic Energy Commission and builders of nuclear reactors projected that nuclear power would produce electricity at such a low cost that it would be "too cheap to meter." Second, the nuclear industry projected that the reactors would have an 80% capacity factor—a measure of the time a reactor is able to produce electricity at its full power potential. Third, the first round of commercial reactors was built with the government paying approximately one-fourth of the cost; these reactors were provided to utilities at a fixed cost with no cost overruns allowed.

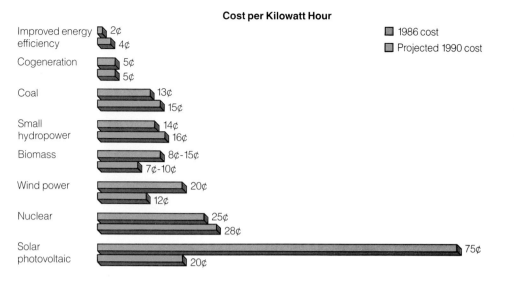

Cost per Kilowatt Hour

Improved energy efficiency: 2¢ / 4¢
Cogeneration: 5¢ / 5¢
Coal: 13¢ / 15¢
Small hydropower: 14¢ / 16¢
Biomass: 8¢-15¢ / 7¢-10¢
Wind power: 20¢ / 12¢
Nuclear: 25¢ / 28¢
Solar photovoltaic: 75¢ / 20¢

■ 1986 cost
■ Projected 1990 cost

Figure 16-12 Costs of electricity generated by improving energy efficiency and by various types of new power plants (including costs of construction, fuel, and operation) in the United States in 1986, with projections to 1990. (Sources of data: Charles Komanoff and Worldwatch Institute)

Fourth, Congress passed legislation that protected the nuclear industry and utilities from significant liability to the general public in case of accident. It was an offer utility company officials could not resist. Today many wish they had.

Since the construction of the first nuclear power plants, it has become increasingly clear that nuclear power is an extraordinarily expensive way to produce electricity, even when it is heavily subsidized as partial protection from free market competition. Construction costs for nuclear power plants in the United States in terms of dollars per kilowatt of electricity have risen more than tenfold since 1960, and this does not even include the costs of storing radioactive wastes and decommissioning plants. By 1986, producing electricity in a new nuclear plant in the United States had a total cost significantly higher than those of coal and cogeneration, as well as the cost of improved energy efficiency (Figure 16-12). By 1990 nuclear power is expected to be even less competitive with other methods for producing electricity in the United States.

Operating costs have also been higher than projected because in 1983 U.S. reactors operated an average of only 56% of their capacity—far below the 80% capacity factor projected by proponents of nuclear power in the 1950s. In contrast, reactors in West Germany operated at an average of 71% of their capacity. Major reasons for this difference include the lack of standardization of U.S. plant design and poorer design, construction quality, and management than in West Germany.

Banks and other lending institutions have become skeptical about financing new U.S. nuclear power plants after the Three Mile Island accident showed that utility companies could lose $1 billion or more in equipment in an hour, plus $1 billion to $1.5 billion in cleanup costs, even without any known serious public health effects. Despite these facts nuclear industry officials continue to push for the building of more nuclear power plants in the United States and hope to restore public confidence in nuclear power (see Spotlight on p. 284).

Several governments with major nuclear power programs—especially Japan, France (see Spotlight on p. 285), the United Kingdom, West Germany, and the Soviet Union—remain strongly committed to using nuclear power to produce a significant amount of their electricity by the end of the century. But public opposition to nuclear power is growing in these and other countries, costs have been much higher than projected (except in France), and plans have been scaled back sharply. By 1987 over two-thirds of the people in most European countries were opposed to the construction of any additional nuclear power plants in their countries and half of the people favored shutting down existing plants.

Advantages and Disadvantages of Conventional Nuclear Fission Using conventional nuclear fission to produce electricity has a number of advantages. Nuclear plants do not release carbon dioxide, particulate matter, sulfur dioxide, or nitrogen oxides into the atmosphere like coal-fired plants. Water pollution and disruption of land are low to moderate if the entire nuclear fuel cycle operates normally. Because of multiple safety systems, a catastrophic accident causing deadly radioactive material to be released into the environment is extremely unlikely.

There are also some disadvantages. Construction and operating costs in the United States and most countries are high and rapidly rising, even with massive government and consumer subsidies. Electricity

Since the Three Mile Island accident the U.S. nuclear industry and utility companies have financed a $40 million advertising campaign by the Committee for Energy Awareness to improve the industry's image, resell nuclear power to the American public, and downgrade the use of solar and other alternatives to nuclear power. These magazine and television ads do not let readers know they are paid for by the nuclear industry and many repeatedly use the misleading argument that nuclear power is needed in the United States to reduce dependence on imported oil.

In fact, since the oil embargo of 1973, the reduction of oil use and oil imports has come not from increased use of nuclear power but mostly from improvements in energy efficiency, increased use of wood as a fuel in homes and businesses, and increased use of coal to produce electricity. Since 1979 only about 5% of the electricity in the United States has been produced by burning oil; thus phasing in the 27 nuclear power plants still under construction in 1986 or building additional plants will not save the country any significant amount of domestic or imported oil.

can be produced by many other, less controversial methods at a cost equal to or lower than that of nuclear power. Although large-scale accidents are unlikely, a combination of mechanical and human errors, sabotage, or shipping accidents could again result in the release of deadly radioactive materials into the environment. The net useful energy yield of nuclear-generated electricity is low (Figure 3-15, p. 46). There is considerable disagreement over how high-level radioactive wastes should be stored, and some scientists doubt that an acceptably safe method can ever be developed. Military and commercial nuclear energy programs commit future generations to safely storing radioactive wastes for thousands of years even if nuclear fission power is abandoned. Furthermore, the existence of nuclear power technology helps spread knowledge and materials that could be used to make nuclear weapons.

16-3 NONRENEWABLE BREEDER NUCLEAR FISSION

At present rates of use the world's supply of uranium should last for at least 100 years and perhaps 200 years. However, some scientists believe that if there is a sharp rise in the use of nuclear fission to produce electricity after the year 2000, breeder nuclear fission reactors can be developed to increase the present estimated lifetime of the world's affordable uranium supplies for at least 1,000 years and perhaps several thousand years.

A **breeder nuclear fission reactor** produces within itself new fissionable fuel in the form of plutonium-239 from nonfissionable uranium-238, an isotope that is in plentiful supply. Radioactive waste and spent fuel elements from conventional fission reactors is taken to a fuel-reprocessing plant (Figure 16-7), where the plutonium-239 is separated and purified for use as fuel in a breeder reactor. In the breeder reactor fast neutrons are used to fission nuclei of plutonium-239 and convert the nonfissionable uranium-238 into enough fissionable plutonium-239 to start up another breeder reactor after 30 to 50 years. Because these devices use fast-moving neutrons for fissioning, they are often called *fast breeder reactors*.

A breeder reactor looks something like the reactor in Figure 16-4, except that its core contains a different fuel mixture and its two heat-exchanger loops contain liquid sodium instead of water. Under normal operation a breeder reactor is considered to be much safer than a conventional fission reactor. But in the unlikely event that all its safety systems failed and the reactor lost its sodium coolant, there would be a runaway fission chain reaction, and perhaps a small nuclear explosion with the force of several hundred pounds of TNT. Such an explosion could blast open the containment building, releasing a cloud of highly radioactive gases and particulate matter. A more common problem, which could lead to temporary shutdowns but poses no significant health hazards, is leakage of molten sodium, which ignites on exposure to air and reacts violently with water. Because the breeder requires little if any mining of uranium, it would have a much

France has the world's most ambitious and cost-efficient plan for using nuclear energy. By 1986 it had 40 reactors—providing 65% of its electricity—and 15 more under construction, with the goal of generating 75% of its electricity by 1990. France builds its reactors in less than 6 years—compared to 12 years in the United States. The plant construction cost per kilowatt of electricity in France is about half that in Japan and most European countries and one-third that in the United States.

The major factor responsible for France's ability to build plants cheaper and quicker than most countries is rigid centralized government control. All plants have standardized designs, and the overall responsibility for design, construction, and operation of all nuclear plants is in the hands of a single government-run national utility company. The public has little opportunity to criticize the program and most French citizens accept the deeply embedded tradition of government secrecy. France's nuclear program survives largely through government subsidies and an increasing national debt (already the third largest in the world), rather than through open-market economic competition with coal and other energy alternatives.

However, because of a massive national debt and a glut of electricity, France reduced its orders for new reactors from six in 1980 to one in 1986 and 1987. The French Planning Ministry concluded that the only reason for not halting all orders for new reactors for several years is to preserve jobs at the government-run company that designs, constructs, and operates all French nuclear plants. In 1986 polls revealed that 52% of of the French public—which for years had supported nuclear power by a large majority—were opposed to construction of any additional nuclear plants in France.

smaller environmental impact on the land than conventional fission.

Since 1966 several small- and intermediate-scale experimental breeder reactors have been built in the United States, the United Kingdom, the Soviet Union, West Germany, and France. In 1986 France put into operation a full-sized commercial breeder reactor, the Superphénix, which cost three times the original estimate to build and produces electricity at a cost per kilowatt over twice that of conventional fission reactors. In 1987 it sprung a sodium leak and was shut down temporarily for expensive repairs.

Tentative plans to build full-size commercial breeders in West Germany, the Soviet Union, and the United Kingdom may be canceled because of the excessive cost of France's reactor, an excess of electric generating capacity, and because studies indicate that breeders will not be economically competitive with conventional fission reactors for at least 50 years.

In 1983, after 13 years of political and scientific debate and a government expenditure of $1.7 billion for planning, the proposed Clinch River intermediate demonstration breeder reactor in Tennessee was canceled. It would have cost 5 to 12 times the original estimate, and it was based on designs outdated by other demonstration breeders in France and the United Kingdom. Numerous studies have also shown that the United States will not need breeder reactors until at least 2025, given the slowdown in the building of conventional nuclear fission reactors.

16-4   NONRENEWABLE AND PERPETUAL NUCLEAR FUSION

Controlled Nuclear Fusion   In the distant future—probably no sooner than 2050, if ever—scientists in the United States, the Soviet Union, Japan, and a consortium of European nations hope to use *controlled nuclear fusion* to provide an essentially inexhaustible source of energy for producing electricity. Current research is focused on the fusion of two isotopes of hydrogen, deuterium (D) and tritium (T) (see Figure 3-6, p. 35), because this nuclear reaction has the lowest ignition temperature, about 100 million degrees.

Deuterium is found in about 150 out of every million molecules of water (150 ppm) and can be separated from ordinary hydrogen atoms fairly easily. Thus the world's oceans provide an almost inexhaustible supply of this isotope. Although there is no significant natural source of tritium, an extremely small but sufficient quantity can be extracted from seawater. The two isotopes can be used as an initial charge in a fusion reactor, with neutrons emitted in the D-T fusion reaction used to bombard a surrounding blanket of lith-

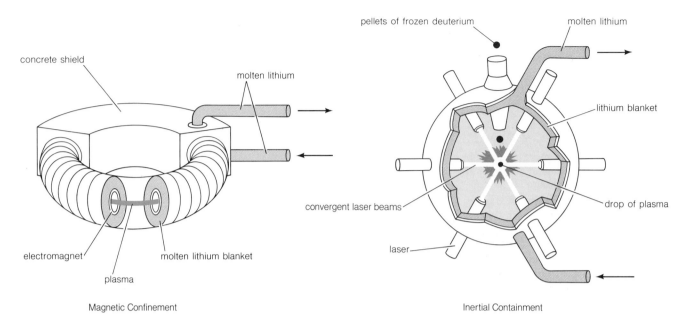

Figure 16-13 Magnetic containment and laser inertial confinement methods for possible ignition of a controlled nuclear fusion reaction.

ium to breed additional tritium fuel. The scarcity of lithium will eventually limit the use of *nonrenewable* D-T fusion, but the earth's estimated supply should last for one thousand to several thousand years, depending on rate of use.

Another possibility is the D-D fusion reaction, in which the nuclei of two deuterium atoms would be fused together to form a helium nucleus (Figure 3-6, p. 35). But this reaction requires an ignition temperature about ten times higher than that for D-T fusion and thus is not being pursued at this time. If controlled D-D nuclear fusion were developed, deuterium in the ocean could supply the world with energy at many times present consumption rates for 100 billion years—making this type of nuclear fusion an essentially *perpetual* source of energy. Few scientists, however, expect D-D fusion to become a major source of energy until 2100, if ever.

Achieving Controlled Nuclear Fusion The development of controlled nuclear fusion to produce thermal energy that can be converted into electricity is still at the laboratory stage after 36 years of research. To bring about a self-sustaining, controlled nuclear fusion reaction, the D-T fuel must be heated to about 100 million degrees and then squeezed together long enough and at a high enough density to ensure that sufficient numbers of the nuclei collide and fuse. No physical walls can be used to confine the hot fuel, known as *plasma*, not only because any known material would be vaporized but also because the walls

would contaminate the fuel and instantly cool it below its ignition temperature.

Two approaches being used in attempts to bring about controlled nuclear fusion are magnetic containment and inertial confinement (Figure 16-13). *Magnetic containment* involves using powerful electromagnetic fields to confine and force the fuel nuclei together within a vacuum. In one promising approach, electromagnetic fields produced by superconducting electromagnets cooled by liquid helium squeeze the fuel into the shape of a large toroid or doughnut (Figure 16-13, left). Such a reactor is known as a *tokamak* (after the Russian words for "toroidal magnetic chamber"), a design pioneered by Soviet physicists. In 1987 important breakthroughs in achieving superconductivity at higher temperatures may eliminate the need for expensive liquid helium as a coolant and make this approach more promising in the near future.

A second approach to nuclear fusion is *inertial confinement*, in which a marble-size, perfectly symmetrical pellet crammed with deuterium and tritium is bombarded from all sides with powerful laser beams, beams of charged particles (light ions), or beams of subatomic particles called *muons*. The beams would drive the fuel inward, compressing and heating it until the fuel nuclei fuse. Charged particle beams carry enough energy for fusion but spread out too much to focus this energy on the fuel pellets because the like-charged particles repel one another. Lasers can be better focused but so far are not powerful enough to bring about fusion.

In 1986 some promising preliminary results were

obtained in a third approach, known as *muon-catalyzed fusion,* that would allow fusion to take place at temperatures even as low as room temperature. If successful, this approach, which is only in the very early stages of development, could represent a major breakthrough.

By late 1987 none of the several laboratory test reactors using any approach throughout the world had been able to reach the *break-even point,* where the energy pumped into the reactor equals the energy it produces. When—and if—the energy break-even point is reached in the laboratory, the next, even more difficult step will be to achieve the *burning point,* at which the D-T nuclear fusion reaction becomes self-sustaining and releases more energy than is put in.

Building a Commercial Nuclear Fusion Reactor Assuming that the burning point can be reached, the next step is to build a small demonstration fusion reactor and then scale it up to commercial size. This task is considered one of the most difficult engineering problems ever undertaken. For example, massive electromagnets cooled to low temperatures would be located only a few meters from the plasma at the highest temperature produced on earth. Protecting the extremely sensitive cooled electromagnets from heat and radiation damage would be like trying to preserve an ice cube next to a blazing fire, only much harder. New developments in superconductivity, however, might help overcome this problem.

Another engineering problem is the necessity of maintaining a near perfect vacuum in the interior section of the reactor containing the plasma. More mind-boggling still, the inner walls of the reactor surrounding the lithium blanket must resist constant baths of highly reactive liquid lithium at a temperature of 1,000°C (1,800°F) and steady bombardment by fast-moving neutrons released when deuterium and tritium fuse. Since neutron bombardment eventually destroys or alters the composition of all currently known reactor wall materials, the walls would have to be replaced about every five years, at such enormous cost that some scientists doubt whether fusion will ever be economically feasible. Scientists and engineers hope to overcome some of these problems by developing special new alloys, but some of their elements may be unaffordably scarce. The estimated cost of a commercial fusion reactor based on currently known approaches is at least two times that of a comparable breeder fission reactor and at least four times that of a comparable conventional fission reactor.

If everything goes as planned—which some believe may be one of the biggest *ifs* in scientific and engineering history—the first U.S. commercial fusion reactor could be completed between 2010 and 2025. If this happens, then between 2050 and 2150 nuclear fusion might produce as much as 18% of U.S. annual commercial energy needs.

The United States, the country that led the world into the age of nuclear power, may well lead it out.
Lester R. Brown

DISCUSSION TOPICS

1. Explain why you agree or disagree with the following statements:
   a. Dry steam, wet steam, and hot water geothermal deposits can provide most of the electricity the United States needs by the year 2010.
   b. Molten rock (magma) geothermal deposits should be able to supply the United States with all the electricity and high-temperature heat it needs by 2025.
   c. Although geothermal energy may not be a major source of energy for the United States over the next few decades, it can supply a significant fraction of energy needs in selected areas where high-quality deposits are found.

2. What method should be used for the long-term storage of high-level nuclear wastes? Defend your choice.

3. Do you favor a U.S. energy strategy based on greatly increased use of conventional nuclear fission reactors to produce electricity between 1988 and 2020? Explain.

4. Explain why you agree or disagree with each of the following proposals made by President Ronald Reagan, utility companies, and the nuclear power industry:
   a. Licensing time of new nuclear power plants in the United States should be halved (from an average of 12 to 6 years) so that these facilities can be built more economically and compete more effectively with coal and other renewable energy alternatives.
   b. A major program for developing the nuclear breeder fission reactor should be developed and funded by the federal government to conserve uranium resources and eventually keep the United States from being dependent on other countries for uranium supplies.

5. Do you believe that the United States and other industrialized countries should try to reduce the risk of nuclear war by pledging not to sell or give any additional nuclear power plants or any forms of nuclear technology to other countries? Explain.

17

Perpetual and Renewable Energy Resources: Conservation, Sun, Wind, Water, and Biomass

GENERAL OBJECTIVES

1. What are the advantages and disadvantages of reducing energy waste as a perpetual source of energy?

2. What are the advantages and disadvantages of capturing and using some of the earth's direct input of perpetual solar energy for heating buildings and water and for producing electricity?

3. What are the advantages and disadvantages of using perpetual indirect solar energy stored in falling and flowing water (hydropower) for producing electricity?

4. What are the advantages and disadvantages of using perpetual indirect solar energy in the form of heat stored in water for producing electricity and heating buildings and water?

5. What are the advantages and disadvantages of using perpetual indirect solar energy stored in winds to produce electricity?

6. What are the advantages and disadvantages of using renewable indirect solar energy stored in plants and organic waste (biomass) for heating buildings and water and for converting biomass to transportation fuels (biofuels)?

7. What are the best present and future energy options for the United States, and what should be the country's long-term energy strategy?

8. What are the advantages and disadvantages of producing and using hydrogen gas and fuel cells to produce electricity, heat buildings and water, and propel vehicles when oil runs out?

Throughout most of human history, people have relied on renewable resources—sun, wind, water, and land. They got by well enough, and so could we.

Warren Johnson

Saving energy by using it more efficiently has been the largest source of energy in MDCs such as the United States since 1978, and it is also the largest potential future source of energy. Some experts project that the contribution of energy from the sun, wind, water, and biomass to global energy use will increase significantly by the year 2000 and then expand even more rapidly. Others believe that these forms of perpetual and renewable energy will grow at a much slower rate because of uncertain costs, loss of momentum during the temporary oil glut of the 1980s, and sharp cutbacks in federal funds and tax breaks for their development.

17-1 ENERGY CONSERVATION: DOING MORE WITH LESS

Reducing Unnecessary Energy Waste: An Offer We Can't Afford to Refuse Our greatest energy resource now and in the future is **energy conservation,** the reduction or elimination of unnecessary energy use and waste. There are three general methods of energy conservation: (1) We can reduce energy consumption at essentially no cost by changing energy-wasting habits. Examples include walking or riding a bicycle for short trips, wearing a sweater indoors in cold weather to allow a lower thermostat setting, and turning off unneeded lights. (2) We can use less energy to do the same amount of work. This requires small, ultimately money-saving investments such as installing more building insulation, keeping car engines tuned, and switching to more energy-efficient cars, houses, heating and cooling systems, appliances, and industrial processes. (3) We can use less energy to do more work by developing new devices that waste less energy than existing ones. Examples include more efficient solar cells for conversion of solar energy directly to electricity, new aerodynamic vehicle designs that reduce fuel consumption, and efficient heating and cooling systems, appliances, and vehicle engines.

Energy conservation is the largest and cheapest source of energy available (Figure 16-12, p. 283), has a high net useful energy yield, reduces the environmental impacts of all other energy resources by reducing overall energy use and waste, adds no carbon dioxide to the atmosphere, extends domestic and world supplies of nonrenewable fossil fuels, buys time for phasing in new perpetual and renewable energy resources, reduces dependence on imported oil and other energy resources, and generally provides more jobs and promotes more economic growth per unit of energy gained than other energy resources. The first two conservation methods also do not require development of new forms of technology, and can be implemented in a very short time, usually days or weeks— compared to 6 to 12 years for building new coal, nuclear, and hydroelectric power plants.

Energy conservation has fewer serious disadvantages than any other present or foreseeable energy alternative. One disadvantage is that improving energy efficiency by replacing houses, industrial equipment, and cars as they wear out with more energy-efficient ones takes a fairly long time—30 to 50 years for buildings and industrial equipment and 10 to 12 years for cars. Although most energy conservation measures improve average life quality, certain ones require lifestyle changes that some people do not like. For example, improved automobile gas mileage depends mainly on smaller, lighter, and less powerful cars, which some drivers don't like. These cars also provide occupants with less protection in accidents as long as consumers do not insist that all cars be equipped with air bags and other safety devices.

Between 1975 and 1985, conservation measures in the United States reduced total consumption of all major types of commercial energy, provided much more new energy (by reducing waste) than all other alternatives combined, and cut national energy bills by about $150 billion a year. On the average, American houses in 1985 used 25% less space-heating energy per square foot than they did in 1975; many new, energy-efficient houses used 75% less. In 1985 new U.S. cars averaged about 26 miles per gallon, nearly double that of new cars in 1973. New refrigerators are now about 72% more efficient than they were in 1972.

Despite these important improvements in energy efficiency, massive amounts of energy are still unnecessarily wasted in the United States. Average gas mileage for new cars and for the entire fleet of cars is below

that in other MDCs. Most U.S. houses and buildings are still underinsulated and leaky and most new houses and buildings do not take advantage of available energy-efficient construction techniques. Even with its low net energy efficiency, electric heating is installed in over half the new homes in the United States (as well as in many European countries).

Studies have shown that fully implementing the first two methods of energy conservation described above would decrease average energy use per person in the United States by 50% between 1986 and 2000, saving about $200 billion a year, enough to pay off the entire national debt and at the same time stimulate the economy. Continuing these improvements and funding crash research programs to create more efficient energy-producing processes could reduce average energy use per person in the United States by 80% to 90% between 1986 and 2020.

Improving Industrial Energy Efficiency In the United States industrial processes consume more commercial energy than transportation, residences, and commercial buildings (Figure 17-1). U.S. industry has led the way in instituting energy conservation measures to save money since 1973 and has the greatest long-term potential to switch to new energy-saving processes. Japan has the highest overall industrial energy efficiency in the world. France, Italy, Spain, and West Germany also have relatively high industrial energy efficiencies.

An important way most U.S. industries can save energy and money in a relatively short time is to install *cogeneration units*, which use excess high-temperature steam and high-temperature heat to produce electricity, which can be used by the industry or sold to utility companies. By 1985 industrial cogeneration in the United States provided electricity equivalent to the output of fifteen 1,000-megawatt power plants. New projects, with an additional output of electricity equal to that of 17 large power plants, were under construction or planned in 1985. The U.S. Office of Technology Assessment estimates that cogeneration has the potential to provide electricity equal to that from 200 large power plants by the year 2000.

Industry accounts for close to half the worldwide use of electricity. Aluminum production is one of the most energy-intensive processes, requiring 1% of the world's commercial energy. Improvements in energy efficiency in the aluminum industry include a new process that reduces electricity use by 25% and another, using recycled aluminum, that reduces electricity use by 90%. Despite the potentially enormous energy savings, the average world aluminum-recycling rate is only 28%; this could easily be doubled or tripled.

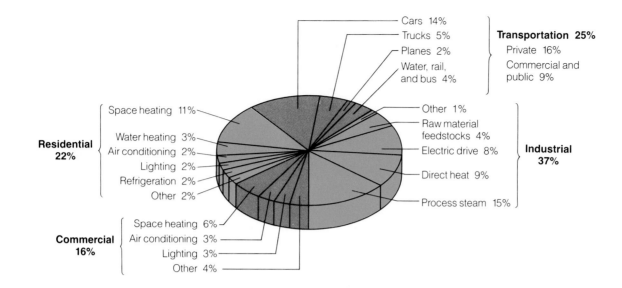

Figure 17-1 Distribution of commercial energy use in the United States among various sectors in 1984. (Data from U.S. Department of Energy)

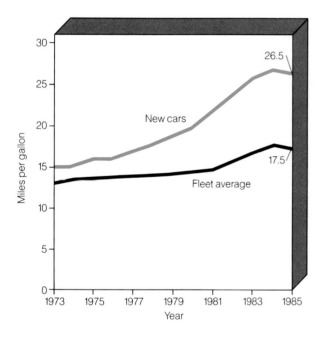

Figure 17-2 Increase in the average fuel efficiency of new cars and the entire fleet of cars in the United States between 1973 and 1985. (Data from U.S. Department of Energy and the Environmental Protection Agency)

Electric industrial motors consume almost two-thirds of the electricity used in industry worldwide and 80% in the United States. Phasing in new motor designs that require 30% to 50% less electricity could reduce overall electricity use in the United States by at least 10%, enough to eliminate the need for two-thirds of the existing U.S. nuclear power plants.

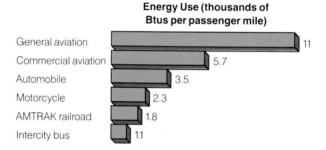

Figure 17-3 Energy efficiency of various modes of domestic transportation.

Improving Transportation Energy Efficiency Transporting people and goods accounts for one-fourth of the commercial energy use in the world and the United States (Figure 17-1)—and a much higher percentage of oil use. About one-tenth of the oil consumed in the world each day is used by American motorists on their way to and from work, two-thirds of them driving alone. Thus the largest savings in oil can come from improved vehicle fuel economy (Figure 17-2), greater use of mass transit (Figure 17-3), and more efficient hauling of freight.

In 1985 Japan had the highest average fuel economy for new and existing cars, and Canada and the United States had the lowest. The U.S. Office of Technology Assessment estimates that new cars produced in the United States could easily reach an average of 45 miles per gallon (mpg) by 2000. But so far Congress has not required U.S. car manufacturers to go beyond the 27.5 mpg set for 1985; even this standard was allowed to decrease to 26.5 mpg (Figure 17-2). Experts

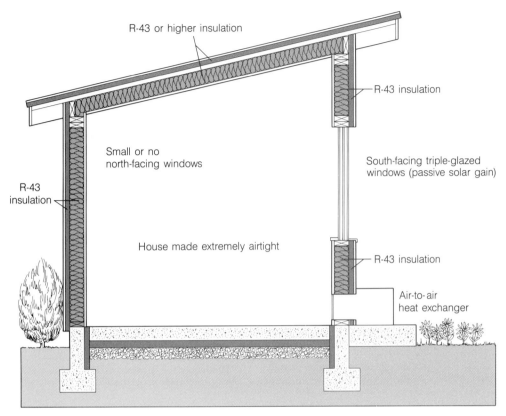

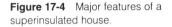

Figure 17-4 Major features of a superinsulated house.

R-43 or higher insulation

R-43 insulation

Small or no
north-facing windows

South-facing triple-glazed
windows (passive solar gain)

R-43
insulation

House made extremely airtight

R-43 insulation

Air-to-air
heat exchanger

estimate that with adequate laws average fuel economy levels of 30 mpg for entire fleets could be achieved everywhere by the end of the century and levels of 50 mpg shortly thereafter. Based on prototype models already in operation, the fuel economy for new cars and light trucks could easily be increased to between 60 and 100 mpg through reduced auto weight, more energy-efficient engines and drivetrains, and reduced aerodynamic and rolling resistance.

Other measures that would conserve the world's diminishing supply of oil include shifting more freight from trucks and airplanes to trains (Figure 17-3), increasing the efficiency of transport trucks by improving aerodynamic design and using turbocharged diesel engines and radial tires, and not allowing trucks to return empty after reaching their destination.

Improving the Energy Efficiency of Commercial and Residential Buildings Most commercial and residential buildings in the United States consume about 50% to 90% more energy than they would if they were designed to use energy more efficiently. A monument to energy waste is the 110-story, twin-towered World Trade Center in Manhattan, which uses as much electricity as a city of 100,000 persons. Windows in its walls of glass cannot be opened to take advan-

tage of natural warming and cooling, and its heating and cooling systems must run constantly, even when no one is in the building.

By contrast, Atlanta's 17-story Georgia Power Company building uses 60% less energy than conventional office buildings. Energy-saving features include an extension of each floor over the one below to create an overhang that allows heating by the low winter sun while blocking out the higher summer sun to reduce air conditioning costs, a computer programmed to turn off all lights at 6 P.M. unless instructed otherwise, energy-efficient lights that focus on desks rather than illuminate entire rooms, and an adjoining three-story building where employees can work at unusual hours so that the entire larger structure does not have to be heated or cooled when few people are at work.

Building a **superinsulated house** is the most effective way to improve the efficiency of residential space heating and save on lifetime costs, especially in cold climates (Figure 17-4). Such a house is massively insulated and made extremely airtight, and can get all its space heating without a conventional backup system from a combination of direct solar gain, waste heat from appliances, and the body heat of the occupants. Such houses retain most of this heat input for at least 100 hours, and inside temperatures probably never fall below 10°C (50°F) even in extremely cold weather. An air-to-air heat exchanger prevents buildup of humidity and indoor air pollution.

The energy-efficient house of the near future will be controlled by microprocessors. Each will be programmed to do a different job such as controlling the thermostat, opening and closing windows and insulated shutters to take advantage of solar energy and breezes, operating fans to control temperature and air distribution, and controlling the security system.

Windows will have a coating like the light-sensitive glass in some sunglasses, automatically becoming opaque to keep the sunlight out when the house gets too hot. Glass with high insulating values (R-10 to R-15)* will be available so that a house can have as many windows as the owner wants in any climate without much heat loss. Thinner insulation material will allow roofs to be insulated to R-100 and walls to R-40, far higher than today's best superinsulated houses (Figure 17-4).

By 1987 small-scale cogeneration units that run on natural or LP-gas were available; they can supply a home with all its space heat, hot water, and electricity needs. The units are no larger than a refrigerator, make less noise than a dishwasher, and, except for an occasional change of oil filters and spark plugs, are nearly maintenance-free. In most cases, this home-use power and heating plant will pay for itself in four to five years.

Homeowners can also get all the electricity they need from rolls of solar cells attached like shingles to a roof or applied to window glass as a coating (already developed by Arco). More people in rural and suburban areas will be drilling wells and using greatly improved heat pumps to extract geothermal heat from warm underground water. Currently, most heat pumps used for space heating and air conditioning last only about five years and must be supplemented in moderately cold climates by expensive and energy-wasting electric-resistance heaters when temperatures fall below $-7°C$ (20°F). As technology improves, however, heat pumps should become more energy-efficient and economical.

---

*The higher the R-value of a material, the greater its insulating ability.

Adding $5,000 to $10,000 to the cost of a new house for energy-saving measures can save the homeowner $50,000 to $100,000 over a 40-year period by cutting lifetime heating and cooling bills by 50% to 100%. Combining currently known conservation measures with emerging technology could greatly increase the energy efficiency of new houses in the not-too-distant future (see Spotlight above).

Many energy-saving features can be added to existing homes, a process called *retrofitting*. For example, simply increasing insulation above ceilings can drastically reduce heating and cooling loads, typically recovering costs in 2 to 10 years. Caulking and weatherstripping around windows, doors, pipes, vents, ducts, and wires saves energy and money quickly. Switching to new gas furnaces with energy efficiencies of 90% to 95%, compared to 60% to 65% for most conventional gas furnaces, also saves energy and money on a lifetime-cost basis.

Legal approaches can also be used to increase energy conservation in homes and buildings. Building codes could be changed to require that all new houses use 80% less energy than conventional houses of the same size, as has been done in Davis, California. Laws can require that any existing house be insulated and weatherproofed to certain standards before it can be sold, as is now done in Portland, Oregon.

Using the most energy-efficient appliances available can also save homeowners energy and money.* About one-third of the electricity generated in the United States and other industrial countries is used to power household appliances. Many American homes have 20 to 50 light bulbs, and burning one 100-watt light bulb just six hours a day each year consumes energy equivalent to fifteen 55-gallon barrels of oil. Socket-type fluorescent light bulbs that use one-fourth as much electricity as conventional bulbs are now available. Although they are expensive, they last 13 times longer than conventional bulbs, and return 3 times more than the original purchase price through reduced electricity bills. Switching to these bulbs would save one-third of the electric energy now produced by all U.S. coal-fired plants or eliminate the need for all electricity produced by the country's 101 nuclear power plants.

Developing a Personal Energy Conservation Plan
Individuals can develop their own plans for saving energy and money (see Appendix 5 for suggestions). Four basic guidelines should be used: (1)

---

*Each year the American Council for an Energy-Efficient Economy publishes a list of the most energy-efficient appliances. For a copy, send $2 to the council at 1001 Connecticut Ave., N.W., Suite 530, Washington, D.C. 20036.

Table 17-1 Energy Use and Conservation in the United States and Sweden

Use or Method	United States	Sweden
Average per capita use	230,000 kcal/day	150,000 kcal/day
Transportation energy use	High	One-fourth of U.S.
Country size	Large	Small
Cities	Dispersed	Compact
Mass transit use	Low	High
Average car fuel economy	Poor	Good
Gasoline taxes	Low	High to encourage conservation
Tariffs on oil imports	Low	High to encourage conservation
Industrial energy efficiency	Fairly low	High
Nationwide energy-conserving building codes	No	Yes
Municipally owned district heating systems	None	30% of population
Emphasis on electricity for space heating	High (one-half of new homes)	High (one-half of new homes)
Domestic hot water	Most kept hot 24 hours a day in large tanks	Most supplied as needed by instant tankless heaters
Refrigerators	Mostly large, frost-free	Mostly smaller, non-frost-free using about one-third the electricity of U.S. models
Long-range national energy plan	No	Yes
Government emphasis and expenditures on energy conservation and renewable energy	Low	High
Government emphasis and expenditures on nuclear power	High	Low (to be phased out)

Don't use electricity to heat space or water; (2) insulate new or existing houses heavily and caulk and weatherstrip to reduce air infiltration and heat loss; (3) get as much heat and cooling as possible from natural sources—especially sun, wind, geothermal energy, and trees for windbreaks and natural shading; and (4) buy the most energy-efficient homes, cars, and appliances available and evaluate them only in terms of lifetime cost.

Energy Efficiency Differences Between Countries Japan, Sweden, and most industrialized western European countries with average standards of living at least equal to that in the United States use only one-third to two-thirds as much energy per person as Americans. This is due to a combination of factors including greater emphasis on energy conservation and fewer passenger miles traveled per person (mainly because cities are more compact in these countries). However, energy use patterns in countries vary, and approaches that save energy in one country can't always be used in others. Table 17-1 compares energy practices between the United States and Sweden.

17-2 DIRECT PERPETUAL SOLAR ENERGY FOR PRODUCING HEAT AND ELECTRICITY

Passive Solar Systems for Low-Temperature Heat A **passive solar system** captures sunlight directly within a structure and converts it to heat (Figure 17-5). Its major design features include (1) large areas of south-facing double- or triple-paned glass or a south-facing greenhouse or solarium to collect solar energy; (2) walls and floors of concrete, adobe, brick, stone, or tile, or water-filled glass or plastic columns or black-painted barrels, or panels or cabinets containing chemicals to store collected solar energy and release it slowly throughout the day and night; (3) few

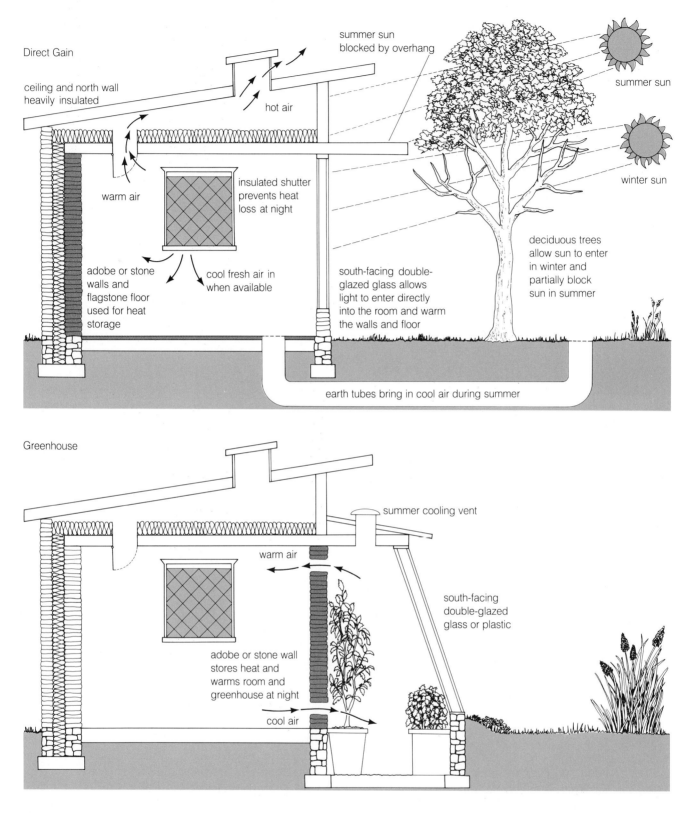

Figure 17-5 Three examples of passive solar design.

or no north-facing windows, thus decreasing heat loss; (4) movable insulated shutters or curtains on windows to reduce heat loss at night; and (5) heavy insulation. A series of roof-mounted passive solar water heaters

(Figure 17-6) can also be used to provide hot water for a house at an installed cost of $3,500 to $4,000 in the United States and $1,000 in Israel and Japan.

Houses with passive solar systems often have an

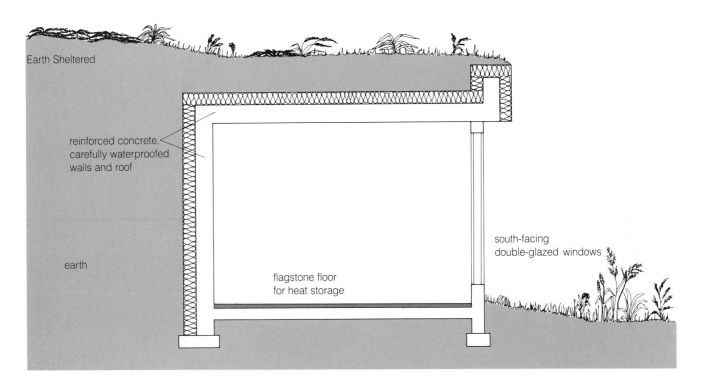

Figure 17-5 Three examples of passive solar design. (Continued)

open design to allow the collected and stored heat to be distributed by natural air flows or fans. An air-to-air heat exchanger is used to provide fresh air without significant heat loss or gain and to prevent buildup of moisture and indoor air pollutants. When necessary, a small backup heating system is also used. A passive solar system can be even more efficient in an earth-sheltered house (see Spotlight on p. 298 and Figure 17-5). A well-designed passive solar system is the simplest, cheapest (on a lifetime-cost basis), most energy-efficient (Figure 3-12, p. 44), most maintenance-free, and least environmentally harmful way to provide 50% to 100% of the space heating of a home or small building, with an added construction cost of only 5% to 10%.

In hot climates the most difficult problem with passively designed buildings is to keep them cool in summer. Passive cooling can be provided by using deciduous trees, window overhangs, or awnings on the south side to block the high summer sun and by using windows and fans to take advantage of breezes and keep air moving (Figure 17-5). Earth tubes, buried 3 to 6 meters (10 to 20 feet) underground, where the temperature remains around 13°C (55°F) all year long in cold northern climates and about 19°C (67°F) in warm southern climates, can also be used to bring in cool and partially dehumidified air (Figure 17-5).

In areas with dry climates, such as the southwestern United States, evaporative coolers can be used to remove interior heat by evaporating water. In hot and humid areas a small dehumidifier or a solar-assisted geothermal heat pump may be needed to reduce humidity to acceptable levels. Solar-powered air conditioners have been developed but so far are too expensive for residential use. According to some experts, climate-sensitive, passive solar cooling and natural ventilation should make it possible to construct all but the largest buildings without air conditioning in most parts of the world.

Active Solar Systems for Low-Temperature Heat

An **active solar system** uses a series of specially designed collectors to concentrate solar energy, pumps to store it as heat in large insulated tanks of water or rock, and thermostat-controlled fans to distribute the stored heat as needed, usually through conventional heating ducts. A series of active solar collectors for space heating and heating water are usually mounted on a roof with an unobstructed southern exposure.

A typical flat-plate solar collector used in an active system consists of a coil of copper pipe attached to a blackened metal base and covered with two layers of glass separated by an insulating layer of air and encased in an aluminum frame (Figure 17-6). Radiant energy from the sun passes through the transparent glass cover, is absorbed by the blackened surface, transferred as heat to water, an antifreeze solution, or air pumped through the copper pipe, and transferred to an insulated water or rock heat storage system. In

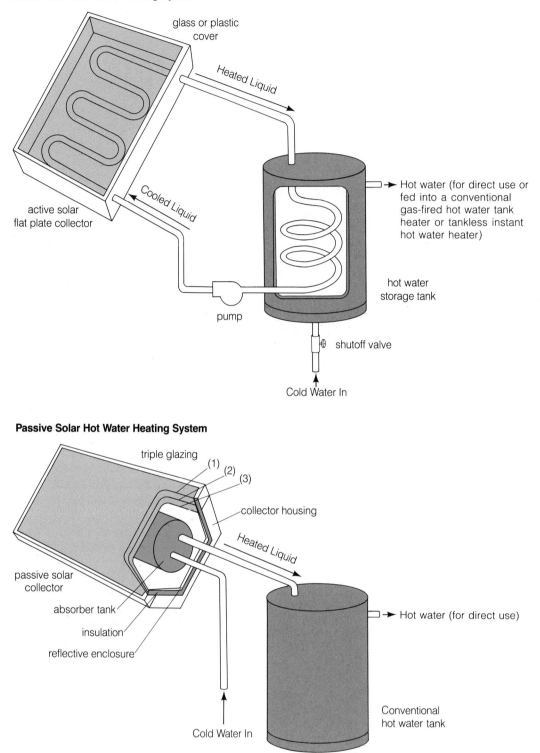

Active Solar Hot Water Heating System

glass or plastic cover

Heated Liquid

Cooled Liquid

active solar flat plate collector

Hot water (for direct use or fed into a conventional gas-fired hot water tank heater or tankless instant hot water heater)

hot water storage tank

pump

shutoff valve

Cold Water In

Passive Solar Hot Water Heating System

triple glazing

(1) (2) (3)

collector housing

Heated Liquid

passive solar collector

absorber tank

insulation

reflective enclosure

Hot water (for direct use)

Conventional hot water tank

Cold Water In

Figure 17-6 Active and passive solar water heaters.

middle and high latitudes where the availability of sun is not as high (Figure 17-7) a small backup heating system is needed during prolonged cold or cloudy periods.

Active systems are more expensive than passive systems on a lifetime basis because they require more materials to build, need more maintenance, and eventually deteriorate and must be replaced. However,

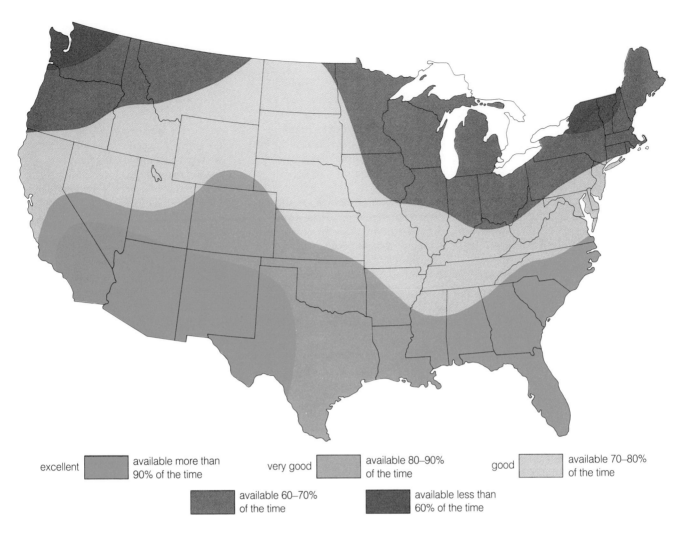

excellent ▨ available more than 90% of the time   very good ▨ available 80–90% of the time   good ▨ available 70–80% of the time

▨ available 60–70% of the time   ▨ available less than 60% of the time

Figure 17-7 Availability of solar energy during the day in the continental United States. (Data from U.S. Department of Energy and the National Wildlife Federation)

retrofitting an existing house with an active solar system is often easier than adding a passive system. About one-third of the existing buildings in the United States have enough southern exposure and sufficient rooftop or other space for active solar retrofitting that should pay for itself in energy savings in three to seven years.

Most solar experts agree that passive solar design is a more effective and cheaper way to heat homes, whereas active systems are usually better for heating large apartment and commercial buildings, which have large rooftops for mounting the solar collectors. Active solar collectors can also be used exclusively for providing hot water. By 1986 over 1 million active solar hot water systems had been installed in the United States, especially in California, Florida, and southwestern states with ample sunshine (Figure 17-7). The main barrier to their widespread use in the United States is a high initial cost of $3,500 to $5,000.

In Israel, which gets 50% of its hot water from the sun, and Japan, where 12% of residences have solar water heaters, mass production and marketing have brought the cost down to about $1,000. Mass production of new designs using thin-film plastic instead of glass and nationwide marketing by large firms could bring costs in the United States down to $750.

Advantages and Disadvantages of Solar Energy for Providing Low-Temperature Heat Using active or passive systems to collect solar energy for low-temperature heating of buildings and water has a number of advantages. The energy supply is free and readily available on sunny days. The technology is well developed and can be installed quickly. Net useful energy yield is moderate to high, carbon dioxide is not added to the atmosphere, and environmental impacts from air pollution, water pollution, and land disturbance are low because the systems themselves produce no pollutants during operation and do not take up valu-

An increasing number of people in the United States are building passively heated and cooled *earth-sheltered* houses (Figure 17-5) and commercial buildings partially or completely underground. Most designs cost about 15% more to build initially than a comparable above-ground structure, primarily because of the large amount of concrete needed to bear the heavy load of earth and the need for careful waterproofing to prevent leaks which would be difficult and expensive to repair. However, a new design using preinsulated curved wooden panels that form underground arches has reduced the cost to that of a comparable above-ground structure.

Even with higher initial costs, earth-sheltered homes are cheaper than conventional above-ground houses of the same size on a lifetime-cost basis because of reduced heating and cooling requirements, elimination of exterior maintenance and painting, and reduced fire insurance rates. Earth-sheltered structures also provide more privacy, quiet, and security from break-ins, fires, hurricanes, tornadoes, earthquakes, and storms than conventional above-ground buildings. The interior of an earth-sheltered house can look like that of any ordinary house, and south-facing solar-collecting windows, an attached greenhouse, and skylights can provide more daylight than is found in most conventional dwellings.

---

able land. A passive solar system is the cheapest way to provide space heating for a home in most places on a lifetime-cost basis. An active solar system is cost-effective in most places for providing hot water for homes and buildings and space heating for fairly large buildings on a lifetime-cost basis.

However, there are disadvantages. The energy supply is not available at night and on cloudy days, so that heat storage systems and small conventional backup heating systems are necessary except in superinsulated houses. Although lifetime costs are highly favorable, high initial costs discourage buyers not used to considering lifetime costs or buyers who move every few years to change jobs. State and local laws must also be in place to guarantee that others cannot build structures that block a user's access to sunlight—legislation often opposed by builders of high-density developments. Most passive solar systems in use today require that owners open and close windows and shades to regulate heat flow and distribution—tasks that in the near future might be done by cheap microprocessors.

Concentrating Solar Energy to Produce High-Temperature Heat and Electricity In experimental systems huge arrays of computer-controlled mirrors track the sun and focus sunlight on a central heat collection point, usually atop a tall tower. This highly concentrated sunlight can produce temperatures high enough for industrial processes or for making high-pressure steam to run turbines and produce electricity.

The world's largest *solar furnace,* the Odeillo Furnace, has been in operation high in the Pyrenees Mountains in southern France since 1970 (Figure 17-8). This system, which produces temperatures up to 2,000°C (5,000°F), is used in the manufacture of pure metals and other substances; the excess heat is used to produce steam and generate electricity fed into the public utility grid. Smaller units are being tested in France, Italy, Spain, and Japan.

Several private and government-financed experimental *solar power towers,* which produce electricity, have been built in the United States. One, known as Solar One, is located in the Mojave Desert in southern California and produces enough electricity to meet the needs of 3,000 homes (Figure 17-9). The main use of such relatively small plants will be to provide reserve power to meet daytime peak electricity loads, especially in California and southwestern states with high electrical load peaks in the summer because of air conditioning demands.

Many observers believe that these systems will make little contribution to overall energy supplies because they have a low net useful energy yield, are costly to build (four to six times the cost of expensive new nuclear power plants on a per kilowatt basis), and produce electricity at a cost several times that of hydroelectric, wind, coal-burning, and conventional nuclear fission power plants. Their environmental impact on air and water is low but land disruption is high because of the large area required for solar collection. They are also usually built in sunny, arid, ecologically fragile desert biomes, where sufficient water may not be available for use in cooling towers to recondense spent steam.

Peter Menzel/Stock, Boston

Figure 17-8 Solar furnace located near Odeillo in the Pyrenees Mountains of southern France.

Sandia National Laboratories, Livermore, California

Figure 17-9 Solar One power tower used to generate electricity in the Mojave Desert near Barstow, California.

Converting Solar Energy Directly to Electricity: Photovoltaic Cells   The earth's direct input of solar energy can be converted by **photovoltaic cells,** commonly called *solar cells* or *photovoltaics,* directly into electrical energy in one simple step. A solar cell consists of a thin wafer of purified silicon (which can be made from inexpensive, abundant sand) to which trace amounts of other substances (such as gallium arsenide or cadmium sulfide) have been added so that the wafer emits electrons and produces a small amount of electrical current when struck by sunlight (Figure 17-10).

Since 1958, expensive solar cells have been used to power space satellites and to provide electricity for at least 12,000 homes worldwide (6,000 in the United States), located mostly in isolated areas where the cost of running electrical lines to individual dwellings is extremely high. Some scientists have also proposed putting billions of solar cells on large orbiting satellites and beaming the energy back to earth in the form of microwaves, but such schemes are considered far too costly.

Because the amount of electricity produced by a single solar cell is very small, many cells must be wired together in a solar panel to provide a generating capacity of 30 to 100 watts. A number of these panels are then wired together and mounted on a roof or on a

Figure 17-10 Use of photovoltaic (solar) cells to provide DC electricity for a home; any surplus can be sold to the local power company.

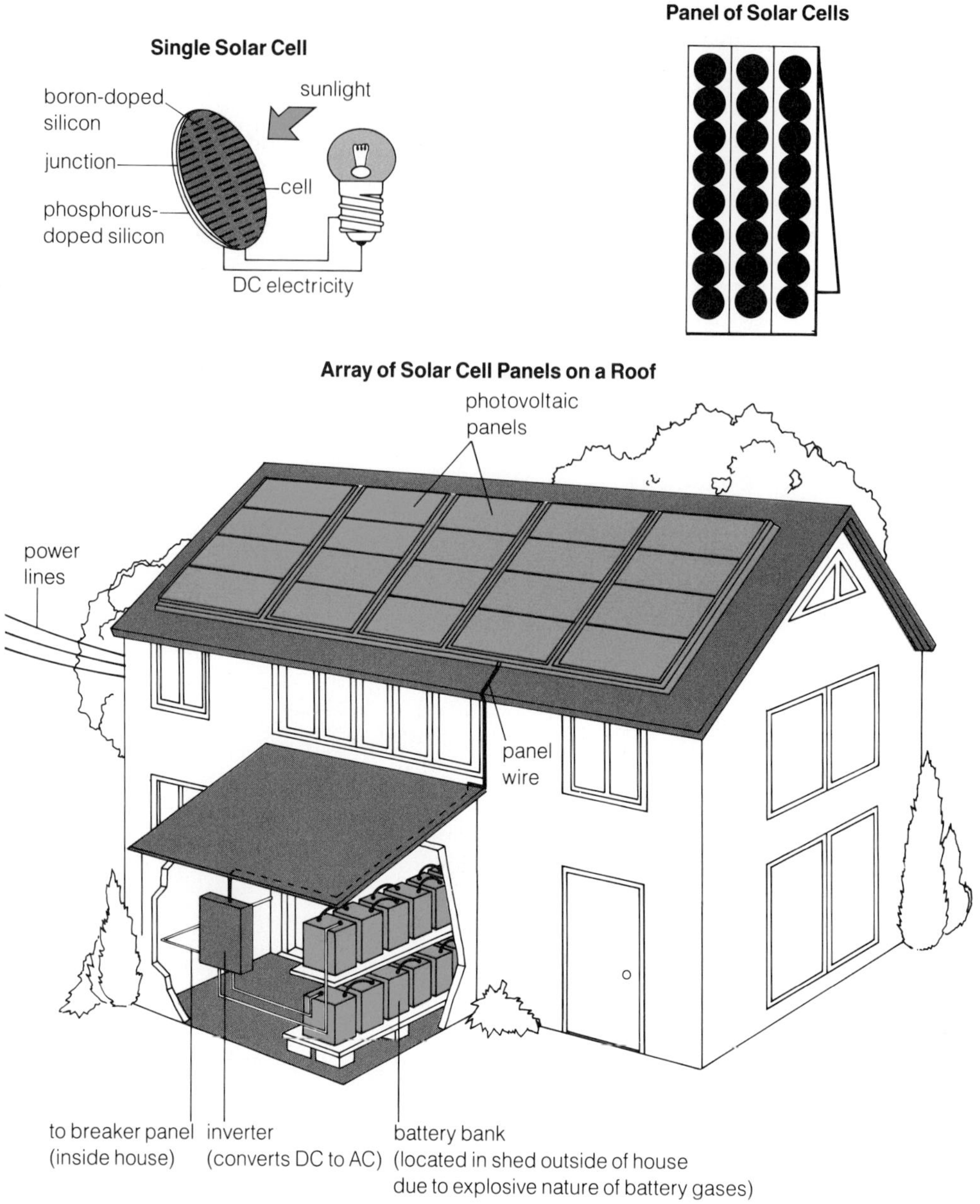

Single Solar Cell

boron-doped silicon

junction

phosphorus-doped silicon

sunlight

cell

DC electricity

Panel of Solar Cells

Array of Solar Cell Panels on a Roof

photovoltaic panels

power lines

panel wire

to breaker panel (inside house)

inverter (converts DC to AC)

battery bank (located in shed outside of house due to explosive nature of battery gases)

rack that tracks the sun to produce electricity for a home or building. Massive banks of such cells can also be used to produce electricity at a small power plant.

The resulting electricity is in the form of direct current (DC), not the alternating current (AC) commonly used in households. One option is to use the electricity directly to power lights and appliances that run on DC, such as those in most recreational vehicles. Any excess energy produced during daylight can be sold to the local utility company or stored for use at night and on cloudy days in long-lasting rechargeable DC batteries such as those used in boats and golf carts. Another alternative is to use an electronic inverter to convert direct current to alternating current. In the future, DC electricity produced by solar cells could also be used to decompose water to produce hydrogen

gas, which can be stored in a pressurized tank and burned in fuel cells to provide space heat, hot water, and electricity and to run cars (Section 17-7).

U.S. Department of Energy and solar cell researchers and manufacturers in the United States and Japan project that development of more energy-efficient cells and cost-effective mass-production techniques should allow solar cells to produce electricity at a competitive price almost everywhere by the mid-1990s or shortly after the turn of the century. This goal came closer to reality in 1986 when a group of electrical engineers at Stanford University developed solar cells with an energy efficiency of 27.5%—not much below the efficiency of nuclear and coal-fired power plants.

If cost-effective solar cells can be mass-produced, their use could spread rapidly in MDCs for buildings and houses and in LDCs, where electricity in rural areas is either unavailable or supplied by expensive diesel generators. Since 1981 the U.S. federal research and development budget for solar cells has been cut sharply while Japanese government expenditures in this area have tripled. Some analysts argue that federal and private research efforts on photovoltaics in the United States should be increased. Otherwise the United States might one day find much of its capital being drained to pay for imports of massive numbers of photovoltaic cells from Japan and other Far Eastern countries, thus losing out on a major global economic market.

Solar cells have a number of important advantages. If used according to projections, they could provide 20% to 30% of the world's electricity by 2050, thus eliminating the need to build large-scale power plants of any type and allowing the phaseout of many existing nuclear and coal-fired power plants. They are reliable and quiet, have no moving parts, can be installed quickly and easily, need little maintenance other than occasional washing to prevent dirt from blocking the sun's rays, and should last 20 to 30 years if encased in glass or plastic. Most are made from silicon, the second most abundant element in the earth's crust. They do not produce carbon dioxide, air and water pollution during operation is low, air pollution from manufacture is low, and land disturbance is very low for roof-mounted systems.

However, there are some drawbacks. Present costs of solar cell systems are high but are projected to become competitive in 7 to 15 years. The net useful energy yield is moderate to low. Depending on design, the widespread use of solar cells may be limited eventually by supplies of expensive or rare elements such as gallium and cadmium used to produce types of cells currently considered to be the most efficient. Their widespread use could also cause economic disruption from the bankruptcy of utilities with unneeded large-scale power plants. Without effective control, solar cell manufacture produces moderate water pollution from hazardous chemical wastes.

17-3 INDIRECT PERPETUAL SOLAR ENERGY: PRODUCING ELECTRICITY FROM FALLING AND FLOWING WATER

Types of Hydroelectric Power Since the 1700s the kinetic energy in the falling and flowing water of rivers and streams has been used to produce electricity in small- and large-scale hydroelectric plants. *Large-scale hydropower projects* tap into the kinetic energy of falling water. High dams are built across large rivers to create large reservoirs. The stored water is then allowed to flow at controlled rates, spinning turbines and producing electricity as it falls downward to the river below the dam. Although based indirectly on the perpetual solar energy that drives the hydrologic cycle, all large hydroelectric dams have finite lives because the reservoirs usually fill with silt and become useless in 30 to 300 years, depending on the rate of natural and human-accelerated soil erosion from land above the dam.

Small-scale hydropower projects tap into the kinetic energy of flowing water. A low dam is built across a small river or stream with no reservoir or only a small one behind the dam, and natural water flow is used to generate electricity. However, electricity production can vary with seasonal changes in stream flow and under drought conditions.

Falling water can also be used to produce electricity in *pumped-storage hydropower systems*, primarily to provide supplemental power during times of peak electrical demand. When electricity demand is low, usually at night, electricity from a conventional power plant is used to pump water uphill from a lake or reservoir to a specially built reservoir at a higher elevation, usually on a mountain. When a power company temporarily needs more electricity than its conventional plants can produce, water in the upper reservoir is released and passed through turbines to generate electricity as it returns to the lower reservoir.

Present and Future Use In 1985 hydropower supplied about one-fourth of the world's electricity—twice that from nuclear power—and almost 7% of the world's total commercial energy. Countries or areas with mountainous regions have the greatest hydropower potential. Hydropower supplies Norway and several countries in Africa with essentially all their electricity, Switzerland 74%, and Austria 67%. Canada gets over 70% of its electricity from hydropower and exports electricity to the United States.

In 1985, LDCs got about 42% of their electricity from hydropower, and their total hydropower capacity is expected to double by 1990. China, with one-tenth of the world's hydropower potential, is likely to become the world's largest producer of hydroelectricity. Work has begun on a hydropower dam across the Yangtze River that will be capable of producing electricity equal to that from 25 large nuclear or coal-fired power plants. This project, however, will force 2 million people to leave their homes.

The United States is the world's largest producer of electricity from hydropower. Hydropower produced at almost 1,600 sites provided 14% of the electricity and about 5% of the total commercial energy used by the United States in 1984. U.S. hydroelectric power plants produce electricity cheaper than any other source. One reason is that most large-scale projects were built from the 1930s to the 1950s, when costs were low. In addition, hydroelectric energy efficiency is high (83% to 93%), plants produce full power 95% of the time (compared to 55% for nuclear plants and 65% for coal plants), and have life spans two to ten times those of coal and nuclear plants. As a result, regions such as the Pacific Northwest, where most of the electricity can be produced by hydropower, enjoy the lowest electric rates in the country.

Despite these important advantages, it is projected that by 2000 hydroelectric power will be supplying only about 5% of the commercial energy used in the United States—the same percentage as in 1985. High construction costs for new large-scale dams is one factor hindering further development. In addition, most suitable sites have already been used, are located in areas far from where electricity is needed so that transmission costs are high, or are located on rivers protected from development by federal and state laws.

In 1983 the U.S. Army Corps of Engineers identified 1,407 abandoned small and medium-size hydroelectric dams that could be retrofitted to supply electricity equal to that from 19 large nuclear or coal power plants, as well as 541 currently undeveloped small to medium hydropower sites, which could supply electricity equivalent to that of 26 large nuclear or coal plants. Rehabilitating existing dams does not require any new technology and has little environmental impact. Once rebuilt, such units have a long life, need minimal operating crews, require little maintenance, and produce electricity at a cost no higher than that of coal and nuclear power.

By 1986 nearly 200 retrofitted small hydroelectric sites were generating power equivalent to that from two large-scale coal or nuclear power plants. Since then, however, development of new projects has fallen sharply because of low oil prices, loss of federal tax credits, and growing opposition by local residents and environmentalists. These groups contend that by reducing stream flow, small hydroelectric projects threaten recreational activities and animal life, disrupt scenic rivers, destroy wetlands, and restrict fish movement. Environmentalists also argue that most of the electricity produced by these projects can be obtained at a lower cost and with less environmental disruption by industrial cogeneration, energy conservation, and importing more hydroelectricity from Canada.

Advantages and Disadvantages of Hydropower

Hydropower has a number of advantages. Many LDCs have large, untapped potential sites, although many are remote from points of use. Hydropower has a moderate to high net useful energy yield and fairly low operating and maintenance costs. Plants rarely need to be shut down, and they produce no emissions of carbon dioxide or other air pollutants during operation. Large dams provide some degree of flood control and a regulated flow of irrigation water for areas below the dam.

There are some drawbacks, however. Construction costs for new large-scale systems are high, and there is a lack of suitable sites for large-scale projects in the United States and Europe. Large-scale projects flood large areas of land to form reservoirs, decrease the natural fertilization of prime agricultural land in river valleys below the dam, and lead to a decline in fishing below the dam (Section 10-4). Without proper land-use control, large-scale projects can also result in greatly increased soil erosion and sediment water pollution near the reservoir above the dam, reducing the effective life of the reservoir. Small-scale projects can disrupt river flows, valleys, and wetlands and restrict fish movement.

Tidal Power

Tidal Power   Another potential source of energy for producing electricity from flowing water is the daily oscillation of ocean water levels as a result of gravitational attraction among the earth, moon, and sun. Twice a day a large volume of water flows in and out of inland bays or other bodies of water near the coast to produce high and low tides. If a bay has an opening narrow enough to be obstructed by a dam with gates that can be opened and closed, and if there is a large difference in water height between high and low tides, the kinetic energy in the daily tidal flows can be used to spin turbines to produce electricity.

Unfortunately, suitable sites occur only at about two dozen places in the world. Since 1968 a small 160-megawatt commercial tidal power plant has been in operation on the north coast of France near St. Malo on the Rance River, which has tides up to 13.5 meters (44 feet), and another small plant is in operation in the Soviet Union. Although operating costs are fairly low, the French project cost about 2.5 times more than a conventional hydroelectric power plant station with the same output built further up the Rance River.

Since 1984, Canada has been operating a 20-megawatt experimental tidal power plant at Annapolis Royal, Nova Scotia, on the Bay of Fundy, which has the largest tidal fluctuation in the world—16 meters (52 feet). If this project is successful, the plant may be expanded in the future. Two possible locations for experimental tidal power stations in the United States are Cook Inlet in Alaska and Passamaquoddy Bay on the Maine coast.

Advantages of tidal energy for producing electricity include a free source of energy (tides), low operating costs, moderate net useful energy yield, low air pollution, no addition of carbon dioxide to the atmosphere, and little land disturbance. However, most analysts expect tidal power to make only a small contribution to world electricity supplies because of the lack of suitable sites, high construction costs, irregular output that varies daily, and extensive seawater corrosion and possible storm damage to dams and power plants.

Wave Power   The kinetic energy in ocean waves, created primarily by wind, is another potential source of energy for producing electricity from moving water. Scientists in Japan, Norway, and France are exploring ways to harness this form of hydropower. So far none of these experiments has led to the production of electricity at an affordable price. In addition, sites with sufficient wave heights are limited, electrical output varies with differences in wave height, construction and operating costs are high, net useful energy yield is low, and equipment is damaged or destroyed by saltwater corrosion and severe storms. In 1986 the United Kingdom abandoned its research on ocean wave power, and most analysts expect this alternative to make little contribution to world electricity production.

17-4 INDIRECT PERPETUAL SOLAR ENERGY: PRODUCING ELECTRICITY FROM HEAT STORED IN WATER

Ocean Thermal Energy Conversion   Ocean water stores immense amounts of heat from the sun. Experiments are under way to use the large temperature differences between the cold bottom waters and the sun-warmed surface waters of tropical oceans to produce electricity. This would be done by a gigantic floating **ocean thermal energy conversion (OTEC)** power plant anchored in suitable tropical ocean areas no more than 80 kilometers (50 miles) offshore (Figure 17-11). Some 62 countries, mostly LDCs in South America and Africa, have suitable sites for such plants. Favorable U.S. sites include portions of the Gulf of Mexico and offshore areas near southern California and the islands of Puerto Rico, Hawaii, and Guam.

In a typical plant, warm surface water would be pumped through a large heat exchanger and used to evaporate and pressurize a low-boiling fluid such as liquid ammonia. The pressurized ammonia gas would drive turbines to generate electricity. Then, cold bottom water as deep as 900 meters (3,000 feet) below the plant would be pumped to the surface through massive pipes 30 meters (100 feet) in diameter and used to cool and condense the ammonia back to the liquid state to begin the cycle again (Figure 17-11).

The pumps in a moderate-size 250-megawatt plant would have to be capable of pumping more water each second than the average flow rate of the Mississippi River; the electricity used by these pumps would reduce the net useful energy yield for the system by one-third. A large cable would transmit the electricity to shore. Other possibilities include using the electricity produced to desalinate ocean water, extract minerals and chemicals from the sea, or decompose water to produce hydrogen gas, which could be piped or transported to shore for use as a fuel. Japan and the United States have been conducting experiments to evaluate the technological and economic feasibility of using this energy resource to produce electricity.

Advantages and Disadvantages of OTEC   The OTEC approach has several advantages. The source of energy is free and perpetual at suitable sites, and there is no need for a costly energy storage and backup system. No air pollution is produced during operation, and the floating power plants require no land area. Nutrients brought up when water is pumped from the ocean bottom might be used to nourish schools of fish and shellfish. Advocates believe that with enough research and development funding, large-scale OTEC plants could be built to produce electricity equivalent to that from ten 1,000-megawatt coal or nuclear power plants by the year 2000.

However, many energy analysts believe that large-scale extraction of energy from ocean thermal gradients may never compete economically with other energy alternatives because of high construction costs (two to three times those of comparable coal-fired plants), high operating and maintenance costs as a result of seawater corrosion of metal parts and fouling of heat exchangers by algae and barnacles, a low net useful energy yield (energy efficiency is only about 3%), the limited number of suitable sites, the potential for severe damage from hurricanes and typhoons, the release of dissolved carbon dioxide gas into the atmosphere when large volumes of deep ocean water are pumped to the surface, and possible disruption of aquatic life.

Inland Solar Ponds   A **solar pond** is a solar energy collector consisting of at least 0.5 hectare (1 acre) of relatively shallow saline water or of fresh water enclosed in large black plastic bags. *Saline solar ponds* can be used to produce electricity and are usually located near inland saline seas or lakes near deserts with ample sunlight. The bottom layer of water in such ponds remains on the bottom when heated because it has a higher salinity and density (mass per unit volume) than the top layer. Heat accumulated during daylight in the bottom layer can be used to produce electricity

Figure 17-11 Possible design of a large-scale ocean thermal electric plant (OTEC) for generating electricity from the temperature gradient in a tropical area of ocean.

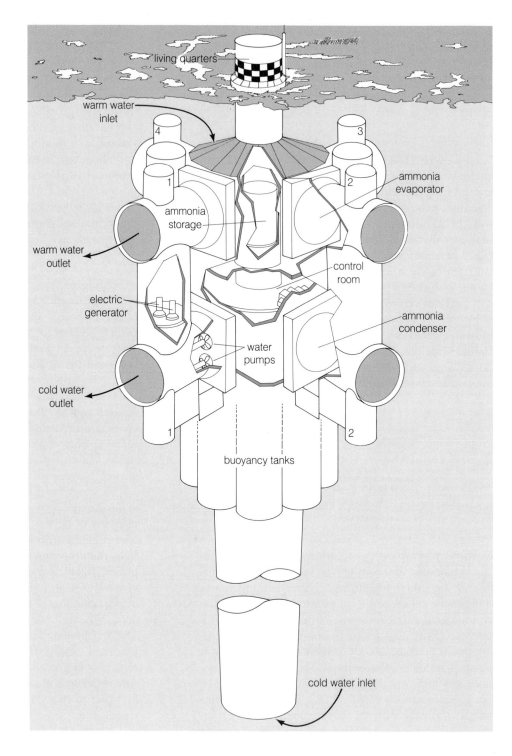

in a manner similar to that used to tap into the temperature difference between bottom and surface waters in a tropical ocean (Figure 17-11).

An experimental saline water solar pond power plant on the Israeli side of the Dead Sea has been operating successfully for several years. By 2000, Israel plans to build a cluster of plants around the Dead Sea to provide most of the electricity it needs for air conditioning and desalinating water. By 1986, more than a dozen experimental saline water solar ponds had

been built in the United States, mostly in desert areas near the Salton Sea in California and the Great Salt Lake in Utah.

Freshwater solar ponds can be used as a source of hot water and space heating. A very shallow hole is dug, lined with concrete, and covered with insulation. A series of large, black plastic bags, each filled with several inches of water, is placed in the hole and covered with fiberglass-reinforced, transparent panels to allow sunlight penetration and to prevent most of the heat stored in the water during daylight from being lost to the atmosphere. A computer-controlled monitoring system determines when water in the bags reaches its peak temperature in the afternoon and turns on pumps to transfer the hot water to large, insulated tanks for distribution as hot water or for space heating. The world's largest freshwater solar pond went into operation at Fort Benning, Georgia, in 1985. This $4 million 11-acre pond provides hot water for 6,500 service personnel and is expected to save up to $10 million over a 20-year period.

Advantages and Disadvantages of Solar Ponds Saline and freshwater solar ponds have the same major advantages as OTEC systems. In addition, they have a moderate net useful energy yield, moderate construction and operating costs, and fairly low maintenance. Freshwater solar ponds can be built in almost any sunny area and may be particularly useful for supplying hot water and space heating for large buildings and small housing developments. Enthusiasts project that with adequate research and development support, solar ponds could supply 3% to 4% of U.S. commercial energy by the year 2000.

However, saline solar ponds are feasible only in areas with ample sunlight, usually ecologically fragile deserts. Operating costs can be high because of extensive saltwater corrosion of pipes and heat exchangers. Unless lined, the ponds can become ineffective when various compounds leach from bottom sediment, darken the water, and reduce sunlight transmission. Freshwater solar ponds require large land areas and are too expensive for providing hot water and space heating for individual homes.

17-5 INDIRECT PERPETUAL SOLAR ENERGY: PRODUCING ELECTRICITY FROM WIND

Wind Power: Past and Present Since the 1600s, prevailing winds, produced indirectly by solar energy, have been harnessed to propel ships, grind grain, pump water, and power many small industrial shops. In the 1700s settlers in the American West used wind to pump groundwater for farms and ranches; in the 1930s and 1940s, small farms beyond the reach of electric utility lines obtained electricity from small wind turbines. By the 1950s, cheap hydropower, fossil fuels, and rural electrification had replaced most wind turbines.

Since the 1970s an array of small to large modern wind turbines, usually consisting of a blade or other device that spins and a generator mounted on a tower, have been developed. Experience has shown that these machines can produce electricity at a reasonable cost for use by small communities and large utility companies in areas with average wind speeds of at least 10 miles per hour and ideally from 14 to 24 miles per hour. Many parts of the continental United States have such winds. Hawaii and parts of Alaska also have very favorable winds.

Small (10 to 100 kilowatt) and intermediate-size (200 to 1,000 kilowatt) wind turbines are the most widely used because they are easier to mass-produce, are less vulnerable to stress and breakdown, and can produce more power in light winds and thus remain in operation longer than large turbines. They are also easier to locate close to the ultimate users, thus reducing electricity transmission costs and energy loss.

Use of wind power in the United States has grown more rapidly since 1981 than any other new source of electricity. Starting near zero in 1981, nearly 13,000 wind machines with a total generating capacity equal to a 1,000-megawatt coal or nuclear plant were in operaton by the end of 1985—enough to meet the electricity needs of 200,000 homes. Most of these were installed by small private companies in California, where 90% of all U.S. wind power electricity is produced, in **wind farms** consisting of clusters of small to intermediate-size wind turbines in windy mountain passes (Figure 17-12). Wind farms can be put into operation within two years and are connected to existing utility lines.

More wind development has taken place in California than the rest of the world because of favorable wind conditions not far from urban areas, high prices paid by utilities for electricity produced by private companies, and very favorable state and federal tax credits (which by 1986 were phased out). The California Energy Commission projects that by 1990 wind power produced by private companies and utility companies will be the state's second least expensive source of electricity—next to hydropower—and will produce 8% of the state's electricity by 2000. However, some of this projected capacity may not be realized because of the elimination of federal and state tax credits.

A smaller number of wind farms, producing a total of about 100 megawatts of electricity, had been installed in other parts of the United States by 1985, especially in the Pacific Northwest, the northern Great Plains, the Northeast, and Hawaii. By 1985 the island of Hawaii obtained about 7% of its electricity from wind and expects to obtain this amount for the entire state by 2000. Other countries planning to make increasing use of wind energy include Denmark (with over 2,000 machines installed by 1985), Canada, Argentina, the United Kingdom, Sweden, West Germany, Australia, the Netherlands, and the Soviet Union.

A 1980 study by the Solar Energy Research Institute indicated that 3.8 million homes and hundreds of

Figure 17-12 A California wind farm consisting of an array of modern wind turbines in a windy mountain pass.

George Gerster/Photo Researchers, Inc.

thousands of farms in the United States are located in areas with sufficient wind speeds to make economical use of small wind generators to provide all or most of their electricity. However, a small, 5-kilowatt wind turbine costs $8,000 to $20,000, including installation, and often needs repairs requiring parts not readily available. Homeowners would have to add an expensive battery, pumped-storage, flywheel or other system to store energy for use when the wind dies down. A cheaper alternative is to use the power company as a backup and sell it surplus electricity, which utilities are currently required by federal law to buy (although prices vary and many utilities throw up roadblocks). At present, it makes more sense economically to build wind farms to serve entire communities or groups of homeowners than for most individual homeowners to install wind power systems.

Wind Power in the Future Wind power experts project that with a vigorous development program, wind energy could provide at least 5% and perhaps as much as 10% to 19% of the projected demand for electricity in the United States and as much as 13% of the world's electricity by the end of the century. Many analysts believe that a key to increasing the use of wind power in the United States is reinstatement (for at least five years) of federal and state tax credits, which reduced the high initial cost of wind installations by 40% to 70%. This would help increase demand and allow small and financially insecure U.S. wind turbine companies to invest in mass production and market-

ing techniques needed to bring the price down to the point where tax credits would no longer be needed.

Advantages and Disadvantages of Wind Power
Wind power has numerous important advantages. It is a perpetual source of energy at favorable sites, and large wind farms with low material requirements can be built within two years. Wind power systems have a moderate net useful energy yield, do not emit carbon dioxide or other air pollutants during operation, and have no cooling water requirements. Their manufacture and use produce little water pollution. The moderate amount of land occupied by wind farms can be used at the same time for grazing and other purposes. Wind farms are projected to have an economic advantage over coal and nuclear power plants in the United States and the world by the 1990s (Figure 16-12, p. 283). This projection, however, assumes that the wind industry can use tax credits to increase demand and can obtain adequate financing to allow construction of mass-production facilities.

Wind power has some disadvantages. It can only be used in areas with sufficient winds, and requires backup electricity from a utility company or from a fairly expensive energy storage system when the wind dies down. Present systems have moderate to high initial costs and high operating costs, and operate at full capacity only about 23% of the time because of variations in wind availability and speed. The industry expects to achieve 35% capacity with improved designs, but this is still well below the 55% capacity

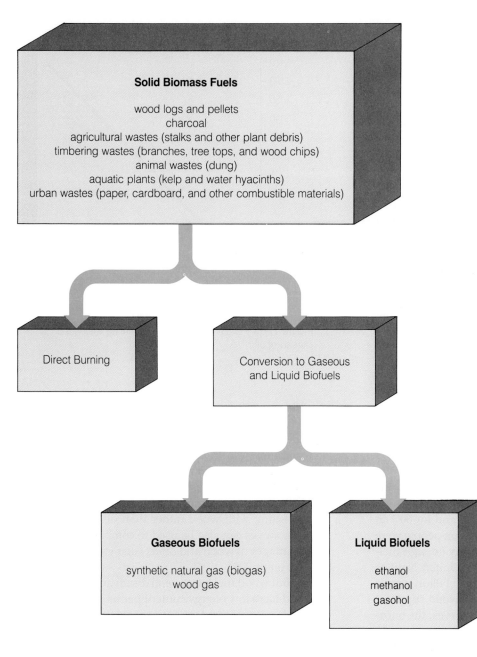

Figure 17-13 Major types of biomass fuel.

Solid Biomass Fuels

wood logs and pellets
charcoal
agricultural wastes (stalks and other plant debris)
timbering wastes (branches, tree tops, and wood chips)
animal wastes (dung)
aquatic plants (kelp and water hyacinths)
urban wastes (paper, cardboard, and other combustible materials)

Direct Burning

Conversion to Gaseous
and Liquid Biofuels

Gaseous Biofuels

synthetic natural gas (biogas)
wood gas

Liquid Biofuels

ethanol
methanol
gasohol

of nuclear power plants and the 65% capacity of coal-fired plants.

Widespread use of wind turbines in the heavily populated Northeast, where the best sites are along heavily used coasts and highly visible mountain ridges, could cause unacceptable visual pollution. Some wind experts have proposed that this problem could be overcome by building large floating wind farms (such as one in operation in Denmark) off the eastern coast to make use of strong offshore winds to produce electricity or hydrogen gas for transmission to shore. Excessive noise and interference with local television reception have been problems with large turbines but can be overcome with improved design and use in isolated areas. Large wind farms might also interfere with flight patterns of migratory birds in certain areas.

17-6 INDIRECT RENEWABLE SOLAR ENERGY: BIOMASS

Renewable Biomass as a Versatile Fuel Produced by solar energy through the process of photosynthesis, **biomass fuel** is organic plant matter that can be burned directly as a solid fuel or converted to a more convenient gaseous or liquid *biofuel* by processes such as distillation and pyrolysis (heating in the absence of air) (Figure 17-13). In 1984, biomass, mostly from the direct burning of wood and animal wastes to heat buildings and cook food, supplied about 11% of the world's commercial energy (4% to 5% in Canada and the United States). By 2000, biomass (including biofuels) is projected to provide 15% of the world's commercial energy and 11% to 20% of the commercial energy in the United States.

All biomass fuels have several advantages in common. They can be used to provide a variety of solid, liquid, and gaseous fuels for space heating, water heating, producing electricity, and propelling vehicles. Biomass is a renewable energy resource as long as trees and plants are not harvested faster than they grow back. There is no net increase in atmospheric levels of carbon dioxide as long as the rate of removal and burning of trees (which releases CO_2) and plants does not exceed their rate of replenishment (living plants remove CO_2 from the air). Burning of biomass fuels adds much less sulfur dioxide and nitric oxide to the atmosphere per unit of energy produced than the uncontrolled burning of coal and thus requires fewer pollution controls.

Biomass fuels also share some disadvantages. Without effective land-use controls and replanting, extensive removal of trees and plants can deplete soil nutrients and cause excessive soil erosion, water pollution, flooding, and loss of wildlife habitat. Biomass resources also have a high moisture content (15% to 95%), which reduces their net useful energy. The added weight makes collection and transport expensive. Each type of biomass fuel has additional specific advantages and disadvantages.

Burning Wood and Wood Wastes In 1985 wood was the primary source of energy for cooking and heating for one-half of the world's population and for 80% in LDCs. However, in 1985 about 1.1 billion people were either unable to find or too poor to buy enough fuelwood to meet minimum needs. The United Nations projects that by the year 2000 about 2.5 billion people will live in areas with inadequate fuelwood supplies.

In MDCs with adequate forest reserves, the burning of wood and wood wastes to heat homes and to produce steam and electricity in industrial boilers has increased rapidly because of price increases in heating oil and electricity. Sweden leads the world in using wood as an energy source, mostly for district heating plants. In 1985 the forest products industry (mostly paper companies and lumber mills) accounted for almost two-thirds of U.S. wood energy consumption, with homes and small businesses burning the rest.

By 1985, one single-family home in nine (one in six in nonmetropolitan areas) in the United States relied entirely on wood for heating—about the same percentage that relied on oil for heating. The percentage of homes heated with wood in different regions of the United States varies; the greatest amount of use is in New England, where wood is plentiful and expensive oil is the primary heating fuel.

Burning wood and wood wastes has several specific advantages in addition to those associated with

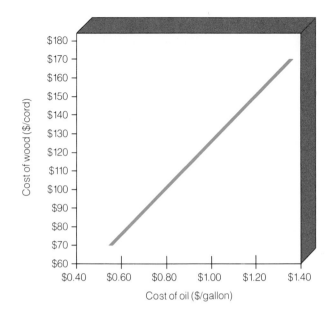

Figure 17-14 Break-even prices for heating with wood instead of oil.

all biomass fuels. Wood has a moderate to high net useful energy yield when collected and burned directly and efficiently near its source. Three-fourths of the wood fuel used in U.S. residences is cut or collected by people, mostly in rural areas, who have access to wood at little or no cost. Figure 17-14 shows the prices at which a homeowner breaks even or saves money by heating with wood rather than oil. Solar-assisted water stoves can also be used to combine some of the advantages of wood and direct solar energy (see Spotlight on p. 309).

But wood fuel has several disadvantages. Because wood has a low energy content per unit of weight and a high water content, it is expensive to harvest, transport, store, and burn. As a result, in urban areas where wood must be hauled long distances, it can cost homeowners more per unit of energy produced than oil and electricity. A considerable number of accidents occur as a result of using wood as a fuel. In 1980, there were an estimated 123,000 chainsaw accidents that required medical attention and 22,000 house fires and 350 deaths caused by wood stoves in the United States.

Residential wood burning also leads to greatly increased outdoor and indoor air pollution, causing as many as 820 cancer deaths a year in the United States, according to the EPA. The estimated 12 million wood stoves in the United States account for over 15% of the nation's emissions of solid particulate matter. They also emit large quantities of carbon monoxide and polycyclic organic matter (unburnt residues containing cancer-causing substances).

This air pollution can be reduced 75% by a $100 to $250 catalytic combustor (a honeycombed ceramic

One way to avoid most of the problems associated with a conventional wood stove is to use a *solar-assisted water stove* (Figure 17-15). Several roof-mounted active solar collectors are used to heat water stored in a large, well-insulated 250 to 1,000 gallon steel tank (depending on the size of the space to be heated), which has a built-in firebox. Wood, paper, coal, or any combustible material is burned in the firebox in very cold weather and at night to supplement the stored solar energy.

The system is placed outdoors in an open lean-to or small enclosed shed, thus eliminating the indoor smoke and soot associated with conventional wood stoves and further increasing the safety of the system. A catalytic converter can be added

to reduce outdoor air pollution. A pump connected to the storage tank circulates the heated water through underground insulated pipes to a heat exchanger (much like the radiator in a car) before the water is returned to the tank. The exchanger transfers heat in the water to air, which is blown by a fan through ducts in the building to be heated. With a conventional thermostat to control indoor temperature, the system eliminates the uneven heating provided by a conventional wood stove.

A separate copper coil placed in the firebox and a second pump allow the stove to meet all hot water needs. Additional coils and pumps can be used to heat a swimming pool, hot tub, sauna, or greenhouse.

A thermostat-controlled oil or gas system can also be attached to the tank as a backup. Maintenance is simple: Occasionally checking a water-level gauge and turning a valve to add water to the tank to replace water lost by slow evaporation (a minor chore that could be automated) and adding a bottle of rust inhibitor once a year is all that is required.

The installed cost for a 500-gallon system with four solar collectors—sufficient to provide all the hot water and space heating for a typical 1,800-square-foot house—is about $5,000 to $6,000 (less if the system is installed by the user). Such a system should pay for itself in several years and then lead to considerable annual savings.

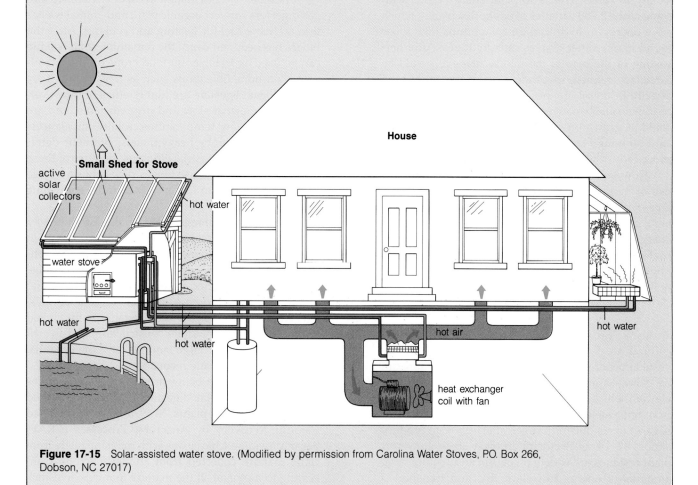

Figure 17-15 Solar-assisted water stove. (Modified by permission from Carolina Water Stoves, P.O. Box 266, Dobson, NC 27017)

chamber containing a chemical that speeds up the burning of emission products). These units also make it more economical to burn wood by increasing the energy efficiency of a typical airtight wood stove from about 55% to as high as 75%, and they reduce the need for chimney cleaning and the chance of chimney fires.

By 1985, Oregon, Colorado, and Montana had passed laws requiring such devices and in some cases forbidding wood burning when particulate matter in the atmosphere reaches certain levels; at least 20 other states were considering such laws. In London, England, and in South Korean cities wood fires have been banned to reduce air pollution. In 1987 the EPA proposed that all wood stoves manufactured after 1988 be required to meet certain air pollution emission standards designed to reduce emissions by 70%.

Energy Plantations Another approach to obtaining plentiful amounts of biomass fuel is to establish large *biomass-energy plantations,* where fast-growing trees, grasses, or other crops would be grown and harvested by automated methods and then burned directly or converted to liquid or gaseous biofuels. Another possibility in some areas is to plant energy farms with some of the 2,000 varieties of euphorbia plants, which store energy in hydrocarbon compounds (like those found in oil) rather than as carbohydrates. After harvesting of the plants, the oil-like material would be extracted, refined to produce gasoline, or burned directly in diesel engines. The unused woody plant residues could be converted to alcohol fuel. Such plants could be grown on semiarid, currently unproductive land, although lack of water might limit the amount produced.

However, energy plantations usually require heavy use of herbicides and pesticides, occupy large land areas, can compete with food crops for prime farmland, and are likely to have low or negative net useful energy yields like conventional crops grown by industrialized agricultural methods. In addition, the economic feasibility of this approach has not been determined.

Burning Agricultural and Urban Wastes In agricultural areas, crop residues (the inedible, unharvested portions of food crops) and animal manure can be collected and burned or converted to biofuels. Hawaii, which plans to produce 90% of its electricity from renewable energy by 2005, was burning enough bagasse (a residue from sugarcane) in 1983 to produce almost 8% of its electricity. In most areas, however, plant residues are widely dispersed and require large amounts of energy to collect, dry, and transport— unless collected along with harvested crops. In addition, ecologists argue that it makes more sense to use

these valuable nutrients to feed livestock, retard soil erosion, and fertilize the soil.

Japan and many European companies have built a number of cogeneration incinerator plants, in which collected urban waste (much of it consisting of paper, cardboard, and plastics) is burned to produce high-temperature steam and electricity, mostly for use by industry or local power companies (Section 20-3). Burning such wastes reduces the volume of urban waste by 90%, extending the lives of existing landfills and reducing the need for new ones. But some analysts argue that more energy is saved by composting or recycling paper and other organic wastes than by burning them.

Converting Solid Biomass to Liquid and Gaseous Biofuels Plants, organic wastes, sewage, and other forms of solid biomass can be converted by bacteria and various chemical processes into gaseous and liquid biofuels such as *biogas* (a gaseous mixture containing about 60% methane and 40% carbon dioxide), *liquid methanol* (methyl, or wood, alcohol), and *liquid ethanol* (ethyl, or grain, alcohol), and other liquid fuels.

In China, bacteria in millions of devices called biogas digesters convert organic plant and animal wastes into methane fuel for heating and cooking. After the biogas has been removed, the remaining solid residue can be used as fertilizer on food crops or, if contaminated, on nonedible crops such as trees. When they work, biogas digesters are highly efficient; however, they are somewhat slow and unpredictable, and are vulnerable to low temperatures, acidity imbalances, and contamination by heavy metals, synthetic detergents, and other industrial effluents. Because of these problems, few MDCs use biogas digestors on a large scale and their use in LDCs such as India and China is declining.

Methane containing biogas is also produced by the underground decomposition of organic matter in the absence of air (anaerobic digestion) in the estimated 17,000 landfill sites around the United States. The gas can be collected simply by inserting pipes into landfills. By 1985, some 40 landfill gas recovery systems (mostly in California) were operating in the United States, 40 others were under construction, and an estimated 2,000 to 3,000 large landfills have the potential for large-scale methane recovery.

Methane is also being produced in some places through the anaerobic digestion of sludge produced at sewage treatment plants. In 1976, a company called Calorific Recovery by Anaerobic Processes (CRAP) began providing Chicagoans with methane made from cattle manure collected from animal feedlots. Converting to methane all the manure that U.S. livestock produce each year could provide nearly 5% of the country's total natural gas consumption at 1985 levels.

But collecting and transporting manure for long distances would require a large energy input. Recycling this manure to the land to replace artificial fertilizer, which requires large amounts of natural gas to produce, would probably save more natural gas.

Anticipating the depletion of crude oil supplies, the world must find a liquid fuel substitute for gasoline and diesel fuel. Some analysts see the biofuels methanol and ethanol as the answer to this problem, since both alcohols can be burned directly as fuel without requiring additives to boost octane ratings.

Currently most of the emphasis is on ethanol as an automotive fuel. Ethanol can be produced from a variety of sugar and grain crops (sugarcane, sugar beets, sorghum, and corn) by the well-established process of fermentation and distillation used in making alcoholic beverages. Pure ethanol can be burned in today's cars with little engine modification. Gasoline can also be mixed with 10% to 23% ethanol to make *gasohol*, now known as a form of super unleaded or ethanol-enriched gasoline, which burns in conventional gasoline engines.

By 1984, ethanol accounted for 43% of automotive fuel consumption in Brazil, where more than 1.2 million of the country's 10 million cars ran on pure ethanol and the remaining 8.8 million cars used an unleaded gasoline mixture containing 23% ethanol produced from sugarcane. This ambitious program helped Brazil cut its oil imports in half between 1978 and 1984. The country plans to triple its production of ethanol fuel by 1993, but this may be delayed by the temporary oil glut of the 1980s, which has reduced conventional gasoline prices.

By 1986, super unleaded gasoline containing 90% gasoline and 10% ethanol accounted for 7% of gasoline sales in the United States. Most is produced from corn in 150 ethanol production facilities built between 1980 and 1985. Ethanol-enriched gasoline is currently exempt from the federal gasoline tax and from varying amounts of state gasoline taxes in at least 30 states in order to stimulate the development of this industry. Despite these and other tax benefits (amounting to almost $1 a gallon), ethanol-enriched gasoline costs between $1.15 and $1.50 a gallon. Until gasoline prices rise again, the use of this form of unleaded gasoline could level off. Sales, however, could increase when leaded gasoline is banned in the United States (supposedly scheduled to occur after 1988). New, energy-efficient distilleries are reducing the costs of producing ethanol, and soon this fuel may be able to compete with other forms of unleaded gasoline without federal tax breaks (scheduled to expire in 1992).

The distillation process used to produce ethanol produces large volumes of a toxic waste material known as "swill," which if allowed to flow into waterways would kill algae, fish, and plants. Another problem is that the net useful energy yield for producing ethanol for use as a fuel is moderate for recently completed distilleries using modern technology and powered by coal, wood, or solar energy, low if the tractor fuel used in growing grain for conversion to ethanol is included, and zero or negative when produced in older oil- or natural-gas-fueled distilleries.

Another alcohol—methanol—can be produced from wood, wood wastes, agricultural wastes, sewage sludge, garbage, coal, and natural gas. High concentrations of methanol corrode conventional engines, but in a properly modified engine burn cleanly without any problems. The use of pure methanol as a fuel, however, will not be economically feasible until gasoline prices reach about $3 or $4 a gallon—already a reality in many European countries, where gasoline taxes are high. *Diesohol*, a mixture of diesel fuel with 15% to 20% methanol by volume, is being tested and could lower emissions of nitrogen oxide pollutants, a drawback of regular diesel fuel.

17-7 HYDROGEN AS A POSSIBLE REPLACEMENT FOR OIL

Some scientists have suggested the use of hydrogen gas (H_2) to fuel cars, heat homes, and provide hot water when oil and natural gas run out. Although hydrogen gas does not occur in significant quantities in nature, it can be produced by chemical processes from nonrenewable coal or natural gas or by using heat, electricity, or perhaps sunlight to decompose fresh water or some of the world's massive supply of seawater (Figure 17-16).

Hydrogen gas can be burned in a reaction with oxygen gas in a power plant, a specially designed automobile engine, or in a fuel cell that converts the chemical energy produced by the reaction into direct-current electricity. Hydrogen burns cleanly in pure oxygen, yielding only water vapor and no air pollutants. In addition, hydrogen can be combined with various metals to produce solid compounds that can be heated to release hydrogen on demand in a small automobile fuel-generating system.

The major problem with using hydrogen as a fuel is that only trace amounts of the gas occur in nature. Thus it must be produced with energy from another source such as nuclear fission, direct solar, or wind. Depending on the source of energy used to decompose water, this raises the cost. Because of the first and second energy laws, hydrogen production by any method will require more energy to produce it than is released when it is burned. Thus its net useful energy yield will always be negative so that its widespread use depends on an ample and affordable supply of some other type of energy. Another problem is that hydrogen gas is highly explosive. However, most ana-

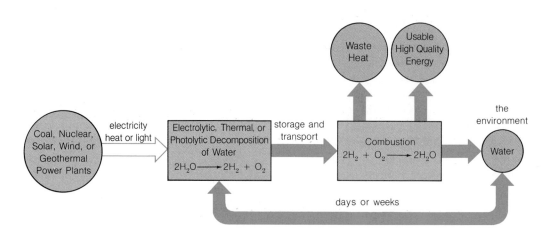

Figure 17-16 The hydrogen energy cycle. Hydrogen gas must be produced by using electricity, heat, or solar energy to decompose water, thus leading to a negative net useful energy yield.

lysts believe we could learn how to handle it safely, as we have for highly explosive gasoline and natural gas.

Although burning hydrogen does not add carbon dioxide to the atmosphere, carbon dioxide would be added if electricity from coal-burning or other fossil fuel-burning power plants were used to decompose water. No carbon dioxide would be added if direct or indirect solar energy or nuclear power were used. Scientists are trying to develop special cells that use ordinary light or solar energy to split water molecules into hydrogen and oxygen gases with reasonable efficiency. But even if affordable materials are used and reasonable efficiencies are obtained, it may be difficult and expensive to develop large-scale commercial cells for producing hydrogen gas. According to the most optimistic projections, affordable commercial cells for using solar energy to produce hydrogen will not be available until after 2000.

17-8 DEVELOPING AN ENERGY STRATEGY FOR THE UNITED STATES

Overall Evaluation of U.S. Energy Alternatives
Table 17-2 summarizes the major advantages and disadvantages of the energy alternatives discussed in this and the two preceding chapters, with emphasis on their potential in the United States. Energy experts argue over these and other futuristic projections, and new data may affect some information in this table. But it does provide a useful framework for making decisions based on currently available information. Four major conclusions can be drawn:

1. The best short-term, intermediate, and long-term alternative for the United States and other countries is to reduce unnecessary energy waste by reducing energy consumption and improving the efficiency of energy use.

2. Total systems for future energy alternatives in the world and in the United States will probably have low to moderate net useful energy yields and moderate to high development costs. Since enough financial capital may not be available to develop all alternative energy systems, projects must be carefully chosen so that capital will not be depleted on systems that yield too little net useful energy or prove to be economically or environmentally unacceptable.

3. Instead of depending primarily on one nonrenewable energy resource like oil, the world and the United States probably must shift to much greater dependence on improving energy efficiency and a combination of perpetual and renewable energy sources over the 50 years it typically takes to develop and phase in new energy resources.

4. As improvements in energy efficiency are made and dependence on perpetual and renewable energy resources is increased, commercial energy production will become more localized and variable, depending on local climatic conditions and availability of usable energy resources.

Economics and National Energy Strategy   Cost is the major factor in determining which commercial energy resources are widely used by consumers. Governments throughout the world use three major economic and political strategies to stimulate or dampen the short- and long-term use of a particular energy resource: (1) *not attempting to control prices,* so that its use depends on open, free market competition (assuming all other alternatives also compete in the same way); (2) *keeping prices artificially low* to encourage its use and development; and (3) *keeping prices artificially high* to discourage its use and development. Each approach has certain advantages and disadvantages. Although effective energy strategies for a particular country should involve a delicately balanced combination of these approaches, most countries place primary emphasis on one.

Free Market Competition Leaving it to the marketplace without any government interference is appealing in principle. However, it rarely exists in practice because business leaders are in favor of it for everyone but their own company. Most energy industry executives work hard to achieve control of supply, demand, and prices for their particular energy resources while urging free market competition for any competing energy resources. They try to influence elected officials and help elect those who will give their business the most favorable tax breaks and other government subsidies. This distorts and unbalances the marketplace.

An equally serious problem with the open marketplace is its emphasis on today's prices to enhance short-term economic gain. This greatly inhibits the long-term development of new energy resources, which can rarely compete effectively in their initial development stages without government-supported research and development and economic subsidies.

Keeping Energy Prices Artificially Low: The U.S. Strategy Many governments provide tax breaks and other subsidies, underwrite expensive long-term research and development, and use price controls to maintain artificially low prices for a particular energy resource. This is the main approach in the United States and in the Soviet Union (where all resources and means of production—and thus prices—are controlled by the central government).

This approach encourages the development and use of energy resources receiving favorable treatment, helps protect consumers (especially the poor) from sharp price increases, can help reduce inflation, and often helps the reelection chances of leaders in democratic societies. At the same time, however, this approach encourages waste and rapid depletion of an energy resource (such as oil) by making its price lower than it should be relative to long-term supply. This strategy discourages development of new energy alternatives not receiving at least the same level of

subsidies and price control. Once energy industries such as the fossil fuel and nuclear power industries receive government subsidies, they usually have the power to maintain this support long after it becomes unnecessary and they often fight efforts to provide equal or higher subsidies for development of new energy alternatives.

In 1984 federal tax breaks and other subsidies for development of energy conservation and perpetual solar-based energy resources in the United States amounted to $1.7 billion; the tax breaks involved were eliminated a year later. In contrast, during 1984 the nuclear power industry received $15.6 billion, the oil industry $8.6 billion, the natural gas industry $4.6 billion, and the coal industry $3.4 billion in federal tax breaks and subsidies, and such subsidies have not been eliminated.

Although energy is the lifeblood of the U.S. economy, 49% of the $4.9 billion allocated for research and development in the Department of Energy (DOE) was apportioned to military weapons programs. Meanwhile, allocations for nondefense energy research and development dropped from 65% to 35% of the DOE budget between 1971 and 1987. About 61% of the remaining nondefense research and development portion of the 1987 DOE budget was allocated to the continued development of nuclear fission and fusion to produce electricity, a form of energy that can only be used for certain purposes. But energy conservation, the country's largest potential source of energy, received only 5% of this portion of the budget, an amount equivalent to about one-fourth the cost of a B-1 bomber. Grants to states and localities for insulation and weatherization of low-income dwellings were virtually eliminated, and direct and indirect solar energy received only 5% of the nondefense research and development budget.

Keeping Energy Prices Artificially High: The Western European Strategy Governments keep the price of a particular energy resource artificially high by withdrawing existing tax breaks and other subsidies or by adding taxes on its use. This approach encourages improvements in energy efficiency, reduces dependence on imported energy, and decreases use of an energy resource (like oil) whose future supply will be limited.

However, such price increases can increase inflation, dampen economic growth, and put a heavy economic burden on the poor unless some of the energy tax revenues are used to help low-income families offset increased energy prices and to stimulate labor-intensive forms of economic growth such as energy conservation. High gasoline and oil import taxes have been imposed by many European governments. This

Table 17-2 Evaluation of Energy Alternatives for the United States (shading indicates favorable conditions)						
Energy Resource	Estimated Availability			Estimated Net Useful Energy of Entire System	Projected Cost of Entire System	Actual or Potential Overall Environmental Impact of Entire System
	Short Term (1989–1999)	Intermediate Term (1999–2009)	Long Term (2009–2039)			
Nonrenewable Resources						
Fossil fuels						
Petroleum	High (with imports)	Moderate (with imports)	Low	High but decreasing	High for new domestic supplies	Moderate
Natural gas	High (with imports)	Moderate (with imports)	Low to moderate	High but decreasing	High for new domestic supplies	Low
Coal	High	High	High	High but decreasing	Moderate but increasing	Very high
Oil shale	Low	Low to moderate	Low to moderate	Low to moderate	Very high	High
Tar sands	Low	Fair? (imports only)	Poor to fair (imports only)	Low	Very high	Moderate to high
Biomass (urban wastes for incineration)	Low	Low	Low	Low to moderate	High	Moderate to high
Synthetic natural gas (SNG) from coal	Low	Low to moderate	Low to moderate	Low to moderate	High	High (increases use of coal)
Synthetic oil and alcohols from coal and organic wastes	Low	Low	Low	Low to moderate	High	High (increases use of coal)
Nuclear energy						
Conventional fission (uranium)	Low to moderate	Low to moderate	Low to moderate	Low to moderate	Very high	Very high
Breeder fission (uranium and thorium)	None	None to low (if developed)	Moderate	Unknown, but probably moderate	Very high	Very high
Fusion (deuterium and tritium)	None	None	None to low (if developed)	Unknown	Very high	Unknown (probably moderate)
Geothermal energy (trapped pockets)	Poor	Poor	Poor	Low to moderate	Moderate to high	Moderate to high
Perpetual and Renewable Resources						
Conservation (improving energy efficiency)	High	High	High	Very high	Low	Decreases impact of other sources

one factor accounts for much lower average energy use per person and greater energy efficiency in these countries than in the United States (Table 17-1, p. 293).

One popular myth is that higher energy prices would cause widespread unemployment. Actually, *low* energy prices increase unemployment because farm-ers and industries find it cheaper to substitute machines run on cheap energy for human labor. On the other hand, *raising* energy prices stimulates employment because building solar collectors, adding insulation, and carrying out other forms of energy conservation are labor-intensive activities.

Energy Resource	Estimated Availability			Estimated Net Useful Energy of Entire System	Projected Cost of Entire System	Actual or Potential Overall Environmental Impact of Entire System
	Short Term (1989–1999)	Intermediate Term (1999–2009)	Long Term (2009–2039)			
Perpetual and Renewable Resources (continued)						
Water power (hydroelectricity)						
New large-scale dams and plants	Low	Low	Very low	Moderate to high	Moderate to very high	Low to moderate
Reopening abandoned small-scale plants	Moderate	Moderate	Low	Moderate to high	Moderate	Low
Tidal energy	None	Very low	Very low	Unknown (moderate)	High	Low to moderate
Ocean thermal gradients	None	Low	Low to moderate (if developed)	Unknown (probably low to moderate)	Probably high	Unknown (probably moderate)
Solar energy						
Low-temperature heating (for homes and water)	Moderate	Moderate to high	High	Moderate to high	Moderate to high	Low
High-temperature heating	Low	Moderate	Moderate to high	Moderate	Very high initially (but probably declining fairly rapidly)	Low to moderate
Photovoltaic production of electricity	Low to moderate	Moderate	High	Moderate	High initially (but declining fairly rapidly)	Low
Wind energy						
Home and neighborhood turbines	Low	Moderate	Moderate to high	Moderate	Moderate to high	Low
Large-scale power plants	None	Very low	Probably low	Low	High	Low to moderate?
Geothermal energy (low heat flow)	Very low	Very low	Low to moderate	Low	High	Moderate to high
Biomass (burning of wood, crop, food, and animal wastes	Moderate	Moderate	Moderate to high	Moderate	Moderate	Moderate to high
Biofuels (alcohols and natural gas from plants and organic wastes)	Low to moderate?	Moderate	Moderate to high	Low to moderate	Moderate to high	Moderate to high
Hydrogen gas (from coal or water)	None	Low	Moderate	Unknown	Unknown	Variable

Why the U.S. Has No Comprehensive Long-Term Energy Strategy After the 1973 oil embargo, Congress was prodded to pass a number of laws (see Appendix 3) to deal with the country's energy problems. Most energy experts agree, however, that these laws do not represent a comprehensive energy strategy. Indeed, analysis of the U.S. political system reveals why the United States has not been able and will probably never be able to develop a coherent energy policy.

One reason is the complexity of energy issues as revealed in this chapter and the two preceding chapters. But the major problem is that the American political process produces laws—not policies—and is not designed to deal with long-term problems. Each law reflects political pressures of the moment and a maze of compromises between competing pressure groups representing industry, environmentalists, and consumers. In addition, a law once passed is difficult to repeal or modify drastically until any undesirable long-term consequences reach crisis proportions.

Taking Energy Matters into Your Own Hands
While elected officials, energy company executives, and environmentalists argue over the key components of a national energy strategy, many individuals have gotten fed up and taken energy matters into their own hands. With or without tax credits, they are insulating, weatherizing, and making other improvements to conserve energy and save money. Some are building new, passively heated and cooled solar homes; others are adding passive solar heating to existing homes.

Similarly, local governments in a growing number of cities are developing their own successful programs to improve energy efficiency and to rely more on locally available energy resources. All of these individual and local initiatives are crucial political and economic actions. Increased and amplified, they can help shape a sane national energy strategy with or without help from federal and state governments.

In the long run, humanity has no choice but to rely on renewable energy. No matter how abundant they seem today, eventually coal and uranium will run out. The choice before us is practical: We simply cannot afford to make more than one energy transition within the next generation.

Daniel Deudney and Christopher Flavin

DISCUSSION TOPICS

1. What are the ten most important things an individual can do to save energy in the home and in transportation (see Appendix 5)? Which, if any, of these do you do? Which, if any, do you plan to do? When?

2. Make an energy use study of your school, and use the findings to develop an energy conservation program.

3. Should the United States institute a crash program to develop solar photovoltaic cells? Explain.

4. Criticize each of the following statements:
   a. The United States can meet essentially all of its future electricity needs by developing solar power plants.
   b. The United States can meet essentially all of its future electricity needs by using direct solar energy to produce electricity in photovoltaic cells.
   c. The United States can meet essentially all of its electricity needs by building new, large hydroelectric plants.
   d. The United States can meet essentially all of its future electricity needs by building ocean thermal electric power plants.
   e. The United States can meet essentially all of its future electricity needs by building a vast array of wind farms.
   f. The United States can meet essentially all of its future electricity needs by building power plants fueled by biomass resources.

5. Give your reasons for agreeing or disagreeing with the following propositions, which have been suggested by various energy analysts:
   a. The United States should cut average energy use per person by at least 50% between 1989 and 2010.
   b. A mandatory energy conservation program should form the basis of any U.S. energy policy.
   c. To solve world and U.S. energy supply problems, all we need to do is recycle some or most of the energy we use.
   d. Federal subsidies for all energy alternatives should be eliminated so that all choices can compete in a true free-enterprise market system.
   e. All government tax breaks and other subsidies for conventional fuels (oil, natural gas, coal), synthetic natural gas and oil, and nuclear power should be removed, and limited subsidies granted for the development of energy conservation and solar, wind, and biomass energy alternatives.
   f. Development of solar and wind energy should be left up to private enterprise without help from the federal government, but nuclear energy should continue to receive federal subsidies.
   g. To solve present and future U.S. energy problems, all we need to do is find more domestic supplies of oil and natural gas and increase our dependence on nuclear power.
   h. The United States should not worry about heavy dependence on foreign oil imports because they improve international relations and prevent the United States from depleting domestic supplies.
   i. A heavy federal tax should be placed on gasoline and imported oil used in the United States.
   j. Between 2000 and 2020 the United States should phase out all nuclear power plants.

6. Explain how a government policy of keeping heating oil, gasoline, and electricity prices artificially low by providing subsidies to fossil fuel and nuclear industries and not imposing higher taxes on gasoline and imported oil can (a) discourage exploration for domestic supplies of fossil fuels; (b) increase or at least not significantly decrease dependence on imported oil; (c) lead to higher-than-necessary unemployment; (d) discourage the development of direct and indirect sources of solar energy; and (e) discourage improvements in energy efficiency.

PART FIVE

Pollution

U.S. Department of Interior, Bureau of Reclamation

Humans of flesh and bone will not be much impressed by the fact that a few of their contemporaries can explore the moon, program their dreams, or use robots as slaves, if the planet Earth has become unfit for everyday life. They will not long continue to be interested in space acrobatics if they have to watch them with their feet deep in garbage and their eyes half-blinded by smog.

René Dubos

18

Air Pollution

GENERAL OBJECTIVES

1. What are the major types and sources of air pollutants?

2. What are industrial smog, photochemical smog, an urban heat island, and acid deposition?

3. What undesirable effects can air pollutants have on people?

4. What undesirable effects can air pollutants have on other species and on materials?

5. What undesirable effects can certain air pollutants have on the ozone layer and global climate?

6. What legal and technological methods can be used to reduce air pollution?

Tomorrow morning when you get up take a nice deep breath. It will make you feel rotten.

Citizens for Clean Air, Inc. (New York)

To stay alive we must inhale about 20,000 liters (21,200 quarts) of air each day. Along with the nitrogen and oxygen gases that make up 99% of the atmosphere, each breath also contains small amounts of other gases, as well as minute droplets of various liquids and tiny particles of a variety of solids known as **particulate matter.** Some of these chemicals come from natural sources, but most come from cars, trucks, power plants, factories, cigarettes, and other sources related to human activities in urban areas. Repeated exposure to even trace amounts of many of these chemicals, known as air pollutants, can damage lung tissue, plants, buildings, metals, and other materials.

18-1 TYPES AND SOURCES OF OUTDOOR AND INDOOR AIR POLLUTION

Our Air Resource: The Atmosphere The atmosphere, a gaseous envelope surrounding the earth, is divided into several zones (Figure 18-1). About 95% of the mass of the air is found in the innermost layer of the atmosphere known as the **troposphere,** extending only 8 to 12 kilometers (5 to 7 miles) above the earth's surface. If the earth were an apple, our vital air supply would be no thicker than the apple's skin.

About 99% of the volume of clean, dry air consists of two gases: nitrogen (78%) and oxygen (21%). The remaining 1% consists of small amounts of other gases such as argon and carbon dioxide. Air also holds water vapor in amounts varying from 0.01% at the frigid poles to 5% in the humid tropics.

Major Types of Outdoor Air Pollutants As clean air moves across the earth's surface, it collects addi-

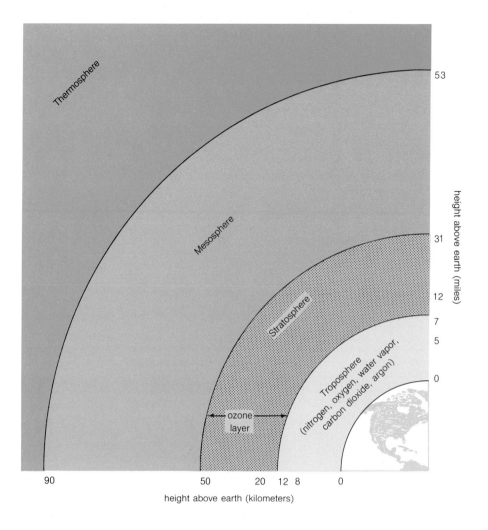

height above earth (miles)

53

31

12

7

5

0

Thermosphere

Mesosphere

Stratosphere

ozone layer

Troposphere (nitrogen, oxygen, water vapor, carbon dioxide, argon)

90   50   20 12 8   0

height above earth (kilometers)

tional loads of chemicals produced by natural events and human activities. Once in the troposphere, potential air pollutants mix vertically and horizontally and often react chemically with each other or with natural components of the atmosphere. When the concentration of a normal component of air or a new chemical added to or formed in the air builds up to the point of causing harm to humans, other animals, vegetation, or materials such as metals and stone, that chemical is classified as an **air pollutant.** Worldwide, each year air pollution causes at least 150,000 premature deaths (53,000 in the U.S.), causes or aggravates debilitating respiratory diseases for tens of millions of people, and results in at least $100 billion in damages to crops, trees, buildings, and other objects.

Although there are hundreds of potential air pollutants, most air pollution results from six major classes of substances (Table 18-1). About 90% of all air pollution problems are caused by five groups of pollutants: carbon monoxide, nitrogen oxides, sulfur oxides, volatile organic compounds (mostly hydrocarbons), and suspended particulate matter (Figure 18-2).

Sources of Outdoor Air Pollutants Natural sources of air pollutants include forest fires started by lightning, pollen dispersal, wind erosion of soil, volcanic eruptions, evaporation of volatile organic compounds from leaves, bacterial decomposition of organic matter, sea spray (sulfate particles), and natural radioactivity (radon-222 gas from deposits of uranium, phosphate, and granite). But emissions from natural sources are dispersed throughout the world and rarely reach concentrations high enough to cause serious damage. Exceptions include massive injections of sulfur dioxide and suspended particulate matter (SPM) from volcanic eruptions and buildup of radon-222 gas inside buildings.

Most potential pollutants are added to the troposphere as a result of human activities (see Spotlight on p. 322): mainly the burning of fossil fuels in power and industrial plants (stationary sources), and in motor vehicles (mobile sources), the sources of 90% of the air pollutants in the United States (Figure 18-2).

The air pollution capital of the world may be Cubato, near Sao Paulo in Brazil. The air in this heavily

Table 18-1 Major Types of Air Pollutants

Class of Pollutants	Major Members of the Class
Carbon oxides (CO_x)	Carbon monoxide (CO), carbon dioxide (CO_2)
Sulfur oxides (SO_x)	Sulfur dioxide (SO_2), sulfur trioxide (SO_3)
Nitrogen oxides (NO_x)	Nitric oxide (NO), nitrogen dioxide (NO_2), nitrous oxide (N_2O)
Volatile organic compounds (VOCs) Hydrocarbons (HCs)—gaseous and liquid compounds containing carbon and hydrogen	Methane (CH_4), butane (C_4H_{10}), ethylene (C_2H_4), benzene (C_6H_6), benzopyrene ($C_{20}H_{12}$)
Other organic compounds	Formaldehyde (CH_2O), chloroform ($CHCl_3$), methylene chloride (CH_2Cl_2), ethylene dichloride ($C_2H_2Cl_2$), trichloroethylene (C_2HCl_3), vinyl chloride (C_2H_3Cl), carbon tetrachloride (CCl_4), ethylene oxide (C_2H_4O)
Suspended particulate matter (SPM) Solid particles	Dust (soil), soot (carbon), asbestos, lead (Pb), cadmium (Cd), chromium (Cr), arsenic (As), beryllium (Be), nitrate (NO_3^-) and sulfate (SO_4^{2-}) salts
Liquid droplets	Sulfuric acid (H_2SO_4), nitric acid (HNO_3), oil, pesticides such as DDT and malathion
Photochemical oxidants formed in the atmosphere by the reaction of oxygen, nitrogen oxides, and VOCs under the influence of sunlight	Ozone (O_3), PANs (peroxyacyl nitrates), formaldehyde (CH_2O), acetaldehyde (C_2H_4O), hydrogen peroxide (H_2O_2), hydroxy radical (HO)

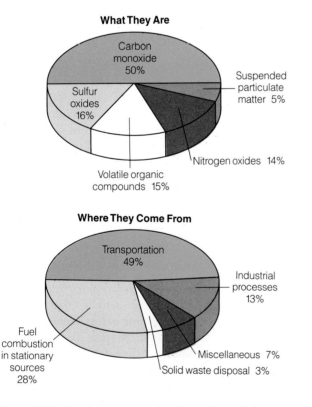

Figure 18-2 Emissions of major air pollutants in the United States. (Data from Environmental Protection Agency)

industrialized city contains twice the level of SPM considered lethal by the World Health Organization. Essentially no birds or insects remain, most trees are blackened stumps, more babies are born deformed there than anywhere in Latin America, air pollution monitoring machines break down from contamination, and the mayor refuses to live in the city.

Primary and Secondary Air Pollutants Air pollutants can be classified as either primary or secondary (Figure 18-3). A **primary air pollutant** is a harmful chemical that directly enters the air as a result of natural events or human activities. For example, carbon monoxide and carbon dioxide are primary pollutants formed when any carbon-containing substance such as coal, oil, natural gas, or wood is burned completely ($C + O_2 \rightarrow CO_2$) or partially ($2C + O_2 \rightarrow 2CO$). In the United States 71% of all carbon monoxide emissions come from motor vehicles. Another primary pollutant, sulfur dioxide (SO_2), is emitted into the air by volcanic eruptions and the burning of oil and coal, which contain sulfur impurities ($S + O_2 \rightarrow SO_2$). In the United States 83% of all SO_2 emissions come from coal and oil-burning electric power plants (68%) and industrial plants (15%).

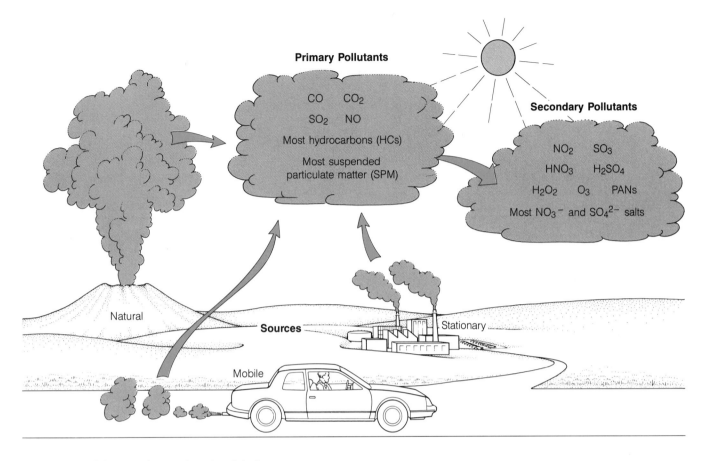

Primary Pollutants

CO   CO₂

SO₂   NO

Most hydrocarbons (HCs)

Most suspended particulate matter (SPM)

Secondary Pollutants

NO₂   SO₃

HNO₃   H₂SO₄

H₂O₂   O₃   PANs

Most NO₃⁻ and SO₄²⁻ salts

Natural

Sources   Stationary

Mobile

Figure 18-3 Primary and secondary air pollutants.

A **secondary air pollutant** is a harmful chemical that forms in the air because of a chemical reaction between two or more air components. For example, the primary pollutant sulfur dioxide reacts with oxygen gas in the atmosphere to form the secondary pollutant sulfur trioxide ($2SO_2 + O_2 \rightarrow 2SO_3$). The sulfur trioxide can then react with water vapor in air to form droplets of sulfuric acid ($SO_3 + H_2O \rightarrow H_2SO_4$), another secondary air pollutant.

Indoor Air Pollution High concentrations of air pollutants can also build up indoors, where people spend 85% to 90% of their time, and in other enclosed spaces such as underground mines, where air is slowly replenished. Indoor air today is generally much cleaner than that found decades ago, when most houses and other buildings were heated with leaky coal-burning furnaces, but there is still cause for concern. In recent years, scientists have found that the air inside some homes, schools, and office buildings is more polluted and dangerous than outdoor air on a smoggy day (Figure 18-4). Indeed, in 1985 the EPA reported that toxic chemicals found in almost every home are three times more likely to cause some type of cancer than outdoor air pollutants. Other air pollutants found in buildings produce dizziness, headaches, coughing, sneezing, burning eyes, and flulike symptoms in many people.

Air pollutants can accumulate in any building. But levels tend to be higher in energy-efficient, relatively airtight houses that do not use air-to-air heat exchangers to bring in sufficient fresh air, and in the more than 5 million mobile homes found in the United States. Mobile homes have a smaller volume of air and lower air-exchange rates than conventional homes; they also consist of a larger proportion of plywood, particle board, and other materials containing volatile organic compounds such as formaldehyde. According to the EPA and public health officials, the most serious indoor air pollution threat is from radioactive radon-222 (see Spotlight on p. 324).

Control of Indoor Air Pollution Despite the seriousness of indoor air pollution, Congress, the EPA, and state legislatures have been reluctant to establish mandatory indoor air quality standards. Part of the problem with monitoring and controlling indoor air pollution is that there are over a hundred million homes and buildings involved. In addition, many home and building owners would resent having their indoor air tested and being required to reduce excessive pollu-

Humans probably first experienced harm from air pollution when they built fires in poorly ventilated caves. As cities grew during the agricultural revolution, air pollution from the burning of wood and later of coal became an increasingly serious problem. In A.D. 1273 King Edward I of England banned the burning of coal and reinstated wood as the primary fuel in order to reduce air pollution. In 1911, at least 1,150 Londoners died from the effects of coal smoke. The author of a report on this disaster coined the word **smog** for the mixture of smoke and fog that often hung over London. An even worse London air pollution incident killed 4,000 people in 1952, and further disasters in 1956, 1957, and 1962 killed a total of about 2,500 people. As a result, London has taken strong measures against air pollution and has much cleaner air today.

In the United States the Industrial Revolution brought air pollution as coal-burning industries and homes filled the air with soot and fumes. In the 1940s air in industrial centers like Pittsburgh and St. Louis became so thick with smoke that automobile drivers sometimes had to use their headlights at midday. The rapid rise of the automobile, especially since 1940, brought new forms of pollution such as photochemical smog, which causes the eyes to sting and water, and toxic lead compounds from the burning of leaded gasoline.

The first known U.S. air pollution disaster occurred in 1948, when fog laden with sulfur dioxide fumes and suspended particulate matter stagnated over the town of Donora in Pennsylvania's Monongahela Valley for five days. About 6,000 of the town's 14,000 inhabitants fell ill and 20 of them died. This killer fog resulted from a combination of mountainous terrain surrounding the valley and stable weather conditions that trapped and concentrated deadly pollutants emitted by the community's steel mill, zinc smelter, and sulfuric acid plant. In 1963, high concentrations of air pollutants accumulated in the air over New York City, killing about 300 people and injuring thousands. Other episodes during the 1960s in New York, Los Angeles, and other large cities led to much stronger air pollution control programs in the 1970s.

---

tion levels, even if their indoor air was making them sick or threatening them and other family members with premature death.

One way to control indoor air pollution is to install air-to-air heat exchangers, which maintain a flow of fresh air without causing major heating or cooling losses, at prices ranging from $500 to $1,500. A 1984 study showed that indoor levels of formaldehyde and several other toxic gases can also be sharply reduced by house plants such as the spider plant (the most effective), golden pathos, syngonium, and philodendron.

18-2 INDUSTRIAL AND PHOTOCHEMICAL SMOG, URBAN HEAT ISLANDS, AND ACID DEPOSITION

Industrial Smog   Various groups of air pollutants found in the air over cities can be classified as either industrial smog or photochemical smog. Although both types of smog are found to some degree in most urban areas, one type often predominates during at least part of the year as a result of differences in climate and major sources of air pollution.

Industrial smog consists mostly of a mixture of sulfur dioxide and SPM, including a variety of solid particles and droplets of sulfuric acid formed from some of the sulfur dioxide. These substances form a grayish haze, explaining why cities where this type of smog predominates are sometimes called *gray-air cities*. This type of air pollution tends to predominate during the winter (especially in the early morning) in older, heavily industrialized cities like London, Chicago, Philadelphia, St. Louis, and Pittsburgh, which typically have cold, wet winters and depend heavily on coal and oil for heating, manufacturing, and producing electric power.

Photochemical Smog: Cars + Sunlight = Tears   A combination of primary pollutants such as carbon monoxide, nitric oxide, and hydrocarbons and secondary pollutants such as nitrogen dioxide, nitric acid, ozone, hydrogen peroxide, peroxacyl nitrates (PANs),

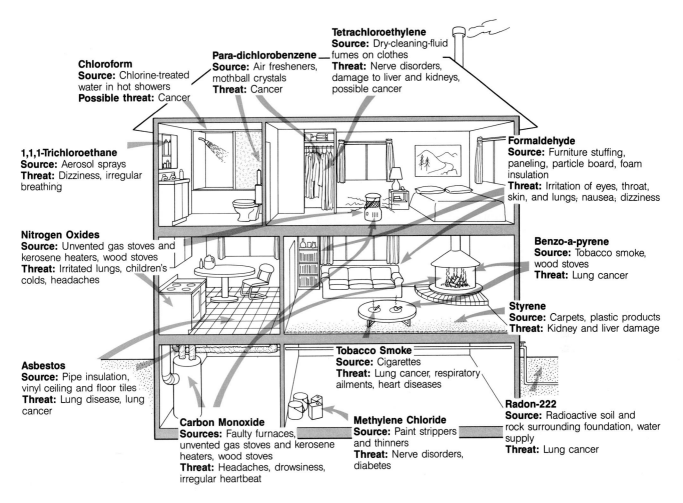

Chloroform
Source: Chlorine-treated water in hot showers
Possible threat: Cancer

Para-dichlorobenzene
Source: Air fresheners, mothball crystals
Threat: Cancer

Tetrachloroethylene
Source: Dry-cleaning-fluid fumes on clothes
Threat: Nerve disorders, damage to liver and kidneys, possible cancer

1,1,1-Trichloroethane
Source: Aerosol sprays
Threat: Dizziness, irregular breathing

Formaldehyde
Source: Furniture stuffing, paneling, particle board, foam insulation
Threat: Irritation of eyes, throat, skin, and lungs; nausea; dizziness

Nitrogen Oxides
Source: Unvented gas stoves and kerosene heaters, wood stoves
Threat: Irritated lungs, children's colds, headaches

Benzo-a-pyrene
Source: Tobacco smoke, wood stoves
Threat: Lung cancer

Styrene
Source: Carpets, plastic products
Threat: Kidney and liver damage

Asbestos
Source: Pipe insulation, vinyl ceiling and floor tiles
Threat: Lung disease, lung cancer

Tobacco Smoke
Source: Cigarettes
Threat: Lung cancer, respiratory ailments, heart diseases

Radon-222
Source: Radioactive soil and rock surrounding foundation, water supply
Threat: Lung cancer

Carbon Monoxide
Sources: Faulty furnaces, unvented gas stoves and kerosene heaters, wood stoves
Threat: Headaches, drowsiness, irregular heartbeat

Methylene Chloride
Source: Paint strippers and thinners
Threat: Nerve disorders, diabetes

Figure 18-4 Some major indoor air pollutants.

and formaldehyde, produced when some of the primary pollutants interact under the influence of sunlight, is called **photochemical smog** (Figure 18-6). Cities in which photochemical smog predominates usually have sunny, warm, dry climates. They are generally newer cities with few polluting industries and large numbers of motor vehicles, which are the major source of air pollution. Examples include Los Angeles, Denver, Salt Lake City (see photo on p. 317), as well as Sydney, Australia; Mexico City, Mexico; and Buenos Aires, Argentina. The worst episodes from this type of smog tend to occur in summer months between noon and 4 P.M.

The first step in the formation of photochemical smog occurs during the early morning traffic rush hours, when NO from automobiles builds up and reacts with O_2 to produce NO_2, a yellowish-brown gas with a pungent, choking odor. This gas produces a characteristic brownish haze, explaining why cities such as Los Angeles, where photochemical smog predominates, are sometimes called *brown-air cities*. Then, as

the sun rises, its ultraviolet rays cause a series of complex chemical reactions that produce the other components of this type of smog (Figure 18-6, p. 326). The mere traces of ozone, PANs, and aldehydes that build up to their peak levels around noon and in the early afternoon on a sunny day can irritate people's eyes and respiratory tracts. During the summer months most industrial smog cities also experience photochemical smog.

Local Climate, Topography, and Smog The frequency and severity of industrial and photochemical smog in an urban area depend on local climate and topography, density of population and industry, and major fuels used in industry and for heating and transportation. In areas with high average annual precipitation, rain and snow help cleanse the air of pollutants. Winds also help sweep pollutants away and bring in fresh air. However, hills and mountains tend to reduce the flow of air in valleys below and allow pol-

Radon-222 is a colorless, odorless, naturally occurring radioactive gas produced by the radioactive decay of uranium-238. Small amounts of radon-producing uranium-238 are found in most soil and rock, but it is much more highly concentrated in underground deposits of uranium, phosphate, and granite rock. Figure 18-5 shows the general locations of such rock deposits in the lower 48 states. When radon gas from such deposits percolates upward to the soil and is released outdoors, it disperses quickly in the atmosphere and decays to harmless levels. However, when the gas is released inside mines or seeps into buildings or water in underground wells over such deposits, it can build up to high levels.

Radon-222 quickly decays into solid particles of other radioactive elements that can be inhaled into the lungs. There they expose lung tissue to a large amount of alpha-ionizing radiation—especially in smokers, because the particles tend to adhere to tobacco tar deposits in the lungs and upper respiratory tract. Repeated exposure to these radioactive particles over 20 or 30 years can cause lung cancer.

On the basis of preliminary data from a nationwide survey of indoor radon levels, the EPA and several scientists estimate that at least one of every nine American homes (perhaps as many as one of every five) may harbor harmful or potentially harmful levels of this gas. Although limited testing has revealed potentially harmful radon levels in 30 states, so far the worst radon hot spots have been found in Pennsylvania, Colorado, Kansas, Wisconsin, and Wyoming.

According to the EPA, prolonged exposure to high levels of radon may be responsible for 5,000 to 20,000 of the 136,000 lung cancer deaths each year in the United States; 100 to 1,000 of these premature deaths may be related to radon released from hot water obtained from groundwater near radon-laden rock and used for showers and washing clothes.

Individuals can measure radon levels in their homes or other buildings by using radon detectors available free or at a low cost from health or environmental agencies in some states (check with local officials). Otherwise, they can be purchased for about $15 to $50 (depending on the type of detector and the number purchased) from private testing firms.

If unacceptable levels are detected, the EPA recommends several ways to reduce radon levels and health risks. The first is to stop all indoor smoking. Ventilation fans (costing about $300) or heat exchangers ($500 to $1,500) can be installed to remove radon and most other indoor air pollutants if infiltration levels are not too high. For houses with serious radon gas problems, special venting systems usually have to be installed below the foundations at a cost of $1,000 to $5,000. To remove radon from contaminated well water, a special type of activated carbon filter can be added to holding tanks at a cost of about $1,000.

In Sweden no house can be built until the lot has been tested for

lutant levels to build up at ground level. Buildings in cities also slow wind speed and impede dilution and removal of pollutants.

During the day the sun warms the air near the earth's surface. Normally, this heated air expands and rises during the day, diluting low-lying pollutants and carrying them higher into the troposphere. Air from surrounding high-pressure areas then moves down into the low-pressure area created when the hot air rises (Figure 18-7, left). This continual mixing of the air helps keep pollutants from reaching dangerous levels in the air near the ground.

But sometimes a layer of dense, cool air is trapped beneath a layer of less dense, warm air in an urban basin or valley. This is called a temperature or **thermal inversion** (Figure 18-7, right, and Figure 18-8). In effect, a warm-air lid covers the region and prevents pollut-

ants from escaping in upward-flowing air currents. Usually these inversions last for only a few hours, but sometimes they last for several days when a high-pressure air mass stalls over an area. When this happens, air pollutants at ground level accumulate to harmful and even lethal levels. Most air pollution disasters—such as those in London and in Donora, Pennsylvania, occurred during lengthy thermal inversions during fall or winter in industrial smog areas.

Thermal inversions occur more often and last longer over towns or cities located in valleys surrounded by mountains, on the leeward sides of mountain ranges, and near coasts. A city with several million people and automobiles in an area with a sunny climate, light winds, mountains on three sides, and the ocean on the other possesses the ideal conditions for photochemical smog worsened by frequent ther-

radon. If the reading is high, the builder must follow government-mandated construction procedures to ensure that the house won't be contaminated with radon from soil or water supplies. Environmentalists urge enactment of a similar program for all new construction in the United States. They also suggest that before purchasing a lot to build a new house, individuals should have the soil tested for radon. Similarly, no one should purchase an existing house unless it has been tested for radon by certified personnel, just as houses must now be tested for termites.

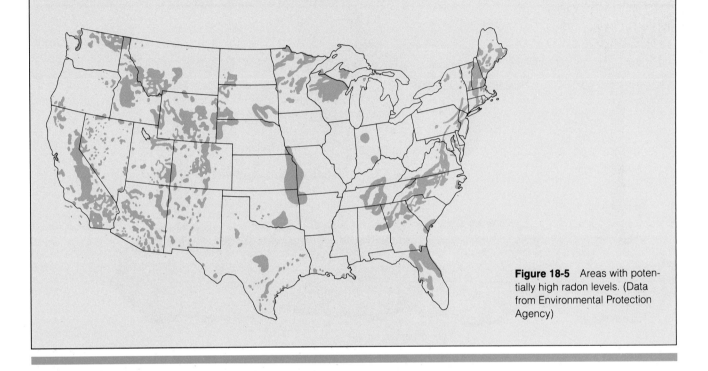

Figure 18-5 Areas with potentially high radon levels. (Data from Environmental Protection Agency)

mal inversions. This describes the Los Angeles basin, which experiences almost daily inversions, many of which are prolonged during the summer months.

Urban Heat Islands In accordance with the second energy law (Section 3-4), when energy is converted from one form to another, low-quality heat is added to the atmosphere. In the United States, energy use is so high that the average continuous heat load per person injected into the atmosphere is equivalent to that from a hundred 100-watt light bulbs.

The effect of all this atmospheric heating is evident in large cities and urban areas, which are typically like huge islands of heat surrounded by cooler suburban and rural areas, a climatic effect known as the **urban heat island** (Figure 18-9). This dome of heat helps trap pollutants, especially SPM, and creates a **dust dome** above urban areas. As a result, concentrations of SPM over urban-industrial areas may be a thousand times higher than those over rural areas. If wind speeds increase, this dust dome elongates downwind to form a *dust plume* that spreads the city's pollutants to rural areas and other urban areas tens to hundreds of miles away. As urban areas grow and merge into vast urban regions, the heat and dust domes from a number of cities can combine to form regional heat islands, which affect regional climates and prevent polluted air from being effectively diluted and cleansed.

Acid Deposition One way to decrease ground-level air pollution from sulfur dioxide, SPM, and nitrogen

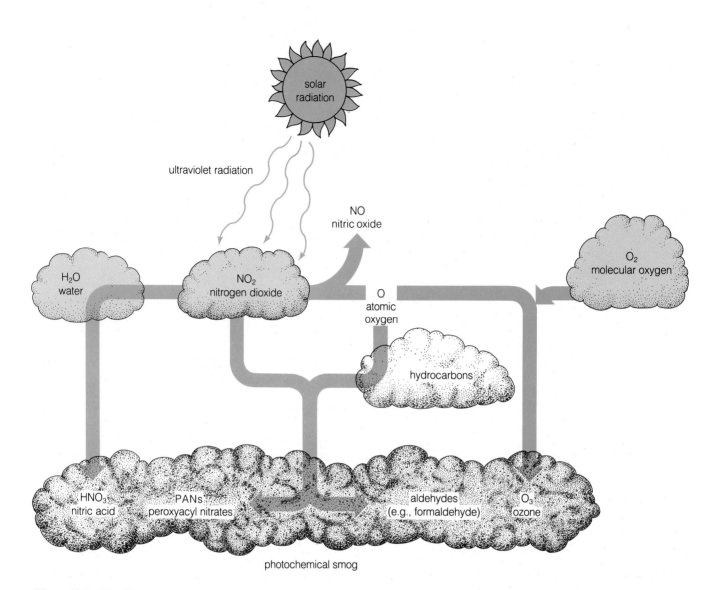

Figure 18-6 Simplified scheme of the formation of photochemical smog.

oxides when coal and oil are burned in electric power plants, metal smelters, and other industrial plants is to discharge these emissions from smokestacks tall enough to pierce the thermal inversion layer (Figure 18-7, right). Use of tall smokestacks in the United States, Canada, and western Europe has led to considerable reduction of ground-level pollution in many urban areas.

This approach, however, leads to increased levels of these pollutants and various secondary pollutants in downwind rural and urban areas. As emissions of sulfur dioxide and nitric oxide are transported over long distances by wind currents, they are chemically transformed into a variety of secondary pollutants such as nitrogen dioxide, droplets of sulfuric and nitric acids, and solid particles of sulfate and nitrate salts.

These chemicals fall or are washed out of the atmosphere onto downwind land and bodies of water. *Wet deposition* occurs when some of the suspended

droplets of sulfuric acid and nitric acid return to the earth as acid rain or its variants, consisting of these acids and snow, sleet, hail, fog, or dew. *Dry deposition* occurs when solid particles of sulfate and nitrate salts and gases such as sulfur dioxide fall or are washed out of the atmosphere, usually near the original pollution sources. These deposited solids can then react with water in soil and bodies of water to form sulfuric and nitric acids. The combined wet and dry deposition of acids or acid-forming substances onto the surface of the earth is known as **acid deposition** (Figure 18-10). This phenomenon is commonly called *acid rain*, but this is a misleading term because these acids and acid-forming substances are deposited not only in rain but also in snow, sleet, fog, and dew and as dry particles and gas.

The relative levels of acidity and basicity of water solutions of substances are commonly expressed in terms of **pH** (Figure 9-6, p. 146). The lower the pH

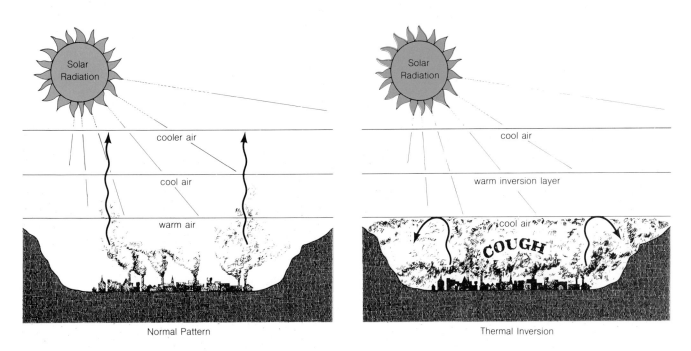

Figure 18-7 Thermal inversion traps pollutants in a layer of cool air that cannot rise to carry the pollutants away.

Figure 18-8 Two faces of New York City. The almost clear view was photographed on a Saturday afternoon (November 26, 1966). The effect of more cars in the city and a thermal inversion is shown in the right-hand photograph, taken the previous day.

value, the higher the acidity, with each whole-number decrease in pH representing a tenfold increase in acidity. Natural precipitation has an average pH value of 5.1 (with a range of 5.0 to 5.6 depending on location), caused when carbon dioxide and traces of natural sulfur and nitrogen compounds and organic acids in the atmosphere dissolve in atmospheric water. This slight acidity of natural precipitation helps water deposited

on soil to dissolve minerals for use by plants and animals. It also deposits some sulfur and nitrogen used as plant nutrients.

However, deposition of acids and acid-forming substances with higher levels of acidity (pH values of 5.0 and less) than those in natural precipitation can damage materials; leach certain nutrients from soil; and kill fish, aquatic plants, and microorganisms in

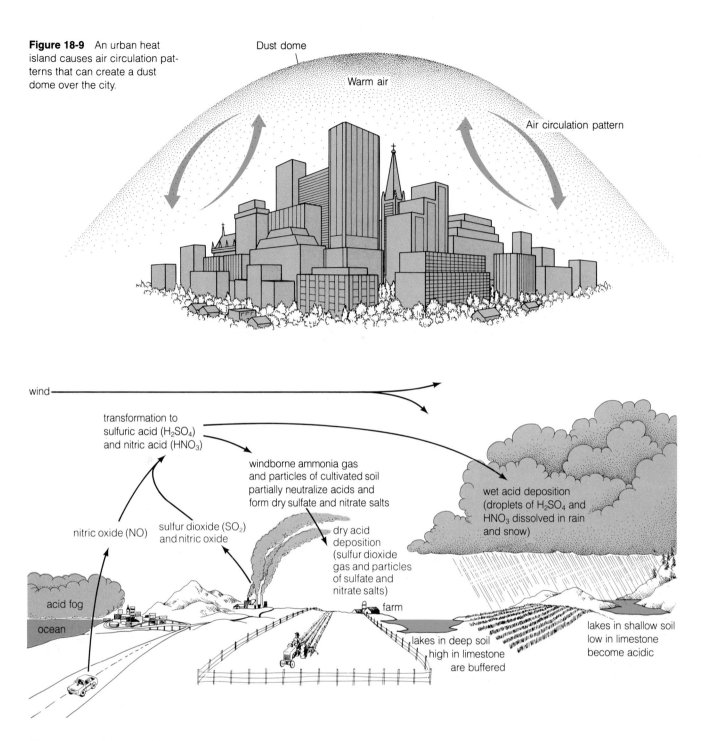

Figure 18-9 An urban heat island causes air circulation patterns that can create a dust dome over the city.

Dust dome

Warm air

Air circulation pattern

wind

transformation to sulfuric acid (H₂SO₄) and nitric acid (HNO₃)

windborne ammonia gas and particles of cultivated soil partially neutralize acids and form dry sulfate and nitrate salts

wet acid deposition (droplets of H₂SO₄ and HNO₃ dissolved in rain and snow)

nitric oxide (NO)

sulfur dioxide (SO₂) and nitric oxide

dry acid deposition (sulfur dioxide gas and particles of sulfate and nitrate salts)

acid fog

ocean

farm

lakes in deep soil high in limestone are buffered

lakes in shallow soil low in limestone become acidic

Figure 18-10 Acid deposition.

lakes and streams. Acid deposition, in combination with other air pollutants such as ozone, sulfur dioxide, and nitrogen oxides, can damage trees, crops, and other plants. It can also affect human health.

Acid deposition as a result of human activities is already a serious problem in western and central Europe, Scandinavia, the northeastern United States, southeastern Canada, and southeastern China, and is expected to become a problem in other areas. Much of the acid-producing chemicals generated in one country are exported to others by prevailing winds. For example, over three-fourths of the acid deposition found in Norway, Switzerland, Austria, Sweden, the Netherlands, and Finland is blown there from indus-

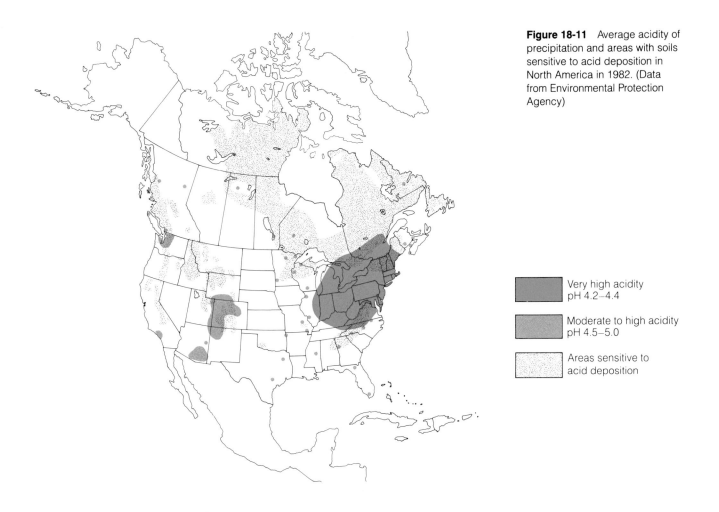

Figure 18-11 Average acidity of precipitation and areas with soils sensitive to acid deposition in North America in 1982. (Data from Environmental Protection Agency)

	Very high acidity pH 4.2–4.4
	Moderate to high acidity pH 4.5–5.0
	Areas sensitive to acid deposition

trialized areas of western and eastern Europe; more than half of the acid deposition in Canada comes from the United States (Figure 18-11).

Almost half of all SO_2 emissions and about one-fourth of NO emissions in the U.S. come from the heavy concentration of coal- and oil-burning power and industrial plants in seven central and upper Midwest states—Ohio, Indiana, Pennsylvania, Illinois, Missouri, West Virginia, and Tennessee. Much of the emission from these states, which comprise the country's industrial heartland, is blown northeastward by prevailing winds, accounting for most of the moderate to very high levels of acid deposition in the northeastern U.S. and southeastern Canada, where 86% of all Canadians and 50% of all Americans live (Figure 18-11). There is also increasing concern over acid deposition in the western United States, especially from sulfur and nitrogen oxides released by smelters on both sides of the Mexico–U.S. border and by large amounts of NO released primarily from automobiles in California.

Once acid deposition reaches the ground, its acidity can be increased or decreased up to tenfold as it passes through local soils before running off into nearby lakes and streams or percolating into groundwater. Soils in some areas contain limestone ($CaCO_3$) and other alkaline (basic) substances that can react with and neutralize the acids, thus reducing their harmful effects on vegetation and aquatic life. But poor, thin soils, such as those overlying granite and some types of sandstone, are already acidic and have little buffering capacity to neutralize additional acids. Such soils are found in much of Scandinavia, parts of Canada and the United States (Figure 18-11), and large portions of Brazil, southern India, Southeast Asia, and eastern China. Acid runoff in these areas can kill many forms of aquatic life in nearby lakes and streams.

18-3 EFFECTS OF AIR POLLUTION ON HUMAN HEALTH

Damage to Human Health Air pollutants have numerous harmful effects on human health. The types and severity of these effects depend on the particular

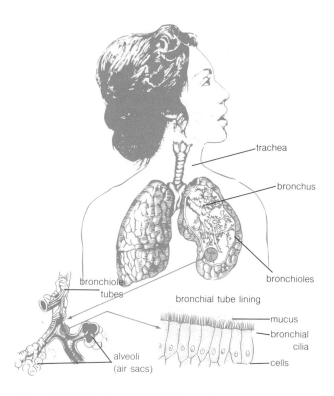

Figure 18-12 The human respiratory system.

true but meaningless. Instead of establishing absolute truth or proof, science establishes only a degree of probability or confidence in the validity of an idea, usually based on statistical or circumstantial evidence.

Body Defenses Against Air Pollution   Fortunately, the human respiratory system has a number of defense mechanisms that help protect us from air pollution (Figure 18-12). When we inhale air, hairs in the nose filter out large particles, and when pollutants irritate the nose, sneezing expels the air in the upper respiratory tract. The linings of the nose; the trachea (windpipe); the bronchi (each **bronchus** is a branch of the trachea); and thousands of minute ducts, or **bronchiole tubes**—which carry air throughout the lungs—are covered with a sticky mucus that captures small particles and dissolves some gaseous pollutants. Most of the upper respiratory tract is lined with hundreds of thousands of tiny mucus-coated hairs, called **cilia,** which continually wave back and forth, transporting mucus and the pollutants it traps to the mouth, where it is either swallowed or expelled. If the lungs become irritated, mucus flows more freely to help remove the irritants and stimulates coughing, which expels the dirty air and some of the contaminated mucus.

Overloading and Degrading the Body's Defense Mechanisms   Despite their effectiveness, our natural defenses against air pollution can be overloaded or impaired. For example, exposure to high levels of an air pollutant or more prolonged exposure to lower concentrations can saturate the mucus so that it can't dissolve any more gases. When this happens, gases penetrate deeper into the respiratory tract and cause persistent irritation and coughing.

Fine particles are particularly hazardous to human health because they are small enough to penetrate the lung's natural defenses; they can also bring with them droplets or other particles of toxic or cancer-causing pollutants that become attached to their surfaces.

Chronic exposure to chemicals in cigarette smoke and other air pollutants such as ozone, sulfur dioxide, nitrogen dioxide, and some types of particulate matter apparently destroy, stiffen, or slow the cilia, and thus make them less effective in removing harmful substances. As a result, bacteria and tiny particles penetrate the **alveoli,** or air sacs, increasing the chances of respiratory infections and lung cancer.

Years of smoking and exposure to air pollutants can trigger so much mucus flow that air passages become blocked, causing coughing. As muscles surrounding the bronchial tubes weaken from prolonged coughing, more mucus accumulates and breathing becomes progressively more difficult. If this condition persists, it indicates **chronic bronchitis**—a persistent

chemicals involved, their concentration in the air, and exposure time. Evidence suggests that air pollution emitted by burning fossil fuels contributes to the premature death of at least 53,000 Americans each year—more than the total number of Americans killed during the nine-year Vietnam War. Groups particularly sensitive to air pollution include the elderly, especially those with lung and heart disorders; infants, whose respiratory systems are not fully developed; active children, who breathe more than most adults; and people with frequent colds or chronic nasal congestion, who breathe through the mouth and thus bypass the filtering mechanism of the nose.

Although tens of thousands of statistical studies provide massive evidence that air pollution harms and sometimes kills people, it is difficult to establish that a specific pollutant causes a particular disease or death. Reasons for this include the large number and variety of air pollutants people are exposed to over decades, synergistic interactions between various pollutants that can lead to more harm than that of one acting alone, and the multiple causes and lengthy incubation times of diseases such as emphysema, chronic bronchitis, lung cancer, and heart disease.

Largely because of these difficulties and a misunderstanding of the nature of science, many people are misled when they hear statements such as, "Science has not proven absolutely that smoking (or any air pollutant) has killed anyone." Such a statement is

inflammation of the mucous membranes of the trachea and bronchi that now affects one out of every five American men between the ages of 40 and 60.

Emphysema occurs when such a large number of the lung's alveoli become so damaged that a person is unable to expel most of the air from the lung. It begins when prolonged irritation of the lungs by chronic exposure to cigarette smoke and other air pollutants causes the bronchioles to close. Some of the trapped air then expands and fuses clusters of alveoli together, and they lose their ability to expand and contract and may even tear. This reduces the surface area available for transferring oxygen to the blood so that walking or the slightest exertion causes acute shortness of breath (Figure 18-13). Eventually the victim may die of suffocation or heart failure.

Emphysema kills far more people than lung cancer and is the fastest rising cause of death in the United States. It is incurable and basically untreatable. Although chronic smoking and exposure to air pollutants can cause emphysema in anyone, about one-fourth of the population is highly susceptible because of a hereditary condition characterized by the absence of a protein that gives the air sacs their elasticity. Anyone with this condition, for which testing is available, should seriously consider not smoking and not living or working in a highly polluted area.

Lung cancer is the abnormal, accelerated growth of cells in the mucous membranes of the bronchial passages. Smoking is considered the leading cause, but lung cancer has also been linked to inhalation of other air pollutants, including particles of radioactive polonium (produced by the decay of radon gas) and plutonium-239, benzopyrene found in cigarette and other types of smoke, and particulate matter—especially particles of asbestos, beryllium, arsenic, chromium, and nickel. Some air pollutants increase the risk of lung cancer by impairing the action of the cilia so that other, carcinogenic pollutants are not effectively removed.

Miners, mill workers, construction workers, and others whose occupations subject them to chronic exposure to high levels of SPM can eventually develop lung disease, which is usually named for the types of particulate matter involved. These diseases, which scar lung tissue, include *black lung* from prolonged inhalation of coal dust, *brown lung* from cotton dust, *silicosis* from quartz dust generated during mining, and *asbestosis* from asbestos fibers. Victims usually experience coughing and shortness of breath and eventually may develop pneumonia, chronic bronchitis, emphysema, or lung cancer.

Considerable evidence indicates that exposure to even a small amount of asbestos fibers can cause lung cancer or mesothelioma (a cancer of the chest and abdominal lining) 15 to 40 years later. The EPA estimates that exposure to asbestos causes 3,000 to 12,000

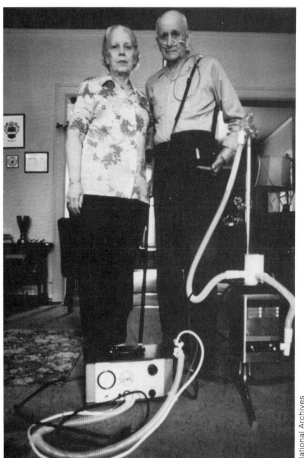

National Archives

Figure 18-13 People who suffer emphysema may have so much difficulty in breathing that they must breathe pure oxygen, carrying the equipment everywhere they go.

cancer cases a year in the U.S., almost all of which are fatal. Some of the asbestos sprayed on ceilings and walls in 45,000 older elementary and secondary schools nationwide is crumbling—a potential threat to 15 million students and 1.4 million employees. In 1987 the EPA issued rules requiring school officials to have buildings inspected and asbestos hazards eliminated. Cleanup costs are estimated at $3.2 billion over 30 years. Financially strapped schools cannot afford such expenditures without increased local taxes or help from state and federal governments.

18-4 EFFECTS OF AIR POLLUTION ON PLANTS AND MATERIALS

Damage to Plants Some forms of air pollution, such as sulfur dioxide, ozone, nitrogen oxides, and PANs, cause direct damage to leaves of crop plants and trees (Figure 18-14) when these gases enter leaf pores (sto-

Figure 18-14 Leaves exposed to sulfur dioxide can take on a bleached look due to destruction of chlorophyll. The leaf on the right is healthy.

U.S. Department of Agriculture

mata). Chronic exposure of leaves and needles to air pollutants can break down the waxy coating that helps prevent excessive water loss and damage from diseases, pests, drought, and frost. Such exposure can also inhibit photosynthesis and plant growth, reduce nutrient uptake, and cause leaves or needles to turn yellow or brown and drop off. Coniferous trees—cone-bearing trees, most of which have needle-shaped leaves—are highly vulnerable to the effects of pollution because of their long life spans and the year-round exposure of their needles to polluted air.

In addition to causing direct leaf damage, acid deposition can leach vital plant nutrients such as calcium from the soil and kill essential soil microorganisms. It also releases aluminum ions, which are normally bound to soil particles, into soil water, where they damage fine root filaments and reduce the uptake of water and nutrients from the soil (Figure 18-15). Prolonged exposure to high levels of air pollutants can kill all trees and vegetation in an area (Figure 14-10, p. 239).

The effects of exposure of trees to multiple air pollutants may not be visible for decades, when suddenly large numbers begin dying off because of soil nutrient depletion and increased susceptibility to pests, diseases, and drought (Figure 18-15). This is what is happening to many forests in parts of Europe. For example, 8% of the trees in West German forests were found to be dead or damaged in 1982. One year later the figure was 34%, and by 1985 the toll stood at 52%. In addition to a $10 billion loss of commercially impor-

tant trees, these diebacks have eliminated habitats for many types of wildlife. Similar damage is occurring to forestlands in at least 15 other European countries; one-quarter to one-half of the total forest area of Luxembourg, the Netherlands, Austria, Switzerland, and Czechoslovakia was damaged by 1985.

So far, similar diebacks from exposure to multiple air pollutants in the United States have occurred primarily to stands on higher-elevation slopes facing moving air masses—especially slopes shrouded in pollution-laden clouds or fog much of the time (Figure 18-16). Measurements taken in 15 eastern states have shown a 40% reduction in growth between 1960 and 1984 for 34 tree species found at high elevations. Some tree species at lower elevations are also beginning to show subtle signs of ill health. Many scientists fear that elected officials will continue to delay implementing more stringent controls on all major forms of air pollution until it is too late to prevent a severe loss of valuable forest resources in the United States and Canada like that taking place in much of Europe.

Acid deposition has a severe impact on aquatic life of freshwater lakes in areas where surrounding soils have little acid-buffering capacity. Much of this damage to aquatic life in the Northern Hemisphere is a result of *acid shock*, which occurs when large amounts of highly acidic water (along with toxic aluminum leached from the soil) suddenly run off into lakes from spring snowmelt or when heavy rains follow a period of drought.

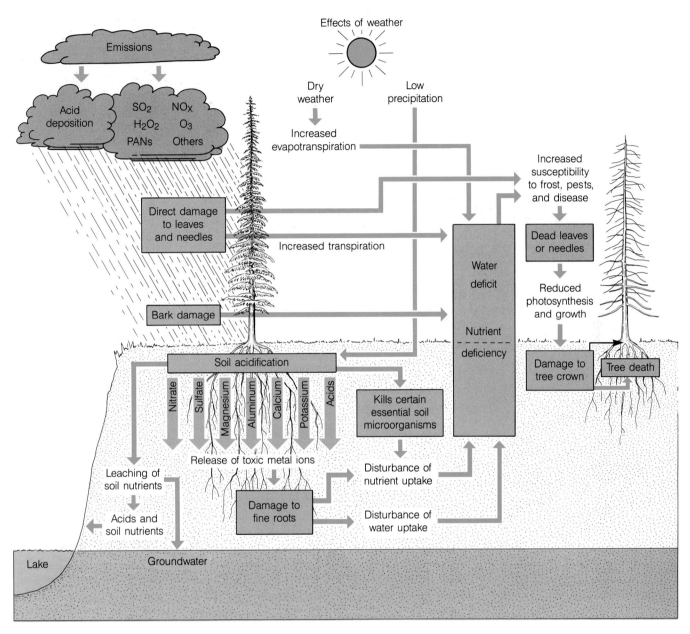

Figure 18-15 Harmful effects of air pollutants on trees.

About 4,000 of Sweden's 40,000 acid-sensitive lakes contain no fish because of excess acidity, 18,000 are partly acidified, and the remainder are at risk if emissions of SO_2 and NO from other parts of Europe are not sharply curbed. The country also has 90,000 kilometers (56,250 miles) of acidified streams. At least 1,000 acid-sensitive lakes in the eastern United States, especially in New England, are highly acidic (pH below 5.0) and have suffered sharp declines in fish populations. Another 3,000 are moderately acidified. In Ontario, Canada at least 1,600 lakes are fishless because of excess acidity. Aquatic life in 48,000 more is threatened unless the U.S. and Canada can agree upon and implement a joint program to sharply reduce acid deposition and other air pollutants that remain in the atmosphere long enough to be carried long distances.

Damage to Materials Each year air pollutants cause tens of millions of dollars in damage to various materials (Table 18-2). Atmospheric fallout of soot and grit on statues, buildings, cars, and clothing requires costly cleaning. Irreplaceable marble statues, historic buildings, and stained-glass windows throughout the world are pitted and discolored by air pollutants (Figure 18-17).

Figure 18-16 This dead coniferous forest on Mount Mitchell, North Carolina, is believed to be the result of long-term exposure to multiple air pollutants.

Figure 18-17 This marble monument in Rome has been damaged by exposure to acidic air pollutants.

18-5 EFFECTS OF AIR POLLUTION ON THE OZONE LAYER AND GLOBAL CLIMATE

Chlorofluorocarbons and Ozone Layer Depletion
In the lower atmosphere, ozone is a pollutant that in trace amounts can damage plants and human health. In the stratosphere, however, the **ozone layer** protects life on earth by screening out more than 99% of the sun's harmful ultraviolet (UV) radiation. Many scientists are concerned that the average concentration of ozone in the stratosphere is being decreased by **chlorofluorocarbons (CFCs),** commonly called by the trade name *Freons,* a group of nontoxic, nonflammable, and cheaply produced chemicals. Since 1955 these chemicals have been widely used as propellants in aerosol spray cans; coolants in refrigerators and air conditioners; industrial solvents; and Styrofoam and other plastic foams for insulating houses, keeping coffee and fast-food hamburgers warm, and as packaging to prevent damage to eggs and shipped items.

Spray cans, discarded or leaking refrigeration and air conditioning equipment, and burning of plastic foam products release these highly unreactive gases into the atmosphere, where they remain up to 110 years. Over several decades they gradually move up to the stratosphere, where under the influence of high-energy UV radiation they break down, releasing chlorine atoms, which speed up the breakdown of ozone into oxygen gas. About 95% of the CFCs released into the atmosphere between 1955 and 1987 are still making their way up to the stratosphere. Since 1978 the use of CFCs in aerosol spray cans has been banned in the United States, Canada, and most Scandinavian countries, but worldwide nonaerosol uses have risen sharply along with aerosol use in western Europe.

The general consensus, based on theoretical models of chemical reactions taking place in the stratosphere, is that continuing CFC emissions at 1987 levels will reduce average levels of ozone in the stratosphere by 3% to 5% over the next 100 years, although the U.S. National Aeronautics and Space Administration projects a 10% depletion of the ozone layer by 2050.

Ozone depletion may be occurring more rapidly and more extensively than these projections indicate. Satellite images have revealed that since 1983 a "hole"—that is, a thinning—in the ozone layer has appeared in the stratosphere over the South Pole each September and October; the hole, covering an area the size of the United States, contains 40% less ozone than normal. A smaller hole has been observed over the North Pole. It is not known whether this loss of ozone during part of each year is caused by CFCs, large volcanic eruptions, natural climatic processes such as cyclic changes in solar output, or some combination of these factors.

Some Effects of Ozone Depletion Less ozone in the stratosphere would allow more UV radiation to reach the earth's surface. The EPA estimates that a 5%

Table 18-2 Harmful Effects of Air Pollution on Materials

Material	Effects	Principal Air Pollutants
Stone and concrete	Surface erosion, discoloration, soiling	Sulfur dioxide, sulfuric acid, nitric acid, solid particulates
Metals	Corrosion, tarnishing, loss of strength	Sulfur dioxide, sulfuric acid, nitric acid, solid particulates, hydrogen sulfide
Ceramics and glass	Surface erosion	Hydrogen fluoride, solid particulates
Paints	Surface erosion, discoloration, soiling	Sulfur dioxide, hydrogen sulfide, ozone, solid particulates
Paper	Embrittlement, discoloration	Sulfur dioxide
Rubber	Cracking, loss of strength	Ozone
Leather	Surface deterioration, loss of strength	Sulfur dioxide
Textile fabrics	Deterioration, fading, soiling	Sulfur dioxide, nitrogen dioxide, ozone, solid particulates

ozone depletion would cause an additional 940,000 cases annually of nonmelanoma skin cancer (a disfiguring but usually not fatal cancer, if treated in time) and 30,000 more cases annually of often-fatal melanoma skin cancer. In addition, humans would be subject to increases in eye cataracts, severe sunburn, and suppression of the immune system.

There would also be a 10% increase in eye-burning photochemical smog. Acid deposition would increase near areas where sulfur dioxide and nitrogen oxides are produced because of an estimated 80% increase in hydrogen peroxide (which speeds up the formation of sulfuric and nitric acids in the atmosphere) for each 1% decrease in stratospheric ozone. Other effects include eye cancer in cattle, damage to many species of terrestrial plants (including some important food crops such as corn, rice, and wheat), damage to aquatic plant species essential to ocean food chains, and a loss of perhaps $2 billion a year from degradation of plastics and other polymer materials.

Protecting the Ozone Layer   Theoretically, the ozone problem is easier to resolve than almost any global pollution issue. CFCs, the primary offender, can be controlled by an international agreement between the 31 major CFC producers and users—the United States, the USSR, Japan, and the Scandinavian and West European countries. CFCs are not necessary for the functioning of society, and substitutes are either available or can be developed.

However, models indicate that just to keep atmospheric CFCs at 1987 levels would require an immediate 85% drop in total CFC emissions throughout the world. Analysts believe that the first step toward this goal should be a total ban on the use of CFCs in aerosol spray cans, egg crates, fast-food containers, and insulation—all nonessential uses for which cost-effective substitutes are available. The next step would be to phase out all other uses of CFCs over a ten-year period. Although substitutes are available for use of CFC coolants in refrigeration and air conditioning, testing and phasing them in may take 10 years and they could cost 5 to 10 times more than CFCs. But compared to the potential economic and health consequences of ozone depletion, such cost increases would be minor.

However, an effective international agreement is unlikely because each country tends to be interested in continuing its use of CFCs for short-term economic gain. In 1987, 31 countries producing CFCs agreed that by 1992 they would cut back on their production of the most widely used CFCs by 20%, followed by a possible further 30% reduction by 1995. Although such a cutback will slow down ozone depletion, most scientists feel that it is not enough to prevent significant deterioration of the ozone layer.

Increased Global Warming from the Greenhouse Effect   The average temperature of the earth's atmosphere is maintained by a system in which the amount

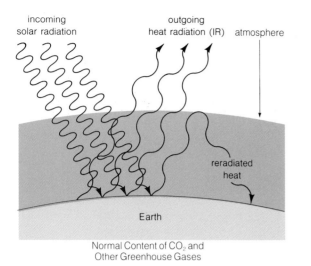

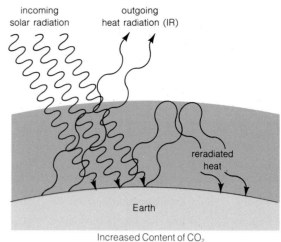

Normal Content of CO_2 and
Other Greenhouse Gases

Increased Content of CO_2
and Other Greenhouse Gases

Figure 18-18 The greenhouse effect.

of energy the earth absorbs from the sun primarily as visible and ultraviolet radiation is balanced by the amount radiated back into space as degraded infrared radiation or heat (Figure 4-13, p. 63). Carbon dioxide, water vapor, and trace amounts of other gases such as ozone in the troposphere, methane, nitrous oxide, and CFCs play a key role in this temperature regulation process.

These gases, known as **greenhouse gases,** acting somewhat like a pane of glass in a greenhouse, let in visible light from the sun but prevent some of the resulting infrared radiation or heat from escaping and reradiate it back toward the earth's surface (Figure 18-18). The resulting heat buildup raises the temperature of the air in the lower atmosphere, a warming action commonly called the **greenhouse effect.** If there were no greenhouse gases in the atmosphere, the earth would be a cold and lifeless planet with an average atmospheric temperature of $-18°C$ (0.4°F).

A buildup of one or several greenhouse gases in the atmosphere would slow down the escape of heat into space and lead to an increase in the average temperature of the earth's atmosphere. This additional heat could affect global climate and food-growing patterns. Buildup of SPM in the atmosphere from natural and human activities might enhance this global warming or cause atmospheric cooling.

Between 1860 and 1986 the average global levels of carbon dioxide in the atmosphere have increased 26% from 275 ppm to 346 ppm. This rise is attributed primarily to the burning of fossil fuels throughout the world. Deforestation, especially wholesale clearing and burning of tropical forests, is also believed to contribute an unknown amount to increased CO_2 levels by drastically reducing the number of plants that absorb carbon dioxide during photosynthesis.

According to the latest climate models, a doubling of the preindustrial CO_2 level of around 275 ppm to 550 ppm would raise the average atmospheric temperature by about 4°C (7°F); temperatures near the poles would rise 2 to 3 times this amount. Although seemingly small, such a change would make the earth warmer than at any time in human history and have profound effects on the earth's climate. During the Ice Age over 20,000 years ago, when vast ice sheets extended to where New York City and Chicago are now, the earth's average temperature was only 5°C (9°F) colder. Even a 1°C or 2°C increase could cause a larger number of violent storms, many more unbearably hot days in summer, and prolonged droughts in many areas.

Depending on the rate of use of fossil fuels, especially coal and oil, CO_2 levels of 550 ppm could be reached sometime between 2040 and 2100. However, a global warming of that magnitude could occur much earlier—between 2010 and 2050—if levels of other greenhouse gases produced by human activities continue to increase at present rates. These include (1) ozone in the lower atmosphere from photochemical smog; (2) methane produced in the digestive tracts of cattle, wetlands, biomass burning, and in the soils of rice paddies; (3) nitrous oxide from fertilization of soils and concentration of animal wastes in feedlots; and (4) CFCs. CFCs alone can contribute about 20% to global warming; thus banning them would not only help protect the ozone layer but also slow the rate of global warming.

Some Effects of Global Climate Changes At first glance a warmer average climate might seem desirable, resulting in lower heating bills in middle and higher latitudes and longer growing seasons in some areas. Crop yields might also increase because more CO_2 in the atmosphere can increase the rate of photosynthesis. Other factors, however, could offset potential gains in crop yields. Damage from insect pests could increase because the warmer temperatures would enhance insect breeding. Significant changes in worldwide precipitation and temperature patterns would cause crop production to decline in some areas and increase in others. Global cooling would cause similar effects.

Since water expands when heated, an increase in average global temperature will raise average sea levels slightly. Warmer temperatures will also cause partial melting of mountain glaciers and parts of the West Antarctic land ice sheet, transferring water from the land to the sea. Present models indicate that raising the average atmospheric temperature by 4°C (7°F) would raise the average global sea level by about 0.6 meters (2 feet) as a result of these processes. This would flood large areas of agricultural lowlands and deltas in Bangladesh, India, and China, where much of the world's rice is grown.

In the United States such an increase would flood major portions of Louisiana and Florida, as well as buildings, roads, tanks storing hazardous chemicals, and other structures along the Gulf and Atlantic coasts, and cause intrusion of saltwater into groundwater supplies. A significant melting of the West Antarctic ice sheet would lead to even larger increases in sea level, but would probably be a gradual process, occurring over 1,000 years or more.

There is widespread agreement among scientists that any significant change in world climate resulting from warming or cooling will disrupt world food production for many years; lead to a sharp increase in food prices; cause considerable economic damage; and require investments of hundreds of billions of dollars in fertilizers, new water transfer and irrigation projects, and dike systems to prevent flooding.

Dealing with the Effects of Global Warming
Basically there are two ways to deal with global warming: slow it down and adjust to its effects. We can slow down the rate of global warming by banning emissions of greenhouse gases such as CFCs and by sharply reducing the use of fossil fuels, especially coal, which emits the most CO_2 per unit of energy produced.

We could achieve significant reductions in fossil fuel use by relying more on a combination of energy conservation and perpetual energy (Chapter 17), or by increased use of nuclear power (Chapter 16). Another approach is to use scrubbers to remove carbon dioxide from the smokestack emissions of coal-burning power and industrial plants and from vehicle exhausts. But present methods remove only about 30%

of the CO_2 and are prohibitively expensive. Planting trees worldwide would also slow global warming by increasing the uptake of CO_2 from the atmosphere, as well as reducing other harmful effects of deforestation.

However, many observers doubt that countries will be able to agree to sharply reduce fossil fuel use and deforestation in time to prevent significant global warming. Countries likely to acquire a more favorable climate would resist severe restrictions, while countries likely to suffer from reduced food-growing capacity may favor taking immediate action. Sharply restricting fossil fuel use, no matter how desirable from a long-term environmental and economic viewpoint, would cause major short-term economic and social disruptions that most countries would find unacceptable.

Thus some analysts suggest that while attempting to reduce fossil fuel use, we should also begin to prepare for the effects of long-term global warming. Efforts should include increased research into the breeding of plants that need less water and plants that can thrive in water too salty for ordinary crops. Improved irrigation methods should be widely used so that less water is wasted (Section 10-5). Dikes should be erected to protect coastal areas from flooding, as the Dutch have done for hundreds of years, and zoning ordinances should prohibit new construction on low-lying coastal areas. Large supplies of key foods stored throughout the world would be insurance against climate changes that could shift the world's major food-growing regions and disrupt food production.

18-6 NOISE POLLUTION

Sonic Assault Because noise travels through the air it can be considered as a form of air pollution. According to the EPA, nearly half of all Americans, mostly urban dwellers, are regularly exposed in their neighborhoods and jobs to levels of noise that interfere with communication or sleep. The American Speech and Hearing Association reports that every day one of every ten Americans lives, works, or plays around noise of sufficient duration and intensity to cause some permanent loss of hearing, and that this number is rising rapidly.

Industrial workers head the list, with 19 million hearing-damaged people out of an industrial work force of 75 million. Workers who run a high risk of temporary or permanent hearing loss include boilermakers, weavers, riveters, bulldozer and jackhammer operators, taxicab drivers, bus and truck drivers, mechanics, machine shop workers, bar and nightclub employees, and performers who use sound systems to amplify their music. Millions of people who listen to music at loud levels using home stereos, portable stereos ("jam

Table 18-3 Effects of Common Sound Pressure Levels

Example	Sound Pressure (dbA)	Effect with Prolonged Exposure
Jet takeoff (25 meters away*)	150	Eardrum rupture
Aircraft carrier deck	140	
Armored personnel carrier, jet takeoff (100 meters away), earphones at loud level	130	
Thunderclap, textile loom, live rock music, jet takeoff (161 meters away), siren (close range), chain saw	120	Human pain threshold
Steel mill, riveting, automobile horn at 1 meter, "jam box" stereo held close to ear	110	
Jet takeoff (305 meters away), subway, outboard motor, power lawn mower, motorcycle at 8 meters, farm tractor, printing plant, jackhammer, garbage truck	100	Serious hearing damage (8 hours)
Busy urban street, diesel truck, food blender, cotton spinning machine	90	Hearing damage (8 hours), speech interference
Garbage disposal, clothes washer, average factory, freight train at 15 meters, dishwasher	80	Possible hearing damage
Freeway traffic at 15 meters, vacuum cleaner, noisy office or party	70	Annoying
Conversation in restaurant, average office, background music	60	Intrusive
Quiet suburb (daytime), conversation in living room	50	Quiet
Library, soft background music	40	
Quiet rural area (nighttime)	30	
Whisper, rustling leaves	20	Very quiet
Breathing	10	
	0	Threshold of hearing

*To convert meters to feet, multiply by 3.3.

boxes") held close to the ear, and earphones are also incurring hearing damage. Studies have shown that 60% of the incoming first-year students at the University of Tennessee have significant hearing loss in the high frequency range. In effect, these and many other young people are entering their twenties with the hearing capability of persons between the ages of 60 and 69.

Measuring and Ranking Noise To determine harmful levels of noise, sound pressure measurements using the **decibel (db)** as the basic unit can be made with a decibel meter. A mathematical equation is used to convert sound pressure measurements to loudness levels.

Sound pressure and loudness, however, are only part of the problem. Sounds also have pitch (fre-quency), and high-pitched sounds seem louder and more annoying than low-pitched sounds at the same intensity. Normally, sound pressure is weighted for high-pitched sounds and reported in dbA units, as shown in Table 18-3. Sound pressure becomes damaging at about 75 dbA, painful at around 120 dbA, and deadly at 180 dbA. Because the db and dbA sound pressure scales are logarithmic, a tenfold increase in sound pressure occurs with each 10-decibel rise. Thus a rise in sound pressure on the ear from 30 dbA (quiet rural area) to 60 dbA (normal restaurant conversation) represents a 1,000-fold increase in sound pressure.

Effects of Noise Excessive noise is a form of stress that can cause both physical and psychological damage. Continued exposure to high sound levels permanently destroys the microscopic hairlike cochlear cells in the fluid-filled inner ear, which wave back and forth to convert sound energy to nerve impulses. Sound experts consider sound levels high enough to cause permanent hearing damage if you need to raise your voice to be heard above the noise, a noise causes your ears to ring, or nearby speech seems muffled.

In addition to hearing damage, sudden noise causes automatic stress reactions, including constricted blood vessels, dilated pupils, tense muscles, increased heartbeat and blood pressure, wincing, holding of breath, and stomach spasms. Constriction of the blood vessels can become permanent, increasing blood pressure and contributing to heart disease. Migraine headaches, gastric ulcers, and changes in brain chemistry can also occur.

What Can Be Done?   Most causes of noise pollution have simple solutions. Industrial employers can control noise by substituting quieter machines and operations. Workers can shield themselves from excessive noise by wearing hearing protectors such as wax or plastic plugs, bulky headsets, and custom-made plastic inserts with valves that close automatically in response to noise. Noisy factory operations can be totally or partially enclosed. Houses and buildings can be insulated to reduce sound transfer (and energy waste). Trucks, motorcycles, vacuum cleaners, and other noisy machines are available in quieter versions, but are often not purchased because consumers falsely equate loudness of engines with their power or effectiveness.

The Soviet Union and many western European and Scandinavian nations are far ahead of the United States in reducing noise and in establishing and enforcing noise control regulations. Europeans have developed quieter jackhammers, pile drivers, and air compressors that do not cost much more than their noisy counterparts. Most European countries also require that small sheds and tents be used to muffle construction noise, and some countries reduce the clanging noises associated with garbage collection by using rubberized collection trucks. Subway systems in Montreal and Mexico City have rubberized wheels to reduce noise. In France, cars are required to have separate highway and city horns.

The government standard for overexposure to noise in any U.S. workplace is 90 dBA for 8 hours a day—still significantly above the standard of 85 dBA considered to be the minimum safe level. Industry officials oppose lowering the standard to 85 dBA because implementation would cost an estimated $20 billion.

In 1972 Congress passed the Noise Control Act, which directed the EPA to set standards for major sources of noise and to support research on the effects of noise and its control. By 1987, 15 years after the act

was passed, the EPA had issued maximum noise standards only for air conditioners, buses, motorcycles, power mowers, some trucks and trains, and some construction equipment. Standards have not been set for aircraft noise, which affects millions. Enforcement of noise regulations has been almost nonexistent because the law merely authorizes fines. In addition, since 1981 the Reagan administration has virtually eliminated the EPA budget for curbing noise pollution.

18-7 CONTROLLING AIR POLLUTION

U.S. Air Pollution Legislation   Air pollution or any other type of pollution can be controlled by laws to establish desired standards and technology to achieve the standards. In the United States little progress was made until Congress passed the Clean Air Acts of 1970 and 1977, which gave the federal government considerable power to control air pollution.

These laws required the EPA to establish **national ambient air quality standards (NAAQS)** for seven major pollutants found in almost all parts of the country: SPM, sulfur oxides, carbon monoxide, nitrogen oxides, ozone, volatile organic compounds (mostly hydrocarbons), and lead. Each NAAQS (listed in Appendix 6) specifies the maximum allowable level, averaged over a specific time period, for a certain pollutant in **ambient** (outdoor) **air.**

The EPA has also established a policy of *prevention of significant deterioration (PSD)* to prevent a decrease in air quality in regions where the air is cleaner than that required by the primary and secondary NAAQS for SPM and sulfur dioxide. Otherwise, industries would move into these areas and gradually degrade air quality to the national standards for these two major pollutants. Any new stationary emission source approved in any PSD area must use the best available technology for controlling emissions of SO_2 and SPM regardless of cost. Once the total allowed increase in emissions of these two pollutants in any PSD area is reached, no further permits are issued.

The Clean Air Acts of 1970 and 1977 required each state to develop an EPA-approved state implementation plan (SIP) showing how it would achieve federal standards fully by 1982, with extensions possible until 1987. Congress gave the EPA the power to halt the construction of major new plants or expansions of existing ones and to cut off federal funds for construction of highways for any state not submitting an acceptable plan.

The Clean Air laws also required the EPA to set uniform national maximum *emission standards* for each industry for newly built plants or major expansions of existing plants. Unlike NAAQs, costs and energy requirements can be considered in setting these standards, which are known as *new source performance standards (NSPS).*

The EPA is also required to identify and set national emission standards for stationary sources emitting hazardous air pollutants that "may cause, or contribute to, an increase in mortality or an increase in serious, irreversible, or incapacitating illness." Scientists have identified at least 600 potentially hazardous air pollutants. However, by 1986 the EPA had either established or proposed national emission standards for only eight hazardous substances: asbestos, arsenic, beryllium, mercury, vinyl chloride, benzene, sulfuric acid, and radioactive isotopes.

Congress set a timetable for achieving certain percentage reductions in emissions of carbon monoxide, hydrocarbons, and nitrogen oxides from motor vehicles. Cars and light trucks were to demonstrate a 96% reduction in hydrocarbon and carbon monoxide emissions and a 76% reduction in nitrogen oxides from 1970 levels by 1982. Although significant progress has been made, a series of legally allowed extensions pushed the deadlines for complete attainment of most of these goals to 1988 or later.

Trends in U.S. Air Quality and Emissions

Between 1975 and 1985 the average ambient concentrations of most major pollutants, except nitrogen oxides, dropped as a result of air pollution control laws, economic recession, and higher energy prices (Figure 18-19). Lead made the sharpest drop because of the gradual phaseout of leaded gasoline. Between 1982 and 1985, however, ambient levels of major pollutants other than lead either remained the same or climbed slightly. Environmentalists allege that these increases are the result of budget cutbacks and efforts by the Reagan administration to relax enforcement of air pollution control regulations.

Averages of air pollutants in several hundred EPA measuring stations across the country do not reveal the severity of air pollution in different major urban areas. The EPA uses a pollution standards index (PSI) to indicate how frequently and to what degree the air quality in a particular city exceeds one or more of the primary health standards (see Appendix 7). Between 1976 and 1983 there was a sharp drop in the number of days in which the air was classified as hazardous, very unhealthful, or unhealthful in most major urban areas. New York, Chicago, and Cleveland showed considerable improvement, while Los Angeles, Houston, and Dallas-Fort Worth showed relatively little improvement.

Methods of Pollution Control

Once a pollution control standard has been adopted, two general approaches can be used to prevent levels from exceed-

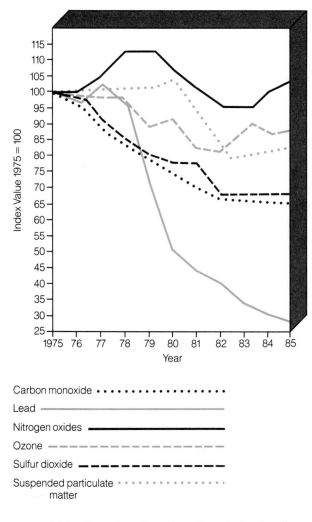

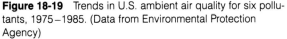

Carbon monoxide · · · · · · · · · · · · · · · · · ·
Lead ————————————————
Nitrogen oxides ━━━━━━━━━━━━━
Ozone – – – – – – – – – – – – –
Sulfur dioxide ▬ ▬ ▬ ▬ ▬ ▬ ▬ ▬ ▬
Suspended particulate · · · · · · · · · · · · · · ·
    matter

Figure 18-19 Trends in U.S. ambient air quality for six pollutants, 1975–1985. (Data from Environmental Protection Agency)

ing it: *input control*, which prevents or reduces the severity of the problem, and *output control*, which treats the symptoms. Output control methods, especially those that attempt to remove the pollutant once it has entered the environment, tend to be expensive and difficult; input methods are usually easier and cheaper in the long run.

There are five major input control methods for reducing the total amount of pollution of any type from reaching the environment: (1) Controlling population growth (Chapter 8); (2) reducing unnecessary waste of metals, paper, and other matter resources through increased recycling and reuse and designing products so that they last longer and are easy to repair (Section 14-5); (3) reducing energy use (Section 17-1); (4) using energy more efficiently (Section 17-1); and (5) switching from fossil fuels to increased reliance on energy from the sun, wind, and water (Chapter 17).

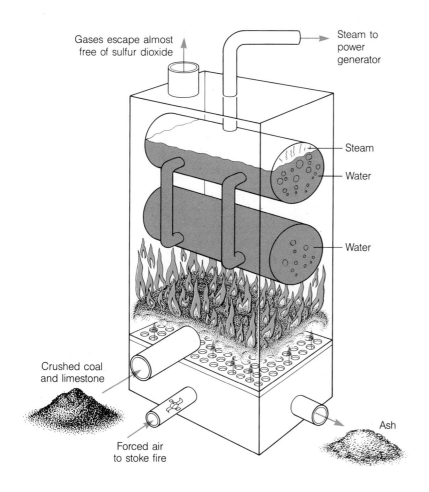

Gases escape almost free of sulfur dioxide

Steam to power generator

Steam

Water

Water

Crushed coal and limestone

Forced air to stoke fire

Ash

Figure 18-20 Limestone injection multiple burning (LIMB). Crushed limestone is injected into a boiler burning powdered coal at a lower temperature than normal burners. The limestone combines with sulfur dioxide to produce a solid material (gypsum).

These are the most effective and least costly ways to reduce air, water, and soil pollution and the only cost-effective methods for reducing the rate of buildup of carbon dioxide in the atmosphere. However, they are rarely given serious consideration in national and international strategies for pollution control.

Control of Sulfur Dioxide Emissions from Stationary Sources In addition to the input control methods mentioned above, the following approaches can lower sulfur dioxide emissions or reduce their effects:

SO₂ Input Control Methods

1. *Burn low-sulfur coal.* Especially useful for new plants located near deposits of such coal; major U.S. supplies are located west of the Mississippi (Figure 15-13, p. 263), far from major eastern power plants and industrial centers to which the coal would have to be transported at high cost and use of much energy; boilers in a number of older power plants cannot burn low-sulfur coal without expensive modifications.

2. *Remove sulfur from coal.* Fairly inexpensive; existing methods remove only 20% to 50% of the sulfur but can be combined with methods to reduce

emissions to meet national air pollution standards; produces large quantities of high-sulfur ash, which can contaminate groundwater; about 10% of the energy content of the coal is lost in the process.

3. *Convert coal to a gas or liquid fuel.* Low net energy yield; too expensive at present coal and oil prices (Section 15-5).

4. *Remove sulfur during combustion by fluidized-bed combustion (FBC)* (Figure 15-17, p. 266). Removes up to 90% of the sulfur dioxide produced during combustion; should be commercially available for small to medium-size plants in the 1990s; less costly to build and operate than scrubbers for new plants, but fairly costly to add to existing plants because a new boiler is required.

5. *Remove sulfur during combustion by limestone injection multiple burning (LIMB)* (Figure 18-20). Still in the development and testing stage, and a number of technical problems remain to be solved to reduce high operating and maintenance costs; reduces SO₂ emissions by 50% to 60%; does not remove as much SO₂ as scrubbers or FBC but is far less costly to install than either of these meth-

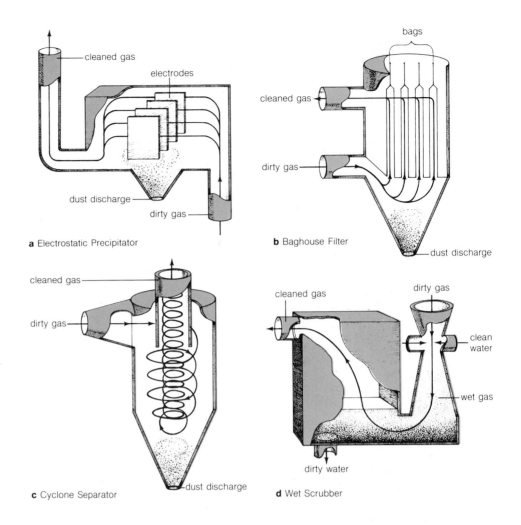

Figure 18-21 Four commonly used methods for removing particulates from the exhaust gases of electric power and industrial plants. The wet scrubber is also used to reduce sulfur dioxide emissions.

a Electrostatic Precipitator

b Baghouse Filter

c Cyclone Separator

d Wet Scrubber

ods; can cost-effectively be retrofitted into existing power and industrial plants; produces more fine particles than other methods.

SO₂ Output Control Methods

1. *Use tall smokestacks* (200 to 1,250 feet high). Can decrease pollution near power or industrial plants but increases pollution levels in downwind areas; favored by industry because it is cheaper than other approaches and passes costs of dealing with harmful effects to taxpayers at large or to another country rather than to the pollution producers; opposed by environmentalists because it does not decrease emissions of of SO_2 and NO, the major contributors to acid deposition.

2. *Remove pollutants after combustion by using flue gas desulfurization (FGD), or scrubbing.* Wet or dry limestone is sprayed into exhaust gases and combines with SO_2 to produce a solid or wet sludge, which can be used as a roadbed filler or for other construction purposes (Figure 18-21*d*); removes up to 95% of SO_2 and 99.9% of solid particulate matter (but not fine particles); can be used in new

plants and retrofitted to most existing large plants; reduces pollution in both local and distant areas; very expensive; so bulky that many small plants do not have enough space to install the equipment; produces large quantities of solid or wet sludge (depending on the method used) as a waste product; opposed by industry because of high costs, part of which must be borne directly by polluting industry rather than indirectly by taxpayers (see Spotlight on p. 343).

3. *Add a tax on each unit emitted.* Encourages development of more efficient and cost-effective methods of emissions control; opposed by industry because it costs more than tall smokestacks and requires polluters to bear more of the costs of control.

4. *Add lime or ground limestone to acidified soil and lakes to neutralize acidity.* Favored by industry over the installation of emission control devices because it shifts costs of pollution control to taxpayers at large; expensive ($100 to $150 per acre) and must be repeated periodically; is a temporary treatment, not a cure, for acid deposition.

The 1977 Clean Air Act required all U.S. coal-burning power plants built since 1978 to use *scrubbers* to remove from 70% to 90% of the sulfur dioxide from smokestack emissions, depending on the sulfur content of the coal being burned. Utility companies have vigorously opposed this requirement, claiming that scrubbers are too costly to build and keep in good operating order, that they are unreliable, that they discourage use of easily mined western low-sulfur coal reserves, and that they could raise the cost of electricity to consumers by as much as 50%.

The EPA argues that scrubber technology is well developed, is used in about 1,000 plants worldwide, and has worked well and remained in operation 90% of the time in Japan and for U.S. utilities that bought and maintained high-quality equipment. The EPA claims scrubbers increase the consumer's cost of electricity by only 5% to 20%—far less than the costs of the harmful effects of sulfur emissions on health, crops, forests, and materials.

Most coal-burning power plants coming on line in the U.S. during the 1980s will have flue gas scrubbers. But utilities are not required to install scrubbers or other SO_2 emission control devices on existing plants, most remodelled plants, and oil-burning plants converted to burn coal. These older plants typically emit seven times as much sulfur dioxide a year as a new plant with a scrubber. To avoid building costly scrubber-equipped plants, some utilities have kept older power plants in operation longer than their normal life spans or renovated them, prolonging pollution from their smokestacks. As a result, only about 12% of the country's 1,800 coal-burning power plants were equipped with scrubbers by 1986.

Executives of utilities and industrial plants continue to oppose more stringent control of sulfur dioxide emissions, especially on existing plants. Environmentalists, however, urge that by 1995 *all* new and existing coal- and oil-burning electric power and industrial plants be required to reduce sulfur dioxide emissions by 90% and emissions of nitrogen oxides by at least 50% by whatever method or combination of methods can achieve this goal. They also believe that all existing tall smokestacks should be either shut down or reduced in size to no more than 30.5 meters (100 feet). What do you think?

Control of Emissions of Nitrogen Oxides from Stationary Sources So far relatively little emphasis has been placed on reducing emissions of nitrogen oxides from stationary sources because control of emissions of sulfur dioxide and particulates has been considered more important. Now it is clear that nitrogen oxides are a major contributor to acid deposition and that they increase tropospheric levels of ozone and other photochemical oxidants that can damage crops, trees, and materials. The following approaches can be used to decrease emissions of nitrogen oxides from stationary sources:

NO$_x$ Input Control Methods

1. *Remove nitrogen oxides during fluidized-bed combustion (FBC).* Removes 50% to 75% of the nitrogen oxides. See SO_2 control methods for advantages and disadvantages.

2. *Remove during combustion by limestone injection multiple burning (LIMB).* Removes 50% to 60% of nitrogen oxides. See SO_2 control methods for advantages and disadvantages.

3. *Reduce by decreasing combustion temperatures.* Removes 50% to 60% of nitrogen oxides; well-established technology; can be used in new plants or retrofitted to existing plants.

NO$_x$ Output Control Methods

1. *Use tall smokestacks.* See SO_2 control methods.

2. *Add a tax for each unit emitted.* See SO_2 control methods.

3. *Remove after combustion through reburning.* Exhaust gases from the primary combustion zone are reburned at a lower temperature in a burner fueled by natural gas or low-sulfur oil; removes 50% or more of nitrogen oxides and up to 90% when combined with input methods; still under development for large plants.

4. *Remove after burning by reacting with isocyanic acid (HCNO).* Still in the laboratory stage and will not be available for at least 10 years; removes up to 99% of nitrogen oxides.

Control of Particulate Matter Emissions from Stationary Sources The following approaches can be used to decrease emissions of SPM from stationary sources:

Figure 18-22 The effectiveness of an electrostatic precipitator in reducing particulate emissions is shown by this stack, with the precipitator turned off (left) and with the precipitator operating.

SPM Input Control Method

1. *Convert coal to a gas or liquid.* See SO$_2$ control methods.

SPM Output Control Methods

1. *Use tall smokestacks.* See SO$_2$ control methods.

2. *Add a tax on each unit emitted.* See SO$_2$ control methods.

3. *Remove particulates from stack exhaust gases.* The most widely used method in electric power and industrial plants. Several methods are in use: (a) electrostatic precipitators (Figures 18-21a and 18-22), which remove up to 99.55% of the total mass of particulate matter (but not most fine particles) by means of an electrostatic field that charges the particles so that they can be attracted to a series of electrodes and removed from exhaust gas; (b) baghouse filters (Figure 18-21b), which can remove up to 99.9% of the particles (including most fine particles) as exhaust gas passes through fiber bags in a large housing; (c) cyclone separators (Figure 18-21c), which remove 50% to 90% of the large particles (but very few medium-size and fine particles) by swirling exhaust gas through a funnel-shaped chamber in which particles collect through centrifugal force; and (d) wet scrubbers (Figure 18-21d), which remove up to 99.5% of the particles (but not most

fine particles). Except for baghouse filters, none of these methods removes many of the more hazardous fine particles; all produce hazardous solid waste or sludge that must be disposed of safely; and except for cyclone separators, all methods are expensive and none prevents particles formed as secondary pollutants in the atmosphere.

Control of Emissions from Motor Vehicles The following approaches can be used to decrease emissions of carbon monoxide, nitrogen oxides, SPM, and lead from motor vehicles:

Motor Vehicle Input Control Methods

1. *Rely more on mass transit and paratransit.*

2. *Shift to less-polluting automobile engines* such as steam or electric engines. At present these engines do not match the internal combustion engine in terms of performance, fuel economy, durability, and cost. Electric cars would increase use of electricity produced by coal-burning or nuclear power plants and trade one set of environmental hazards for another.

3. *Shift to less-polluting fuels* such as natural gas, alcohols, and hydrogen gas. Supplies of natural gas are limited; alcohol is still too costly but may become competitive when oil prices rise (Section

17-6); hydrogen gas has a negative net useful energy yield and requires greatly improved fuel cells (Section 17-7).

4. *Improve fuel efficiency.* A quick and cost-effective approach. Present U.S. fuel-efficiency standards should be greatly increased from 26.5 mpg in 1986 to 40 mpg by 2000.

5. *Modify the internal combustion engine to reduce emissions.* Burning gasoline using a lean, or more air-rich, mixture reduces CO and hydrocarbons but increases NO emissions; however, this method can be combined with output control to reduce overall emissions, and a new lean-burn engine that reduces emissions of nitrogen oxides by 75% to 90% may be available in about 10 years.

Motor Vehicle Output Control Methods

1. *Use emission control devices.* Most widely used approach; positive crankcase ventilation (PCV) recycles hydrocarbons released from the crankcase back into the engine for combustion; exhaust gases are recirculated back through the engine to cool combustion temperature and reduce NO emissions; three-way catalytic converters change carbon monoxide and hydrocarbons in exhaust gas into CO_2 and water vapor and NO into N_2; platinum and palladium catalysts used in catalytic converters to speed up these reactions are easily deactivated by lead in gasoline, but this problem will decrease as leaded gasoline is phased out; engines must be kept well tuned for converters to work effectively; three-way catalytic converters now under development can decrease pollutants by 90% to 95% and should be available within a few years.

2. *Require car inspections twice a year and increase fines* to ensure that emission control devices are not tampered with and are in good working order. About 48% of U.S. cars and light trucks have malfunctioning emissions equipment.

What Needs to Be Done   Since 1981, when the 1977 Clean Air Act was up for revision, industry officials and the Reagan administration have pushed hard to ease federal auto emission standards, relax industrial cleanup goals, allow more pollution in PSD regions, and extend nationwide EPA deadlines for meeting primary air pollution standards. They claim that the benefits of existing air pollution control laws are not worth their high costs, that these laws threaten economic growth, and that they are implemented too inflexibly by the EPA.

Environmentalists, on the other hand, have proposed stricter limits on all polluting emissions; a ban on CFCs; enactment of standards for key indoor air pollutants in homes, factories, and office buildings; and enactment and enforcement of much stricter noise control standards. They recognize that the costs of implementing a much stricter and more comprehensive air pollution control program will be high. But they argue that the long-term costs of not doing so could be astronomical: massive damage to people, livestock, crops, materials, forests, soils, and lakes.

Environmentalists accuse the EPA of too frequently granting industries and cities extensions for meeting standards and of reducing efforts to enforce the existing laws. They point out that these laws have not led to the wave of plant closings and high unemployment predicted by industry and have created numerous jobs in the air pollution control industry.

At the international level, countries—especially MDCs—need to develop and strictly enforce agreements to ban CFCs, reduce emissions of SO_2 and NO to control acid deposition, and reduce emissions of CO_2 to delay global warming. An important start was made when the Soviet Union and 20 European countries agreed to reduce their annual emissions of sulfur dioxide by 30% to 50% by 1995 from 1980 levels. In 1987 countries producing and using most of the world's CFCs also agreed to make a modest 20% reduction in the production of these chemicals by 1992. Meaningful reductions in atmospheric carbon dioxide inputs could be accomplished by just three countries—China, the Soviet Union, and the United States—the world's three largest users of coal.

There is the very real possibility that the human race— through ignorance or indifference or both—is irreversibly altering the ability of the atmosphere to support life.
Sherwood Rowland

DISCUSSION TOPICS

1. Rising oil and natural gas prices and environmental concerns over nuclear power plants could force the U.S. to depend more on coal, its most plentiful fossil fuel, for electric power. Comment on this in terms of air pollution. Would you favor a return to coal instead of increased use of nuclear power? Explain.

2. Evaluate the pros and cons of the statement, "Since we have not proven absolutely that anyone has died or suffered serious disease from nitrogen oxides, automobile manufacturers should not be required to meet the federal air pollution standards."

3. Should all uses of CFCs be banned in the U.S., including their use in refrigeration and air conditioning units? Explain.

4. Should MDCs set up a world food bank to store several years' supply of food to reduce the harmful

effects of a loss in food production from a change in climate? How would you decide who gets this food in times of need?

5. How do levels of major pollutants in your area compare with the NAAQS shown in Appendix 6? What trends in these levels have taken place during the past ten years?

6. What topographical and climate factors either enhance or help decrease air pollution in your community?

7. Do you favor or oppose requiring a 50% reduction in emissions of sulfur dioxide and nitrogen oxides by fossil fuel-burning electric power and industrial plants and a 50% reduction in emissions of nitrogen oxides by motor vehicles in the United States between 1989 and 1999? Explain.

8. Should all tall smokestacks be banned? Explain.

9. Do buildings in your college or university contain asbestos? If so, what is being done about this potential health hazard?

10. Should standards be set and enforced for most major indoor air pollutants? Explain.

11. As a class or group project try to borrow one or more sound pressure decibel meters from the physics or engineering department or from a local stereo or electronics repair shop. Make a survey of sound pressure levels at various times of day and at several locations; record the results on a map. Include a room with a stereo, and take readings at an indoor concert or nightclub at various distances from the sound system speakers. Also measure sound pressure levels from earphones at several different volume settings. Correlate your findings with those in Table 18-3.

19

Water Pollution

GENERAL OBJECTIVES

1. What are the major types, sources, and effects of water pollutants?
2. What are the major pollution problems of rivers and lakes?
3. What are the major pollution problems of the world's oceans?
4. What are the major pollution problems of groundwater?
5. What technological and legal methods can be used to reduce water pollution?

You can use the latest toothpaste,
Then rinse your mouth with industrial waste.
Tom Lehrer

Water pollution is any physical or chemical change in surface water or groundwater that can adversely affect living organisms. The level of purity required for water depends on its use. Water too polluted to drink may be satisfactory for washing steel, producing electricity at a hydroelectric power plant, or cooling the steam and hot water produced by a nuclear or coal-fired power plant. Water too polluted for swimming may not be too polluted for boating or fishing.

19-1 TYPES, SOURCES, AND EFFECTS OF WATER POLLUTION

Major Water Pollutants and Their Effects   For convenience, biological, chemical, and physical forms of water pollution can be broken down into eight major types:

1. disease-causing agents (bacteria, viruses, protozoa, and parasites)
2. oxygen-demanding wastes (domestic sewage, animal manure, and other biodegradable organic wastes that deplete water of dissolved oxygen)
3. water-soluble inorganic chemicals (acids, salts, toxic metals and their compounds)
4. inorganic plant nutrients (water-soluble nitrate and phosphate salts)
5. organic chemicals (insoluble and water-soluble oil, gasoline, plastics, pesticides, cleaning solvents, and many others)
6. sediment or suspended matter (insoluble particles of soil, silt, and other inorganic and organic materials that can remain suspended in water)
7. radioactive substances
8. heat

Table 19-1	Major Water Pollutants		
Pollutant	Sources	Effects	Control Methods
Oxygen-demanding wastes	Natural runoff from land; human sewage; animal wastes; decaying plant life; industrial wastes (from oil refineries, paper mills, food processing, etc.); urban storm runoff	Decomposition by oxygen-consuming bacteria depletes dissolved oxygen in water; fish die or migrate away; plant life destroyed; foul odors; poisoned livestock	Treat wastewater; minimize agricultural runoff
Disease-causing agents	Domestic sewage; animal wastes	Outbreaks of waterborne diseases, such as typhoid, infectious hepatitis, cholera, and dysentery; infected livestock	Treat wastewater; minimize agricultural runoff; establish a dual water supply and waste disposal system
Inorganic chemicals and minerals			
Acids	Mine drainage; industrial wastes; acid deposition (Section 18-2)	Kills some organisms; increases solubility of some harmful minerals	Seal mines; treat wastewater; reduce atmospheric emissions of sulfur and nitrogen oxides (Section 18-7)
Salts	Natural runoff from land; irrigation; mining; industrial wastes; oil fields; urban storm runoff; deicing of roads with salts	Kills freshwater organisms; causes salinity buildup in soil; makes water unfit for domestic use, irrigation, and many industrial uses	Treat wastewater; reclaim mined land; use drip irrigation; ban brine effluents from oil fields
Lead	Leaded gasoline; some pesticides; smelting of lead (see Section 20-5)	Toxic to many organisms, including humans	Ban leaded gasoline and pesticides; treat wastewater
Mercury	Natural evaporation and dissolving; industrial wastes; fungicides	Highly toxic to humans (especially methyl mercury)	Treat wastewater; ban unessential uses (Section 20-5)
Plant nutrients (phosphates and nitrates)	Natural runoff from land; agricultural runoff; mining; domestic sewage; industrial wastes; inadequate wastewater treatment; food-processing industries; phosphates in detergents	Algal blooms and excessive aquatic growth; kills fish and upsets aquatic ecosystems; eutrophication; possibly toxic to infants and livestock; foul odors	Advanced treatment of industrial, domestic, and food-processing wastes; recycle sewage and animal wastes to land; minimize soil erosion
Sediments	Natural erosion, poor soil conservation; runoff from agricultural, mining, forestry, and construction activities	Major source of pollution (700 times solid sewage discharge); fills in waterways, harbors, and reservoirs; reduces shellfish and fish populations; reduces ability of water to assimilate oxygen-demanding wastes	More extensive soil conservation practices (Section 9-4)

Table 19-1 summarizes the major sources, effects, and methods for controlling these eight major types of water pollutants.

Point and Nonpoint Sources   For purposes of control and regulation it is convenient to distinguish between point sources and nonpoint sources of water pollution from human activities. A **point source** is a source that discharges pollutants, or any effluent, such as wastewater, through pipes, ditches, and sewers into bodies of water at specific locations (Figure 19-1). Examples include factories, sewage treatment plants (which remove some but not all pollutants), electric

Table 19-1 Major Water Pollutants *(continued)*

Pollutant	Sources	Effects	Control Methods
Radioactive substances	Natural sources (rocks and soils); uranium mining and processing; nuclear power generation; nuclear weapons testing	Cancer; genetic defects (see Section 3-2)	Ban or reduce use of nuclear power plants and weapons testing; more strict control over processing, shipping, and use of nuclear fuels and wastes (see Section 16-2)
Heat	Cooling water from industrial and electric power plants	Decreases solubility of oxygen in water; can kill some fish; increases susceptibility of some aquatic organisms to parasites, disease, and chemical toxins; changes composition of and disrupts aquatic ecosystems	Decrease energy use and waste; return heated water to ponds or canals or transfer waste heat to the air; use to heat homes, buildings, and greenhouses
Organic chemicals			
Oil and grease	Machine and automobile wastes; pipeline breaks; offshore oil well blowouts; natural ocean seepages; tanker spills and cleaning operation	Potential disruption of ecosystems; economic, recreational, and aesthetic damage to coasts, fish, and waterfowl; taste and odor problems	Strictly regulate oil drilling, transportation, and storage; collect and reprocess oil and grease from service stations and industry; develop means to contain and mop up spills
Pesticides and herbicides	Agriculture; forestry; mosquito control	Toxic or harmful to some fish, shellfish, predatory birds, and mammals; concentrates in human fat; some compounds toxic to humans; possible birth and genetic defects and cancer (see Section 6-4)	Reduce use; ban harmful chemicals; switch to biological and ecological control of insects (Section 21-5)
Plastics	Homes and industries	Kills fish; effects mostly unknown	Ban dumping, encourage recycling of plastics; reduce use in packaging
Detergents (phosphates)	Homes and industries	Encourages growth of algae and aquatic weeds; kills fish and causes foul odors as dissolved oxygen is depleted	Ban use of phosphate detergents in crucial areas; treat wastewater (see Section 19-5)
Chlorine compounds	Water disinfection with chlorine; paper and other industries (bleaching)	Sometimes fatal to plankton and fish; foul tastes and odors; possible cancer in humans	Treat wastewater; use ozone for disinfection and activated charcoal to remove synthetic organic compounds

power plants, active and abandoned underground coal mines, oil tankers, and offshore oil wells. So far most water pollution control efforts have concentrated on reducing discharges to surface water from industrial and municipal point sources because they are easy to identify.

In contrast, a **nonpoint source** of water pollution is a source that is one of many widely scattered sources that discharge pollutants over a large area (Figure 19-1). Examples include runoff into surface water and seepage into groundwater from croplands, livestock feedlots, logged forests, urban and suburban lands, construction areas, parking lots, and roadways. While point sources usually produce regular, year-round dis-

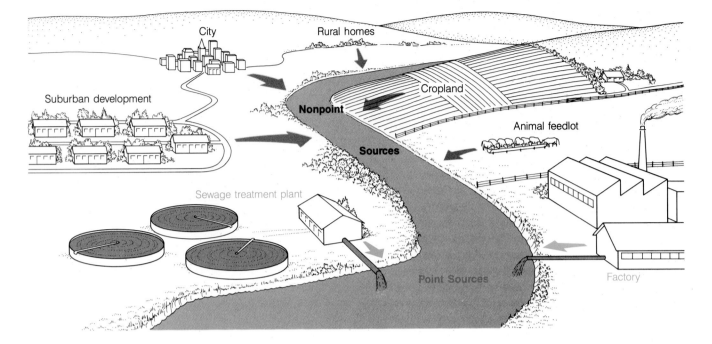

Figure 19-1 Point and nonpoint sources of water pollution.

charges, significant pollution from most nonpoint sources usually occurs only during major storms or when snow melts. Relatively little progress has been made in the control of nonpoint water pollution because of the difficulty and expense of identifying and controlling discharges from so many diffuse sources.

Some Indicators of Water Quality Three indicators of water quality are the coliform bacteria count, concentration of dissolved oxygen (DO), and biological oxygen demand (BOD). Detecting *specific* disease-causing agents in water is a difficult, time-consuming, and costly process. Instead, water used for drinking or swimming is routinely analyzed for the presence of **coliform bacteria,** found in great numbers in the intestines and thus in the feces of humans and other animals. Most coliform bacteria themselves are not harmful. But the presence of large numbers of these bacteria in water indicates recent contamination by untreated feces, which are likely to contain more dangerous bacteria and viruses.

Usually, several samples are taken, and the water is considered safe to drink by EPA standards if the arithmetic mean of the **coliform bacteria count** of all samples does not exceed 1 coliform bacterial colony per 100 milliliters of water, with no single sample having a count higher than 4 colonies per 100 milliliters. When this level is exceeded, a municipal water treatment plant must either add more chlorine to the water or use an alternative source for drinking water.

The EPA-recommended maximum level for swimming water is 200 colonies per 100 milliliters, but some cities and states allow higher levels. When the allowable level is exceeded, the contaminated pool, river, or beach is usually closed to swimming.

One of the most useful indicators of the ability of a body of surface water to support fish and most other forms of aquatic life is its **dissolved oxygen (DO)** content: the amount of oxygen gas dissolved in a given quantity of water at a particular temperature and atmospheric pressure. When surface waters are overloaded with biodegradable wastes, the resulting population explosion of aerobic decomposers can reduce the supply of dissolved oxygen so that aquatic organisms, especially fish and shellfish, die from suffocation. Complete oxygen depletion kills all forms of aquatic life except anaerobic bacteria, which do not require oxygen to break down organic material. The decomposition of organic wastes by these anaerobic decomposers produces toxic and foul-smelling substances such as hydrogen sulfide (recognized by its rotten-egg smell), ammonia, and methane (swamp gas), which bubble to the surface. Figure 19-2 shows the correlation between water quality and its parts per million of dissolved oxygen.

The quantity of oxygen-consuming wastes in water is usually determined by measuring the **biological oxygen demand (BOD):** the amount of dissolved oxygen needed by aerobic decomposers to break down the organic materials in a given volume of water over a 5-day incubation period at 20°C (68°F). Certain chem-

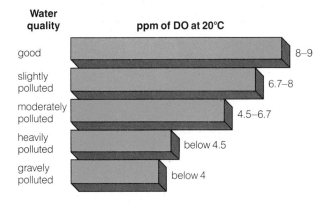

Water quality	ppm of DO at 20°C	
good		8–9
slightly polluted		6.7–8
moderately polluted		4.5–6.7
heavily polluted	below 4.5	
gravely polluted	below 4	

Figure 19-2 Water quality and dissolved oxygen (DO) content.

icals other than organic wastes can react with oxygen dissolved in water. The **chemical oxygen demand (COD)** is a more complete and accurate measurement of the total depletion of dissolved oxygen in water. Water is considered seriously polluted when its BOD or COD causes the dissolved oxygen content to fall below 4.5 ppm, and gravely polluted if the DO level falls below 4 ppm.

19-2 POLLUTION OF RIVERS, LAKES, AND RESERVOIRS

Natural Processes Affecting Pollution Levels in Surface Water The concentrations and chemical forms of most pollutants change once they are added to bodies of surface water as a result of four natural processes: dilution, biodegradation, biological amplification, and sedimentation. The degree of dilution of all pollutants and biodegradation of oxygen-consuming wastes by decomposers taking place in a body of surface water depends on its volume and flow rate. In a large, rapidly flowing river, relatively small amounts of pollutants are quickly diluted to low concentrations and the supply of dissolved oxygen needed for aquatic life and for biodegradation of oxygen-consuming wastes is rapidly renewed. However, such rivers can be overloaded with pollutants. In addition, dilution and biodegradation are sharply reduced when flow is decreased during dry spells; or when large amounts of water are withdrawn for irrigation or cooling and returned in smaller amounts because of evaporation, or returned at high temperatures, which decrease its DO content.

In lakes, reservoirs, estuaries, and oceans, dilution is often less effective than in rivers because these bodies of water frequently contain stratified layers that undergo little vertical mixing. This also reduces the levels of dissolved oxygen, especially in bottom lay-

ers. In addition, lakes and reservoirs have little flow, further reducing dilution and replenishment of DO.

Another problem with relying on dilution to disperse pollution is that some substances, especially synthetic organic chemicals, can have harmful effects on aquatic life and humans at extremely small concentrations. Biodegradation is ineffective in removing nonbiodegradable and persistent pollutants. Also, concentrations of some of these pollutants (such as DDT, PCBs, some radioactive isotopes, and some mercury compounds) are biologically amplified to higher concentrations as they pass through food webs (Figure 4-17, p. 67).

Sedimentation can remove trace amounts of some organic and inorganic pollutants, which become attached to particles that settle and accumulate in the mud at the bottom of lakes, reservoirs, and slow-flowing rivers. However, toxic substances stored in bottom sediments can become resuspended if the bottom is dredged or if it is stirred up by high flow rates during flooding. Although the natural processes of dilution, biodegradation, and sedimentation can reduce many pollutants to harmless levels, we have learned that the solution to most forms of water pollution is to prevent or reduce their entry.

Rivers and Degradation of Oxygen-Consuming Wastes Because they flow, most rivers recover rapidly from some forms of pollution—especially excess heat and oxygen-demanding wastes—as long as they are not overloaded. How long this recovery process takes depends on the river's volume, flow rate, and the volume of incoming biodegradable wastes. Changes in the amount of DO at various points in a river lead to changes in the types and diversity of fish and other aquatic organisms (Figure 19-3).

Rivers are particularly susceptible to being overloaded with oxygen-demanding wastes during hot summer months, when stream flow and turbulence are low and the effects of dilution and oxygen transfer from the air are reduced. DO levels are also reduced because the warm water holds less oxygen and speeds up bacterial decay. In the United States many once-rapid major rivers, such as the Ohio, have been converted into chains of long, slow-flowing lakes by construction of flood control, navigational, and recreational dams. As a result, they are less effective in biodegrading oxygen-demanding wastes and dissipating inputs of heated water from power and industrial plants.

Along many rivers, water for drinking is removed upstream from a town, and industrial and sewage wastes are discharged downstream. This pattern is usually repeated hundreds of times. If a river or stream receives heavy loads of oxygen-demanding wastes along most or all of its path, the river can suffer from

Figure 19-3 The oxygen sag curve (solid) versus oxygen demand (dashes). Depending on flow rates and the amount of pollutants, rivers recover from oxygen-demanding wastes and heat if given enough time and if they are not overloaded.

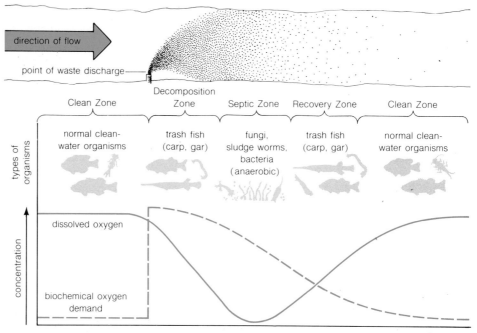

severe oxygen depletion, loss of most fish, and buildup of high levels of disease-carrying human and animal wastes. Requiring each town or city to withdraw its drinking water downstream rather than upstream would lead to a dramatic improvement in the quality of river water. Each town would be forced to clean up its own waste outputs rather than passing them on to downstream areas.

U.S. River Water Quality   Since 1972, about 13% of the 570,000 kilometers (354,000 miles) of U.S. rivers and streams on which the EPA has information have improved in quality in terms of increased DO, diversity of aquatic life, and decreased biological oxygen demand. This improvement in water quality is primarily a result of water pollution control laws, which have greatly increased the number and quality of sewage treatment plants and required industries to reduce or eliminate discharges into surface waters.

Since 1972 about 3% of monitored U.S. stream miles have suffered a decrease in quality. The quality of about 84% of monitored U.S. stream miles has remained about the same since 1972. Primarily by reducing discharges from point sources, the country is holding the line against pollution of most of its rivers and streams—an impressive accomplishment considering the rise in economic activity and population since 1972. Further improvements will require stricter monitoring and enforcement of existing standards for discharges from point sources and massive efforts to reduce inputs from nonpoint sources.

River Water Quality in Other Parts of the World   Pollution control laws have also led to improved water quality in many rivers and streams in Canada, Japan, and most western European countries since 1970. Many rivers in the Soviet Union, however, have become more polluted with industrial wastes as industries have expanded without adequate pollution controls.

A spectacular river cleanup has occurred in Great Britain. In the 1950s the river Thames was little more than a flowing anaerobic sewer. But after more than 30 years of effort, $250 million of British taxpayers' money, and millions more spent by industry, the Thames has made a remarkable recovery. Dissolved oxygen levels have risen to the point where the river now supports increasing populations of at least 95 species of fish, including the pollution-sensitive salmon. Commercial fishing is thriving, and many species of waterfowl and wading birds have returned to their former feeding grounds.

Despite important progress, stretches of some rivers in MDCs are still polluted with excessive quantities of wastes from point and nonpoint sources. Large fish kills and contamination of drinking water still occur because of accidental or deliberate releases of toxic inorganic and organic chemicals by industries (see Spotlight on p. 353), malfunctioning sewage treatment plants, and runoff of pesticides from cropland.

Available data indicate that pollution of rivers in most LDCs is a serious and growing problem. Currently, more than two thirds of India's water resources are polluted. Of the 78 rivers monitored in China, 54

The Rhine River winds 1,320 kilo-meters (820 miles) through Switzerland, France, West Germany, and the Netherlands before emptying into the North Sea (Figure 19-4). Numerous cities and chemical, steel, and other plants border this heavily used and abused river.

In 1970 the Rhine was so heavily polluted with toxic and oxygen-demanding wastes that it was devoid of most fish and contained only about 25 forms of aquatic animal life. Between 1970 and 1986 cleanup efforts increased DO levels by almost 60%, decreased BOD by 50%, and increased the number of types of aquatic animal life to 100, including 15 species of fish reintroduced by scientists.

This progress was set back when a fire broke out on November 1, 1986, near Basel, Switzerland, at a chemical warehouse owned by Sandoz—a large, Swiss-based international chemical and pharmaceutical company. The fire ruptured drums of chemicals, and water hosed onto the flames flushed at least 27 metric tons (30 tons) of toxic chemicals—mostly herbicides, mercury-containing fungicides, and dyes (which break down into cyanide)—into the Rhine. At least nine other spills of toxic chemicals by other industries occurred between November 6 and December 2—causing environmentalists to charge that some were deliberately spilled.

Within a week after their discharge most of the chemicals had flowed into the already polluted North Sea, where it is hoped most will be diluted to harmless levels or broken down into harmless substances. Villages and cities in the four countries depending on the river for drinking water had to temporarily find other supplies. Near Strasbourg, France, sheep that drank water from the river died. Most countries banned all fishing in the river and tributaries until further notice.

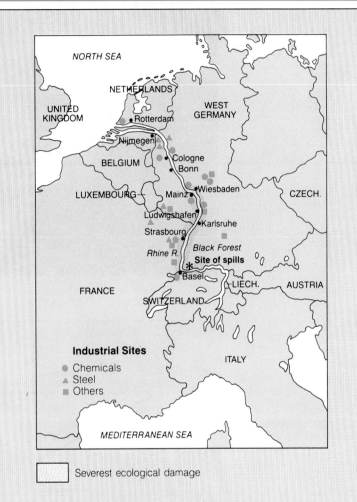

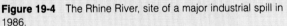
Severest ecological damage

Figure 19-4 The Rhine River, site of a major industrial spill in 1986.

At least half a million fish and large numbers of eels, mussels, snails, aquatic insects, and waterfowl were killed. About 328 kilometers (205 miles) of the river, from Basel to Mainz, West Germany, suffered severe ecological damage. Of greatest concern are the estimated 200 kilograms (440 pounds) of highly toxic mercury compounds, most of which settled to the bottom of the river. Scientists fear that eventually much of this mercury will be biologically amplified in food webs, killing fish-eating birds, and poisoning people who eat large quantities of contaminated fish.

Officials of countries downstream from Switzerland complained that Swiss authorities had failed to notify them of the accident for 24 hours and even then had not warned them about the severity of the spill. Although they broke no laws, Sandoz officials had decided five years earlier not to act on recommendations made by an insurance company to improve warehouse safety. A small investment would have saved the company large amounts of money it is now expected to pay as compensation—and, more important, would have prevented a major ecological disaster.

Figure 19-5 Fish killed by pesticide runoff in Arrowhead Lake, North Dakota.

Nelius B. Nelson/U. S. Fish and Wildlife Service

Figure 19-6 Major sources of nutrient overload from human activities, or cultural eutrophication.

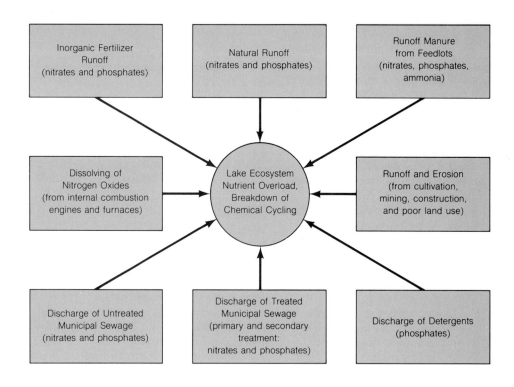

are seriously polluted with untreated sewage and industrial wastes. In Latin America many rivers are severely polluted. The Bogota River, for example, is contaminated by sewage from 5 million people and industrial wastes from Bogota, Columbia. At the town of Tocaima, 120 kilometers (75 miles) downstream, the river has an average coliform bacteria count that is 36,500 times the maximum level allowed for swimming in the United States. Nearly all the children in a

nearby village have sores on their skin from swimming in the river.

Pollution Problems of Lakes and Reservoirs The flushing and changing of water in lakes and large reservoirs can take from one to a hundred years, compared to several days to several weeks for rivers. As a result, these bodies of surface water are more sus-

Figure 19-7 Mass of dead and dying algae from cultural eutrophication of a lake.

National Archives

ceptible than rivers to contamination with plant nutrients, oil, and toxic substances that can destroy bottom life and kill fish (Figure 19-5). Runoff of acids into lakes is a serious and increasing problem in areas subject to acid deposition (Figure 18-11, p. 329) and has killed most fish and other forms of aquatic life in thousands of lakes in western Europe, northwestern Canada, and the northeastern United States.

In 1985 the EPA reported that the water quality in 78% of the area of 3,755 U.S. lakes and reservoirs monitored fully support their designated uses such as boating, commercial and sport fishing, and swimming. Despite this encouraging news, significant water pollution problems remain, especially as a result of runoff of plant nutrients and toxic wastes from nonpoint sources.

The most widely reported problem impairing some uses of relatively shallow lakes and reservoirs, especially those near urban or agricultural centers, is accel-

erated eutrophication as a result of human activities (Figure 19-6). **Eutrophication** is a natural process by which a lake gradually becomes enriched in plant nutrients such as phosphates and nitrates as a result of natural erosion and runoff from the surrounding land basin. The stepped-up addition of phosphates and nitrates as a result of human activities, sometimes called **cultural eutrophication,** can produce in a few decades the same degree of plant nutrient enrichment that takes thousands to millions of years by natural processes.

Overloading of shallow lakes and reservoirs with plant nutrients during the summer produces dense growths of rooted plants like water chestnuts and water hyacinths near the shore. It also causes population explosions, or **algal blooms,** of floating algae, especially the blue-green species, which give the water the appearance of green soup and which release substances that make the water taste and smell bad.

Dissolved oxygen in the surface layer of water near the shore and the bottom layer in areas covered with algae is depleted when large masses of these algae die, fall to the bottom, and are decomposed by aerobic bacteria (Figure 19-7). Then lake trout and other deepwater species of game fish die of oxygen starvation, leaving the lake populated by panfish species like carp, which need less oxygen. Cultural eutrophication may cause the actual number of fish to increase, but there will be fewer of the kinds of game fish that most people prefer. If excess nutrients continue to flow into a lake, the bottom water becomes foul and almost devoid of animals, as anaerobic bacteria take over and produce their smelly decomposition products.

About one-third of the 100,000 medium to large lakes and about 85% of the large lakes near major population centers in the United States suffer from some degree of cultural eutrophication. The Great Lakes, for example, receive massive inputs of plant nutrients and numerous toxic water pollutants from point and nonpoint sources (see Spotlight on p. 356). Severe cultural eutrophication jeopardizes the use of lakes and reservoirs for drinking, sport and commercial fishing, recreation, irrigation, and cooling of electric power and industrial plants.

Control of Cultural Eutrophication The solution to cultural eutrophication is to use input methods to reduce the flow of nutrients into lakes and reservoirs and output methods to clean up lakes that suffer from excessive eutrophication.

Cultural Eutrophication: Input Control Methods

1. Use advanced waste treatment (Section 19-5) to remove 90% of phosphates from effluents of sew-

The five interconnected Great Lakes comprise the largest surface body of fresh water on earth. The massive Great Lakes watershed contains thousands of industries and over 60 million people—one-third of Canada's population and one-eighth of the U.S. population. The lakes also serve as a source of drinking water for 24 million people.

Despite their massive size, the lakes are especially sensitive to pollution from point and nonpoint sources throughout their vast drainage basin because only 1% of the water entering them flows out at the St. Lawrence River in any single year. By the 1960s many areas of the lakes were suffering from severe cultural eutrophication, massive fish kills, and contamination with bacteria and other wastes, forcing the closing of many bathing beaches and severely reducing commercial and sport fishing.

Although all five lakes were affected, the impact on Lake Erie was particularly intense. It is the shallowest of the Great Lakes and has the smallest volume of water, and its drainage basin is heavily industrialized and contains the largest population of any of the lakes. At one time in the mid-1960s, massive algal blooms choked off oxygen to 65% of the lake's bottom, and populations of many of its most desirable species of commercial and game fish were sharply reduced. Lake Ontario's small size and shallowness also make it susceptible to cultural eutrophication and other forms of water pollution.

Since 1972 a joint $15 billion pollution control program by Canada and the United States has led to significant decreases in levels of phosphates, coliform bacteria, and many toxic industrial chemicals, along with decreases in algal blooms and increases in dissolved oxygen, aquatic life, and sport and commercial fishing. Only eight of 516 swimming beaches remained closed because of pollution by 1987. These accomplishments were mainly the result of decreases in discharges from point sources brought about by building and upgrading sewage treatment plants; improving treatment of industrial wastes; and banning or limiting the amount of phosphate in detergents, household cleaners, and water conditioners in many areas of the Great Lakes drainage basin.

Although runoff of phosphates from nonpoint sources is still a problem in some areas, the most serious problem today is contamination from toxic wastes. Most toxic "hot spots" are found in harbors or near the mouths of tributaries emptying into the lakes, especially Lake Erie and Lake Ontario.

---

age treatment and industrial plants before they reach the lake.

2. Ban or set low limits on phosphates in household detergents and other cleaning agents to reduce the amount of phosphate reaching sewage treatment plants and nonpoint sources such as septic tanks.

3. Control land use, use sound soil conservation practices, and clean streets regularly to reduce runoff of fertilizers, manure, and soil from nonpoint sources.

4. Divert wastewater to fast-moving streams or to the ocean. This approach is not possible in most places; where it is possible, it may transfer the problem from a lake to a nearby estuary.

Cultural Eutrophication: Output Control Methods

1. Dredge bottom sediments to remove excess nutrient buildup. Impractical in large, deep lakes; not very effective in shallow lakes; often reduces water quality by resuspending toxic pollutants;

dredged material must go somewhere—usually into the ocean.

2. Remove or harvest excess weeds and debris. Disruptive to some forms of aquatic life and difficult and expensive in large lakes.

3. Control nuisance plant growth with herbicides and algicides. Can pollute water and kill off other plants.

4. Aerate lakes and reservoirs to avoid oxygen depletion (Figure 19-8). Expensive.

As with other forms of pollution, input approaches are the most effective. Input control methods have to be tailored to each situation based on the *limiting factor principle:* When a number of nutrients are needed for the growth of various plant species, the one in smallest supply will limit or stop growth. For example, because phosphorus is the limiting factor in most freshwater lakes, its control should be emphasized.

Fortunately, if excessive inputs of limiting plant nutrients stop, the lake will usually return to its previous state. But there is disagreement over whether

USDA/Soil Conservation Service

phosphorus inputs should be reduced by banning or limiting phosphates in laundry detergents, by removing phosphates from wastewater at sewage treatment plants, or both. Studies of over 400 bodies of water indicate that a reduction of 20% in the total phosphate load must be achieved to produce a discernible effect on water quality. By 1987 eight states—Indiana, Maryland, Michigan, Minnesota, New York, Vermont, Virginia, amd Wisconsin—and many parts of Canada had banned the use of phosphate detergents. Such bans have made a major contribution to reducing cultural eutrophication in the Great Lakes and other areas and have saved consumers and taxpayers money. In some lakes and in coastal waters and estuaries, emphasis should be on reducing inputs of nitrogen because it is the limiting factor.

Thermal Pollution of Rivers and Lakes Almost half of all water withdrawn in the United States each year is for cooling electric power plants. The cheapest and easiest method is to withdraw cool water from a nearby ocean bay or inlet, major river, or large lake, pass it through heat exchangers in the plant, and return the heated water to the same body of water.

Large rivers with rapid flow rates can dissipate heat rapidly and suffer little ecological damage unless their flow rates are sharply reduced during summer months or prolonged drought. However, large inputs of heated water from a single plant or a number of plants using the same lake or slow-moving river can have an adverse effect called **thermal pollution.**

Warmer temperatures lower DO content by decreasing the solubility of oxygen in water. Warmer water also causes aquatic organisms to increase their respiration rates and consume oxygen faster, and it increases their susceptibility to disease, parasites, and toxic chemicals. Although some fish species can survive in heated water, many game fish cannot because they have lower temperature tolerance limits and higher oxygen requirements. Discharge of heated water into shallow water near the shore also disrupts spawning and kills young fish.

Fish and other organisms adapted to a particular temperature range can also be killed from **thermal shock:** sharp changes in water temperature when new power plants open up or when plants shut down for repair. Many fish die on intake screens used to prevent fish and debris from clogging the heat exchanger pipes. Pumping large volumes of cold, nutrient-rich water from the bottom layer of moderate-size lakes also speeds up the eutrophication process.

While some scientists call the addition of excess heat to aquatic systems thermal pollution, others talk about using heated water for beneficial purposes, calling it **thermal enrichment**. They point out that heated water results in longer commercial fishing seasons and reduction of winter ice cover in cold areas, and can be used for irrigation to extend the growing season in frost-prone areas.

Warm water from power plants can also be cycled through aquaculture pens to speed the growth of commercially valuable fish and shellfish. For example, waste hot water is used to cultivate oysters in aquaculture lagoons in Japan and in New York's Long Island Sound, and to cultivate catfish and redfish in Texas. Heated water could also be used to heat nearby buildings and greenhouses, melt snow, desalinate ocean water, and provide low-temperature heat for some industrial processes. However, because of dangers from air pollution and release of radioactivity, most coal-burning and electric power plants are usually not located near enough to aquaculture operations, buildings, and

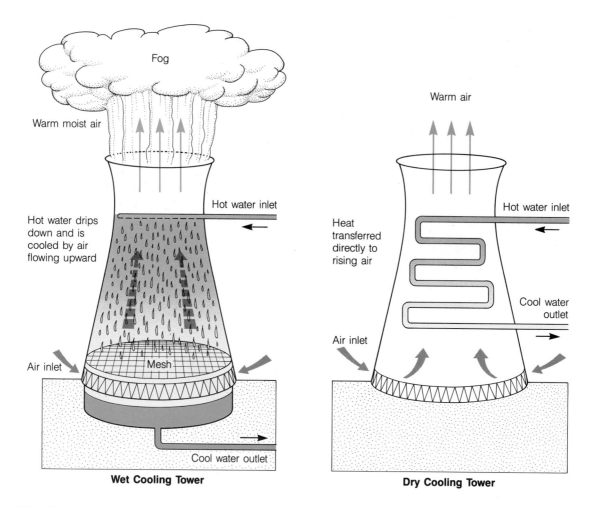

Figure 19-9 Wet and dry cooling towers transfer heat from cooling water to the atmosphere.

industries to make thermal enrichment economically feasible.

Reduction of Thermal Water Pollution There are a number of ways to minimize the harmful effects of excess heat on aquatic ecosystems. We can use and waste less electricity (Section 17-1) and limit the number of power and industrial plants discharging heated water into the same body of water. The heated water can also be returned at a point away from the ecologically vulnerable shore zone. In areas particularly susceptible to thermal pollution, heat in the water can be transferred to the atmosphere by means of wet or dry cooling towers (Figure 19-9). Water can also be discharged into shallow cooling canals, left to transfer its heat to the atmosphere, and then withdrawn for reuse as cooling water (Figure 19-10).

Most new power plants use *wet cooling towers*. This approach, however, has several disadvantages: larger withdrawals of surface water to replace water lost by evaporation, visual pollution from the gigantic cooling towers (Figure 19-11), high construction costs (about

$100 million per tower for a 1,000-megawatt plant), high operating costs, and excessive fog and mist in nearby areas. *Dry cooling towers* are seldom used because they cost two to four times more to build than wet towers. Cooling ponds and canals are useful where enough affordable land is available.

19-3 OCEAN POLLUTION

The Ultimate Sink Oceans are the ultimate sink for natural and human wastes from land runoff, atmospheric fallout, direct dumping, and accidental oil spills from tankers and offshore oil drilling platforms. The major pollution problems of the oceans occur around their edges—the harbors; estuaries; wetlands; and inland seas, such as the Baltic and Mediterranean, near large cities, industrial centers, and the mouths of polluted rivers.

Although the ocean can dilute, disperse, and break down large amounts of sewage and some types of

Figure 19-10 A 2,400-hectare (6,000-acre) canal system is used to transfer heat from cooling water into the atmosphere at the Turkey Point power plant site near Miami, Florida, rather than discharging the heated water into Biscayne Bay. One of the four units—two nuclear and two fossil fuel—is located at the upper right.

Figure 19-11 Wet cooling towers for the Rancho Seco nuclear power plant near Sacramento, California. Compare the size of the towers with the power plant and automobiles. Each tower is more than 120 meters (400 feet) high and could hold a baseball field in its base.

industrial waste, especially in its deep-water areas, it does have limits. The sheer magnitude of discharges, especially near coasts, can overload natural purifying systems (see Spotlight on p. 360). In addition, these natural processes cannot readily degrade many of the plastics, pesticides, and other synthetic chemicals created by human ingenuity. For example, studies indicate that each year one to two million seabirds and more than 100,000 marine mammals, including whales, seals, dolphins, and manatees, die as a result of ingestion of plastic cups, bags, six-pack yokes, fishing gear, and other forms of plastic trash thrown or washed into the ocean.

Ocean Dumping Barges and ships dump large volumes of waste into the sea. About 80% of these wastes are **dredge spoils,** materials scraped from the bottoms of harbors and rivers to maintain shipping channels. Typically, about one-third of these dredged materials are contaminated with effluents from industries and urban areas and runoff from farmlands. Most dredge spoils, too costly to transport to the land for disposal, pose relatively little risk to aquatic ecosystems.

Most of the remaining 20% of the wastes barged out and dumped into the ocean are industrial wastes and **sewage sludge,** a gooey mixture of bacteria- and virus-laden organic matter, toxic metals, synthetic organic chemicals, and settled solids removed from wastewater at sewage treatment plants. As a result of more stringent ocean-dumping laws, the volume of industrial wastes dumped at sea in U.S. coastal waters declined by about 75% between 1975 and 1985. However, the volume of sewage sludge dumped into coastal waters increased by almost 60% during the same period. This increase occurred because of the greatly increased levels of sewage treatment required by water pollution control laws enacted in the 1970s and the lack of suitable and affordable land dumping sites. Great Britain also dumps large quantities of sewage sludge at sea.

Currently, all dredge spoils, industrial wastes, and sewage sludge are dumped at designated sites near the Atlantic, Gulf, and Pacific coasts. The most intensely used ocean dumping site is the New York Bight, a shallow (80 feet deep) area 19 kilometers (12 miles) off the New York–New Jersey coast near the mouth of the Hudson River. After over 60 years of dumping, a 105-square kilometer (40-square mile) area of the ocean bottom in the New York Bight is covered with a black sludge containing high levels of bacteria, long-lived viruses, toxic metals, and organic compounds. During storms some of this sludge has washed ashore on Long Island and New Jersey beaches. It has also contaminated shellfish beds and caused disease outbreaks among people consuming raw clams and oysters illegally harvested from closed beds.

The Chesapeake Bay, the largest estuary in the United States, is an ecosystem in decline from pollution by toxic chemicals and excessive inputs of nitrogen and phosphorus plant nutrients. This generally shallow estuary receives wastes from point and nonpoint sources scattered throughout a massive drainage basin including nine large rivers and parts of six states. The bay becomes a massive pollution sink because only 1% of the waste entering it is flushed into the Atlantic Ocean. Between 1940 and 1986 the number of people living close to the bay grew from 3.7 million to 13.2 million and is projected to reach 14.5 million by 2000.

Harvests of oysters, crabs, and several other commercially important fish such as striped bass (rockfish) have fallen sharply since 1960. Studies have shown that point sources, primarily sewage treatment plants, are the major contributors of phosphorus, while nonpoint sources, primarily runoff from urban and suburban areas and agricultural activities, are the main sources of nitrogen. Between 1983 and 1985 over $550 million in federal and state funds were spent on a cleanup program that could ultimately cost several billion dollars and will require prolonged, cooperative efforts by officials in six states and thousands of cities, towns, and industries and by millions of individual homeowners and farmers.

For 15 years federal officials and environmentalists have been trying to relocate the dumping of all sewage sludge to a deepwater site 170 kilometers (106 miles) east of New York City just beyond the continental shelf. Because of its 610-meter (2,000-foot) depth and nearness to Gulf Stream currents, this site has a much greater capacity to dilute and disperse wastes than the New York Bight. After years of legal maneuvering by urban officials wishing to avoid the extra costs of hauling the sludge further, the EPA and the municipalities agreed on a schedule to phase out all dumping of sewage sludge in the New York Bight by December 1987. Even if this schedule is met, the bight will continue to receive substantial inputs of contaminants from coastal wastewater discharges, urban runoff, atmospheric fallout, and dumping of dredged material.

Although the deep ocean is better equipped than land to handle sewage and some forms of industrial

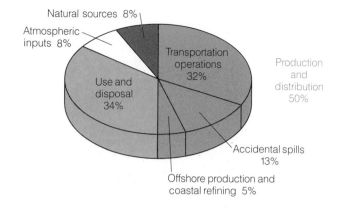

Figure 19-12 Sources of ocean oil pollution. (Data from the National Academy of Sciences)

waste, most scientists believe that the ocean should not be used for the dumping of slowly degradable or nondegradable pollutants such as PCBs, some pesticides and radioactive isotopes, and toxic mercury compounds, which can all be biologically amplified in ocean food webs. Unfortunately, many of these materials are mixed with some types of sewage sludge and industrial waste being dumped into the oceans in large quantities.

Ocean Oil Pollution Crude petroleum (oil as it comes out of the ground) and refined petroleum (fuel oil, gasoline, and other products obtained by distillation and chemical processing of crude petroleum) are accidentally or deliberately released into the environment from a number of sources (Figure 19-12).

The most publicized releases of oil into the oceans are tanker accidents and blowouts (oil escaping under high pressure from a borehole in the ocean floor) at offshore drilling rigs. However, releases of oil from offshore wells during normal operations and during transport of the oil to shore add a much larger volume of oil to the oceans than occasional blowouts and tanker accidents (Figure 19-12).

Effects of Oil Pollution The effects of oil spills are difficult to predict because they depend on a number of factors, including type of oil spilled (crude or refined), amount, distance of the spill from shore, time of year, weather conditions, and ocean and tidal currents. Crude oil and refined oil are collections of hundreds of substances with widely differing properties. After an oil spill, low-boiling, aromatic hydrocarbons are the primary cause of the immediate killing of a number of aquatic organisms, especially in their larval forms. Fortunately, most of these toxic chemicals evaporate

Figure 19-13 This bird was coated with crude oil from an underwater offshore oil leak that occurred near Santa Barbara, California, in 1969. Such birds die unless the oil is removed with a detergent solution.

into the atmosphere within a day or two. Some other chemicals remain on the surface and form small, floating, tarlike globs. These floating substances are gradually broken down by bacteria over several weeks or months, although they persist much longer in cold polar waters.

Floating oil can coat the feathers of marine birds, especially diving birds (Figure 19-13), and the fur of marine mammals such as seals and sea otters. This oily coating destroys the animals' natural insulation and buoyancy, and most drown or die of exposure from loss of body heat. It is estimated that between 150,000 and 450,000 marine birds in the North Sea and North Atlantic regions are killed each year by chronic oil pollution, mostly from routine tanker releases.

Heavy oil components that sink to the ocean floor or wash into estuaries are believed to have the greatest long-term impact on marine ecosystems. These components can kill bottom-dwelling organisms such as crabs, oysters, mussels, and clams or make them unfit for human consumption because of their oily taste and smell.

Oil slicks that wash onto beaches can have serious economic effects on coastal residents, who lose income from fishing and tourist activities. Oil-polluted beaches washed by strong waves or currents are cleaned up fairly rapidly, but beaches in sheltered areas remain contaminated for several years. Oil cleanup is very expensive for oil companies and coastal communities.

Studies of the effects of several crude oil spills in the open sea between 1969 and 1978 found that most forms of marine life recovered nearly completely within 3 years. In contrast, spills of oil, especially refined oil, near shore or in estuarine zones, where sea life is most abundant, have much more damaging and long-lasting effects. For example, damage to estuarine species from the 1969 spill of refined oil at West Falmouth, Massachusetts, was still being detected 10 years later.

Controlling Ocean Oil Pollution The best way to deal with oil pollution is to use various input approaches to prevent it from happening in the first place. Output strategies can be used to deal with oil pollution once it has occurred, but these have not been very effective.

Ocean Oil Pollution: Input Control Methods

1. Use and waste less oil (Section 17-1).

2. Collect used oils and greases from service stations and other sources and reprocess them for reuse.

3. Strictly regulate the building, maintenance, loading and unloading procedures, training of crews, and routing of oil tankers to reduce accidental releases.

4. Require all tankers to have double hulls to reduce chances of a spill and to have separate tanks for the oil cargo and ballast water used to provide stability for empty ships. In older ships, the empty oil tanks were filled with salt water for ballast. When the ships arrived at port to be refilled with oil, the oil-contaminated water was dumped at sea.

5. Eliminate the rinsing of sludge from empty oil tanks and dumping it onto the sea by requiring crude oil to be cleaned before loading to prevent buildup of sludge in the ship's holding tanks.

6. Strictly regulate safety, training, and operation procedures for offshore wells.

7. Strictly regulate safety, operation, and disposal procedures for refineries and industrial plants.

Ocean Oil Pollution: Output Control Methods

1. Use mechanical barriers to prevent oil from reaching the shore and then vacuum it up or soak it up with pillows filled with chicken feathers.

Ineffective in high seas and bad weather conditions and in ice-congested water.

2. Treat spilled oil with detergents so that it will disperse, dissolve, or sink. Experience has shown that the detergents kill more marine life than the oil does.

3. Use genetic engineering techniques to develop bacterial strains that can degrade compounds in oil faster and more efficiently than natural bacterial strains. Possible ecological side effects of the "superbugs" should be carefully investigated before widespread use.

4. Use helicopters equipped with lasers to ignite and burn as much as 90% of an oil spill in a few seconds. Cheaper and more effective than most other approaches and is the only effective method in ice-congested seas. Creates air pollution.

19-4   GROUNDWATER POLLUTION

Is It Safe to Drink the Water?   Groundwater is a vital resource that provides drinking water for one out of two Americans and 95% of those in rural areas. About 75% of American cities depend on groundwater for all or most of their supply of drinking water. The EPA estimates that roughly 1% to 2% of the country's usable groundwater is moderately or severely polluted. Although this estimate may seem small, it is significant because most contaminated aquifers are near heavily populated areas. Thus contaminated aquifers affect 5 million to 10 million people.

Furthermore, according to the Office of Technology Assessment, the EPA estimate is probably low; there has been no uniform or comprehensive testing of the country's groundwater resources. By 1987 only 38 of the 700 different chemicals found in groundwater were covered by federal water quality standards and routinely tested for in municipal drinking water supplies. No testing at all is required for the country's millions of private wells.

In a 1982 survey the EPA found that 45% of the large public water systems served by groundwater were contaminated with synthetic organic chemicals that posed potential health threats. Another EPA survey found that two-thirds of the rural household wells tested violated at least one federal health standard for drinking water.

Vulnerability of Groundwater to Pollution   Some bacteria and most suspended solid pollutants are removed as contaminated surface water percolates through the soil into aquifers. But this process can be overloaded by large volumes of wastes, and its effectiveness varies with the type of soil. No soil is effective in filtering out viruses and most synthetic organic chemicals. Bacterial degradation of oxygen-demanding wastes reaching aquifers does not occur readily because of the lack of dissolved oxygen and sufficient microorganisms in groundwater.

Once contaminants reach groundwater, they are not effectively diluted and dispersed because the rate of movement of most groundwater is very slow. Once an aquifer becomes contaminated, it remains that way for decades or centuries, depending on how rapidly it is replenished. Thus we can see why most analysts believe that groundwater pollution is the most serious U.S. water quality problem today and why they expect it to get much worse.

Sources of Groundwater Contamination   Groundwater can be contaminated from a number of point and nonpoint sources (Figure 19-14). Two major sources of groundwater contamination are leaks of hazardous organic chemicals from underground storage tanks and seepage of such chemicals and toxic heavy metal compounds from landfills, abandoned toxic waste dumps, and lagoons. There are at least 22,000 abandoned hazardous waste disposal sites, 1,500 active hazardous waste landfills, and 15,000 active municipal landfills that are unlined and located above or near aquifers in the United States.

Leaks of gasoline, home-heating oil, industrial-cleaning solvents, and other hazardous chemicals from 1.5 million underground storage tanks may be responsible for as much as 40% of the country's groundwater contamination. EPA surveys indicate that up to 35% of these tanks may be leaking some of their contents into groundwater as a result of improper installation, corrosion (most are steel tanks designed to last only 15 to 20 years), cracking (fiberglass tanks), and overfilling. An additional 350,000 aging tanks are expected to begin leaking between 1987 and 1990. A gasoline leak of just 1 gallon a day can seriously contaminate the water supply for 50,000 persons. Such slow leaks usually remain undetected until someone discovers that a well is contaminated. By 1987 only two states, Florida and Connecticut, required monitoring of underground tanks.

The EPA requires that all underground tanks installed after 1988 have double walls or concrete vaults built around them to help prevent leaks into groundwater. But this does not solve the problem of leaks from existing tanks. Environmentalists believe that monitoring of all underground tanks should be required by law, and that operators should be required to carry

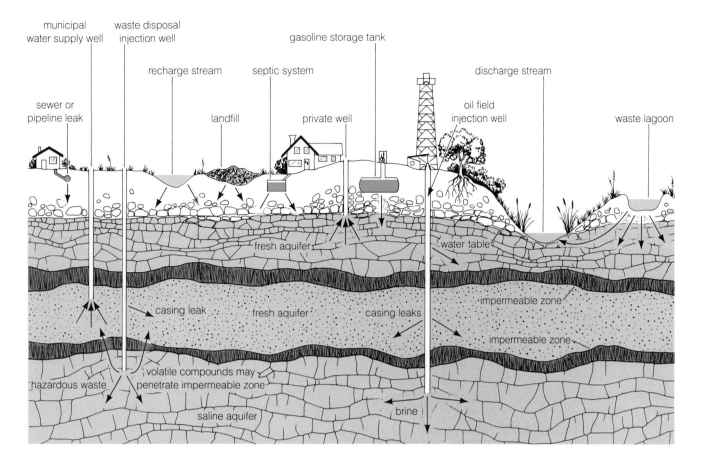

Figure 19-14 Major sources of groundwater contamination in the United States.

sufficient liability insurance to cover cleanup and damage costs and should be liable for leaks from abandoned tanks.

Another concern is leaks into groundwater from deep wells used to inject hazardous wastes deep underground. As a result of stricter regulation of surface water pollution and more recently of disposal of hazardous wastes in landfills, industries have found that deep-well injection of wastes is a much cheaper and less carefully regulated solution. Such wells are bored through aquifers and separated from them by an impermeable layer of rock that theoretically seals the waste in an underground tomb. In practice, however, liquid wastes from some wells have migrated into aquifers and surface waters through other abandoned and unplugged gas and oil wells, blowouts, and cracks caused by earth tremors and by dynamite used to expand the disposal area. Wastes can also escape through faulty casings and joints.

Laws regulating deep-well injection are weak and poorly enforced. There are no limits on the types of wastes that can be injected, no required reporting of the types of wastes injected, and no national inventory of active and abandoned wells. Operators are not required to monitor nearby aquifers and are not liable

for any damages from leaks once the well is abandoned and plugged. By 1986 Alabama, Florida, Louisiana, and California had either banned or were phasing out deep-well disposal. Environmentalists believe that there should be a national ban on all deep-well disposal of hazardous waste since there are other, less risky alternatives (Section 20-6).

Control of Groundwater Pollution Groundwater pollution is much more difficult to detect and control than surface water pollution. Locating and monitoring groundwater pollution is expensive (up to $10,000 per monitoring well) and usually many monitoring wells must be sunk to determine the area of contamination. Pumping water out of a contaminated aquifer, cleaning it up, and then returning it is usually prohibitively expensive.

Some scientists propose that genetic engineering techniques be used to develop strains of anaerobic bacteria that could be injected into polluted aquifers to break down specific pollutants under low-oxygen conditions. Other scientists, however, are concerned about the possible side effects of such bacteria, including mutations into other strains.

Because of these difficulties, it is generally recognized that preventing contamination in the first place is the only effective long-range way to protect groundwater resources. Despite the seriousness of the threat to U.S. drinking water supplies, by 1987 there was no single federal law designed to protect groundwater. Some aspects of groundwater pollution, however, can be controlled by parts of existing water pollution control laws (Section 19-6).

In 1984 the EPA suggested that each state classify its groundwater resources into one of three categories. Class I, consisting of irreplaceable sources of drinking water and ecologically vital sources, should be given the highest protection. Class II includes all other groundwater currently in use or potentially available for drinking water or for other beneficial uses. Class III groundwater is not a potential source of drinking water and is of limited beneficial use, usually because it is already too contaminated.

Protecting groundwater, however, will require enacting and strictly enforcing many unpopular laws at the federal and state levels. Any potentially polluting activities would have to be banned in Class I and II areas, but they could be encouraged in Class III areas. Essentially all disposal of hazardous wastes in landfills and deep wells would have to be banned except perhaps in Class III areas. Expensive monitoring would be required for aquifers near existing landfills, underground tanks, and other potential sources of groundwater contamination.

Much stricter controls would have to be placed on the application of pesticides and fertilizers by millions of farmers and homeowners—regulations that would also be a key factor in protecting surface water from contamination by nonpoint sources. People using private wells for drinking water would probably have to have their water tested once a year. Unless nationwide federal laws are enacted, states that take such measures to protect their water supplies and the health of their citizens might lose industries and jobs to states with less regulation.

19-5 WATER POLLUTION CONTROL

Control of Nonpoint Source Pollution Although most U.S. surface waters have not declined in quality since 1970, they have also not improved. The primary reason is the absence of any national strategy for controlling water pollution from nonpoint sources. Such a strategy will require greatly increased efforts to control soil erosion through conservation and land-use control for farms, construction sites, and suburban and urban areas (Section 9-4).

Fertilizer runoff and leaching can be reduced by avoiding excessive application, applying fertilizer only during the growing season, not using it on steeply sloped land, requiring buffer zones of vegetation between fields and surface water, using slow-release fertilizers, and using crop rotation with nitrogen-fixing plants to reduce the need for fertilizer.

Pesticide runoff and leaching can be reduced by applying no more pesticide than needed and by applying it only when needed; increased reliance on biological methods of pest control can significantly reduce the need for pesticides (Section 21-5). Methods to control runoff and infiltration of animal wastes from feedlots and barnyards include regulating animal density, not locating such operations on land sloping toward nearby surface water, and diverting runoff into lagoons or detention basins from which the nutrient-rich water can be pumped and applied as fertilizer to cropland or forestland.

Control of Point Source Pollution: Wastewater Treatment In many LDCs and parts of MDCs, sewage and waterborne industrial wastes from point sources are not treated and are discharged into the nearest waterway or into a storage basin such as a cesspool or lagoon. Widespread discharge of untreated human and livestock wastes into surface water is a major cause of illness and death from waterborne diseases.

In most MDCs, however, most of the wastes from point sources are purified to varying degrees. In rural and suburban areas with suitable soils, sewage and wastewater from each house is usually discharged into a **septic tank,** which traps greases and large solids and discharges the remaining wastes over a large drainage field for filtration by the soil and biodegradation by soil bacteria (Figure 19-15). In the United States, septic tanks are regulated by all states to ensure that they are installed in soils with adequate drainage, not placed too close together or too near well sites, and installed properly. To prevent backup and overflow, grease and solids must be periodically pumped out of the tank.

In urban areas most waterborne wastes from homes, businesses, factories, and storm runoff flow through a network of sewer pipes to sewage treatment plants. Some urban areas have separate lines for sewage and storm water runoff, but in other areas (such as parts of Boston) lines for these two sources are combined because it is cheaper (Figure 19-16). The problem with a combined sewer system is that during heavy rains the total volume of wastewater and storm runoff flowing through the system usually exceeds—by as much as 100 times—the amount that can be handled by the sewage treatment plant. As a result,

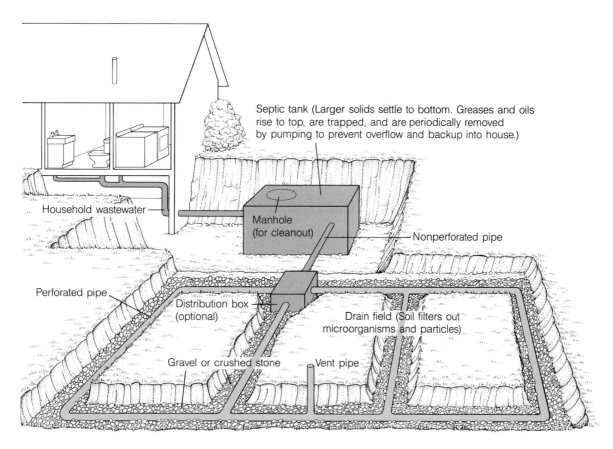

Figure 19-15 Septic tank system used for disposal of domestic sewage and wastewater in rural and suburban areas.

Labels in figure:
- Septic tank (Larger solids settle to bottom. Greases and oils rise to top, are trapped, and are periodically removed by pumping to prevent overflow and backup into house.)
- Household wastewater
- Manhole (for cleanout)
- Nonperforated pipe
- Perforated pipe
- Distribution box (optional)
- Drain field (Soil filters out microorganisms and particles)
- Gravel or crushed stone
- Vent pipe

the overflow, which contains untreated sewage, is discharged directly into surface waters.

When sewage reaches a treatment plant, it can undergo various levels of purification, depending on the sophistication of the plant and the degree of purity desired. **Primary sewage treatment** is a mechanical process that uses screens to filter out debris like sticks, stones, and rags; then suspended solids settle out as **sludge** in a sedimentation tank (Figure 19-17). These operations remove about 60% of suspended solids, 30% of oxygen-demanding wastes, 20% of nitrogen compounds, 10% of phosphorus compounds, and little or none of other chemical pollutants.

Secondary sewage treatment is a biological process that uses aerobic bacteria to remove biodegradable organic wastes (Figure 19-18). It removes up to 90% of the oxygen-demanding wastes by using either **trickling filters,** where aerobic bacteria degrade sewage as it seeps through a large vat bed filled with crushed stones covered with bacterial growths (Figure 19-19), or an **activated sludge process,** in which the sewage is pumped into a large tank and mixed for several hours with bacteria-rich sludge and air bubbles to increase bacterial degradation. The water then

goes to a sedimentation tank, where most of the suspended solids settle out as sludge. The sludge is removed and broken down in an anaerobic digestor, disposed of by incineration, dumped in the ocean or a landfill, or applied to land as fertilizer.

The combined primary and secondary treatments still leave about 10% of the oxygen-demanding wastes, 10% of the suspended solids, 50% of the nitrogen (mostly as nitrates), 70% of the phosphorus (mostly as phosphates), 30% of most toxic metal compounds, 30% of most synthetic organic chemicals, and essentially all the long-lived radioactive isotopes and dissolved persistent organic substances such as some pesticides. In the United States, secondary treatment must be used in all communities served by sewage treatment plants. Preliminary experiments have shown that waste water can also be purified effectively by allowing it to flow slowly through long ponds filled with plants such as water hyacinths. These plants also remove toxic organic chemicals and metals not removed by conventional primary and secondary treatment.

A small-scale, **package sewage treatment plant** is sometimes used for secondary treatment of small quantities of wastes from a shopping center, an apart-

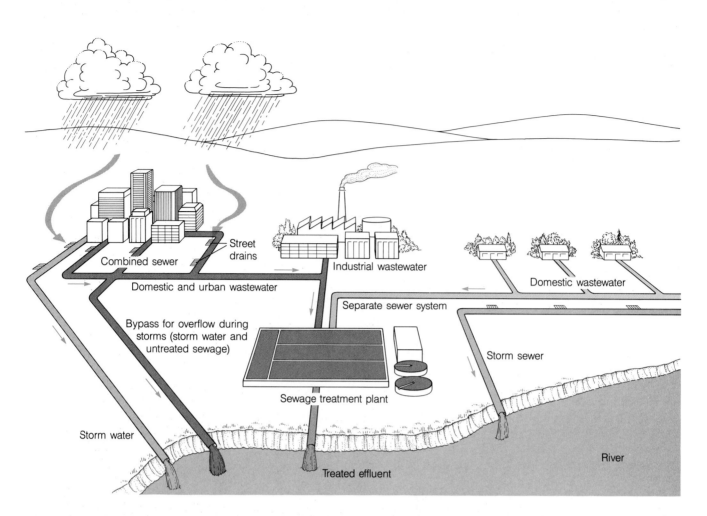

Figure 19-16 Separated and combined storm and sewer systems used in cities.

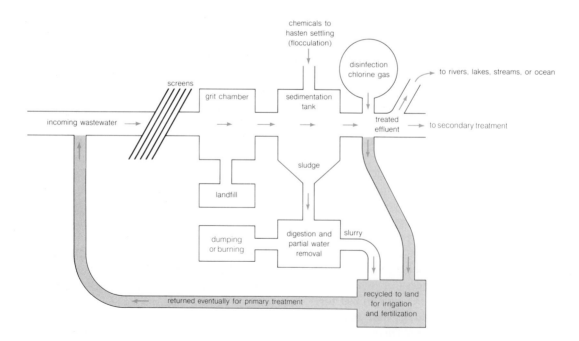

Figure 19-17 Primary sewage treatment. Shaded areas show recycling of plant nutrients in sludge and treated wastewater to land instead of discharge into surface waters.

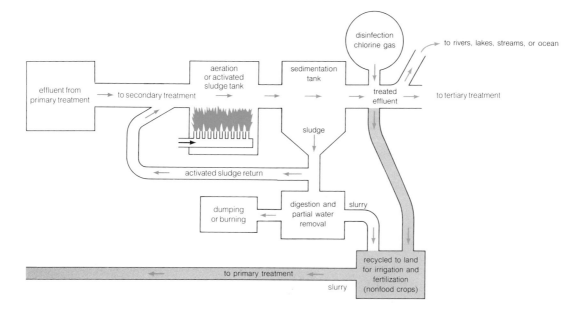

Figure 19-18 Secondary sewage treatment. Shaded areas show recycling of nutrients to land.

Figure 19-19 Trickling filter used as part of secondary sewage treatment.

ment complex, a village, or a small housing subdivision. However, this type of plant often does not work properly and requires considerable attention and maintenance.

Tertiary sewage treatment is a series of specialized chemical and physical processes that reduce the quantity of specific pollutants still left after primary and secondary treatment (Figure 19-20). Types of tertiary treatment vary depending on the contaminants in specific communities and industries. The most common methods of tertiary treatment are precipitation (settling out as insoluble solids) to remove up to 90% of suspended solids and phosphates, filtration with activated (finely powdered) carbon to remove dissolved organic compounds and any remaining suspended solids, and reverse osmosis by passage through a membrane to remove dissolved organic and inorganic substances. Tertiary treatment is rarely used, except in Sweden and Denmark, because the plants are about twice as expensive to build and four times as costly to operate as secondary plants.

One of the last steps in any form of sewage treatment is to disinfect the water before it is discharged into nearby waterways or applied to land for further filtering and use as fertilizer. Disinfection removes water coloration and kills disease-carrying bacteria and some

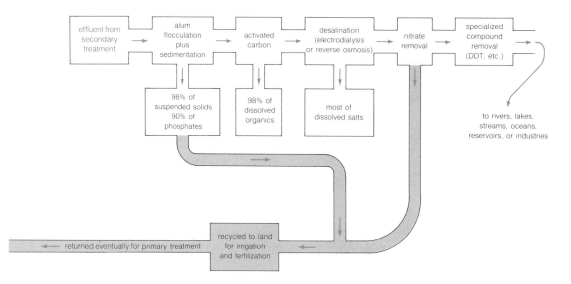

Figure 19-20 Tertiary sewage treatment. Shaded areas show recycling of nutrients to land.

(but not all) viruses. The usual method is **chlorination.** A problem is that chlorine also reacts with organic materials in the wastewater or in surface water to form small amounts of chlorinated hydrocarbons, some of which, such as chloroform, are known carcinogens. Consequently, several other disinfectants are being tested, and one of the leading contenders is ozone. Ozone is more expensive than chlorine but is more effective against viruses.

Treatment of water for drinking by urban residents is much like wastewater treatment. The degree of treatment required varies with the purity of the water coming into the treatment plant. Little, if any, treatment is necessary in areas with very pure sources of groundwater.

Land Disposal of Sewage Effluent and Sludge
Building secondary sewage treatment plants throughout the United States is an important step in water pollution control, especially in reducing oxygen-demanding wastes, suspended solids, and bacterial contamination. However, without more expensive tertiary treatment, sewage plant effluents can overload lakes and slow-moving rivers with phosphates and nitrates, triggering algal blooms and oxygen depletion. Sewage treatment also produces large volumes of sludge that must be disposed of.

An alternative is to spray sewage plant effluent or untreated wastewater or to spread sludge as fertilizer to forestlands, parks, surface-mined lands, croplands, wetlands, and aquaculture ponds. By 1986 about 25% of all U.S. municipal sludge was returned to the land as fertilizer.

One problem with land disposal of wastewater and sludge is that bacteria, viruses, toxic metals, and hazardous synthetic organic chemicals can contaminate food and groundwater. Several treatments have been used to avoid or minimize the problem. Before it is applied, sludge can be heated to kill harmful bacteria, as is done in West Germany and Switzerland, or composted and allowed to decay. Sludge and wastewater can be treated to remove toxic metals and organic chemicals before application, or they can be applied only on land not used to grow crops or raise livestock, such as forests, surface-mined land, and highway medians in areas where groundwater is already contaminated or is not used as a source of drinking water. Land disposal is not a panacea, but with careful management and control it is a useful ecological approach to waste handling in some areas.

Individual Waste Management People can reduce pollution of water by safely disposing of products containing harmful chemicals. These products should never be poured down house or street drains or flushed down the toilet. Waste oil drained from automobiles should be put in a container and taken to a local service station, where it will be turned in for recycling. Antifreeze drained from motor vehicles should be collected and poured onto a porous surface, such as gravel, away from water supplies. Call your local health or water department for information about proper disposal of insecticides, herbicides, paints, lacquers, thinners, brush cleaners, wood preservatives, turpentine, and household cleaners containing organic solvents.

We can also reduce water pollution by using less water and by using low-phosphate or nonphosphate detergents and cleaners. Commercial inorganic fertilizers, pesticides, detergents, bleaches, and other chemicals should be used only if necessary and then in the smallest amounts possible (see Table 20-1 on p. 388).

19-6 U.S. WATER POLLUTION CONTROL LAWS

Protecting Drinking Water   Before 1974 the United States had no enforceable national standards for drinking water. Each state set its own standards, and these varied in range and rigor from state to state. This began changing with passage of the Safe Drinking Water Act of 1974, which required the EPA to establish national drinking water standards, called maximum contaminant levels (MCLs), for any pollutants that "may" have adverse effects on human health.

The first MCLs went into effect in 1977. By 1986 the EPA had set MCLs for 26 water pollutants. According to the EPA, 87% of the country's 59,000 municipal water systems were in compliance with these MCLs in 1985. Environmentalists and health officials, however, have criticized the EPA for not setting MCLs for more of at least 700 potential pollutants found in municipal drinking water supplies. Amendments added to the Safe Drinking Water Act in 1986 require the EPA to set MCLs for 83 new contaminants by 1989 and for 25 more by 1991. These amendments also require disinfection for all public water supplies, ban the use of lead pipe and solder in any new public water system, and require the EPA to establish regulations for monitoring deep waste-injection wells.

Wells for millions of individual homes in suburban and rural areas are not required to meet federal drinking water standards, primarily because of the high cost of testing each well regularly and because of political problems associated with verifying individual compliance. A study conducted for the EPA by Cornell University scientists in 1982 indicated that 39 million rural residents—two out of three of all rural Americans—were drinking water from private wells that did not meet one or more federal water quality standards.

Contaminated wells and concern about possible contamination of public drinking water supplies has led about 1 of every 15 Americans to drink bottled water at an average cost of $.45-$1 a gallon. Bottled water is regulated by the Food and Drug Administration (FDA) based on standards that are required to be equivalent to EPA standards for drinking water. However, the FDA does little testing of bottled water and generally relies on tests by the bottled water industry. Furthermore, there is no assurance that such water won't contain synthetic organic chemicals and other chemicals not regulated by EPA standards.

Most activated-charcoal filter units (costing from $500 to $2,500) for attachment under home sinks remove most synthetic organic chemicals if filters are changed regularly. But they are not effective in removing bacteria, viruses, and toxic metals. Filter units that contain "bacteriocides" prevent bacteria from building up in the filter only—not in the water. Coupling of a reverse osmosis system (costing from $500 to $1,000) with an activated-charcoal filter removes just about all pollutants if both systems are properly maintained at a cost of about $100 a year.

U.S. Control Efforts   The Federal Water Pollution Act of 1972 and the Clean Water Act of 1977, along with amendments in 1981 and 1987, form the basis of U.S. water pollution control efforts, which have the goal of making all U.S. surface waters safe for fishing and swimming. These acts require the EPA to establish national effluent standards that limit the amounts of conventional and toxic water pollutants that can be discharged into surface waters from factories, sewage treatment plants, and other point sources and to set up a nationwide system for monitoring water quality. All municipalities are required to use secondary sewage treatment by 1988—a deadline that will not be met in a number of areas.

Between 1972 and 1986 the federal government provided almost $45 billion to municipalities for the construction, operation, and maintenance of municipal wastewater treatment facilities. State and local governments provided an additional $25 billion in matching grants for these purposes. Amendments to the Clean Water Act in 1987 authorized expenditure of an additional $18 billion between 1987 and 1996.

By 1986 about 67% of the country's municipalities had completed the construction needed to ensure full compliance with national effluent limits, and only 12% of the municipalities where construction had been completed were still in significant noncompliance. The failure of almost one-third of the country's municipalities to complete construction by 1986 is attributed to fraud, overbuilding, bureaucratic and construction delays, and a 37% cut in federal funding for water pollution control between 1981 and 1986 by the Reagan administration.

By 1986 an impressive 94% of all industrial dischargers were officially in compliance with their discharge permits. However, a 1984 study by the General Accounting Office showed that most of the 33 industries studied violated their EPA waste discharge permits; the GAO accused the EPA of doing little to enforce compliance of water pollution control laws.

These water pollution control laws also require states to establish local and regional planning to reduce water pollution from nonpoint sources. But no goals and standards have been established, and relatively little funding has been provided to reduce water pollution from nonpoint sources. The country also has no comprehensive legislation, goals, or funding designed to protect its groundwater supplies from contamination.

Future Water Quality Goals   Since 1972 improved control of discharges of oxygen-demanding wastes from point sources has meant that about 75% of the country's monitored areas of rivers and lakes are fishable and swimmable. Environmentalists believe that future water pollution control efforts in the United States should be focused on three major goals. First, existing legislation controlling the discharge of conventional and toxic pollutants into surface waters from point sources should be strictly enforced and not weakened. Second, new legislation should be enacted and funded to establish national goals and regulations for sharply reducing runoff of conventional and toxic pollutants into surface waters from nonpoint sources. Finally, comprehensive and well-funded legislation should be enacted to protect the country's groundwater drinking supplies from pollution by point and nonpoint sources.

The reason we have water pollution is not basically the paper or pulp mills. It is, rather, the social side of humans—our unwillingness to support reform government, to place into office the best-qualified candidates, to keep in office the best talent, and to see to it that legislation both evolves from and inspires wise social planning with a human orientation.

Stewart L. Udall

DISCUSSION TOPICS

1. Explain why dilution is not always the solution to water pollution. Cite examples and conditions for which this solution is, and is not, applicable.

2. Explain how a river can cleanse itself of oxygen-demanding wastes. Under what conditions will this natural cleansing system fail?

3. Give your reasons for agreeing or disagreeing with the idea that we should deliberately dump most of our wastes in the ocean because it is a vast sink for diluting, dispersing, and degrading wastes, and if it becomes polluted, we can get food from other sources.

4. Should all dumping of wastes in the ocean be banned? Explain. If so, where would you put these wastes? What exceptions, if any, would you permit? Under what circumstances? Explain why banning ocean dumping alone will not stop ocean pollution.

5. Contact local officials to find out the source of drinking water in your area. How is it treated? Has it been analyzed recently for the presence of synthetic organic chemicals, especially chlorinated hydrocarbons? If so, were any found and are they being removed?

6. Contact local officials to determine whether during the past 10 years any swimming areas in your area have been closed because of high coliform bacteria counts. How often are swimming areas tested for coliform bacteria?

7. What are the major nonpoint sources of contamination of surface water and groundwater in your area?

8. Should the injection of hazardous wastes into deep underground wells be banned? Explain.

20

Solid Waste and Hazardous Wastes

GENERAL OBJECTIVES

1. How much urban solid waste is produced in the United States?

2. What options do we have for dealing with this waste?

3. How can we recover some of the valuable resources in this waste?

4. What are the major types, sources, and effects of the hazardous waste we produce?

5. What are some significant types of hazardous waste?

6. What can we do with hazardous waste?

The shift from a throwaway society to a recycling one can help restore a broad-based gain in living standards.
Lester R. Brown and Edward C. Wolf

Since 1970 air and water quality in the United States have improved—or at least not worsened—in most areas. But these improvements have produced large quantities of solid waste, such as fly ash removed from smokestack exhaust and toxic sludge removed from wastewater at sewage treatment plants. To this increasing volume of solid waste are added massive quantities of other solid and hazardous wastes produced every day by individuals, businesses, and factories in our throwaway society. Complicating matters even further are the highly toxic wastes disposed of in thousands of land dump sites before environmental laws were passed—and now threatening the health and lives of entire communities. Thus the control and management of solid and hazardous waste is one of our most urgent environmental problems.

20-1 SOLID WASTE PRODUCTION IN THE UNITED STATES

What Is Solid Waste and How Much Is Produced? Any useless, unwanted, or discarded material that is not a liquid or a gas is classified as **solid waste.** It is yesterday's newspaper and junk mail, today's dinner table scraps, raked leaves and grass clippings, nonreturnable bottles and cans, worn-out appliances and furniture, abandoned cars, animal manure, crop residues, food-processing wastes, sewage sludge, fly ash, mining and industrial wastes, and an array of other cast-off materials.

The total amount of solid waste from all sources produced each year in the United States is staggering—estimated to be at least 4.6 billion metric tons (5.1 billion tons). This amounts to an average of 19 metric tons (21 tons) a year for each American, or 53

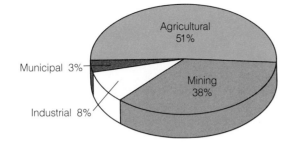

Figure 20-1 Sources of solid waste in the United States. (Data from the Environmental Protection Agency)

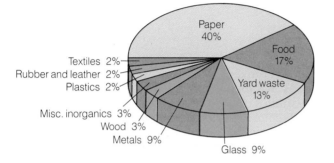

Figure 20-2 Composition of the approximately 1.8 kilograms (4 pounds) of urban solid waste thrown away in a typical day by each American. (Data from National Solid Wastes Management Association)

Figure 20-3 An unsightly and illegal roadside dump.

kilograms (115 pounds) a day. About 89% of this solid waste is produced as a result of agricultural and mining activities (Figure 20-1). Fortunately, over 90% of agricultural solid waste is recycled into the soil by being plowed under or used as fertilizer (manure and some crop residues).

Industrial solid waste makes up about 8% of the total produced each year. Much of this is scrap metal, plastics, slag, paper, fly ash from electrical power plants, and sludge from sewage treatment plants. Fly ash and sewage sludge will increase rapidly as more coal-burning plants are built, older plants are required to install air pollution control equipment, and sewage plants that are now under construction or still being planned are put into operation.

Solid municipal waste produced by homes and businesses in or near urban areas makes up the remaining 3% of the solid waste produced in the United States. Each American produces an average of about 1.8 kilograms (4 pounds) per day, or 657 kilograms (1,460 pounds) per year. About 70% of what the typical American throws away as garbage and rubbish consists of paper, food, and yard waste (Figure 20-2). Because this solid waste is concentrated in highly populated areas, it must be removed quickly and effi-

ciently to prevent health problems, infestations by rats and other disease-carrying organisms, and buildup of massive piles of unsightly trash.

Strategies for Dealing with Solid Waste Today we rely primarily on *throwaway output approaches* to dump these wastes in the ocean (Section 19-3) or on the land or to burn them in incinerators. Environmentalists, however, believe we should begin shifting from this throwaway approach (Figure 3-16, p. 47) to a sustainable-earth or low-waste approach (Figure 3-17, p. 48). With this approach most of what we throw away would not be viewed as solid waste but as wasted solids, which should be reused, recycled, or burned to provide energy.

This *resource recovery output approach* can be coupled with *input approaches* designed to produce less solid waste. Examples include reducing average consumption per person by wasting fewer resources and buying things we really need rather than merely want, increasing the average lifetime of products, decreasing the amount of material used in some products (smaller cars, for example), and designing products for easier repair, reuse, and recycling (Section 14-5).

Because solid waste from agricultural, mining, and industrial activities is discussed elsewhere in this book, the next two sections of this chapter are devoted to evaluating the throwaway output and resource recovery strategies for dealing with urban solid waste.

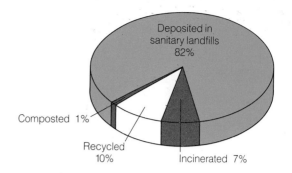

Figure 20-5 Fate of solid waste in the United States. (Data from the Environmental Protection Agency)

Figure 20-4 Burning of solid waste at an open dump near Ralls, Texas. This practice pollutes the air and is now banned, but it is still done illegally in some places.

20-2 DISPOSAL OF URBAN SOLID WASTE: DUMP, BURY, BURN, OR COMPOST?

Littering and Open Dumps   People often casually discard solid waste by throwing bottles, cans, fast-food containers, and other items on the street or out of car windows. In addition to creating visual pollution, this adds to the taxpayers' burden because collecting widely dispersed litter is very expensive. Although it is illegal, some people in rural areas, who must dispose of their own trash, dump it along roadsides rather than haul it to dumpster locations or county landfills (Figure 20-3). Between 1975 and 1986, average urban waste collection and disposal costs doubled and are continuing to rise as many urban areas run out of convenient places to dispose of their refuse and have to haul it farther or find other alternatives. Before passage of the 1976 Resource Conservation and Recovery Act (RCRA), most urban solid waste was merely dumped on the ground at selected sites—sometimes ecologically valuable wetlands. These unsightly **open dumps** supported large populations of disease-carrying rodents and insects and often contaminated groundwater and surface water through leaching and runoff. The dumps also created air pollution when they caught on fire from spontaneous combustion or when they were set on fire to reduce the volume of wastes (Figure 20-4). The RCRA, however, required

that all existing open dumps in the United States be closed or upgraded to sanitary landfills by 1983, and banned the creation of new open dumps.

Sanitary Landfills   Currently 82% of the urban solid waste collected in the United States is deposited in sanitary landfills. The remainder is burned in municipal incinerators, recycled, or composted (Figure 20-5). A **sanitary landfill** is a land waste disposal site that eliminates most of the problems associated with open dumps by spreading wastes in thin layers, compacting them, and covering them with a fresh layer of soil each day (Figure 20-6). No open burning is allowed, odor is seldom a problem, and rodents and insects cannot thrive. In addition, sanitary landfills are supposed to be situated so as to minimize water pollution from runoff and leaching—although this is not always done.

A landfill can be put into operation fairly quickly, has low operating costs, and can handle a massive amount of solid waste (Figure 20-7). Although a filled landfill that has been allowed to settle for a few years is not a suitable building site, it can be regraded and used as a site for a park, golf course, athletic field, wildlife area, or other recreational purpose (Figure 20-8).

Landfills do have some drawbacks. Wind can scatter litter and dust during the day before each day's load of trash is covered with soil. There is a danger that explosive methane gas and toxic hydrogen sulfide gas, produced by anaerobic decomposition of organic wastes, can seep into nearby buildings and cause explosions or asphyxiation. This can be prevented by equipping landfills with vent pipes to collect these gases so they can be burned or allowed to escape into the air. At a large landfill on Staten Island in New York City, enough methane gas is collected to heat 10,000 homes a year. In addition to saving energy, collecting and burning methane gas from all landfills worldwide would reduce annual atmospheric emissions of meth-

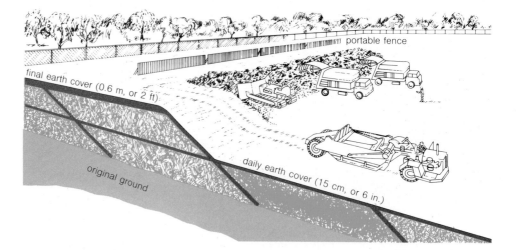

portable fence

a sanitary landfill.
spread in a thin layer
compacted with a bull-
er. A scraper (foreground)
covers the wastes with a fresh
layer of soil at the end of each
day. Portable fences are used to
catch and hold windblown
debris.

final earth cover (0.6 m, or 2 ft)

original ground

daily earth cover (15 cm, or 6 in.)

Figure 20-7 A mountain of solid
waste is deposited in a sanitary
landfill each day before it is covered with soil.

Institute for Local Self Reliance

ane by 6% to 18%—thus helping reduce depletion of
the ozone layer and global warming from greenhouse
gases (Section 18-5). Contamination of groundwater
is also a potential problem without proper siting, construction, and monitoring. This is especially true for
thousands of older landfills filled and abandoned before
stricter landfill siting and operating regulations were
established.

Even if it is eventually converted to a useful purpose, most people do not want a landfill nearby because
of the traffic, noise, and dust that are inevitable during
the years the landfill is being filled. Because of citizen
opposition, escalating land prices, and lack of environmentally acceptable sites, more than half of the
cities in the United States will exhaust their present
landfill capacity and run out of acceptable new sites
by 1990.

Incineration Another way to deal with solid waste
is **incineration,** the burning of combustible materials

and melting of certain noncombustible materials in
municipal incinerators. The ash or residue left after
incineration can then be deposited in landfills or in
the ocean. Incineration kills disease-carrying organisms and reduces the volume of solid waste by 80%
to 90%. Salvaged metals and glass can generate income,
and the waste energy can be used to produce electricity or heat for nearby buildings. Incinerators do not
pollute groundwater and add very little air pollution
if equipped with adequate air pollution control devices.
Sweden, which burns half its solid waste, now requires
such controls.

However, there are drawbacks. Even with air pollution control devices, incinerators emit large quantities of fine particulate matter into the atmosphere.
Construction, maintenance, and operating costs are
much higher for incinerators than for landfills, except
in areas where land prices are high or waste must be
hauled long distances to the landfills. In addition, for
every 10 tons of municipal waste fed into an incinerator, one ton of ash is produced. This resulting ash is

Joe Melena/The Peninsula Times Tribune

usually contaminated with toxic metals and dioxins, and may soon be classified as hazardous waste by the EPA. Disposal of this waste at a hazardous waste facility costs about 15 times more per ton than its disposal at sanitary landfills.

Appropriate incinerator sites are difficult to find because of citizen opposition. But as urban areas run out of acceptable landfills, incineration will become more economically attractive. By the end of this century incinerators are projected to be burning 30% of the country's solid waste, compared to 7% today.

Composting Biodegradable solid waste from slaughterhouses, food processing industries, and kitchens can be mixed with soil and decomposed by aerobic bacteria to produce a material known as **compost,** which can be used as a soil conditioner and fertilizer. Kitchen waste, paper, leaves, and grass clippings can be decomposed in backyard compost heaps and used in gardens and flower beds. With food processing and

other industries, the large supply of organic waste can be collected and degraded in large composting plants, as is done in many European countries such as the Netherlands, West Germany, and Italy. The compost is then bagged and sold.

Composting, however, has some drawbacks. It is not economically feasible to use it with mixed urban waste because sorting out the glass, metals, and plastics is too expensive. Thus composting requires that consumers and plants separate food and yard waste for collection. In some countries, such as the United States, the demand for compost is not great enough to justify its large-scale production.

20-3 RESOURCE RECOVERY FROM SOLID WASTE

The High-Technology Approach By various methods of **resource recovery,** usable materials or energy can be salvaged from solid waste. Whether most resources should be recycled by a centralized high-technology approach or by a decentralized low-technology approach is the subject of debate. In the ideal high-technology approach, large, centralized **resource recovery plants** would shred and automatically separate mixed urban waste to recover glass, iron, aluminum, and other valuable materials, which would be sold to manufacturing industries for recycling (Figure 20-9). The remaining paper, plastics, and other combustible wastes would be incinerated to produce steam, hot water, or electricity, which could then be used in municipal facilities or sold to nearby buildings and manufacturing plants.

By 1987, the United States had 70 resource recovery plants and at least 100 others were under construction or in the planning stage. Although a few of these plants separate and recover some iron, aluminum, and glass for recycling, most are sophisticated incinerators used to produce energy by burning trash. Hundreds of such energy resource recovery plants are in operation in European and Japanese cities. Denmark uses incinerator plants to convert about 60% of its burnable waste to energy, Sweden burns 50% of its burnable waste, and Switzerland and the Netherlands burn 40% and 30%, respectively, compared to only 7% in the United States.

Unlike their European counterparts, most large U.S. energy resource recovery plants have been a disappointment. They have been expensive to build ($50 million to $500 million per plant) and have suffered from delays, breakdowns, high operating and maintenance costs, lack of enough daily waste for economical operation, and continuing financial losses, even though 55% of the costs of these facilities was subsi-

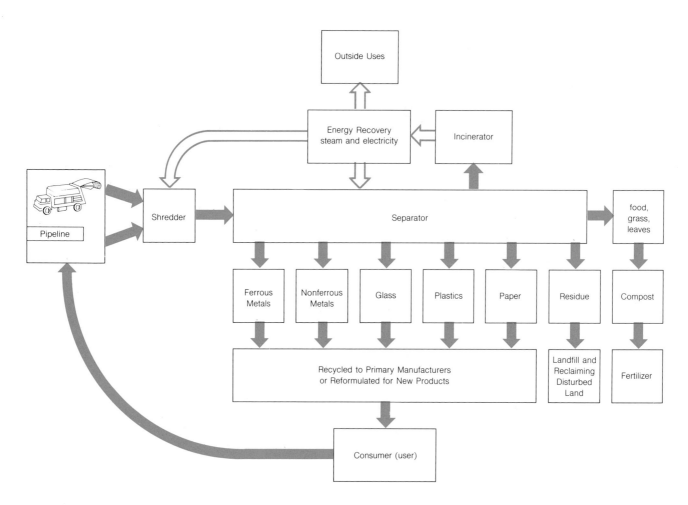

Figure 20-9 Generalized urban resource recovery system. At present, most resource recovery plants are designed primarily to burn paper and other combustible materials in an incinerator; the energy produced by the incinerator is used to produce electricity or for heating nearby buildings.

dized by federal funds. They also cause air pollution if not properly controlled and maintained. Several have gone bankrupt and have been abandoned.

By 1987 about 4% of urban trash in the United States was burned and the energy sold to power companies. Energy produced by burning trash is projected to more than quadruple by the year 2000, based on a new series of plants based mostly on European technology. In Florida, state officials hope to burn most of the state's waste at 18 plants by 1995.

Small-scale, high-tech resource recovery systems can be used for apartment buildings, hospitals, and housing developments. By 1987, over 700 such waste collection systems were in operation in Sweden, the United Kingdom, West Germany, France, the Soviet Union, and other countries. The largest system in the world is at Disney World in Florida.

The Low-Technology Approach Most waste materials recovered in the United States are recycled in a *low-technology approach*. In this simpler, small-scale approach, homes and businesses place waste materials—such as glass, paper, metals, and food scraps—into separate containers. Compartmentalized city collection trucks, private haulers, or volunteer recycling organizations pick up the segregated wastes, clean them if necessary, and sell them to scrap dealers, compost plants, and manufacturers. Studies have shown that this source separation takes only 16 minutes a week for the average American family.

In the United States more than 3,000 municipal or community-based recycling centers together recycle 12 million tons of resources in trash for an annual cash return exceeding $380 million. A number of U.S. cities are using this approach to recycle a significant percentage of their solid waste—Wilton, New Hampshire (46%); Davis, California (25%), Berkeley, California (20%); San Francisco (23%); and Seattle (20%). Oregon, New Jersey, and Rhode Island require reusable household waste to be separated into three or four groups to facilitate recycling.

A comprehensive low-technology recycling program could save 5% of annual U.S. energy use—more

Although paper can be recycled at a fairly high rate, only about 25% of the world's wastepaper is now recycled (Figure 20-10). The Netherlands, Mexico, and Japan have high paper recycling rates primarily because they are sparsely forested. A number of analysts believe that with sufficient economic incentives and laws, at least half the world's wastepaper could be recycled by the end of the century. During World War II the United States recycled about 45% of its wastepaper when paper drives and recycling were national priorities.

Directly or indirectly, each American uses an average of about 275 kilograms (600 pounds) of paper per year—about 8 times the world average and about 40 times the average in LDCs. Product overpackaging—packages inside of packages and oversized containers made to trick consumers into thinking they are getting more for their money—is a major contributor to paper use and waste. Almost three-fourths of U.S. paper produced ends up in the trash. Nearly $1 of every $10 spent for food and drink in the United States goes for throwaway packaging.

In addition to saving trees and land, recycling paper saves about 30% to 55% of the energy needed to produce paper from virgin pulp-

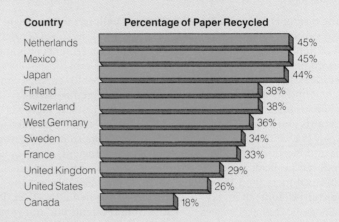

Country	Percentage of Paper Recycled
Netherlands	45%
Mexico	45%
Japan	44%
Finland	38%
Switzerland	38%
West Germany	36%
Sweden	34%
France	33%
United Kingdom	29%
United States	26%
Canada	18%

Figure 20-10 Percentage of paper recycled in various countries. (Data from Organization for Economic Cooperation and Development)

wood and can reduce air pollution from pulp mills by about 95%. If half the discarded paper were recycled, the country would save enough energy to provide 10 million people with electrical power each year.

Having individual homes and businesses sort out paper for recycling is an important key to increased recycling. Otherwise, the paper becomes so contaminated with other trash that wastepaper dealers will not buy it. Slick paper magazines, magazine sections, and advertising supplements cause contamination problems and must not

be included. By 1986, more than 150 U.S. cities required residences and businesses to sort out newspapers and cardboard for pickup and recycling.

Factors hindering wastepaper recycling in the United States include federal tax subsidies and other financial incentives that make it cheaper to produce new paper from trees than from recycling, widely fluctuating prices that make recycling paper a risky financial venture, and increased burning of paper in resource recovery plants to produce energy. What do you think should be done?

than the energy generated by all U.S. nuclear power plants—at perhaps one-hundredth of the capital and operating costs. By contrast, if all combustible urban solid waste was burned in high-technology energy recovery plants, this would provide only 1% of the country's annual commercial energy use. The low-technology approach also produces little air and water pollution, has low start-up costs and moderate operating costs, and saves more energy and provides more jobs for unskilled workers than high-technology resource recovery plants.

However, the low-tech approach requires the

public to separate waste into recyclable categories and requires large and stable markets for the recycled materials. Currently, with discriminatory tax policies and no guaranteed market, prices for recycled materials are subject to sharp and sudden changes, making recycling a highly risky business.

Regardless of what resource recovery methods are used, there is considerable room for improvement in the United States, where only about 8% of salvagable glass, metals, and paper in urban solid waste is presently recycled or burned as a source of energy (see Spotlight above).

In 1977, the residents of a suburb of Niagara Falls discovered that "out of sight, out of mind" often does not apply. Hazardous industrial waste buried decades earlier bubbled to the surface, found its way into groundwater supplies, and ended up in backyards and basements.

Between 1942 and 1953, Hooker Chemicals and Plastics Corporation, which produced pesticides and plasticizers, dumped more than 19,000 metric tons (21,000 tons) of highly toxic, carcinogenic chemical wastes (mostly contained in steel drums) into an old canal excavation, known as the Love Canal, and sealed the dump with a clay cap and topsoil.

In 1953 Hooker Chemicals sold the canal area to the Niagara Falls school board for one dollar on the condition that the company would have no future liability for any injury or property damage caused by the dump's contents. The company says that it warned the school board against carrying out any kind of construction at the disposal site.

An elementary school and a housing project, eventually containing 949 homes, were built in the Love Canal area (Figure 20-11). Residents began complaining to city officials in 1976 about chemical smells and chemical burns received by children playing in the canal, but these complaints were ignored. In 1977 chemicals from badly corroded barrels filled with hazardous waste began leaking into storm sewers, gardens, and basements of homes adjacent to the canal.

Informal health surveys conducted by alarmed residents revealed an unusually high incidence of birth defects; miscarriages; assorted cancers; and nerve, respiratory, and kidney disorders among people who lived near the canal. Complaints to local elected and health officials had little effect. Pressure from residents and unfavorable publicity, however, led state officials to conduct a preliminary health survey and tests. They found that women age 30 to 34 in one area of the canal had a miscarriage rate four times higher than normal; they also found that the air, water, and soil of the canal area and basements of nearby houses were contaminated with a wide range of toxic and carcinogenic chemicals.

In 1978 the state closed the school and permanently relocated the 239 families whose homes were closest to the dump, and fenced this area off. In 1980, after protests from the outraged 710 families still living nearby, President Carter declared Love Canal a federal disaster area and had these families relocated. Federal and New York state funds were then provided to buy the homes of those who wanted to move permanently.

Since that time the homes and the school within a block and a half of the canal have been torn down, and the state has purchased 570 of the remaining homes. In 1987 about 50 families remained in the desolate neighborhood, unwilling or unable to sell their houses and move. The dump site has been covered with a clay cap and surrounded by a drain system that pumps leaking wastes to a new treatment plant. A chain-link fence surrounds the entire contaminated area.

Local officials have pressed the federal government for a clean bill of health so that the state can resell the homes it bought from fleeing homeowners and begin rehabilitating the neighborhood. But cleanup has proved to be quite difficult. The EPA hopes to complete its evaluation of cleanup efforts by 1988.

As yet no definitive study has been made to determine the long-term effects of exposure to these hazardous chemicals on the former Love Canal residents. All studies made so far have been criticized on scientific grounds. Even if the effects of exposure to these chemicals prove to be less harmful than expected, the psychological damage to the evacuated families is enormous. For the rest of their lives, they will wonder whether a disorder will strike and will worry about the possible effects of the chemicals on

20-4   TYPES, SOURCES, AND EFFECTS OF HAZARDOUS WASTE

What Is Hazardous Waste?   Any discarded material that may pose a substantial threat or potential hazard to human health or the environment when managed improperly is a **hazardous waste.** These wastes may be in solid, liquid, or gaseous form, and include a variety of toxic, ignitable, corrosive, or dangerously reactive substances.

Until recently there was little concern over hazardous waste in the United States and most other parts of the world. This changed in 1977, when it was discovered that hazardous chemicals leaking from an abandoned waste dump had contaminated homes in a suburban development known as Love Canal, located in Niagara Falls, New York (see Spotlight above). This event triggered the realization that one of the country's primary environmental problems is dealing with the large amounts of hazardous waste we are producing today, and that we must also solve the problem of

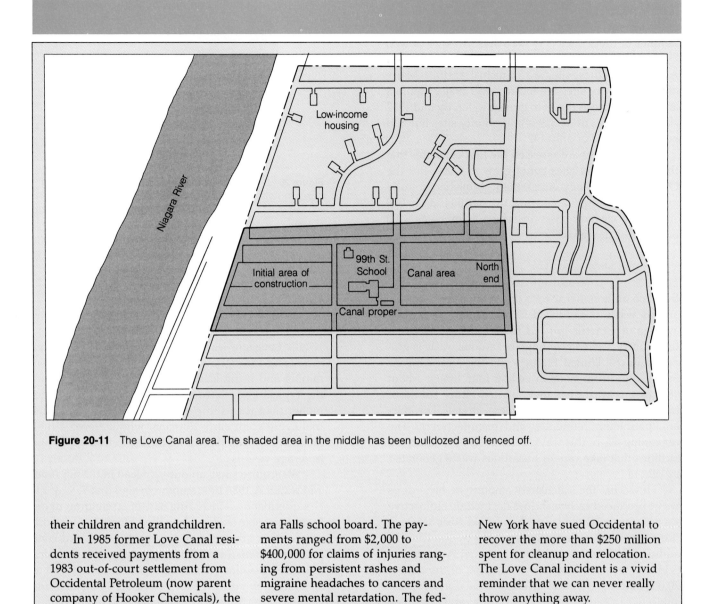

Figure 20-11 The Love Canal area. The shaded area in the middle has been bulldozed and fenced off.

their children and grandchildren.

In 1985 former Love Canal residents received payments from a 1983 out-of-court settlement from Occidental Petroleum (now parent company of Hooker Chemicals), the city of Niagara Falls, and the Niagara Falls school board. The payments ranged from $2,000 to $400,000 for claims of injuries ranging from persistent rashes and migraine headaches to cancers and severe mental retardation. The federal government and the state of New York have sued Occidental to recover the more than $250 million spent for cleanup and relocation. The Love Canal incident is a vivid reminder that we can never really throw anything away.

what to do with what was dumped in the past in up to 50,000 sites before any laws were established for its disposal.

How Much Has Been Dumped on the Land in the Past? Between 1950 and 1975 an estimated 5 trillion kilograms (6 billion tons) of hazardous waste was deposited on or under land throughout the United States. The EPA estimates that there are at least 26,000 U.S. sites where hazardous materials were dumped before present laws regulating disposal of such materials were enacted in 1976. The full extent of the problem is unknown because no one knows where they all are or what is in them.

Every country in Europe (except Sweden and Norway) also contains a number of abandoned and active hazardous waste sites needing urgent attention. For example, small and densely populated Holland has 5,000 identified hazardous waste sites; at least 350 pose an immediate danger to public health.

By early 1987 the EPA had placed 951 U.S. sites

on a priority cleanup list because of their threat to nearby populations from actual or potential pollution of the air, surface water, and groundwater. The largest number of these sites is in New Jersey (94), followed by Michigan (66), New York (64), and California (53). At least 180 of these sites are municipal landfills. Many of these priority sites are located over major aquifers and pose a serious threat to groundwater. By mid 1987 cleanup had begun at 819 of the priority sites but had been completed at only 14.

As more sites are assessed the EPA estimates that the list of priority sites could grow to 2,000. But the Office of Technology Assessment published a study in 1985 estimating that the final list may include 10,000 sites, with cleanup costs absorbing as much as $100 billion over the next 50 years.

How Much Is Produced Today? Each year more than 265 million metric tons (292 million tons) of hazardous waste is produced in the United States—an average of 1 metric ton (1.1 tons) for each person in the country. Each day enough hazardous waste is produced in the United States to fill the New Orleans Superdome from floor to ceiling four times. About 96% of this waste is generated and either stored or treated on site by large companies—chemical producers, petroleum refineries, and manufacturers. The remaining 4% of this waste is handled by commercial facilities that take care of hazardous waste generated by others.

However, the calculated amount of hazardous waste produced each year does not include radioactive waste, sewage sludge, and household toxic waste—none of which is currently regulated by the EPA. In addition, a 1984 study by the National Academy of Sciences revealed that only about 20% of the almost 70,000 different chemicals in commercial use have been subjected to extensive toxicity testing, and one-third of these chemicals have never been tested at all for toxicity. If such tests were run, many of these chemicals would be classified as hazardous waste.

About 93% of the hazardous wastes produced in the U.S. come from the chemical, petroleum, and metal-related industries (Figure 20-12). Although all states produce hazardous waste, about 65% of the volume is produced, in decreasing order, by Texas, Ohio, Pennsylvania, Louisiana, Michigan, Indiana, Illinois, Tennessee, West Virginia, and California.

Transporting hazardous waste, mostly by truck and train, is another area of increasing concern. According to the EPA, between 1980 and 1985 there were 7,000 accidents involving the release of 191,000 metric tons (210,000 tons) of hazardous chemicals in the United States. These accidents killed 139 people, injured 1,478, led to the evacuation of 217,000 people, and caused at least $50 million in property damage.

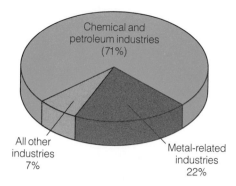

Figure 20-12 Major sources of hazardous waste in the United States in 1985. (Data from the Environmental Protection Agency)

20-5 SOME EXAMPLES OF HAZARDOUS WASTES

Lead Levels of lead in the environment have been increasing throughout the world since people began mining and using lead in about 800 B.C. As a result, the typical body burden of lead today is 500 times higher than it was in people living before the industrial age (except early Romans, who suffered a high incidence of lead poisoning linked to lead pipes and beverage vessels).

We acquire small amounts of lead in the air, food, and water. A 1986 EPA study revealed that 77% of the U.S. population—including 88% of all children under 5—have unsafe lead levels in their blood. Some of this comes from inhalation of tiny particles of lead compounds emitted into the atmosphere from the burning of leaded gasoline (which contains tetraethyl lead as an antiknock additive) and from lead smelters and steel factories. Vehicle emissions of lead have declined due to the gradual reduction of the lead allowed in gasoline in the United States. Studies show that the 68% drop in gas lead content between 1977 and 1982 caused lead concentrations in the atmosphere and in human blood to drop by almost two-thirds. A total ban on leaded gasoline in the United States is supposed to take place by 1988.

Brazil is phasing out gasoline entirely in favor of ethanol, and leaded gasoline has been prohibited in all large cities in the Soviet Union since 1959. European Common Market countries, Japan, and Canada are also sharply reducing the lead content of gasoline but have not banned it altogether.

Another source of lead is ingesting food contaminated with airborne particles that have settled on agricultural and grazing areas, especially those near highways. Some of this lead is absorbed by plants, but most can be removed by careful washing. Solder used

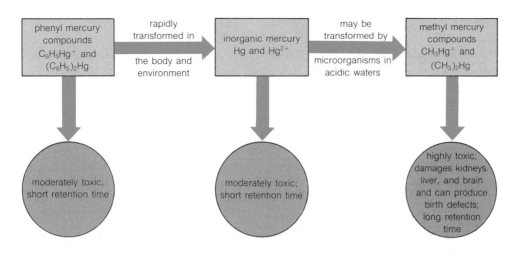

Figure 20-13 Some chemical forms of mercury and how they may be transformed in the environment.

phenyl mercury compounds $C_6H_5Hg^+$ and $(C_6H_5)_2Hg$ — rapidly transformed in the body and environment → inorganic mercury Hg and Hg^{2+} — may be transformed by microorganisms in acidic waters → methyl mercury compounds CH_3Hg^+ and $(CH_3)_2Hg$

moderately toxic; short retention time

moderately toxic; short retention time

highly toxic; damages kidneys, liver, and brain and can produce birth defects; long retention time

to seal the seams on food cans is another source of lead in food, especially in acidic foods such as tomatoes and citric juices.

As leaded gasoline is phased out, the remaining source of lead for many people is drinking water. According to the EPA, nearly one in five Americans drinks tap water containing excess levels of lead, which enters plumbing systems when acidic "soft" water erodes lead conduits and the lead solder used to join copper pipe. Since 1987, the use of pipes and solder containing lead in public water systems has been banned. However, a large percentage of existing houses and buildings contain pipes and solder joints containing lead.

For about $40 homeowners can have private laboratories test tapwater for lead and other contaminants (check with local health department officials). Running the water for two to three minutes before using will flush out most lead. Home water filtration systems using activated charcoal filters don't remove lead or other toxic metals. An alternative is to drink bottled water, most of which is spot-checked for lead content by the bottler. In building or remodeling, homeowners should use plastic pipes or ask plumbers to use lead-free solder, which costs only a few dollars more.

Before 1950 lead oxide and other lead compounds were added to interior and exterior paint to make it shinier and longer-lasting and to fix colors. Levels of lead in paint were decreased in the 1950s, but it was not until 1976 that the U.S. Consumer Products Safety Commission reduced the amount of lead in paint sold for home use to 0.06%.

In the United States it is estimated that 40 million houses built before 1950 and 20% of those built between 1960 and 1975 are potential sources of heavily leaded paint. These houses are a major source of lead poisoning for children between ages 1 and 3, who crawl around the floor and inhale harmful amounts of lead dust from cracking and peeling paint or ingest it by

sucking their thumbs or putting toys in their mouths. Many infants also eat chips of leaded paint that have peeled off, apparently because they taste sweet.

Another common household source of lead is in burning certain types of paper. Homeowners who burn used paper in wood stoves should not burn comic strips, gift-wrapping paper, and painted wood, which can be sources of lead contamination indoors and outdoors.

Children up to about age 9 are particularly vulnerable to lead poisoning because their bodies absorb lead very readily. Pregnant women can also transfer dangerous levels of lead to unborn children. Studies indicate that 15% to 20% of all preschool children in the United States may suffer some degree of lead poisoning. About 200 American children die each year from lead poisoning, especially from ingesting large quantities of leaded paint chips. Each year thousands of others suffer from some brain damage, hearing disorders, and behavioral disorders as a result of lead poisoning.

Mercury Mercury enters the air and water when we burn coal (which contains mercury as a contaminant) and through industrial discharges into sewers and surface waters. Yet these discharges are small compared to natural inputs of mercury vaporizing from the earth's crust and from the vast amounts stored as bottom sediments in the ocean.

It is dangerous to eat large amounts of tuna, swordfish, and other large ocean species that contain high levels of mercury. But most, if not all, of this mercury comes from natural sources, and the danger has probably always been present. Although the human input of mercury into the ocean is insignificant compared to natural sources, dangerous levels of mercury compounds are sometimes discharged by industries into lakes, rivers, bays, and estuaries.

Figure 20-13 shows the major forms of mercury and the ways they are transformed. Metallic mercury is dangerous when inhaled but it isn't as dangerous when swallowed. The major threat posed by mercury is in an extremely toxic organic mercury compound known as methyl mercury (CH_3Hg^+). It can remain in the body for months; attack the central nervous system, kidneys, liver, and brain tissue; and cause birth defects.

Under acidic conditions, anaerobic bacteria dwelling in the bottom mud of lakes and other surface waters can convert elemental mercury and mercury salts into methyl mercury. Most surface waters apparently aren't acidic enough to cause this, but acidification of an increasing number of lakes from acid deposition may aggravate the problem (Section 18-2).

One tragic episode occurred in the late 1950s when 649 people died and 1,385 suffered mercury poisoning from methyl mercury discharged into Minamata Bay, Japan, from a nearby chemical plant. Most victims in this seaside area had eaten mercury-contaminated fish and shellfish three times a day. Another tragedy occurred in 1972 when Iraqi villagers, who had received a large shipment of seed grain fumigated with methyl mercury, fed it to their animals and baked bread with it instead of planting it. Reportedly, 459 people died and 6,530 others became ill.

Dioxins   In 1971, dirt roads in several St. Louis, Missouri, suburbs were sprayed with waste automobile oil to control dust, a widely used procedure in many states. The day after the oil was sprayed, the owner of a nearby horse ranch found dozens of dead sparrows on the floor of a barn near the road. Most dogs and cats on the ranch died within a month of the spraying. Within a year, 43 horses that had regularly exercised in an arena near the road died, and most of the pregnancies of the horses bred in 1971 resulted in spontaneous abortions. During 1971, one of the ranch owners suffered from headaches, diarrhea, and chest pains. One of his daughters suffered from bad sores and severe headaches, and another daughter had to be hospitalized as a result of severe internal bleeding.

Several years later, tests on the oil sprayed on the road revealed that it was contaminated with several highly toxic compounds, including a number of toxic chlorinated hydrocarbon compounds known as **dioxins**. Further investigation revealed that the contaminated oil had been sold to a chemical company that was supposed to clean it up for reuse but instead sprayed it on roads at several sites in Missouri. Soil in the Times Beach suburb was found to be so contaminated that in 1983 the EPA bought out the entire town at a cost of $36.7 million and had to relocate 2,200 people. Twenty-six other sites in Missouri are known to be contaminated with dioxins, and 75 more are suspected. Significant levels of dioxins have also been found in rivers in Michigan, fish taken from the Great

Lakes, and flooded basements of homes near the Love Canal in Niagara Falls.

Dioxins are a family of 75 different chlorinated hydrocarbon compounds. One form in particular, usually referred to as TCCD, has been shown to be extremely toxic and to cause liver cancer, birth defects, and death in laboratory animals at extremely low levels. This chemical also persists in the environment, especially in soil and fatty tissue in the human body, and can apparently be biologically amplified to higher levels in food webs. Workers and others exposed to TCCD in industrial plant accidents have complained of headaches, weight and hair loss, liver disorders, irritability, insomnia, nerve damage in the arms and legs, loss of sex drive, and chloracne (a severe, painful, and often disfiguring form of acne).

TCCD and several other dioxins form in trace amounts during the high-temperature combustion of various organic compounds in incinerators and other combustion processes. In 1981 the EPA concluded that the quantity of TCCD and other dioxins released into the atmosphere during combustion processes is small enough and diffuse enough to be relatively harmless, though this finding is disputed by some scientists. The major potential threat comes from the larger quantities of the chemical that are present primarily in industrial dump sites, many of which are abandoned. In 1986 the U.S. Centers for Disease Control found that through the fat in human breast milk, nursing infants may be exposed to TCCD levels 1,300 times the recomended daily maximum exposure level.

But there is some possible good news. The EPA had assumed that dioxins are highly mobile in soil, thus making disposal of dioxin-contaminated soil and waste on land hazardous. However, in 1986 a preliminary study indicated that dioxins apparently move very slowly in soil, perhaps as little as 1 centimeter in 400 years to 5,000 years. If this finding is substantiated, the EPA hopes to be able to dispose of soils containing low concentrations of dioxins in abandoned surface mines.

Polychlorinated Biphenyls (PCBs)   Since 1966, scientists have found widespread contamination from a widely used group of toxic, oily synthetic organic chemicals known as **polychlorinated biphenyls (PCBs)**. PCBs are mixtures of about 70 different but closely related chlorinated hydrocarbon compounds.

There was little concern about PCBs until 1968, when 1,300 Japanese came down with skin lesions (chloracne) and suffered liver and kidney damage after they had eaten rice oil accidentally contaminated with PCBs that had leaked from a heat exchanger. As a result of this incident, Japan banned all uses of PCBs.

Like DDT and dioxins, PCBs are insoluble in water, soluble in fats, and very resistant to biological and

Binghamton Press and Sun-Bulletin

Figure 20-14 Cleanup begins after a 1981 fire in a basement transformer spread PCBs throughout a state office building in Binghamton, New York.

chemical degradation; thus they are biologically amplified in food webs. PCBs entering the body through food, skin contact, and inhalation accumulate in fatty tissues and body organs. Also, like DDT, dioxins, and other chlorinated hydrocarbons, the long-term health effects on people exposed to low levels of PCBs are unknown. But tests have shown that PCBs produce liver damage, kidney damage, gastric disorders, reproductive disorders, skin lesions, and tumors in laboratory animals.

In 1974 the U.S. chemical industry voluntarily stopped producing PCBs for all uses except in closed systems such as electrical transformers, and in 1976 Congress banned the further manufacture and use of PCBs, except in existing electrical transformers. However, prior to this ban, the EPA estimates that at least 68,000 metric tons (75,000 tons) of PCBs had entered the environment because of indiscriminate dumping at landfills and fields, into sewers, and along roadsides. Traces of PCBs have been found all over the world in soil, surface and groundwater, fish, human breast milk, fatty tissue, and even in Arctic snow.

Until 1981 the major risk of exposure to PCBs was thought to be from leaks and spills and improper disposal of PCBs and PCB equipment. But since 1981, fires involving PCB-filled electrical transformers in several cities have exposed people in office buildings, apartment complexes, shopping malls, and train and subway stations to much more severe risks. During a transformer fire, PCBs and even more toxic by-products produced by their combustion spread throughout buildings and are also flushed into storm sewers and

surface waters by the water used to extinguish the fire (Figure 20-14).

In 1985, the EPA ordered that by 1990 PCBs be removed from all electrical transformers in U.S. apartment and office buildings, hospitals, and shopping malls, and banned the further installation of PCB-filled transformers in or near commercial buildings. This will still leave about 140,000 sealed electrical transformers and capacitors (owned mostly by utility companies) containing some 341 million kilograms (375,000 tons) of PCBs. Each year PCBs are released into the environment when a few of these transformers and capacitors leak, catch fire, or explode.

20-6 CONTROL AND MANAGEMENT OF HAZARDOUS WASTE

Methods for Dealing with Hazardous Waste There are three basic ways of dealing with hazardous waste, as outlined by the National Academy of Sciences in 1983 (Figure 20-15). The first and most desirable approach aims at reducing the total amount of waste produced by modifying industrial or other processes to eliminate or reduce waste output and by reusing or recycling the hazardous wastes that are produced.

Any remaining wastes should then be converted to less hazardous or nonhazardous materials by spreading them on the land where they can decompose biologically, incinerating them on land or at sea

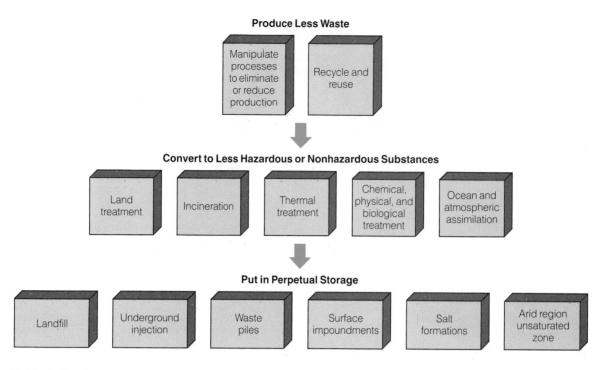

Produce Less Waste

Manipulate processes to eliminate or reduce production

Recycle and reuse

Convert to Less Hazardous or Nonhazardous Substances

Land treatment

Incineration

Thermal treatment

Chemical, physical, and biological treatment

Ocean and atmospheric assimilation

Put in Perpetual Storage

Landfill

Underground injection

Waste piles

Surface impoundments

Salt formations

Arid region unsaturated zone

Figure 20-15 Options for dealing with hazardous waste. (National Academy of Sciences)

Figure 20-16 Fate of hazardous waste in the United States in 1985. (Data from the Congressional Budget Office)

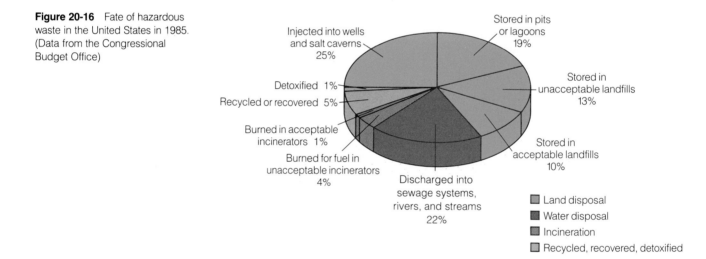

Stored in pits or lagoons 19%

Injected into wells and salt caverns 25%

Stored in unacceptable landfills 13%

Detoxified 1%

Recycled or recovered 5%

Burned in acceptable incinerators 1%

Burned for fuel in unacceptable incinerators 4%

Discharged into sewage systems, rivers, and streams 22%

Stored in acceptable landfills 10%

- Land disposal
- Water disposal
- Incineration
- Recycled, recovered, detoxified

using specially designed incinerators, thermally decomposing them, treating them chemically or physically, or in some cases diluting them to acceptable levels in the ocean or atmosphere. Any waste still remaining after such detoxification processes should then be placed in perpetual storage in a geologically and environmentally secure place that is carefully monitored for leaks.

Present Management of Hazardous Waste So far, most hazardous waste produced in the United States is managed by the third and least desirable option. About 80% of the hazardous wastes produced each

year in the United States is disposed of legally or illegally on the land using environmentally unacceptable methods (deep wells, unlined lagoons, and unlined landfills); discharged into sewage systems, rivers, and streams; or burned in unacceptable incinerators (Figure 20-16). Only 10% is stored in environmentally acceptable landfills and only 7% is recycled, recovered, detoxified, or burned in acceptable incinerators. An unmeasured amount of hazardous waste is illegally dumped on the land or into water supplies (see Spotlight on p. 385).

In most European countries, where vacant land is in short supply, about 50% of hazardous waste is burned in incinerators equipped with sophisticated air

In the United States a large and probably growing amount of hazardous waste is illegally dumped into municipal landfills, rivers, sewer drains, wells, empty lots and fields, old quarries, and abandoned mines, and along roadsides. Some truckers carrying liquid hazardous waste drive along freeways or rural roads, especially at night, with the tank spigots on their trucks open.

In Tennessee, illegal dumpers have sent freight cars loaded with hazardous waste to fictitious addresses, to be paid for upon delivery. Some companies pick up hazardous waste from customers, charge them $50 to $150 a barrel for disposal, and then deposit the bar-

rels in a rented warehouse. Once the warehouse is filled, the disposal company may declare bankruptcy and move out of the state or leave the building owner liable for disposing of the waste and for damages.

Hazardous wastes are also sprayed on ordinary trash, and garbage collectors are bribed to dispose of the trash in landfills not designed to handle such substances. Some are also illegally mixed with heating oil and burned in the boilers of schools, hospitals, and office and apartment buildings, resulting in the release of toxic chemicals into the air.

U.S. law enforcement officials warn that illegal dumping is becom-

ing more frequent as waste generators and haulers try to cut costs. Some officials also warn that the lure of large profits and generally lax law enforcement has led to increased involvement of organized crime in every aspect of the hazardous waste disposal industry. For example, mob-controlled garbage collection companies, principally in New York, New Jersey, Ohio, and Florida, are picking up toxic waste from client firms, mixing it with regular garbage, and then dumping the deadly mixture at municipal landfills and other sites not authorized to receive hazardous waste.

---

pollution controls; much of the energy released by the incinerators is used to produce electricity or heat.

In 1984 Congress amended the 1976 Resource Conservation and Recovery Act to make it national policy to minimize or eliminate land disposal of hazardous waste by 1990 unless the EPA has determined that it is an acceptable or the only feasible approach for a particular hazardous material. Even then, each chemical is to be treated to the fullest extent possible to reduce its toxicity before land disposal of any type is allowed. Although it is unlikely that the 1990 deadline will be met, this national policy represents a much more ecologically sound approach to dealing with hazardous waste.

Recycling, Reuse, and Industrial Process Redesign   In Europe, waste exchanges or clearinghouses are used to transfer about one-third of a firm's waste so that another firm can use it as raw material. By 1986 at least 30 regional waste exchanges in the United States were transferring about 10% of the listed wastes, and this fraction could increase significantly in the future. Although the EPA estimates that at least 20% of the hazardous materials that are currently generated in the United States could be recycled or reused, currently only about 5% of the materials are managed in this manner. Despite the enormous potential for recycling and waste trading, between 1979 and 1985 the

EPA spent almost no money and assigned only one person to promote this approach.

Conversion to Less Hazardous or Nonhazardous Materials   In the geographical center of Denmark, a plant detoxifies as much as 90% of the country's hazardous waste, while providing 30% of the heating needs for the 18,000 residents of a nearby town. In West Germany, 15 regional waste treatment centers detoxify almost 85% of the country's hazardous waste. So far, this approach is not used very widely in the United States.

Some hazardous organic compounds that contain little or no toxic metal compounds, volatile materials, or persistent organic compounds can be detoxified biologically by landfarming. In **landfarming,** wastes are applied onto or beneath surface soil and mixed to expose the contaminated material to oxygen; then microorganisms and nutrients are added as needed to ensure biological decomposition. This method is particularly useful for stimulating forest growth and for reclamation of surface-mined land.

Other biological treatment and decomposition processes include the use of composting, trickling filters, activated sludge, and aerated lagoons. There are also experiments with using mutant bacteria produced by genetic engineering techniques to detoxify specific waste materials. But critics worry that these "superbugs" may get out of control and destroy other useful material before it has a chance to become waste. Hazardous waste is also detoxified by various physical and chemical processes such as neutralization of acidic or alkaline waste and removal of toxic metals and other compounds by precipitation or absorption.

Hazardous waste can also be decomposed by incineration in the presence of oxygen at high temperatures. The Netherlands incinerates about half its hazardous waste, and the EPA estimates that about 60% of all U.S. hazardous waste could be incinerated. With proper air pollution controls, incineration has a number of advantages. It is potentially the safest method of disposal for most types of hazardous waste and is also the most effective method of waste disposal of organic waste material, such as pesticides, solvents, and PCBs.

But there are also some disadvantages. Incineration is the most expensive method, and the ash that is left must be disposed of and often contains toxic metals. Not all hazardous wastes are combustible, and the gaseous and particulate combustion products emitted by incinerators can be health hazards if not controlled.

Because of the upcoming ban on land disposal of most hazardous wastes, and because most people object to living near incinerators, interest has been growing in incinerating liquid hazardous waste at sea in specially designed ships, as several European countries are doing. These ships would burn waste without the expensive smokestack scrubbers required for land-based incinerators. This approach is about one-third cheaper than on-land incineration and minimizes the accidental dangers to people. Since 1977, the United States has conducted several experimental hazardous waste burnings in incinerator ships in the Pacific Ocean and the Gulf of Mexico. Although these test burns were declared successful, several scientists contend that the measurements were inadequate or invalid.

Many environmentalists have opposed burning of hazardous waste at sea, fearing that accidental chemical spills resulting from human error, fog, storms, or reefs—or residue from incomplete destruction of toxic wastes—could threaten marine life. In addition, they suspect that some companies would take money-saving shortcuts or would cover up accidents at sea, far away from scrutiny. Although a land-based incinerator equipped with scrubbers injects only a tiny amount of unburned toxic ash into the atmosphere, an incinerator ship burning the same volume of waste without such controls injects a large amount of unburned toxic particulate matter into the atmosphere, where winds carry it long distances and allow it to settle on land and inland surface waters. Environmentalists also point out that at-sea incineration works only for liquid waste and thus is not appropriate for most of the hazardous waste generated in the United States.

Land Disposal of Hazardous Waste Since 1976, the Resource Conservation and Recovery Act has required that any landfill used for the storage of hazardous waste be a secured landfill, that operators of such landfills show financial responsibility for up to $10 million in accidental damages, and that the facilities be monitored for at least 30 years to minimize the chance of hazardous waste escaping into the environment. A **secured landfill** is a site for the containerized burial and storage of hazardous solid waste (Figure 20-17). Liquid hazardous wastes cannot be stored in secured landfills unless they are solidified to reduce volume and encased in cement, asphalt, glass, or organic polymers to decrease the chance of escape into the air or water.

Ideally, a secured landfill is situated in thick natural clay deposits; isolated from surface or subsurface water supplies; not subject to flooding, earthquakes, or other disruptions; and unlikely to transport leachate to an underground water source. By 1986 there were about 525 EPA-approved hazardous waste land disposal sites in the United States, 49 of them commercial operations.

In 1983, however, the Office of Technology Assessment concluded that sooner or later, any secured landfill will leak from tears in the plastic liners caused by bulldozers or freezing temperatures; leachate disintegration of the liner; crushing of leachate collection pipes by the weight of the waste; clogging of the perforations in leachate collection pipes by debris; or disruption of the protective cover by erosion, new construction, or subsidence.

Federal Legislation and Control of Hazardous Waste The Resource Conservation and Recovery Act of 1976, as amended in 1984, requires the EPA to identify hazardous wastes, set standards for their management, and provide guidelines and some financial aid to establish state programs for managing such wastes. The RCRA also requires all firms that store, treat, or dispose of more than 100 kilograms (220 pounds) of hazardous wastes per month to apply to the EPA for a permit.

To reduce illegal dumping, hazardous waste producers granted disposal permits by the EPA must use a "cradle-to-grave" manifest system to keep track of the 4% of hazardous waste transferred from point of origin to offsite approved disposal facilities. EPA administrators, however, point out that this requirement is almost impossible to enforce because the EPA and state regulatory agencies do not have enough personnel to review the documentation of more than 750,000 hazardous waste generators and 15,000 haulers each year, let alone verify them and prosecute offenders.

Environmentalists argue that the Resource Conservation and Recovery Acts of 1976 and 1984 have

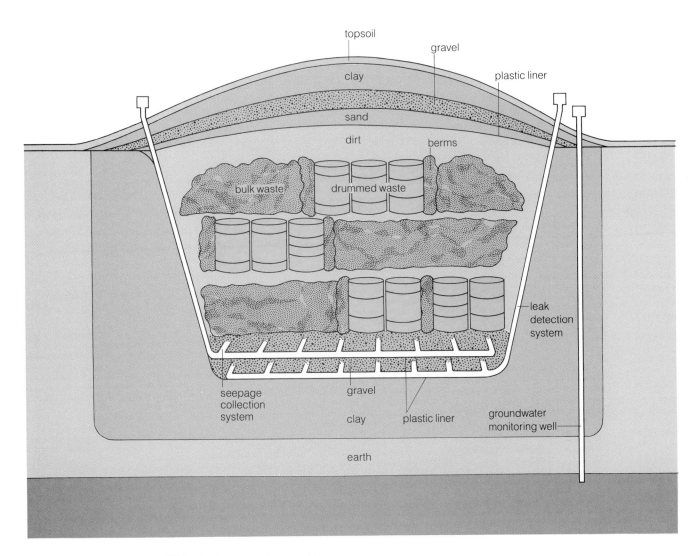

topsoil

gravel

clay

plastic liner

sand

dirt

berms

bulk waste

drummed waste

leak
detection
system

seepage
collection
system

gravel

clay

plastic liner

groundwater
monitoring well

earth

Figure 20-17 A secured landfill for the long-term storage of hazardous wastes.

several serious loopholes. By early 1987 the EPA had listed only 350 of the estimated 35,000 potentially hazardous chemicals as being hazardous. Sampling and testing procedures used by waste producers to determine whether their wastes are classified as hazardous under federal guidelines are inadequate. At least 4.5 million metric tons (5 million tons) of hazardous waste discharged down sewers is not regulated and is a major source of contamination of sewage sludge (which limits its use as a fertilizer) and surface waters.

States are not required to regulate all hazardous wastes identified by the EPA, making states with weaker programs more attractive choices as dumping grounds for certain wastes. In addition, because federal money for inspections ended in 1982, few states regularly monitor city and county landfills to determine whether leachate is percolating into groundwater. Furthermore, in 1985 the EPA budget for drafting and enforcing hazardous waste control regulations decreased by 25% from 1981. Most violators draw only a warning,

and with only 35 EPA criminal investigators, most illegal dumpers have little chance of being caught.

In 1980 Congress passed the Comprehensive Environmental Response, Compensation and Liability Act (CERCLA), known as the Superfund program, and added amendments in 1986. CERCLA was designed primarily to deal with the problems of financing the cleanup of abandoned or illegal hazardous waste sites. The original act established a $1.6 billion fund, financed jointly by federal and state governments and taxes on chemical and petrochemical industries, for cleanup of abandoned or inactive hazardous waste dump sites. The EPA is authorized to collect fines and sue the owners later (if they can be found and held responsible) to recover up to three times the cleanup costs.

Drastic EPA budget cuts since 1981 and lack of sufficient matching funds by states have made it difficult to implement the Superfund legislation. In 1983, critics charged the EPA with letting some noncomply-

Table 20-1 Alternatives for Some Hazardous Household Chemicals

Chemical	Alternative
Oven cleaner	Use baking soda for scouring. For baked-on grease, apply ¼ cup of ammonia in oven overnight to loosen; scrub the next day with baking soda.
Drain cleaner	Pour ½ cup salt down drain, followed by boiling water; flush with hot tap water.
Glass polish	Use ammonia and soap.
Wall and floor cleaners containing organic solvents	Use detergents to clean large areas and then rinse with water.
Toilet bowl, tub, and tile cleaner	Mix borax and lemon juice in a paste. Rub on paste and let set two hours before scrubbing.
Mildew stain remover and disinfectant cleaner	Chlorine bleach
Furniture polish	Melt 1 pound carnauba wax into 1 pint of mineral oil. For lemon oil polish, dissolve 1 teaspoon of lemon oil into 1 pint of mineral oil.
Shoe polish	Use polishes that do not contain methylene chloride, trichloroethylene, or nitrobenzene.
Spot removers	Launder fabrics when possible to remove stains. Also try cornstarch or vinegar.
Carpet and rug shampoos	Cornstarch
Detergents and detergent boosters	Washing soda and soap powder
Water softeners	Washing soda
Pesticides (indoor and outdoor)	Use natural biological controls (Section 21-5), boric acid for roaches
Mothballs	Soak dried lavender, equal parts of rosemary and mint, dried tobacco, whole peppercorns, and cedar chips in real cedar oil and place in a cotton bag

ing firms off too easily and settling for superficial cleanups. Congressional investigations of Superfund mismanagement and alleged inside deals led to the firing of the director of the program and the resignation of the head of the EPA.

In 1986, amendments to the Superfund program authorized $9 billion more to be used for cleanup of sites between 1987 and 1994. The amendments also provided $500 million to clean up leaking underground fuel tanks, required manufacturers to provide citizens with detailed information about any hazardous chemicals produced or stored in their community, and made it easier for citizens to sue polluters for damages.

A 1983 study by the Office of Technology Assessment concluded that in the long run the Superfund program might be ineffective because many wastes are simply moved from one burial site to another, and leakage eventually will occur. A 1985 study by a congressional research team found that of the 1,246 hazardous waste dumps it surveyed, nearly half showed signs of polluting nearby groundwater. This team of investigators charged that the EPA's monitoring of these sites was "inaccurate, incomplete, and unreliable."

Individual Action Individuals can reduce their own exposure to hazardous waste by insisting that existing laws governing hazardous waste be enforced and strengthened and that the EPA and state agencies administering these laws be adequately funded and staffed. Used motor oil should be taken to a local auto service center for recycling. Less hazardous (and usually cheaper) household cleaning products should be used (Table 20-1), and hazardous chemicals such as pesticides should only be used when absolutely necessary and in the smallest amount possible.

Hazardous household chemicals should not be mixed because many react and produce deadly chemicals. For example, when ammonia and household bleach are combined or even get near one another, they react to produce deadly poisonous chloramine gas. Hazardous chemicals should also not be flushed down the toilet, poured down the drain, buried in the yard, or dumped down storm drains. They should not be thrown away in the garbage because they will end up in a landfill, where they can contaminate drinking water supplies. Instead, contact your local health department or environmental agency for information on what do with such chemicals.

Waste is a human concept. In nature nothing is wasted, for everything is part of a continuous cycle. Even the death of a creature provides nutrients that will eventually be reincorporated in the chain of life.

Denis Hayes

DISCUSSION TOPICS

1. How is solid waste collected and disposed of in your community? Is the groundwater near sanitary landfills periodically monitored for contamination? Does the community have any EPA-approved secured landfill sites for disposal of hazardous waste?

2. Keep a list for a week of the solid waste materials you dispose of. What percentage is composed of materials that could be recycled, reused, or burned as a source of energy? How much of this material is a result of unnecessary packaging?

3. List the advantages and disadvantages of the high-technology (resource recovery plant) and the low-technology (source separation) approaches to recycling materials from solid waste. Would you favor requiring all households and businesses to sort recyclable materials for curbside pickup in separate containers? Explain.

4. Determine whether (a) your college and your city have recycling programs; (b) your college and your local government require that a certain fraction of all paper purchases contain recycled fiber; (c) teachers in your college and in local schools expect everyone to write on both sides of paper; (d) your college sells soft drinks in throwaway cans or bottles; and (e) your state has, or is contemplating, a law requiring deposits on all beverage containers.

5. What responsibility, if any, do you feel Hooker Chemicals has for damages and cleanup costs resulting from the leakage of hazardous waste at Love Canal? Explain.

6. Would you oppose locating a secured landfill in your community for the storage of hazardous waste? Would you oppose an incinerator to detoxify such wastes? Explain. If you oppose both of these alternatives, how would you propose that the hazardous waste generated in your community be managed?

7. Give your reasons for agreeing or disagreeing with each of the following proposals for dealing with hazardous waste in the United States:

   a. Burn all liquid hazardous waste in federally approved at-sea incinerator ships.

   b. Reduce the production of hazardous waste and encourage recycling and reuse of such materials by levying a tax or fee on producers for each unit of waste generated.

   c. Ban all land disposal of hazardous waste as a means of encouraging recycling, reuse, and treatment and as a means of protecting groundwater from contamination.

   d. Provide low-interest loans, tax breaks, and other financial incentives for industries that produce hazardous waste to encourage them to recycle, reuse, treat, destroy, and reduce generation of such waste.

21

Pesticides and Pest Control

A weed is a plant whose virtues have not yet been discovered.

Ralph Waldo Emerson

A **pest** is any unwanted organism that directly or indirectly interferes with human activity. Since 1945 vast fields planted with only one crop or only a few crops, as well as home gardens and lawns, have been blanketed with a variety of chemicals called pesticides (or *biocides*). A **pesticide** is a substance than can kill organisms that humans consider to be undesirable. A substance in this arsenal can kill unwanted insects (an **insecticide**), plants (a **herbicide**), rodents such as rats and mice (a **rodenticide**), fungi (a **fungicide**), or other organisms.

Pesticides can improve crop yields and help control populations of disease organisms. However, there is considerable evidence that the widespread use of pesticides can have harmful effects on wildlife, ecosystem structure and function, and human health. Their overuse can even lead to an increase in crop losses and a resurgence of the diseases and pests they are supposed to control. Rachel Carson's 1962 book *Silent Spring* dramatized the potential dangers of pesticides to wildlife and people and set off a controversy between environmentalists and pesticide industry officials that is still raging.

21-1 PESTICIDES: TYPES AND USES

Natural Control of Pests in Diverse Ecosystems About 100 of the at least 1 million catalogued insect species cause about 90% of damage to food crops. Indeed, most insects, fungi, rodents, and soil microorganisms help to cycle essential chemicals, pollinate plant species, and keep populations of potentially harmful species from reaching the levels that cause significant economic loss of food crops or livestock. Furthermore, through natural selection many plants and insects contain or give off chemicals that tend to

Figure 21-1 Throughout recorded history periodic outbreaks of locusts (such as these in Somalia, Africa) have devoured wild and cultivated plants used to feed the world's human population.

protect them from certain predators, parasites, and diseases.

Why the Need for Pest Control Has Increased

Locusts have always ravaged the wild and cultivated plants that people eat (Figure 21-1). Over the past 200 years, however, an increasing number of insects and other pests have become serious threats to crops that feed the world's rapidly increasing population.

Large areas of diverse ecosystems, containing small populations of many species, have been replaced with greatly simplified agricultural ecosystems and lawns that contain large populations of only one or two desired plant species. In such biologically simplified ecosystems, organisms can grow in number and achieve pest status, whereas their populations would have been controlled naturally in more diverse ecosystems. As a result, people have had to spend an increasing amount of time, energy, and money to control pests in crop fields, lawns, and other simplified ecosystems. The most widely used approach has been to spray fields and lawns with certain synthetic chemicals.

The Ideal Pest Control Method

The ideal pest-killing chemical would (1) kill only the target pest; (2) have no short- or long-term health effects on nontarget organisms, including human beings; (3) be broken down into harmless chemicals in a relatively short time; (4) prevent the development of genetic resistance in target organisms; and (5) be more economic than not using pest control. Unfortunately, no known pest control method meets all these criteria.

First-Generation Pesticides

Before 1940 there were only a few dozen pesticides on the market. Many of these *first-generation pesticides* were nonpersistent organic compounds made or extracted from insect poisons found naturally in plants. For example, pyrethrum, a powder obtained from the heads of chrysanthemums (Figure 21-2), was used by the Chinese 2,000 years ago and is still in use today. Caffeine is also an excellent insecticide that can be used to control tobacco hornworms, mealworms, milkweed bugs, and mosquito larvae. Other insecticides derived from natural plant sources include nicotine (as nicotine sulfate) from tobacco, rotenone from the tropical derris plant, and garlic oil and lemon oil, which can be used against fleas, mosquito larvae, houseflies, and other insects.

A second type of first-generation commercial pesticide in use before 1940 consisted of persistent inorganic compounds made from toxic metals such as arsenic, lead, and mercury. Most of these compounds are no longer used because they are highly toxic to people and animals, they contaminate the soil for 100 years or more, and they tend to accumulate in soil to the point of inhibiting plant growth.

Second-Generation Pesticides

A major revolution in insect pest control occurred in 1939 when it was discovered that **DDT** (**d**ichloro**d**iphenyl**t**richloroethane), a chemical known since 1874, was a potent insecticide. Since 1945, chemists have developed many varieties of such synthetic organic chemicals known as *second-generation pesticides.*

DDT and many related second-generation pesticides have been widely used for two reasons: they are easy and fairly cheap to produce, and they kill many

Figure 21-2 The heads of these pyrethrum flowers being harvested in Kenya, Africa, are ground into a powder and used directly as commercial insecticides or converted to other chemically related pyrethroid insecticides.

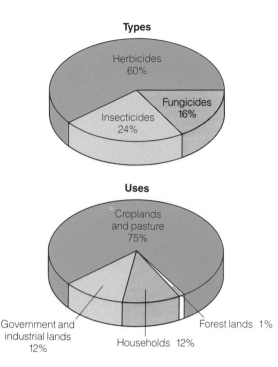

Figure 21-3 Major types of pesticides and their uses in the United States. (Data from the EPA and David Pimentel)

types of pest organisms over a long period of time because they are not broken down readily in the environment and are not dissolved and removed from cropfields by rain or irrigation water.

Worldwide, about 2.3 million metric tons (2.5 million tons) of second-generation pesticides are used each year—amounting to an average of about 0.45 kilogram (1 pound) of pesticide for each person on earth. About 85% of all pesticides are used in MDCs, but use in LDCs is growing rapidly and is projected to increase at least fourfold between 1985 and 2000.

In the United States about 600 active ingredients and 900 inert ingredients are mixed to make some 50,000 individual pesticide products. In 1986 about 1.2 million metric tons (1.3 million tons)—an average of 4.9 kilograms (10.7 pounds) for each American—were used in the United States (Figure 21-3).

Although we usually think of pesticides in terms of cropland, an EPA study showed that 92% of all U.S. households use one or more type of pesticide to control insects, weeds, fungi, or rodents. This study also showed that the average homeowner applies much more pesticide per unit of land area than do farmers; each year more than 250,000 Americans become sick because of pesticides used in the home. Thus everyone—not only farmers—has an important role to play in reducing unnecessary pesticide use and in ensuring that unused pesticides are not thrown in the trash to end up in landfills and pollute groundwater.

Major Types of Insecticides and Herbicides Most of the thousands of different insecticides used today fall into one of four classes of compounds: chlorinated hydrocarbons, organophosphates, carbamates, or pyrethroids (Table 21-1). Most of these chemicals are broad-spectrum poisons that kill most of the target and nontarget insects in the sprayed area by disrupting their nervous systems. They vary widely in their persistence—the length of time they remain active in killing insects (Table 21-1).

Organophosphates are generally much less persistent than DDT and most other chlorinated hydrocarbons now banned or restricted in the United States and many MDCs. However, they are more water soluble and can leach into groundwater; they also are much more toxic to birds, human beings, and other mammals than are chlorinated hydrocarbons. Furthermore, to compensate for their fairly rapid breakdown, farmers usually apply nonpersistent insecticides at regular intervals to ensure more effective insect control. As a result, these chemicals are often present in the environment almost continuously, like persistent pesticides. Use of pyrethroids is growing rapidly because they are generally nonpersistent, effective at low doses, and not highly toxic to mammals.

Herbicides can be placed into three classes based on their effect on plants: contact herbicides, systemic herbicides, and soil sterilants (Table 21-2). Most herbicides are active for only a short time.

Table 21-1 Major Types of Insecticides

Type	Examples	Persistence
Chlorinated hydrocarbons	DDT, DDE, DDD, aldrin, dieldrin, endrin, heptachlor, toxaphene, lindane, chlordane, kepone, mirex	High (2–15 years)
Organophosphates	Malathion, parathion, Azodrin, Phosdrin, methyl parathion, Diazinon, TEPP, DDVP	Low to moderate (normally 1–12 weeks, but some can last several years)
Carbamates	Carbaryl (Sevin), Zineb, maneb, Baygon, Zectran, Temik, Matacil	Usually low (days to weeks)
Pyrethroids	Pyrethrums extracted from flowers (Figure 21-2) and used directly or modified chemically	Usually low (days to weeks)

Table 21-2 Major Types of Herbicides

Type	Examples	Effects
Contact	Triazines such as atrazine	Kills foliage by interfering with photosynthesis
Systemic	Phenoxy compounds such as 2,4-D, 2,4,5-T, and Silvex; substituted ureas such as diuron, norea, and fenuron	Absorption creates excess growth hormones; plants die because they cannot obtain enough nutrients to sustain their greatly accelerated growth
Soil sterilants	Treflan, Dymid, Dowpon, Sutan	Kills soil microorganisms essential for plant growth; most also act as systemic herbicides

21-2 THE CASE FOR PESTICIDES

Using Insecticides to Control Disease During World War II, DDT was sprayed directly on soldiers and war refugees to control body lice, which spread typhus. The World Health Organization (WHO) also used DDT and related second-generation pesticides to control the spread of insect-transmitted diseases such as malaria (carried by the *Anopheles* mosquito), bubonic plague (rat fleas), typhus (body lice and fleas), sleeping sickness (tsetse fly), and Chagas' disease (kissing bugs).

Largely because of DDT, dieldrin, and several other chlorinated hydrocarbon insecticides, more than 1 billion people have been freed from the risk of malaria, and the lives of at least 7 million people have been saved since 1947 (see Spotlight on p. 394). Thus *DDT and other insecticides have probably saved more human lives than any other synthetic chemicals since people have inhabited the earth.*

Although DDT and several other chlorinated hydrocarbon insecticides deserve their reputation as life-givers, they are no longer effective in many parts of the world, primarily because of genetic resistance (see Spotlight on p. 394). As a result, between 1970 and 1987 there was a thirty- to fortyfold increase in malaria in countries where it had been almost eradicated. Despite the increasing ineffectiveness of DDT and other insecticides, the WHO points out that a ban on chlorinated hydrocarbon and organophosphate insecticides would lead to large increases in disease, human suffering, and death.

Using Insecticides and Herbicides to Increase Food Supplies Each year pests and disease consume or destroy about 45% of the world's food supply: One-third of this loss occurs before harvest and 12% occurs after harvest. Even in the United States, which uses vast amounts of pesticides and has sophisticated food storage and transportation networks, the total loss due to pests and disease is estimated to be 42% of the potential yearly production—33% before harvest and 9% after. This leads to annual crop losses worth about $20 billion.

People in the United States and in most MDCs tend to view malaria as a disease of the past. But in the tropical and subtropical regions of the world, malaria is still the single most serious health problem—killing at least 1 million people a year (mostly children under 5) and incapacitating tens of millions. Today more than half the world's population live in malaria-prone regions in about 100 different countries. Even in the United States, each day an average of four people discover they have malaria.

Malaria's symptoms come and go; they include fever and chills, anemia, an enlarged spleen, severe abdominal pain and headaches, extreme weakness, and greater susceptibility to other diseases. Caused by one of four species of protozoa (one-celled organisms) of the genus *Plasmodium*, the disease is transmitted from person to person by a bite from the female of about 60 of the 400 different kinds of *Anopheles* mosquito (Figure 21-4).

Malaria can also be transmitted when a person receives the blood of an infected donor or when a drug user shares a needle with an infected user. For this reason heroin is usually "cut" or diluted with quinine, an antimalarial drug.

One way to control malaria is to administer antimalarial drugs like quinines that protect people against infection from *Anopheles* mosquitoes. Although these drugs are helpful, they cannot be used effectively to rid an area of malaria—people in infected areas would have to take the drugs continuously throughout their lives. In addition, new strains of carrier mosquitoes eventually develop genetic resistance to any widely used antimalarial drug.

Another approach involves trying to eliminate the mosquito carriers by draining swamplands and marshes and by spraying breeding areas with DDT and other pesticides. During the 1950s and 1960s, the WHO made great strides in reducing malaria in many areas,

eliminating it in 37 countries by widespread spraying of DDT and the use of antimalarial drugs. In India malaria cases dropped from 100 million in 1952 to only 40,000 in 1966, and in Pakistan cases were reduced from 7 million in 1961 to 9,500 in 1967.

Since 1970, however, malaria has made a dramatic comeback in many parts of the world. According to the WHO, there are now at least 250 million new cases of the disease each year. Epidemiologists at the U.S. Centers for Disease Control estimate that a more accurate figure may be 800 million new cases a year. Most malaria victims are children under the age of 5.

Major factors contributing to this tragic resurgence are the mosquito's increased genetic resistance to DDT and other insecticides and to antimalarial drugs, rising costs of pesticides and antimalarial drugs, spread of irrigation ditches that provide new mosquito breeding grounds, and reduced budgets for

---

The yearly $3 billion that the United States invests in pest control yields about $11 billion in increased crop yields after deducting the $1 billion in annual losses as a result of social and environmental damages from using pesticides. Thus from an economic viewpoint pesticides represent an excellent investment return. The Department of Agriculture estimates that food prices in the United States are 30% to 50% lower than they would be without the country's widespread use of pesticides and extensive after-harvest storage facilities.

Use of pesticides to help increase crop yields in the United States has meant that more land is available for recreation, forests, and wildlife reserves. By producing more crops per hectare of land, there has been less pressure to expand crop production to lands subject to excessive erosion; an indirect effect of pesticide use is that reduced soil erosion has helped maintain water quality by limiting the flow of sediment into surface waters.

There are alternatives to relying on insecticides and herbicides to control insects and weeds (Section 21-5). But proponents argue that these chemicals have several advantages over other approaches. They can control most insect pests quickly and at a reasonable cost, they have a relatively long shelf life, they are easily shipped and applied, and they are safe when handled properly. When genetic resistance occurs in pest insects and weeds, farmers can usually switch to other pesticides.

21-3   THE CASE AGAINST PESTICIDES

Development of Genetic Resistance   The most serious drawback to using chemicals to control pests is that most pest species, especially insects, can develop

malaria control due to the misbelief that the disease has been eliminated. By early 1985, the WHO reported that 51 of the 60 malaria-carrying species of mosquitoes had become genetically resistant to DDT and one or more of the other chlorinated hydrocarbons widely used to control the disease. At least ten of the species are also resistant to widely used organophosphate pesticides.

Research is being carried out to develop biological controls for *Anopheles* mosquitoes and to develop antimalaria vaccines, but such approaches are in the early stages of development and lack adequate funding. The WHO estimates that only 3% of the money spent each year on biomedical research is spent on malaria and other tropical diseases, even though more people suffer and die from these diseases than from all others combined.

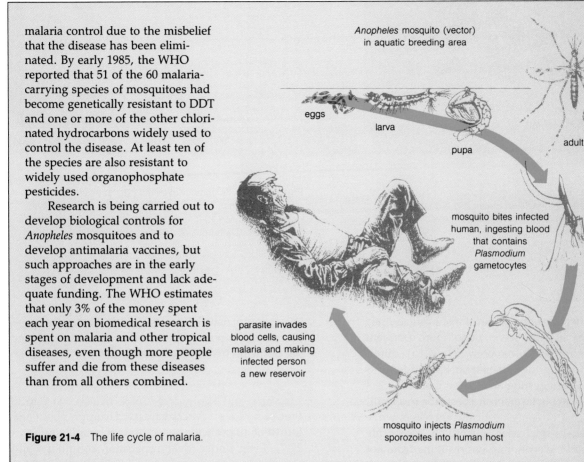

Figure 21-4 The life cycle of malaria.

Anopheles mosquito (vector) in aquatic breeding area

eggs

larva

pupa

adult

mosquito bites infected human, ingesting blood that contains *Plasmodium* gametocytes

Plasmodium develops in mosquito

parasite invades blood cells, causing malaria and making infected person a new reservoir

mosquito injects *Plasmodium* sporozoites into human host

genetic resistance to any chemical poison through natural selection. When an area is sprayed with a pesticide, most of the pest organisms are killed. However, a few organisms in a given population of a particular species survive because they have genes that make them resistant or immune to a specific pesticide.

Because most pest species—especially insects and disease organisms—have short generation times, a few surviving organisms can reproduce a large number of similarly resistant offspring in a short time. For example, the boll weevil (Figure 21-5), a major cotton pest, can produce a new generation every 21 days.

When populations of offspring of resistant parents are repeatedly sprayed with the same pesticide, each succeeding generation contains a higher percentage of resistant organisms. Thus the widespread use of any chemical to control a rapidly reproducing insect pest species typically becomes ineffective within about five years—even sooner in the hot, humid tropics, where insects and disease organisms can adapt

quickly to new environmental conditions. Weeds and plant diseases also develop genetic resistance, but not as quickly as most insects.

Since 1950 there has been a dramatic increase in the number of insect species with genetic resistance to one or more insecticides (Figure 21-6). Worldwide, by 1987 almost 500 species of insects, 150 species of plant pathogens, 50 species of fungi, 50 species of weeds, and 10 species of small rodents (mostly rats) had strains resistant to one or more pesticides. About 20 species of particularly damaging pests have become resistant to virtually every pesticide targeted at them.

According to insect expert Robert Metcalf, by 1995 the number of insect pest species resistant to one or more insecticides could exceed 1,500, and by the turn of the century virtually all insect pest species will probably show some form of genetic resistance. Because half of all pesticides applied worldwide are herbicides, genetic resistance in weeds is also expected to increase significantly.

Figure 21-5 The cotton boll weevil accounts for about 35% of the pesticides used in the United States, but farmers are now increasing their use of natural predators to control this major pest.

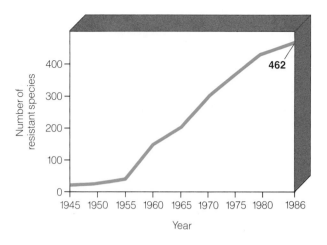

Figure 21-6 The number of insect pest species showing genetic resistance to one or more pesticides has increased dramatically since 1945, when pesticides were first applied in large doses. (Data from World Resources Institute)

When genetic resistance develops, pesticide sale representatives usually recommend more frequent applications, stronger doses, or switching to a different chemical to keep the resistant species under control, rather than suggesting alternative methods (Section 21-5). This can put farmers on a **pesticide treadmill,** in which the cost of using pesticides increases while their effectiveness decreases. Eventually insecticide costs can exceed the economic loss resulting from not using these chemicals. For example, even though insecticide use in the United States increased tenfold between 1940 and 1980, crop losses from insects during the same period almost doubled (from 7.1% to 13%).

Killing of Natural Pest Enemies and Creation of New Pests Most pesticides are broad-spectrum poisons that kill not only the target pest species but also a number of natural predators and parasites that may have been maintaining the pest species at a reasonable level. Without sufficient natural enemies, and with much food available, a rapidly reproducing insect pest species can make a strong comeback a few days or weeks after being initially controlled. This revival of the pest population requires the use of more pesticides, again placing farmers on a pesticide treadmill.

The repeated use of broad-spectrum pesticides also creates new pests and converts minor pests to major pests—the reverse of what pesticides are supposed to do. This occurs when pesticides kill off not only natural predators of the pest species but also the predators of minor pests, since predators, like parasites and

mites, are not usually species-specific. Then minor pests become major pests. Additional pests are needed, and the pesticide treadmill occurs again.

Mobility and Biological Amplification of Persistent Pesticides Aircraft are used to apply about one-fourth of all pesticides used on U.S. cropland (Figure 21-7). Even under ideal application conditions only about 50% of the spray reaches the target area. Most of this spray, as well as pesticides applied by ground sprayers, is washed or leached off the land or blown into the air. As a result, it is estimated that an average of only 1% of pesticides applied to crops reaches the target pests, and this amount may range from 0.1% to 5% depending on the chemical and its method of application. The remaining 99% ends up in the air, drinking water, food, and other nontarget organisms, including people. In 1986 the EPA reported that traces of 17 different pesticides were found in groundwater supplies in 23 states.

DDT and other fat-soluble pesticides can be biologically amplified in food chains and webs to levels hundreds to millions of times higher than those in the soil or water (Figure 4-17, p. 67). The high levels of pesticides stored in the fatty tissues of organisms feeding on pesticide-contaminated organisms that are in the lower trophic levels in the food chain can kill many forms of wildlife outright or interfere with their reproduction. Because of biological amplification, all people born after 1950 have carried traces of DDT and other chlorinated hydrocarbon pesticides in their fatty tissues; long-term harmful effects are unknown.

Figure 21-7 Crop duster spraying a Florida orange grove with fungicide. No more than 1% of the chemical being applied reaches the target organisms.

Figure 21-8 The population of peregrine falcons in the United States was sharply reduced between 1950 and 1970 when high levels of DDT and its breakdown products caused the birds to produce eggs with shells so thin that many chicks died.

Threats to Wildlife Marine organisms, especially shellfish, can be killed by minute concentrations of chlorinated hydrocarbon pesticides. The young offspring of some species are often particularly susceptible to pesticides. Honeybees, which pollinate crops that provide one-third of the food consumed in the United States, are extremely susceptible to pesticide poisoning. Each year an estimated 20% of all honeybee colonies in the U.S. are killed by pesticides and another 15% of the colonies are damaged—causing annual losses of at least $135 million from reduced pollination of vital crops.

During the 1950s and 1960s there were drastic declines in populations of fish-eating birds such as the osprey, cormorant, brown pelican, and bald eagle. There were also sharp declines in populations of predatory birds such as the prairie falcon, sparrow hawk, Bermuda petrel, and peregrine falcon (Figure 21-8); these birds help control populations of rabbits, ground squirrels, and other crop-damaging small mammals.

Research has shown that these population declines occurred because DDE, a chemical produced by the breakdown of high levels of DDT, accumulated in the bodies of the affected bird species and reduced the calcium deposition in the shells of their eggs. The resulting thin-shelled eggs are so fragile that many of them break, and the unborn chicks die before they can hatch normally.

Since the U.S. ban on DDT in 1972, populations of most of these bird species have made a comeback. By 1980, however, it was discovered that levels of DDT and other banned pesticides were beginning to rise again in some areas and in some species such as the peregrine falcon—probably the result of illegal use of banned pesticides.

Short-Term Threats to Human Health The World Health Organization estimates that over 500,000 farm workers, pesticide plant employees, and children

In 1975 state officials found that 70 of the 150 employees in a pesticide manufacturing plant in Hopewell, Virginia (near Richmond) had been poisoned by exposure to high levels of kepone (chlorodecone)--a persistent, chlorinated hydrocarbon pesticide used as an ant and roach poison. In the plant kepone dust filled the air, covered equipment, and was even found in the employees' lunch area. Some workers also brought kepone dust home on their clothes and contaminated family members.

The plant, associated with Allied Chemical Company, was shut down in 1975, and 29 workers were hospitalized with uncontrollable shaking, slurred speech, apparent brain and liver damage, inability to concentrate, joint pain, and, in some cases, sterility. Allied Chemical Company has paid out $13 million in damage suits to the victims and their families.

Further investigation revealed that a large area of the James River, the largest river in Virginia, and its fish and shellfish were contaminated with kepone. Between 1966 and 1975, the manufacturer illegally dumped kepone into the Hopewell, Virginia municipal sewage system. The compound disrupted the bacterial decomposition processes in the sewage treatment plant and led to the discharge of untreated, kepone-laden sewage into the nearby James River. Since 1975 more than 160 kilometers (100 miles) of the river and its tributaries were closed to commercial fishing, resulting in a loss of jobs and millions of dollars.

In 1984 the world's worst industrial accident occurred at a Union Carbide pesticide plant located in Bhopal, India. Almost 2,700 people were killed when highly toxic methyl isocyanate gas, used in the manufacture of pesticides, leaked

from a storage tank. Official Indian government figures indicate that 320,000 of Bhopal's population of about 1 million suffered some sort of illness. At least 14,000 became seriously ill and suffered from blindness, sterility, kidney and liver infections, tuberculosis, brain damage, and other disorders that can lead to premature death. The victims have sued Union Carbide for $3.1 billion in damages in a district court in Bhopal.

This tragedy could probably have been prevented by the expenditure of perhaps no more than a million dollars to ensure more adequate plant safety. This incident has aroused concern about the safety of the almost 11,600 chemical plants located in the United States, especially after a toxic gas leak in 1985 from another Union Carbide plant in Institute, West Virginia, sent 135 nearby residents to the hospital.

---

become seriously ill, and about 5,000 to 20,000 die each year around the world from exposure to toxic insecticides, especially organophosphates. Insecticide-related illnesses and deaths are particularly high among farm workers in LDCs, where educational levels are low, warnings are few, and pesticide regulation and control methods are often lax.

In the United States pesticides cause an estimated 45,000 illnesses and at least 25 deaths each year among the country's 7 million farm workers. Studies by the World Resources Institute (WRI) indicate that pesticide-related illness among farm workers in the United States and throughout the world is probably greatly underestimated because of poor records, lack of doctors and reporting in rural areas, and faulty diagnosis. According to the WRI, the true number of pesticide-related illnesses each year probably runs in the millions, including 300,000 in the United States. Injuries and deaths can also occur from the manufacture of pesticides (see Spotlight above).

In the 1960s a controversy began over the possible health effects from the use of the herbicides 2,4,5-T and 2,4-D. Between 1962 and 1970, Agent Orange, a 50-50 mixture of 2,4-D and 2,4,5-T, was sprayed to

defoliate swamps and forests in South Vietnam to prevent guerrilla ambushes, discourage the movement of troops and supplies through demilitarized zones, clear areas around military camps, and destroy crops that could feed Vietcong and North Vietnamese soldiers.

In 1965 and 1966 a study commissioned by the National Cancer Institute found that low levels of 2,4,5-T caused high rates of birth defects in laboratory animals. This report was not released to the public until 1969. Because of the resulting pressure from environmentalists and health officials, however, the Vietnam defoliation program was halted in 1970. Investigations revealed that the birth defects in laboratory animals were probably caused by a highly toxic dioxin called TCCD (Section 20-5), formed in minute quantities as an unavoidable contaminant during the manufacture of 2,4,5-T.

In the late 1970s as many as 40,000 previously healthy Vietnam veterans began experiencing a variety of medical disorders including dizziness, blurred vision, insomnia, fits of uncontrollable rage, nausea, chloracne on large areas of their skin, and depression. An abnormally high percentage of these veterans fathered infants who were aborted prematurely, still-born, or had multiple birth defects. Other veterans had higher than expected incidences of leukemia, lymphoma, and rare testicular cancer.

By 1980, more than 1,200 Vietnam veterans had filed claims with the Veterans Administration for disabilities allegedly caused by exposure to Agent Orange. The VA and chemical manufacturers of Agent Orange, however, continue to deny any connection between the medical disorders and Agent Orange and attribute the problems to the post-Vietnam stress syndrome. In 1984, the companies making Agent Orange agreed to an out-of-court settlement with the Vietnam veterans, without admitting any guilt or connection between the disorders and the use of the herbicide.

In 1986 a National Cancer Institute study indicated a strong statistical link between 2,4-D—a component of Agent Orange and the active ingredient in more than 1,500 herbicide products used by farmers and home gardeners—and a rare form of cancer known as non-Hodgkin's lymphoma. A study found that farmers who used 2,4-D were more than twice as likely as nonfarmers to develop this form of cancer. Earlier, Swedish researchers had shown an association between 2,4,5-T and 2,4-D and two other cancers—soft-tissue sarcoma and Hodgkin's disease.

Long-Term Threats to Human Health Many scientists are concerned about the possible long-term effects on people of long-term, low-level exposure to DDT and other persistent pesticides. Such effects, if any, won't be known for several decades because the people who have carried these chemicals in their bodies the longest were only 43 years old by 1988. The results of this long-term worldwide experiment, with people involuntarily playing the role of guinea pigs, may never be known, because it is almost impossible to determine that a specific chemical such as DDT caused a particular cancer or other harmful effect.

However, some disturbing but inconclusive evidence has emerged. DDT, aldrin, dieldrin, heptachlor, mirex, endrin, and 19 other pesticides have all been found to cause cancer in test animals, especially liver cancer in mice. In addition, autopsies have shown that the bodies of people who died from cancers, cirrhosis of the liver, hypertension, cerebral hemorrhage, and softening of the brain contained fairly high levels of DDT or its breakdown products DDD and DDE.

21-4 PESTICIDE REGULATION IN THE UNITED STATES

Is the Public Adequately Protected? Because of the potentially harmful effects of pesticides on wildlife and people, Congress passed the Federal Insecticide,

Fungicide, and Rodenticide Act (FIFRA) in 1972. This act, which was amended in 1975 and 1978, requires that all commercially available pesticides be registered with the Environmental Protection Agency. Using information provided by the pesticide manufacturer, the EPA may refuse to approve the use of the pesticide or may classify it for general or restricted use.

Since its passage in 1972 environmentalists have considered the FIFRA the weakest environmental law on the books because of strong lobbying by the powerful agricultural chemicals industry and because the committee that draws up the legislation is controlled by pro-industry elected officials from farm states. Unlike other environmental laws, the FIFRA authorizes the EPA to allow a dangerous chemical to stay on the market if the supposed economic benefits (substantiated by the pesticide industry) outweigh the risk to human health or the environment.

Since 1972 the EPA has used the FIFRA to ban the use, except for emergency situations, of DDT and several other persistent chlorinated hydrocarbon pesticides such as aldrin, dieldrin, heptachlor, lindane, chlordane, and toxaphene. These pesticides were banned because their persistence and biological amplification in food chains and webs threatens some forms of wildlife and because laboratory tests show they cause birth defects, cancer, and neurologic disorders in laboratory test animals. Between 1975 and 1985 the EPA has also banned or restricted the use of 28 other pesticides, including the herbicides Silvex and 2,4,5-T, because of their potential hazards to human health.

However, even when the health and environmental effects are shown to outweigh the economic benefits of continued use of a particular pesticide, the procedural labyrinth the EPA must follow to cancel or restrict the use of a pesticide can take ten years. During this time consumers continue to be exposed to health hazards from the product. For example, the National Cancer Institute established that low dosages of ethylene dibromide (EDB)—widely used for decades as a fumigant on grains and citrus fruits—caused stomach cancer and genetic mutations in all test animals in a relatively short time. But the EPA did not ban this chemical until 1984. Even this action came only after several states, fed up with inaction by the EPA, began removing EDB-contaminated products from grocery store shelves after finding alarming levels of EDB in groundwater supplies in fruit-growing areas and in some cake and muffin mixes and other grain products.

Most Americans might assume that the ingredients approved for use in pesticide products in the United States have been carefully examined by the EPA and have passed rigorous health and safety tests to determine their potential for producing cancer, genetic mutations, and birth defects in humans. However, according to a 1984 study by the National Academy of Sciences, only 10% of the ingredients used in pesticides in 1984 had sufficient health and safety data for an adequate assessment of their dangers to human health.

The FIFRA required the EPA to reevaluate the 600 active ingredients approved for use in pesticide products before 1972 to determine whether any of these substances caused cancer, birth defects, or other health risks. The EPA was supposed to complete this analysis by 1975. Yet, by 1987 the EPA had only completed its evaluation for three of these ingredients. In addition, during the late 1970s, the EPA discovered that data had been falsified by a now-defunct laboratory test on over 200 pesticides approved for use (including at least 90 used on food crops). Under the FIFRA, the EPA is not required to ban a pesticide even if can be shown that its registration was based on faulty data. By 1987 repeated attempts by environmentalists to have Congress strengthen the FIFRA had failed.

Export of Banned Pesticides In response to a slowdown in the pesticide use rate in the United States since 1981, the U.S. chemical industry has increased exports of pesticides or their basic ingredients to countries (mostly LDCs) where they have not been banned. In most LDCs, up to 70% of these pesticides are applied to crops such as cotton, coffee, cocoa, and bananas destined for export to Europe, Japan, and the United States. About half of the produce consumed in the United States each winter is provided by Mexico.

A 1979 report by the General Accounting Office concluded that a large proportion of food and natural fiber imported into the United States from LDCs may contain unsafe residues of pesticides that have been banned here. In 1984 tests by the Food and Drug Administration found that 30% to 50% of the imported coffee was tainted with residues of pesticides banned in the United States.

Although high levels of residues on imported foods are illegal, enforcement is lax. Each year the Food and Drug Administration inspectors check less than 1% of imported food products for pesticide contamination. Even then in 1985 only 3% of the foods found to be contaminated were not allowed to be sold.

In 1983, the United Nations passed a resolution calling for stringent restrictions on the export of products whose use has been banned or severely restricted in the exporting nation; in addition, the UN asked for widespread publication of a list of such products. Under orders from President Reagan, the United States was the only country opposing this resolution. It was argued that countries receiving exports of banned or unregistered pesticides, drugs, and other chemicals from the United States should be free to use these chemicals if they so desire and that if U.S. companies do not sell these substances to them, someone else will.

Has DDT Really Been Banned? A 1983 study showed that 44% of the fruits and vegetables grown in California contained higher than expected residues of 19 different pesticides, including DDT and other banned substances. Investigators suspect that this is due to a combination of illegal smuggling of relatively large amounts of DDT and some other banned insecticides into the U.S. (used mostly in western states) from Mexico, and a loophole in the FIFRA that allows the sale in the United States of up to 50 insecticides such as dicofol (usually sold under the trade name Kelthane) that can consist of much as 15% DDT by weight. This DDT is allowed because it is classified by the EPA as an "unintentional impurity" that occurs when certain insecticides are manufactured.

21-5 ALTERNATIVE METHODS OF INSECT CONTROL

Modifying Cultivation Procedures A number of alternatives to relying on conventional pesticides to control pest populations are available. For centuries, farmers have used cultivation methods that discourage or inhibit pests. Rotating the type of crop grown in a given field from year to year can be used to reduce populations in the soil or crop residues of nonmigrating pests that feed on a particular crop. Planting times can also be adjusted to ensure that most of the pest population starves to death before the crop is available or adjusted to favor natural predators over the pests. Rows of hedges around or through fields can serve as barriers to insect invasion and create refuges for enemies of pests. Crops can also be grown in areas where their major pests do not exist. Unfortunately, to increase profits and in some cases to avoid bankruptcy, many farmers in MDCs such as the United States have abandoned these cultivation methods.

Biological Control For decades, agricultural scientists have been using **biological control** to regulate the populations of insects, weeds, rodents, and other pests by reintroducing effective natural predators, parasites, and pathogens (disease-causing bacteria and viruses) or by importing new ones. Worldwide, there have been more than 300 successful biological control projects, especially in China and the Soviet Union.

Examples of biological control include using ladybugs and praying mantises (Figure 21-9) to control aphids, and using speck-sized parasitic wasps to con-

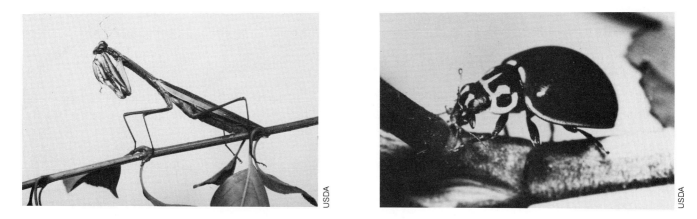

Figure 21-9 The praying mantis (left) and the ladybug (on the right, eating an aphid) are used to control insect pests.

trol various crop-eating moths and flies such as the leaf miner. A bacterial agent (*Bacillus thuringiensis*), sold commercially as a dry powder, is effective in controlling many strains of leaf-eating caterpillars, mosquitoes, and gypsy moths. As pesticide prices continue to climb, it will become economically feasible to mass-produce many biological control agents.

Biological control has a number of important advantages. It normally affects only the target species and is nontoxic to other species, including people. Once a population of predators or parasites is established, control is often self-perpetuating and does not have to be reintroduced each year. Development of genetic resistance is minimized because both pest and predator species usually undergo natural selection to maintain a stable interaction (coevolution). In the United States biological control has saved farmers an average of $30 for every $1 invested, compared to $4 saved for every $1 invested in pesticides.

No method of pest control, however, is perfect. Typically, 10 to 20 years of research may be required to understand how a particular pest interacts with its various enemies and to determine the best control agent, and mass production is often difficult. Farmers find that pesticides are faster-acting and simpler to apply than biological agents. Biological agents must also be protected from pesticides sprayed in nearby fields, and there is a chance that some can also become pests. In addition, some pest organisms can develop genetic resistance to viruses and bacterial agents used for biological control.

Genetic Control by Sterilization   Males of an insect species can be raised in the laboratory, sterilized by radiation or chemicals, and then released in an infested area to mate unsuccessfully with fertile wild females.

If sterile males outnumber fertile males by 10 to 1, a pest species in a given area can be eradicated in about four generations, provided reinfestation does not occur. This sterile male technique works best if the females mate only once; if the infested area is isolated so that it can't be periodically repopulated with nonsterilized males; and if the insect pest population has already been reduced to a fairly low level by weather, pesticides, or other factors.

The screwworm fly is a major livestock pest in South America, Central America, and the southeastern and southwestern United States. This metallic blue-green insect, about two to three times the size of the common housefly, deposits its eggs in open wounds of warm-blooded animals such as cattle and deer. Within a few hours the eggs hatch into parasitic larvae that feed on the flesh of the host animal (Figure 21-10). A severe infestation of this pest can kill a mature steer within ten days.

The Department of Agriculture used the sterile-male approach to essentially eliminate the screwworm fly from the southeastern states between 1962 and 1971. In 1972, however, the pest made a dramatic comeback, infesting 100,000 cattle and causing serious losses until 1976, when a new strain of the males was developed, sterilized, and released to bring the situation under temporary control. To prevent resurgences of this pest, new strains of sterile male flies will have to be developed, sterilized, and released every few years.

Major problems with this approach include ensuring that sterile males are not overwhelmed numerically by nonsterile males, knowing the mating times and behavior of each target insect, risking that laboratory-produced strains of sterile males will not be as sexually active as normal wild males, preventing reinfestation with new nonsterilized males, and high costs.

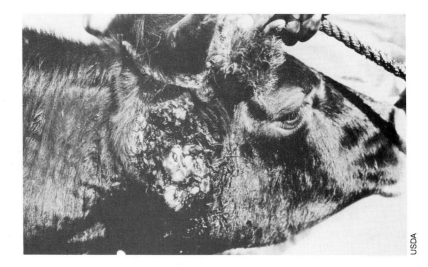

Figure 21-10 Infestation of screwworm fly larvae in the ear of a steer. A fully grown steer can be killed in ten days from thousands of maggots feeding on a single wound.

Genetic Control by Breeding Resistant Crops and Animals For many decades agricultural scientists have used artificial selection, cross-breeding, and genetic engineering to develop new varieties of plants and animals resistant to certain insects, fungi, and diseases. For example, scientists developed a number of wheat strains resistant to the Hessian fly, a major wheat pest that was accidentally introduced into the United States in the straw bedding of German mercenaries during the Revolutionary War.

Breeding new resistant strains of crops and animals is expensive and can require from 10 to 20 years of painstaking work by highly trained scientists. The new strains must also produce high yields. Furthermore, insect pests and plant diseases can develop new strains that attack the once-resistant varieties, forcing scientists to continually develop new resistant strains. In the future, genetic engineering techniques may be used to develop resistant crops and animals more rapidly.

Chemical Control Using Natural Sex Attractants and Hormones Various pheromones and hormones can be used to control populations of insect pest species. Some observers believe that these two new types of chemical agents, sometimes called *third-generation pesticides,* may eventually replace the present use of less desirable second-generation pesticides.

In many insect species, when a virgin female is ready to mate she releases a minute amount (typically about one-millionth of a gram) of a species-specific chemical sex attractant called a *pheromone.* Males of the species up to a half-mile away can detect the chemical with their antennal receptors and follow the scent upwind to its source.

Pheromones extracted from an insect pest species or synthesized in the laboratory can be used to lure pests such as Japanese beetles into traps containing toxic chemicals. An infested area can also be sprayed with the appropriate pheromone or covered with millions of tiny cardboard squares impregnated with the substance so that the males become confused and are unable to find a mate because they detect the smell of virgin females everywhere. Recent research indicates that instead of using pheromones to trap pests, it is more effective to use them to lure the pests' natural predators into fields and gardens. Pheromones are now commercially available for use against 30 major pests.

Pheromones have a number of advantages: they work on only one species, they are effective in extremely minute concentrations, they usually break down within a week, they have relatively little chance of causing genetic resistance, and they are not poisonous to animals, thus not affecting wildlife and not creating new pests. However, the difficulties with pheromones include identifying and isolating the specific sex attractant for each pest species, determining the mating behavior of the target insect, and coping with periodic reinfestation from surrounding areas. Pheromones have also failed for some pests because only adults are drawn to the traps; for most species, the juvenile forms—such as caterpillars—do most of the damage. The major problem with pheromones is their lack of availability for most pests or natural predators.

Hormones are chemicals produced in an organism's cells that travel through the bloodstream and control various aspects of the organism's growth and development. Each step in the life cycle of a typical insect is regulated by the timely release of juvenile hormones (JH) and molting hormones (MH) (Figure 21-11). Extracted or laboratory-synthesized juvenile hormones or molting hormones applied at certain stages in an insect's life cycle can produce abnormalities that cause the insect to die before it can reach maturity and reproduce (Figure 21-12).

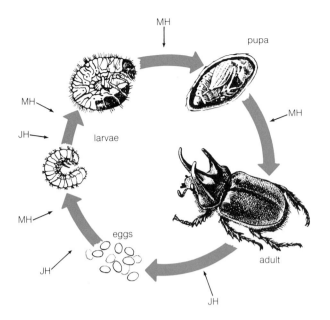

Figure 21-11 For normal growth, development, and reproduction certain juvenile hormones (JH) and molting hormones (MH) must be present at genetically determined stages in the typical life cycle of an insect. If applied at the right time, synthetic hormones can be used to disrupt the life cycle of insect pests.

Figure 21-12 Chemical hormones can prevent insects from maturing completely, thus making it impossible for them to reproduce. Compare a normal mealworm (left) with one that failed to develop an adult abdomen after being sprayed with a synthetic hormone.

Insect hormones have the same advantages as pheromones, except that they sometimes affect natural predators of the target insect species and other nonpest species. However, hormones require weeks rather than minutes to kill, are often ineffective with a large infestation, sometimes break down chemically in the environment before they can act, must be applied at the right time in the life cycle of a target insect, and are difficult and costly to isolate and produce.

Irradiation of Foods Exposing certain foods to various levels of radiation is being touted by the nuclear industry and the food industry as a means of killing and preventing insects from reproducing in certain foods after harvest, extending the shelf life of some perishable foods, and destroying parasites such as trichinae and bacteria such as salmonella that each year kill 2,000 Americans. In 1986 the FDA approved use of low doses of ionizing radiation on fruits, vegetables, and fresh pork, and it may soon be approved for use on poultry and seafood. Irradiated foods are already sold in 33 countries, including the Soviet Union, Japan, Canada, Brazil, Israel, and many west European countries.

Because tests show that consumers will not buy food if it is labeled as being irradiated, foods exposed to radiation sold in the United States bear a characteristic logo and a label stating that the product has been "picowaved"—information that will be meaningless to many consumers, but better than no label. The label, however, is only required on foods that have been directly irradiated, not on those that contain irradiated components such as various spices used in processed foods, cheese spreads, and luncheon meats.

Exposure to higher doses—generally those above what is presently allowed by the FDA—also could extend the shelf life of some perishable foods. A food does not become radioactive when it is irradiated, just as being exposed to X rays does not make the body radioactive.

The FDA and the WHO claim that over 1,000 studies show that foods exposed to low radiation doses are safe for human consumption. However, critics of irradiation argue that not enough animal studies have been done and that tests of the effects of irradiated foods on people have been too few and brief to turn up any long-term effects, which typically require 30 to 40 years to be evaluated. The focus of this controversy is the fact that irradiation produces trace amounts of at least 65 chemicals in foods, some of which cause cancer in test animals.

The FDA estimates that only 10% of the chemicals produced in irradiated foods are probably unique—that is, they are chemicals not found in nonirradiated foods—and assumes that the concentrations of these chemicals will be too small to affect human health. Opponents of food irradiation say this assumption is

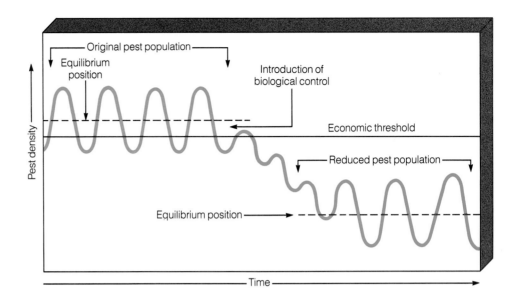

Figure 21-13 The goal of biological control and integrated pest management is not to eliminate pests but to keep the pest population beneath the level of economic damage.

unwarranted until these chemicals have been identified and thoroughly tested.

Opponents also fear that more people might die of deadly botulism because present levels of irradiation do not destroy the spore-enclosed bacteria that cause this disease, but they do destroy the microbes that give off the rotten odor that warns of the presence of botulism bacteria. They also note that irradiation can be expensive, adding as much as five cents a pound to the price of some fresh vegetables. Proponents, however, respond that irradiation of food is likely to reduce health hazards to people by decreasing the use of some potentially damaging pesticides and food additives.

Integrated Pest Management An increasing number of experts believe that in many cases our present, eventually self-defeating pesticide-based approach to pest control can be replaced with a carefully designed ecological approach called **integrated pest management (IPM).** In IPM, each crop and its major pests are considered as an ecological system, and a control program is developed that uses a variety of biological, chemical, and cultivation methods in proper sequence and timing.

The overall aim of IPM is not eradication but keeping pest populations just below the level of economic loss (Figure 21-13). Fields are carefully monitored to check whether pests have reached an economically damaging level. When such a level is reached, farmers first use biological and cultural controls. Pesticides are applied only when absolutely necessary, and in small amounts, and different chemicals are cycled to retard development of genetic resistance.

Over the past 30 years, more than three dozen IPM programs have been used successfully, especially on cotton. These experiments have shown that a properly designed IPM program can reduce preharvest pest-induced crop losses by 50%, pesticide use and pesticide control costs by 50% to 75%, and fertilizer and irrigation needs, and can at the same time increase crop yields and reduce costs.

Integrated pest management has a number of advantages. It requires minimal use of pesticides and therefore reduces their harmful side effects, including genetic resistance. Also, it decreases soil erosion and use of fertilizer and irrigation water, and it is cheaper on a long-term basis than the increasingly expensive pesticide treadmill.

However, there are some drawbacks to IPM. It requires expert knowledge about each pest-crop situation and is slower-acting and more labor-intensive than the use of conventional pesticides. Methods developed for a given crop in one area may not be applicable to another area with slightly different growing conditions. Although long-term costs are typically lower than those using conventional pesticides, initial costs may be higher.

A massive switch to IPM in the United States is very difficult, primarily because of the political clout of the agricultural chemical companies that see little profit in such methods. As a result, IPM will have to be developed through financial and technical support from federal and state agencies. Environmentalists also urge that its use be promoted by providing government subsidies and perhaps government-backed crop-loss insurance to farmers who use IPM or other approved alternatives to the widespread use of pesticides.

Changing the Attitudes of Consumers and Farmers Two attitudes tend to increase the use of pesticides and lock us into the pesticide treadmill. First, many people believe that the only good bug is a dead bug. Second, most consumers insist on buying only perfect, unblemished fruits and vegetables, even though a few holes or frayed leaves do not significantly affect the taste, nutrition, or shelf life of such produce. Educating farmers and consumers to change these attitudes would significantly help reduce the use of pesticides.

Individuals can also reduce the dangers of pesticides by using the smallest amount possible only when absolutely necessary and by disposing of unused pesticides in a safe manner. (Call your local health department for advice.) Elected representatives should also be pressured to significantly strengthen the Federal Insecticide, Fungicide, and Rodenticide Act in order to better protect human health and the environment.

Homeowners should seriously consider planting all or most of the land around a house site not used for gardening with a diverse mix of wildflowers, herbs (for cooking), low-growing ground cover, small bushes, and other forms of vegetation natural to the area. Create paths covered with wood chips or stones through the area so that little, if any, mowing is ever required.

This approach, based on natural diversity rather than the traditional monoculture of grass, saves time, energy, money (no lawn mower, gasoline, or frequent lawn mower repairs), and strife over why the lawn hasn't been cut. This type of yard also reduces infestations by mosquitoes and other insects by providing a diversity of habitats for their natural predators. The largest homesite infestations by insect pests and weeds such as crabgrass usually occur in yards planted with grass that is kept cut below three inches in height. In other words, work with, not against, nature by establishing a diversity of vegetation and letting grass grow at least three inches.

We need to recognize that pest control is basically an ecological, not a chemical problem.
Robert L. Rudd

DISCUSSION TOPICS

1. Should DDT and other pesticides be banned from use in malaria control? Explain.

2. Should the United States abandon or sharply decrease pesticide use and replace it with integrated pest management? Explain. What might be the consequences for LDCs? For the United States? For you?

3. Explain how the use of insecticides can actually increase the number of insect pest problems and threaten some carnivores and omnivores, including people?

4. How does genetic resistance to a particular pesticide occur? What major advantages do insects have over humans in this respect?

5. Debate the following resolution: Because DDT and the other banned chlorinated hydrocarbon pesticides pose no demonstrable threat to human health and have probably saved more lives than any other chemicals in history, they should again be approved for use in the United States.

6. Should certain types of foods used in the United States be irradiated? Explain.

7. List the major advantages and disadvantages of pest control by (a) biological control; (b) sterilization of male insects; (c) sex attractants; (d) juvenile hormones; (e) resistant crop varieties; (f) cultivation practices; (g) integrated pest management; and (h) conventional pesticides.

Epilogue

Achieving a Sustainable-
Earth Society

The frontier or throwaway mentality sees the earth as a place of unlimited room and resources where ever-increasing production, consumption, and technology inevitably lead to a better life for everyone. If we pollute one area, we merely move to another or eliminate or control the pollution through technology. This mentality represents an attempt to dominate nature.

Many environmentalists argue that over the next 50 years or so we must change from our present *throwaway* or *frontier* rules to a new set of *sustainable-earth* or *conserver* rules designed to maintain the earth's vital life-support systems not only for our own sustenance, well-being, and security, but also to fulfill our obligation to pass this vital legacy along to future generations.

In contrast to the old or new frontier mind-set, a sustainable-earth or conserver mentality sees that the earth is a place of limited room and resources and that ever-increasing production and consumption can put severe stress on the natural processes that renew and maintain the air, water, and soil upon which we depend. Sustaining the earth calls for cooperating with nature, rather than blindly attempting to dominate it. Achieving a sustainable-earth world view involves working our way through four levels of environmental awareness summarized in the Spotlight on p. 407.

But the sustainable-earth view by itself is not enough. It is unrealistic to expect poor people living at the margin of existence to think about the long-term survival of the planet. When people need to burn wood to keep from freezing, they will cut down trees. When their livestock face starvation, they will overgraze grasslands. For this reason, analysts argue that an equally important element in the transition to a sustainable-earth society is generous and effective aid from the MDCs to LDCs that are in need of of grants, loans, and technical advice. According to these observers, such aid should help the poorer nations to become more self-reliant rather than causing them to become increasingly dependent on the MDCs for goods and services.

We can read and talk about environmental problems, but finally it comes down to what you and I are willing to do individually and collectively. We can begin at the individual level and work outward by joining with others to amplify our actions. This is the way the world is changed. Envision the world as made up of all kinds of cycles and flows in a beautiful and diverse web of interrelationships and a kaleidoscope of patterns and rhythms whose very complexity and multitude of potentials remind us that cooperation, honesty, humility, and love must be the guidelines for our behavior toward one another and the earth.

It is not too late. There is time to deal with the complex environmental problems we face if enough of us really care. It's not up to "them," it's up to "us." Don't wait.

First Awareness Level: Pollution and Environmental Degradation

We must discover the symptoms. At this level, we must point out and try to stop irresponsible acts of pollution and environmental degradation by individuals and organizations and resist being duped by slick corporate advertising. But we must at the same time change our own lifestyles. We have all been working toward our own destruction by "drilling holes in the bottom of the boat." Arguing over who is drilling the biggest hole only diverts us from working together to keep the boat from sinking. The problem of remaining at the first awareness level is that individuals and industries see their own impact as too tiny to matter, not realizing that millions of individual impacts acting together threaten our life-support systems. Remaining at this first level of awareness also leads people to see the crisis as a problem comparable to a "moon shot," and to look for a quick technological solution: "Have technology fix us up, send me the bill at the end of the month, but don't ask me to change my way of living."

Second Awareness Level: Consumption Overpopulation

We must recognize that the cause of pollution is not just people but their level of consumption and the environmental impact of various types of production (Figure 1-9, p. 10). At this second level the answers seem obvious. We must control the world's population, but we must also reduce wasteful consumption of matter and energy resources—especially in MDCs, which, with less than 26% of the world's population, account for about 80% of the world's resource consumption and environmental pollution.

Third Awareness Level: Spaceship Earth (Shallow Ecology)

We must become aware that population and resource use will not be controlled until enough world leaders and citizens stress that protecting and preserving the environment must be our primary purpose. The goal at this level is to use technology, economics, and conventional politics to control population growth, pollution, and resource depletion to prevent ecological overload. Some argue that the Spaceship Earth metaphor is a sophisticated expression of our arrogance toward nature—the idea that through technology we can control nature and create artifical environments to avoid environmental overload. They point out that this approach eventually poses a dire threat to individual freedom because to protect the life-support systems that are necessary in space, a centralized authority (ground control) must rigidly control astronauts' lives. Instead of novelty, spontaneity, joy, and freedom, the spaceship model is based on cultural homogenization, social regimentation, artificiality, monotony, and gadgetry. In addition, it is argued that this approach can cause environmental overload in the long run because it is based on the false idea that we have essentially complete understanding of how nature works. Furthermore, this awareness level does not seriously question the economic, political, social, and ethical foundations of modern industrial society, which some see as the major causes of our environmental problems.

Fourth Awareness Level: Sustainable Earth (Deep Ecology)

We must recognize that (1) all living species are interconnected; (2) the role of human beings is not to rule and control nature but to work with nature and to meet meet human needs on the basis of ecological understanding; (3) because the earth's organisms and their interactions are so diverse, attempts at excessive control will sooner or later backfire; (4) our major goal should be to preserve the ecological integrity, sustainability, and diversity of the life-support systems for all species; (5) the forces of biological evolution, not technological control, should determine which species live or die; and (6) human beings have no right to interfere destructively with nonhuman life except to satisfy vital needs.

Periodicals

The following publications can help you keep well informed and up to date on environmental and resource problems. Those marked with an asterisk are recommended as basic reading. Subscription prices, which tend to change, are not given.

Acid Precipitation Digest, Center for Environmental Information, 33 S. Washington St., Rochester, NY 14608. Summarizes current news and research related to acid deposition.

Alternate Sources of Energy, Alternate Sources of Energy, Inc., 107 S. Central Ave., Milaca, MN 56353. Useful source of information on renewable energy alternatives.

American Demographics, American Demographics, Inc., P.O. Box 68, Ithaca, NY 14851. Basic information.

*American Forests, American Forestry Association, 1319 18th St., N.W., Washington, DC 20036. Popular treatment, "seeks to promote an enlightened public appreciation of natural resources."

*Annual Energy Review, National Energy Information Center, Energy Information Administration, Forrestal Building, Room 1F-048, 1000 Independence Ave., S.W., Washington, DC 20585. Useful summary of data.

*Audubon, National Audubon Society, 950 Third Ave., New York, NY 10022. Conservationist viewpoint; covers more than bird watching. Good popularizer of environmental concerns; well-produced, sophisticated graphics.

*Audubon Wildlife Report, National Audubon Society, 950 Third Ave., New York, NY 10022. Superb annual summary of wildlife conservation.

BioScience, American Institute of Biological Sciences, 1401 Wilson Blvd., Arlington, VA 22209. Official monthly publication of AIBS; gives major coverage to biological aspects of the environment, including population. Style ranges from semipopular to technical. Features and news sections attentive to legislative and governmental issues.

*Bulletin of the Atomic Scientists, 935 E. 60th St., Chicago, IL 60637. Includes coverage of environmental issues, particularly in relation to nuclear power and nuclear testing and fallout.

Catalyst for Environmental/Energy, 274 Madison Ave., New York, NY 10016. High-level, popular treatment; substantial articles on all aspects of environment, including population control. Reviews books and films suited to environmental education.

Ceres, Food and Agriculture Organization of the United Nations (FAO), UNIPUB, Inc., 650 First Avenue, P.O. Box 433, New York, NY 10016. Contains articles on the population–food problem.

*The CoEvolution Quarterly, P.O. Box 428, Sausalito, CA 94965. Covers a wide range of environmental and self-sufficiency topics.

*Conservation Biology, Blackwell Scientific Publications, Inc., 52 Beacon St., Boston, MA 02108. Excellent coverage of wildlife conservation.

*Conservation Foundation Letter, Conservation Foundation, 1717 Massachusetts Ave., N.W., Washington, DC 20036. Usually 12 pages long. Good summaries of key issues.

*Conservation News, National Wildlife Foundation, 1412 16th St., N.W., Washington, DC 20036. Good coverage of wildlife issues.

Demographic Yearbook, Department of International Economic and Social Affairs, Statistical Office, United Nations Publishing Service, United Nations, NY 10017. Excellent source of population data.

Design and Environment, 355 Lexington Ave., New York, NY 10017. Useful for architects, engineers, and city planners.

*Earth Island Journal, Earth Island Institute, 300 Broadway, Suite 28, San Francisco, CA 94133. Excellent summaries of national and global environmental issues.

*The Ecologist, Ecosystems Ltd., 73 Molesworth St., Wadebridge, Cornway PL27 7DS, United Kingdom. Wide range of articles on environmental issues from an international viewpoint.

Ecology, Ecological Society of America, Dr. Ralph E. Good, Business Manager, Department of Biology, Rutgers University, Camden, NJ 08102. Good source of information on more technical ecology research.

The Energy Consumer, U.S. Department of Energy, Office of Consumer Affairs, Forrestal Building, 1000 Independence Ave., S.W., Washington, DC 20585. Free monthly summaries of energy information.

*Environment, Heldref Publications, 4000 Albemarle St., N.W., Suite 504, Washington, DC 20016. Seeks to put environmental information before the public. Excellent in-depth articles on key issues.

Environmental Abstracts, Environment Information Center, Inc., 48 W. 38th St., New York, NY 10018. Compilation of environmental abstracts; basic bibliographic tool. Too expensive for individual subscription but should be available in university libraries.

*Environmental Action, 1525 New Hampshire Ave., N.W., Washington, DC 20036. Political orientation. Excellent coverage of environmental issues from legal, political, and social action viewpoints.

Environmental Ethics, Department of Philosophy, University of Georgia, Athens, GA 30602. Major journal in the field.

*The Environmental Professional, Editorial Office, Department of Geography, University of Iowa, Iowa City, IA 52242. Excellent discussion of environmental issues.

*Environmental Quality, Council on Environmental Quality, 722 Jackson Place, N.W., Washington, DC 20006. Annual report on environmental problems and progress in environmental protection and improvement.

Environmental Science & Technology, American Chemical Society, 1155 16th St., N.W., Washington, DC 20036. Emphasis on water, air, and solid waste chemistry. Basic reference on current technological developments and research.

EPA Journal, Environmental Protection Agency. Order from Government Printing Office, Washington, DC 20402. Broad coverage of environmental issues and updates on EPA activities.

Family Planning Perspectives, Planned Parenthood–World Population, Editorial Offices, 666 Fifth Ave., New York, NY 10019. Excellent coverage of population issues and latest information on birth control methods.

FDA Consumer, U.S. Department of Health and Human Services, Public Health Service, 5600 Fishers Lane, Rockville, MD 20857. Useful source of information on health issues and food additives.

The Futurist, World Future Society, P.O. Box 19285, Twentieth Street Station, Washington, DC 20036. Covers wide range of societal problems, including environmental, population, and food issues. A fascinating and readable journal.

Impact of Science on Society, UNESCO, 317 East 34th St., New York, NY 10016. Essays on the social consequences of science and technology.

Journal of the Air Pollution Control Association, 4400 Fifth Ave., Pittsburgh, PA 15213. Technical research articles.

Journal of the American Public Health Association, 1015 18th St., N.W., Washington, DC 20036. Some coverage of environmental health issues.

Journal of Environmental Education, Heldref Publications, 4000 Albemarle St., N.W., Suite 504, Washington, DC 20016. Good for teachers.

Journal of Environmental Health, National Environmental Health Association, 1600 Pennsylvania Ave., Denver, CO 80203. Good coverage of technical research.

Journal of the Water Pollution Control Federation, 2626 Pennsylvania Ave., N.W., Washington, DC 20037. Technical research articles.

Journal of Wildlife Management, Wildlife Society, Suite 611, 7101 Wisconsin Ave., N.W., Washington, DC 20014. Good coverage of basic issues and information.

Living Wilderness, Wilderness Society, 1901 Pennsylvania Ave., N.W., Washington, DC 20006. Strong statement of "wild areas" viewpoint.

Monthly Energy Review, National Energy Information Center, Energy Information Administration, Forrestal Building, Room 1F-048, 1000 Independence Ave., S.W., Washington, DC 20585. Useful monthly summaries of U.S. energy production and consumption.

Mother Earth News, P.O. Box 70, Hendersonville, NC 28739. Superb articles on organic farming, alternative energy systems, and alternative lifestyles.

National Parks and Conservation Magazine, National Parks and Conservation Association, 1701 18th St., N.W., Washington, DC 20009. Good coverage of parks and wildlife issues.

National Wildlife, National Wildlife Federation, 8925 Leesburg Pike, Vienna, VA 22180. Good summaries of issues with wildlife emphasis. Action-oriented, with a "Washington report."

Natural History, American Museum of Natural History, Central Park West at 79th St., New York, NY 10024. Popular; wide school and library circulation. Regularly concerned with environment.

Nature, 711 National Press Building, Washington, DC 20045. British equivalent to *Science*; enjoys outstanding reputation.

New Scientist, 128 Long Acre, London, WC 2, England. Excellent general science journal with extensive coverage of environmental issues.

Not Man Apart, Friends of the Earth, 1245 Spear St., San Francisco, CA 94105. Excellent capsule summaries and a few in-depth articles on national and international environmental issues.

Organic Gardening & Farming Magazine, Rodale Press, Inc., 33 E. Minor St., Emmaus, PA 18049. The best guide to organic gardening.

Pollution Abstracts, Data Courier, Inc., 620 S. 5th St., Louisville, KY 40202. Basic bibliographic tool. Too expensive for individual subscription but should be available in university libraries.

Population and Vital Statistics Report, UN Publications Sales Section, New York, NY 10017. Latest world figures.

Population Bulletin, Population Reference Bureau, 2213 M St., N.W., Washington, DC 20037. Nontechnical articles on population issues. Highly recommended.

Population Bulletin, UN Publications Sales Section, New York, NY 10017. Statistical summaries. English and French editions.

Population Reports, Population Information Program, Johns Hopkins University, Hampton House, 624 N. Broadway, Baltimore, MD 21205. Useful information on population issues and birth control.

Practical Homeowner, Rodale Press, Inc., 33 E. Minor St., Emmaus, PA 18049. Excellent nontechnical articles for homeowners on energy conservation and alternative energy systems.

PV News, PV Energy Systems, 2401 Childs Lane, Alexandria, VA 22308. Monthly newsletter summarizing developments in solar photovoltaic cell technology.

Renewable Energy News, Solar Vision, Inc., 7 Church Hill, Harrisville, NH 03450. Useful summary of latest technological and political developments.

Resource Recycling, Resource Recycling, P.O. Box 10540, Portland, OR 97210. Useful summaries of ways in which consumers and companies can profit from recycling, reuse, and waste reduction.

Resources, Resources for the Future, Inc., 1755 Massachusetts Ave., N.W., Washington, DC 20036. Free on request; summarizes information and research on natural resources.

Science, American Association for the Advancement of Science, 1515 Massachusetts Ave., N.W., Washington, DC 20036. Useful source of information. Formerly an excellent source of key environmental articles, under new editorship a technical research journal with greatly decreased coverage of environmental and interdisciplinary issues.

Science News, Science Service, Inc., 1719 N St., N.W., Washington, DC 20036. Good popular summaries of scientific developments, including environmental topics.

Scientific American, 415 Madison Ave., New York, NY 10017. Outstanding journal for the intelligent citizen who wants to keep up with science. Many general articles on environment and ecology.

Sierra, 530 Bush St., San Francisco, CA 94108. Excellent coverage of a wide range of environmental problems and of citizen action. Beautiful photographs.

Solar Age, Solar Vision, Inc., 7 Church Hill, Harrisville, NH 03450. Summary of advances in solar technology.

State of the World, Worldwatch Institute, 1776 Massachusetts Ave., N.W., Washington, DC 20036. Superb annual summary of environment and resource issues.

Statistical Yearbook, Department of International Economic and Social Affairs, Statistical Office, United Nations Publishing Service, United Nations, NY 10017. Useful annual summary of data on population, food production, resource production and consumption, energy, housing, and forestry.

Sun Times, Solar Lobby, Suite 510, 1001 Connecticut Ave., N.W., Washington, DC 20036. Quarterly newsletter summarizing political developments related to alternative energy options.

*Technology Review, Room E219–430, Massachusetts Institute of Technology, Cambridge, MA 02139. Not specialized or always technical, but addressed to a sophisticated audience. Devotes more than half its pages to environment-related material; strong on issues of science policy.

Transition, Laurence G. Wolf, ed., Department of Geography, University of Cincinnati, Cincinnati, OH 45221. Quarterly journal of the Socially and Ecologically Responsible Geographers.

UNESCO Courier, UNESCO Publications Center, 317 E. 34th St., New York, NY 10016. A magazine for the general reader; frequently attentive to environmental issues.

World Development Report, World Bank, Publications Department, 1818 H Street, N.W., Washington, DC 20433. Useful annual summary.

*World Resources, World Resources Institute, 1735 New York Ave., N.W., Washington, DC 20006. Superb annual summary of environment and resource problems published jointly by the World Resources Institute and the International Institute for Environment and Development.

*Worldwatch Papers, Worldwatch Institute, 1776 Massachusetts Ave., N.W., Washington, DC 20036. A series of reports designed to serve as an early warning system on major environmental problems. Worldwatch also publishes an annual *State of the World* in book form. Highly recommended.

Yearbook of World Energy Statistics, Department of International Economic and Social Affairs, Statistical Office, United Nations Publishing Service, United Nations, NY 10017. Useful annual summary of data on worldwide energy production.

Units of Measurement

Length
Metric
1 kilometer (km) = 1,000 meters (m)
1 meter (m) = 100 centimeters (cm)
1 meter (m) = 1,000 millimeters (mm)
1 centimeter (cm) = 0.01 meter (m)
1 millimeter (mm) = 0.001 meter (m)
English
1 foot (ft) = 12 inches (in)
1 yard (yd) = 3 feet (ft)
1 mile (mi) = 5,280 feet (ft)
Metric-English
1 kilometer (km) = 0.621 mile (mi)
1 meter (m) = 39.4 inches (in)
1 inch (in) = 2.54 centimeters (cm)
1 foot (ft) = 0.305 meter (m)
1 yard (yd) = 0.914 meter (m)
1 nautical mile = 1.85 kilometers (km)

Area
Metric
1 square kilometer (km^2) = 1,000,000 square meters (m^2)
1 square meter (m^2) = 1,000,000 square millimeters (mm^2)
1 hectare (ha) = 10,000 square meters (m^2)
1 hectare (ha) = 0.01 square kilometer (km^2)
English
1 square foot (ft^2) = 144 square inches (in^2)
1 square yard (yd^2) = 9 square feet (ft^2)
1 square mile (mi^2) = 27,880,000 square feet (ft^2)
1 acre (ac) = 43,560 square feet (ft^2)
Metric-English
1 hectare (ha) = 2.471 acres (ac)
1 square kilometer (km^2) = 0.386 square mile (mi^2)
1 square meter (m^2) = 1.196 square yards (yd^2)
1 square meter (m^2) = 10.76 square feet (ft^2)
1 square centimeter (cm^2) = 0.155 square inch (in^2)

Volume
Metric
1 cubic kilometer (km^3) = 1,000,000 cubic meters (m^3)
1 cubic meter (m^3) = 1,000,000 cubic centimeters (cm^3)
1 liter (L) = 1,000 milliliters (mL) = 1,000 cubic centimeters (cm^3)
1 milliliter (mL) = 0.001 liter (L)
1 milliliter (mL) = 1 cubic centimeter (cm^3)
English
1 gallon (gal) = 4 quarts (qt)
1 quart (qt) = 2 pints (pt)
Metric-English
1 liter (L) = 0.265 gallon (gal)
1 liter (L) = 1.06 quarts (qt)
1 liter (L) = 0.0353 cubic foot (ft^3)
1 cubic meter (m^3) = 35.3 cubic feet (ft^3)
1 cubic meter (m^3) = 1.307 cubic yards (yd^3)
1 cubic kilometer (km^3) = 0.24 cubic mile (mi^3)
1 barrel (bbl) = 159 liters (L)
1 barrel (bbl) = 42 U.S. gallons (gal)

Mass
Metric
1 kilogram (kg) = 1,000 grams (g)
1 gram (g) = 1,000 milligrams (mg)
1 gram (g) = 1,000,000 micrograms (μg)
1 milligram (mg) = 0.001 gram (g)
1 microgram (μg) = 0.000001 gram (g)
1 metric ton (mt) = 1,000 kilograms (kg)
English
1 ton (t) = 2,000 pounds (lb)
1 pound (lb) = 16 ounces (oz)
Metric-English
1 metric ton (mt) = 2,200 pounds (lb) = 1.1 tons
1 kilogram (kg) = 2.20 pounds (lb)
1 pound (lb) = 454 grams (g)
1 gram (g) = 0.035 ounce (oz)

Energy and Power
Metric
1 kilojoule (kJ) = 1,000 joules (J)
1 kilocalorie (kcal) = 1,000 calories (cal)
1 calorie (cal) = 4.184 joules (J)
Metric-English
1 kilojoule (kJ) = 0.949 British thermal unit (Btu)
1 kilojoule (kJ) = 0.000278 kilowatt-hour (kW-h)
1 kilocalorie (kcal) = 3.97 British thermal units (Btu)
1 kilocalorie (kcal) = 0.00116 kilowatt-hour (kW-h)
1 kilowatt-hour (kW-h) = 860 kilocalories (kcal)
1 kilowatt-hour (kW-h) = 3,400 British thermal units (Btu)
1 quad (Q) = 1,050,000,000,000,000 kilojoules (kJ)
1 quad (Q) = 2,930,000,000,000 kilowatt-hours (kW-h)
Approximate crude oil equivalent
1 barrel (bbl) crude oil = 6,000,000 kilojoules (kJ)
1 barrel (bbl) crude oil = 2,000,000 kilocalories (kcal)
1 barrel (bbl) crude oil = 6,000,000 British thermal units (Btu)
1 barrel (bbl) crude oil = 2,000 kilowatt-hours (kW-h)
Approximate natural gas equivalent
1 cubic foot (ft^3) natural gas = 1,000 kilojoules (kJ)
1 cubic foot (ft^3) natural gas = 260 kilocalories (kcal)
1 cubic foot (ft^3) natural gas = 1,000 British thermal units (Btu)
1 cubic foot (ft^3) natural gas = 0.3 kilowatt-hour (kW-h)
Approximate hard coal equivalent
1 ton (t) coal = 20,000,000 kilojoules (kJ)
1 ton (t) coal = 6,000,000 kilocalories (kcal)
1 ton (t) coal = 20,000,000 British thermal units (Btu)
1 ton (t) coal = 6,000 kilowatt-hours (kW-h)

Temperature Conversions
Fahrenheit (°F) to Celsius (°C): $°C = \dfrac{(°F - 32.0)}{1.80}$

Celsius (°C) to Fahrenheit (°F): $°F = (°C \times 1.80) + 32.0$

Major U.S. Environmental Legislation

General

National Environmental Policy Act of 1969 (NEPA)

Energy

National Energy Acts of 1978 and 1980

Water Quality

Federal Water Pollution Control Act of 1972

Ocean Dumping Act of 1972

Safe Drinking Water Act of 1974, 1984

Toxic Substances Control Act of 1976

Clean Water Acts of 1977, 1987

Air Quality

Clean Air Acts of 1965, 1970, 1977

Noise Control

Noise Control Act of 1972

Quiet Communities Act of 1978

Resources and Solid Waste Management

Solid Waste Disposal Act of 1965

Resources Recovery Act of 1970

Toxic Substances

Toxic Substances Control Act of 1976

Resource Conservation and Recovery Act of 1976

Comprehensive Environmental Response, Compensation, and Liability (Superfund) Acts of 1980, 1986

Wildlife

Species Conservation Act of 1966

Federal Insecticide, Fungicide, and Rodenticide Control Act of 1972

Marine Protection, Research, and Sanctuaries Act of 1972

Endangered Species Act of 1973

Land Use

Multiple Use Sustained Yield Act of 1960

Wilderness Act of 1964

Wild and Scenic River Act of 1968

National Coastal Zone Management Acts of 1972 and 1980

Federal Land Policy Management Act of 1976

Forest Reserves Management Acts of 1974, 1976

National Forest Management Act of 1976

Surface Mining Control and Reclamation Act of 1977

Endangered American Wilderness Act of 1978

Alaskan National Interest Lands Conservation Act of 1980

4

How to Save Water and Money

Bathroom (65% of residential water use; 40% for toilet flushing)

■ For existing toilets, reduce the amount of water used per flush by putting a tall plastic container weighted with a few stones into each tank, or buy (for about $10) and insert a toilet dam made of plastic and rubber; bricks also work but tend to disintegrate and gum up the water.

■ In new houses, install water-saving toilets or, where health codes permit, waterless or composting toilets. Flush only when necessary, using the advice found on a bathroom wall in a drought-stricken area: "If it's yellow, let it mellow—if it's brown, flush it down."

■ Take short showers—showers of less than 5 minutes use less water than a bath. Shower by wetting down, turning off the water while soaping up, and then rinsing off. If you prefer baths, fill the tub well below the overflow drain.

■ Use water-saving flow restrictors, which cost less than a dollar and can be easily installed, on all faucets and showerheads.

■ Check frequently for toilet, shower, and sink leaks and repair them promptly. A pinhole leak anywhere in a household water system can cost $25 a month in excess water and electricity charges; a fast leak, $50 or more.

■ Don't keep water running while brushing teeth, shaving, or washing.

Laundry Room (15%)

■ Wash only full loads; use the short cycle and fill the machine to the lowest possible water level.

■ When buying a new washer, choose one that uses the least amount of water and fills up to different levels for loads of different sizes.

■ Check for leaks frequently and repair all leaks promptly.

Kitchen (10%)

■ Use an automatic dishwasher only for full loads; use the short cycle and let dishes air-dry to save energy.

■ When washing many dishes by hand, don't let the faucet run. Instead use one filled dishpan for washing and another for rinsing.

■ Keep a jug of water in the refrigerator rather than running water from a tap until it gets cold enough to drink.

■ While waiting for faucet water to get hot, catch the cool water in a pan and use it for cooking or to water plants.

■ Check for sink and dishwasher leaks frequently and repair them promptly.

■ Try not to use a garbage disposal or water-softening system—both are major water users.

Outdoors (10%; higher in arid areas)

■ Don't wash your car, or wash it less frequently. Wash the car from a bucket of soapy water; use the hose only for rinsing.

■ Sweep walks and driveways instead of hosing them off.

■ Reduce evaporation losses by watering lawns and gardens in the early morning or in the evening, rather than in the heat of midday or when windy. Better yet, landscape with pebbles, rocks, sand, wood chips, or native plants adapted to local average annual precipitation so that watering is not necessary.

■ Use drip irrigation systems and mulch on home gardens to improve irrigation efficiency and reduce evaporation.

5

How to Save Energy and Money

Transportation (50% of average personal energy use)

- Walk or ride a bike for short trips (100% savings).

- Use a car pool or mass transit as much as possible (50% or more).

- Use a bus or train for long trips (50% to 75%).

- Buy an energy-efficient car (30% to 70%).

- Consolidate trips to accomplish several purposes (up to 50%).

- Keep engine tuned and replace air filter regularly (20% to 50%).

- Obey speed limits (20% or more).

- Accelerate and brake gently and don't warm up the engine for more than a minute (15% to 20%).

- Use steel-belted radial tires and keep tire pressure at the recommended level (2% to 5%).

Home Space Heating (25%)

- Build a superinsulated or highly energy-efficient house (50% to 100% savings).

- Dress more warmly, humidify air, and use fans to distribute heat so that thermostat setting can be lowered without loss of comfort (saves 3% for each °F decrease).

- Install the most energy-efficient heating system available (15% to 50%).

- Install an electronic ignition system in furnace, have furnace cleaned and tuned once a year, and clean or replace intake filters every two weeks (15% to 35%).

- Do not heat closets and unused rooms (variable savings).

- Insulate ceilings and walls (20% to 50%).

- Caulk and weatherstrip cracks (10% to 30%).

- Install storm windows and doors or insulated drapes or shutters (5% to 25%).

Hot Water Heating (9%)

- Install the most energy-efficient system available, such as active solar, instant tankless, or high-efficiency-gas water heaters (15% to 60%).

- Turn down thermostat on water heater (5% to 25%).

- Insulate hot water pipes and water heater (10% to 15%).

- Use less hot water by taking two- to five-minute showers instead of baths, washing dishes and clothes only with full loads, washing clothes with warm or cold water, repairing leaky faucets, installing flow reducers on faucets and showerheads, and not letting water run while bathing, shaving, brushing teeth, or washing dishes (10% to 25%).

Cooking, Refrigeration, and Other Appliances (9%)

- Buy the most energy-efficient stove, refrigerator, and other appliances available—ideally powered by natural or LP gas, not electricity (25% to 60%).

- Install electronic ignition systems on all gas stoves and other appliances (10% to 30%).

- Use a chest freezer rather than an upright model to prevent unnecessary loss of cool air when door is opened, and keep it almost full (variable).

- Do not locate refrigerator or freezer near a stove or other source of heat and keep condenser coils on back clean (variable).

- Don't use oven for space heating (very expensive).

Cooling, Air Conditioning, and Lighting (7%)

- Buy the most energy-efficient air conditioning system available (30% to 50%).

- Increase thermostat setting (3% to 5% for each °F).

- Close off and do not air condition closets and unused rooms (variable).

- Use small floor fans and whole-house window or attic fans to eliminate or reduce air conditioning needs (variable).

- Close windows and drapes on sunny days and open them on cool days and at night (variable).

- Close bathroom doors and use an exhaust fan or open window to prevent transfer of heat and humid air to rest of house (variable).

- Try to schedule heat- and moisture-producing activities such as bathing, ironing, and washing during the coolest part of the day (variable).

- Cover pots while cooking (variable).

- Use fluorescent and other energy-saving bulbs wherever possible (15% to 25%).

- Use natural lighting whenever possible (variable).

- Turn off lights and appliances when not in use and reduce lighting levels by using dimmers and lower wattage (variable).

National Ambient Air Quality Standards for the United States

Pollutant	Averaging Time	Primary Standard Levels [micrograms (μg) or milligrams (mg) per cubic meter (m^3) and parts per million (ppm)]	Secondary Standard Levels [micrograms (μg) or milligrams (mg) per cubic meter (m^3) and parts per million (ppm)]
SPM	Annual (geometric mean)	75 μg/m^3	60 μg/m^3
	24 hours*	260 μg/m^3	150 μg/m^3
Sulfur dioxide	Annual (arithmetic mean)	80 μg/m^3 (0.03 ppm)	—
	24 hours*	365 μg/m^3 (0.14 ppm)	—
	3 hours*		1300 μg/m^3 (0.5 ppm)
Carbon monoxide	8 hours*	10,000 μg/m^3 (9 ppm)	same as primary
	1 hour	40 mg/m^3 (35 ppm)*	same as primary
Nitrogen dioxide	Annual (arithmetic mean)	100 μg/m^3 (0.05 ppm)	same as primary
Ozone	1 hour	235 μg/m^3 (0.12 ppm)	same as primary
Volatile organic compounds[†]	3 hours (6 to 9 A.M.)	160 μg/m^3 (0.24 ppm)	160 μg/m^3 (0.24 ppm)
Lead	3 months	1.5 μg/m^3	same as primary

Source: Environmental Protection Agency.
*Not to be exceeded more than once a year.
†A non-health-related standard used as a guide for ozone control. Does not include methane.

U.S. Pollutant Standard Index (PSI) Values
(Data from the Environmental Protection Agency)

PSI Index Value	Air Quality Level	Pollutant Levels (micrograms per cubic meter)					Health Effect Description	General Health Effects	Suggested Action
		SPM (24 hour)	SO_2 (24 hour)	CO (8 hour)	O_3 (1 hour)	NO_2 (1 hour)			
500	Significant harm	1,000	2,620	57.5	1,200	3,750		Premature death of ill and elderly. Healthy people will experience adverse symptoms that affect their normal activity.	All persons should remain indoors, keeping windows and doors closed. All persons should minimize physical exertion and avoid traffic.
400	Emergency	875	2,100	46.0	1,000	3,000	Hazardous (300 and above)	Premature onset of certain diseases. Significant aggravation of symtoms in the ill; decreased exercise tolerance in healthy persons.	Elderly and persons with existing diseases should stay indoors and avoid physical exertion. General population should avoid activity.
300	Warning	625	1,600	34.0	800	2,260	Very Unhealthful (200–299)	Significant aggravation of symptoms and decreased tolerance in persons with heart or lung disease with widespread symptoms in the healthy population.	Elderly and persons with existing heart or lung disease should stay indoors and reduce physical activity.
200	Alert	375	800	17.0	400	1,130	Unhealthful (100–199)	Mild aggravation of symptoms in susceptible persons with irritation symptoms in the healthy population.	Persons with existing heart or respiratory ailments should reduce physical exertion and outdoor activity.
100	NAAQS	260	365	10.0	240	—	Moderate		
50	50% of NAAQS	75*	80*	5.0	120	—	Moderate		
0		0	0	0	0	—	Good		

*Annual primary NAAQS.

Further Readings

For a more comprehensive set of references, see Further Readings in the expanded version of this book: *Living in the Environment*, 5th ed.: Belmont, Calif.: Wadsworth, 1988.

Chapter 1/Population, Resources, Environmental Degradation, and Pollution

Brown, Lester R., et al. Annual. *State of the World*. New York: W. W. Norton. Superb overview published annually since 1984 by the Worldwatch Institute, Washington, D.C.

Council on Environmental Quality. *Annual Report*. Washington, D.C.: Government Printing Office. Annual summaries since 1970 of environmental problems and progress. Useful source of information.

Council on Environmental Quality and U.S. Department of State. 1980. *The Global 2000 Report to the President*. 3 vols. Washington, D.C.: Government Printing Office. Outstanding summary of global population, resource, environmental degradation, and pollution problems with projections to the year 2000.

Dahlberg, Kenneth A., et al. 1985. *Environment and the Global Arena*. Durham, N.C.: Duke University Press. Superb overview of problems, policies, and possible environmental futures.

Durrell, Lee. 1986. *State of the Ark: An Atlas of Conservation in Action*. Garden City, N.Y.: Doubleday. Superb overview.

Hardin, Garrett. 1968. "The Tragedy of the Commons." *Science*, vol. 162, 1243–1248. Classic environmental article describing how land and other resources are abused when they are shared by everyone.

Hardin, Garrett. 1985. *Filters Against Folly*. East Rutherford, N.J.: Viking. Thought-provoking series of essays by a prominent neo-Malthusian.

Myers, Norman, ed. 1984. *Gaia: An Atlas of Planet Management*. Garden City, N.Y.: Anchor Press/Doubleday. Outstanding overview of the planet's resources, environmental problems, and possible solutions. Excellent illustrations.

Schumacher, E. F. 1973. *Small Is Beautiful: Economics As If People Mattered*. New York: Harper & Row. Eloquent presentation of the need for appropriate technology. An environmental classic.

Simon, Julian L. 1981. *The Ultimate Resource*. Princeton, N.J.: Princeton University Press. Very good presentation of the cornucopian position. However, the author bases some of his conclusions on the false idea that energy can be recycled.

Simon, Julian L., and Herman Kahn, eds. 1984. *The Resourceful Earth: A Response to "Global 2000."* Useful collection of articles by a group of mostly cornucopian thinkers who argue that increased economic growth and technology can solve the world's population, resource, and pollution problems and that these problems were exaggerated in *The Global 2000 Report to the President* (1980).

World Resources Institute and International Institute for Environment and Development. Annual. *World Resources*. Excellent annual report published since 1986.

Chapter 2/Human Impact on the Earth

Carson, Rachel. 1962. *Silent Spring*. Boston: Houghton Mifflin. An environmental classic.

Congressional Quarterly. 1983. *The Battle for Natural Resources*. Washington, D.C.: Congressional Quarterly, Inc. Excellent overview of the history of federal land use.

Lash, Jonathan, et al. 1984. *A Season of Spoils*. New York: Pantheon. Well-researched analysis of Ronald Reagan's environmental policies during most of his first term.

Leopold, Aldo. 1949. *A Sand County Almanac*. New York: Oxford University Press. An environmental classic describing Leopold's ecological land-use ethic.

Marsh, George Perkins. 1864. *Man and Nature*. New York: Scribners. An environmental classic considered to be one of the greatest American works on the environment.

Nash, Roderick. 1968. *The American Environment: Readings in the History of Conservation*. Reading, Mass.: Addison-Wesley. Excellent collection of articles.

Osborn, Fairfield. 1948. *Our Plundered Planet*. Boston: Little, Brown. An environmental classic that attempted to alert readers to the environmental problems we face today.

Sears, Paul B. 1980. *Deserts on the March*. Norman: University of Oklahoma Press. Probably the best account of how human activities have contributed to the spread of deserts. Originally published in 1935.

Shanks, Bernard. 1984. *This Land Is Your Land*. San Francisco: Sierra Club Books. Outstanding history of the use and abuse of public lands in the United States.

Vig, Norman J., and Michael J. Craft. 1984. *Environmental Policy in the 1980s*. Washington, D.C.: Congressional Quarterly Press. Useful summary of President Reagan's environmental policy during his first term.

Zaslowsky, Dyan, and Wilderness Society. 1986. *These American Lands*. New York: Henry Holt. Superb history of use of federal lands.

Chapter 3/Matter and Energy Resources: Types and Concepts

Bent, Henry A. 1971. "Haste Makes Waste: Pollution and Entropy." *Chemistry*, vol. 44, 6–15. Excellent and very readable account of the relationship between the second law of thermodynamics and environmental problems.

Clark, Wilson, and Jake Page. 1981. *Energy, Vulnerability, and War: Alternatives for America*. New York: W. W. Norton. Excellent analysis of vulnerability of centralized U.S. energy system to nuclear attack and to cutoffs of imported oil. Based on a study carried out by energy expert Clark for the Department of Defense.

Colorado Energy Research Institute. 1976. *Net Energy Analysis: An Energy Balance Study of Fossil Fuel Resources*. Golden, Colo.: Colorado Energy Research Institute. Excellent source of data.

Fowler, John M. 1984. *Energy and the Environment*. 2d ed. New York: McGraw-Hill. Excellent overview.

Gofman, John W. 1981. *Radiation and Human Health*. San Francisco: Sierra Club Books. An expert's detailed and controversial evaluation of the health effects of exposure to low-level radiation.

Lovins, Amory B. 1977. *Soft Energy Paths*. Cambridge, Mass.: Ballinger. Superb analysis of energy alternatives. See also Nash (1979).

Nash, Hugh, ed. 1979. *The Energy Controversy: Soft Path Questions and Answers*. San Francisco: Friends of the Earth. Pros and cons of the soft energy path.

Odum, Howard T., and Elisabeth C. Odum. 1980. *Energy Basis for Man and Nature*. New York: McGraw-Hill. Outstanding discussion of energy principles and energy options, with emphasis on net energy analysis.

Pochin, Edward. 1985. *Nuclear Radiation: Risks and Benefits*. New York: Oxford University Press. Excellent overview.

Rifkin, Jeremy. 1980. *Entropy: A New World View*. New York: Viking. Superb nontechnical description of the need to develop a sustainable-earth society based on the second law of thermodynamics.

Chapter 4/Ecosystems: What Are They and How Do They Work?

Colinvaux, Paul A. 1978. *Why Big Fierce Animals Are Rare*. Princeton, N.J.: Princeton University Press. Fascinating and very readable description of major ecological principles.

Ehrlich, Paul R. 1986. *The Machinery of Life: The Living World Around Us and How It Works*. New York: Simon & Schuster. Superb nontechnical description.

Ehrlich, Paul R., Anne H. Ehrlich, and John P. Holdren. 1977. *Ecoscience: Population, Resources and Environment*. San Francisco: W. H. Freeman. Excellent, more detailed text at a higher level than *Environmental Science*.

Kormondy, Edward J. 1984. *Concepts of Ecology*. 3d ed. Englewood Cliffs, N.J.: Prentice-Hall. First-rate introduction at a slightly higher level than *Environmental Science*.

Lovelock, James E. 1979. *Gaia: A New Look at Life*. Oxford, England: Oxford University Press. Superb analysis of interactions in the biosphere.

Odum, Eugene P. 1983. *Basic Ecology*. Philadelphia: Saunders. Outstanding textbook on ecology by a prominent ecologist.

Ramadé, François. 1984. *Ecology of Natural Resources*. New York: John Wiley. Excellent introduction.

Rickleffs, Robert E. 1976. *The Economy of Nature*. Portland, Ore.: Chiron Press. Beautifully written introduction to ecology.

Smith, Robert L. 1985. *Elements of Ecology*. 2d ed. New York: Harper & Row. Outstanding basic text.

Watt, Kenneth E. F. 1982. *Understanding the Environment*. Newton, Mass.: Allyn & Bacon. Excellent introduction.

Chapter 5/Ecosystems: What Are the Major Types and What Can Happen to Them?

See also the readings for Chapter 4.

Attenborough, David. 1984. *The Living Planet*. Boston: Little, Brown. Superb survey of life in the biosphere.

Brown, J. H., and A. C. Gibson. 1983. *Biogeography*. St. Louis: C. V. Mosby. Excellent analysis of factors affecting distribution of plants and animals in various parts of the earth.

Clapham, W. B., Jr. 1984. *Natural Ecosystems*. 2d ed. New York: Macmillan. Useful introduction.

Ehrlich, Paul R. 1980. "Variety Is the Key to Life." *Technology Review*. March/April. Excellent summary of the need to preserve biological diversity.

Maltby, Edward. 1986. *Waterlogged Wealth*. Washington, D.C.: Earthscan. Excellent overview of destruction of wetlands in the United States.

Marx, Wesley. 1981. *The Oceans: Our Last Resource*. San Francisco: Sierra Club Books. Excellent discussion of how to preserve the ocean's resources.

Odum, Eugene P. 1969. "The Strategy of Ecosystem Development." *Science*, vol. 164, 262–270. Excellent summary of succession.

Pilkey, Orin H., Sr., et al. 1984. *Coastal Design, A Guide for Builders, Planners, & Homeowners*. New York: Van Nostrand Reinhold. Excellent analysis.

Teal, J., and M. Teal. 1969. *Life and Death of a Salt Marsh*. New York: Ballantine. A classic work on estuaries.

Woodwell, G. M. 1970. "Effects of Pollution on the Structure and Physiology of Ecosystems." *Science*, vol. 168, 429–433. Excellent analysis.

Chapter 6/Risk-Benefit and Cost-Benefit Analysis

Andrews, Richard N. L. 1981. "Will Benefit-Cost Analysis Reform Regulations?" *Environmental Science and Technology*, vol. 15, no. 9, 1016–1021. Excellent summary of cost-benefit and cost-effective analysis.

Chandler, William U. 1986. *Banishing Tobacco*. Washington, D.C.: Worldwatch Institute. Superb analysis.

Conservation Foundation. 1985. *Risk Assessment and Risk Control*. Washington, D.C.: Conservation Foundation. Superb overview.

Crutzen, Paul J. 1985. "The Global Environment After Nuclear War." *Environment*, vol. 27, no. 8, 6–11, 34–37. Excellent overview of research on the nuclear winter effect.

Dotto, Lydia. 1986. *Planet Earth in Jeopardy: Environmental Consequences of Nuclear War*. New York: Wiley. Excellent overview.

Douglas, Mary, and Aaron Wildavsky. 1982. *Risk and Culture*. Berkeley: University of California Press. Excellent overview.

Imperato, P. J., and Greg Mitchell. 1985. *Acceptable Risks*. New York: Viking. Excellent nontechnical overview.

National Academy of Sciences. 1985. *The Effects on the Atmosphere of a Major Nuclear War*. Washington, D.C.: National Academy Press. Authoritative summary of nuclear winter/autumn effect.

National Academy of Sciences. 1982. *Diet, Nutrition, and Cancer*. Washington, D.C.: National Academy Press. Authoritative review of relationships between diet and cancer.

Olsen, Steve. 1986. *Biotechnology*. Washington, D.C.: National Academy Press. Useful summary of a conference of experts on the nature and possible regulation of genetic engineering.

Rifkin, Jeremy. 1985. *Declaration of a Heretic*. Boston: Routledge & Kegan Paul. Excellent summary of potential problems with genetic engineering by its most active and effective critic.

Thompson, Starley L., and Stephen H. Schneider. 1986. "Nuclear Winter Reappraised." *Foreign Affairs*, Summer. Research indicating that effects of nuclear war might be a less severe nuclear autumn.

Chapter 7/Population Dynamics and Distribution

Bouvier, Leon F. 1984. "Planet Earth 1984–2034: A Demographic Vision." *Population Bulletin*, vol. 39, no. 1, 1–39. Outstanding overview of future population trends and problems.

Brown, Lester R., and Jodi Jacobson. 1987. "Assessing the Future of Urbanization." In *State of the World 1987*, by Lester R. Brown et al. New York: W. W. Norton, pp. 38–56. Excellent overview.

Chandler, William U. 1985. *Investing in Children*. Washington, D.C.: Worldwatch Institute. Excellent analysis of how to reduce infant mortality.

Dantzig, George B., and Thomas L. Saaty. 1973. *Compact City: A Plan for a Liveable Environment*. San Francisco: W. H. Freeman. Outstanding analysis of urban design for a more ecologically sound and self-reliant city.

Fabos, Julius G. 1985. *Land-Use Planning*. New York: Chapman and Hall. Excellent examples.

Haupt, Arthur, and Thomas T. Kane. 1978. *The Population Handbook*. Washington, D.C.: Population Reference Bureau. Superb introduction to demographic terms and concepts.

McHarg, Ian L. 1969. *Design with Nature*. Garden City, N.Y.: Natural History Press. A beautifully written and illustrated description of an ecological approach to land-use planning. Also available in paperback from Doubleday.

Merrick, Thomas W. 1986. "World Population in Transition," *Population Bulletin*, vol. 41, no. 2, 1–51. Superb overview.

National Academy of Sciences. 1983. *Future Directions of Urban Public Transportation*. Washington, D.C.: National Academy Press. Excellent analysis.

Population Reference Bureau. Annual. *World Population Data Sheet*. Washington, D.C.: Population Reference Bureau. This concise annual summary is the source for most of the population data used in this book.

Population Reference Bureau. 1982. "U.S. Population: Where We Are; Where We're Going." *Population Bulletin*, vol. 37, no. 2. Excellent summary.

Teitelbaum, Michael, and Jay M. Winter. 1985. *The Fear of Population Decline*. Orlando, Fla.: Academic Press. Excellent overview of this problem feared by some MDCs.

UNICEF. 1987. *The State of the World's Children*. New York: United Nations. Excellent overview.

Chapter 8/Population Control

Brown, Lester R. 1981. *Building a Sustainable Society*. New York: W. W. Norton. An outstanding discussion of the need for population control and the conservation of renewable and nonrenewable resources.

Brown, Lester R., and Jodi L. Jacobson. 1986. *Our Demographically Divided World*. Washington, D.C.: Worldwatch Institute. Superb analysis of demographic transition.

Croll, Elisabeth, et al. 1985. *China's One-Child Family Policy*. New York: St. Martin's Press. Excellent overview.

Goliber, Thomas J. 1985. "Sub-Saharan Africa: Population Pressures on Development." *Population Bulletin*, vol. 40, no. 1, 1–45. Excellent summary of problems and possible solutions.

Grupte, Pranay. 1984. *The Crowded Earth: People and the Politics of Population*. New York: W. W. Norton. Superb summary of population problems and what is being done about them in selected countries.

Hardin, Garrett. 1982. *Naked Emperors: Essays of a Taboo Stalker*. Los Altos, Calif.: William Kaufmann. Thought-provoking essays on a variety of subjects including overpopulation and abortion.

Lamm, Richard D., and Gary Imhoff. 1985. *The Immigration Time Bomb*. New York: Dutton. Excellent analysis arguing for greater control of legal and illegal immigration in the United States.

Menken, Jane, ed. 1986. *World Population and U.S. Policy: The Choices Ahead*. New York: W. W. Norton. Eight experts provide an insightful analysis of population policy.

NARAL Foundation. 1984. *Legal Abortion: Arguments Pro & Con*. Washington, D.C.: National Abortion Rights Action League. Excellent summary.

Population Reference Bureau. 1986. *Women in the World: The Women's Decade and Beyond*. Washington, D.C.: Population Reference Bureau. Excellent summary of condition of women throughout the world.

Religious Coalition for Abortion Rights. 1985. *Point, Counterpoint*. Washington, D.C.: Religious Coalition for Abortion Rights. Excellent summary of arguments for and against abortion by a group of 31 national religious organizations.

Chapter 9/Soil Resources

Batie, Sandra S. 1983. *Soil Erosion: Crisis in America's Croplands?* Washington, D.C.: Conservation Foundation. Useful and objective analysis.

Brady, Nyle C. 1974. *The Nature and Properties of Soils*. New York: Macmillan. Excellent introductory text.

Brown, Lester R., et al. 1984. *State of the World 1984*. New York: W. W. Norton. See Chapter 4 of this annual publication for a superb summary of world and U.S. soil erosion.

Dale, Tom, and V. G. Carter. 1955. *Topsoil and Civilization*. Norman: University of Oklahoma Press. Classic work describing soil abuse throughout history.

Sophen, C. D., and J. V. Baird. 1982. *Soils and Soil Management.* Reston, Va.: Reston Publishing. Excellent introductory text that is easy to read and scientifically sound.

Swanson, Earl R., and Earl O. Heady. 1984. "Soil Erosion in the United States," In *The Resourceful Earth,* edited by Julian L. Simon and Herman Kahn. New York: Basil Blackwell, pp. 202–222. Useful overview.

Chapter 10/Water Resources

Ashworth, William. 1982. *Nor Any Drop To Drink.* New York: Summit Books. Outstanding overview of the water crisis in the United States.

Brown, Lester R., et al. 1985. *State of the World 1985.* New York: W. W. Norton. See Chapter 3 of this annual publication for an excellent discussion of water resources.

Dregnue, H. E. 1983. *Desertification of Arid Lands.* New York: Academic Press. Excellent overview by an expert.

Fradkin, Phillip L. 1981. *A River No More: The Colorado and the West.* New York: Alfred A. Knopf. Overview of politics and water resources in the western United States.

Goldsmith, Edward, and Nicholas Hidyard, eds. 1986. *The Social and Environmental Effects of Large Dams,* 3 vols. Detailed analysis. Useful source of data.

Postel, Sandra. 1985. *Conserving Water: The Untapped Alternative.* Washington, D.C.: Worldwatch Institute. Excellent overview.

Pringle, Laurence. 1982. *Water—The Next Great Resource Battle.* New York: Macmillan. Superb overview of the water crisis in the United States.

Sheaffer, John, and Leonard Stevens. 1983. *Future Water.* New York: William Morrow. Excellent overview of U.S. water resource problems with suggested solutions.

Wijkman, Anders, and Lloyd Timberlake. 1984. *Natural Disasters: Acts of God or Acts of Man?* Washington, D.C.: Earthscan. Excellent overview of effects of human activities on damages from floods and drought.

World Resources Institute and International Institute for Environment and Development. Annual. *World Resources.* New York: Basic Books. Excellent annual summary of water resources published since 1986.

Chapter 11/Food Resources and World Hunger

Brown, Lester R., et al. Annual. *State of the World.* New York: W. W. Norton. Excellent annual summary published since 1984.

Calder, Nigel. 1986. *The Green Machines.* New York: Putnam. Excellent summary of the potential of genetic engineering and other forms of biotechnology.

Crosson, Pierre R. 1984. "Agricultural Land: Will There Be Enough?" *Environment,* vol. 26, no. 7, 17–20, 40–45. Excellent analysis.

Freydberg, N., and W. Gortner. 1982. *The Food Additives Book.* New York: Bantam. Useful source of information.

Huessy, Peter. 1978. *The Food First Debate.* San Francisco: Institute for Food and Development Policy. Pros and cons of the proposals made by Lappé and Collins (1977).

Hunger Project. 1985. *Ending Hunger: An Idea Whose Time Has Come.* New York: Praeger. Superb discussion of world food problems and possible solutions.

Lappé, Francis M., and Joseph Collins. 1977. *Food First.* Boston: Houghton Mifflin. Provocative discussion of world food problems.

Murphy, Elaine M. 1984. *Food and Population: A Global Concern.* Washington, D.C.: Population Reference Bureau. Excellent overview.

Pimentel, David. 1987. "Down on the Farm: Genetic Engineering Meets Ecology." *Technology Review,* January, pp. 24–30. Excellent analysis of usefulness and potential harmful effects of genetic engineering.

Pimentel, David, and Marcia Pimentel. 1979. *Food, Energy, and Society.* New York: Wiley. Outstanding discussion of food problems and possible solutions, with emphasis on energy use and food production.

Reichert, Walt. 1982. "Agriculture's Diminishing Diversity." *Environment,* vol. 24, no. 9, 6–11, 39–43. Excellent summary.

Timberlake, Lloyd. 1985. *Africa in Crisis.* Washington, D.C.: Earthscan. Excellent overview.

Todd, Nancy J., and John Todd. 1984. *Bioshelters: Ocean Arks, City Farming: Ecology as a Basis for Design.* San Francisco: Sierra Club Books. Outstanding description of how individuals can practice sustainable-earth agriculture.

United States Department of Agriculture. 1980. *Report and Recommendations on Organic Farming.* Washington, D.C.: U.S. Department of Agriculture. Excellent summary of research.

Wijkman, Anders, and Lloyd Timberlake. 1984. *Natural Disasters: Acts of God or Acts of Man?* Washington, D.C.: Earthscan. Excellent overview of the causes of famine and the pros and cons of famine relief.

Wolf, Edward C. 1986. *Beyond the Green Revolution: New Approaches for Third World Agriculture.* Washington, D.C.: Worldwatch Institute. Excellent analysis.

Chapter 12/Land Resources: Wilderness, Parks, Forests, and Rangelands

See also the readings for Chapter 2.

Allin, Craig W. 1982. *The Politics of Wilderness Preservation.* Westport, Conn.: Greenwood Press. Excellent political history of efforts to preserve wilderness.

Brown, Lester R., et al. Annual. *State of the World.* New York: W. W. Norton. Excellent annual overviews.

Caufield, Catherine. 1985. *In the Rainforest.* New York: Alfred Knopf. Excellent summary of tropical forest problems.

Conservation Foundation. 1985. *National Parks and the New Generation.* Washington, D.C.: Conservation Foundation. Excellent analysis of problems, with proposed solutions.

Ferguson, Denzel, and Nancy Ferguson. 1983. *Sacred Cows at the Public Trough.* Bend, Ore.: Maverick Publications. Excellent investigative reporting of environmental degradation of western public rangelands by the cattle industry.

Frome, Michael. 1983. *The Forest Service.* Boulder, Colo.: Westview Press. Excellent overview.

Libecap, Gary D. 1986. *Locking Up the Range: Federal Land Control and Grazing.* San Francisco: Pacific Institute for Public Policy Research. Excellent analysis.

McNeely, Jeffery A., and Kenton R. Miller, eds. 1984. *National Parks, Conservation, and Development*. Washington, D.C.: Smithsonian Institution Press. Excellent analysis.

Minckler, Leon S. 1980. *Woodland Ecology*. 2d ed. Syracuse, N.Y.: Syracuse University Press. Superb introduction to ecological management of forests.

Myers, Norman. 1984. *The Primary Source: Tropical Forests and Our Future*. New York: W. W. Norton. Superb analysis by an expert.

Nash, Roderick. 1982. *Wilderness and the American Mind*. 3d ed. New Haven, Conn.: Yale University Press. Outstanding book on American attitudes toward wilderness and conservation.

Office of Technology Assessment. 1984. *Technologies to Sustain Tropical Forest Resources*. Washington, D.C.: Government Printing Office. Excellent analysis.

Sierra Club. 1982. *Our Public Lands: An Introduction to the Agencies and Issues*. San Francisco: Sierra Club Books. Excellent overview.

Smith, D. M. 1982. *The Practice of Silviculture*. New York: Wiley. Excellent standard text.

U.S. Department of Interior. 1984. *50 Years of Public Land Management: 1934–1984*. Washington, D.C.: Bureau of Land Management. Useful overview of range management.

World Resources Institute and International Institute for Environment and Development. Annual. *World Resources*. New York: Basic Books. Excellent annual summary since 1986 of land resources.

Chapter 13/Wild Plant and Animal Resources

Anderson, S. H. 1985. *Managing Our Wildlife Resources*. Columbus, Ohio: Charles Merrill. Excellent textbook.

Baker, Ron. 1985. *The American Hunting Myth*. New York: Vantage Press. Useful analysis of arguments for and against sport hunting.

Credlund, Arthur G. 1983. *Whales and Whaling*. New York: Seven Hills Books. Excellent overview.

Durrell, Lee. 1986. *State of the Ark: An Atlas of Conservation in Action*. Garden City, N.Y.: Doubleday. Superb overview.

Ehrlich, Paul, and Anne Ehrlich. 1981. *Extinction*. New York: Random House. One of the best treatments of the value of wildlife and the causes of extinction, with suggestions for preventing extinction.

Huxley, Anthony. 1984. *Green Inheritance*. Garden City, N.Y.: Anchor/Doubleday. Superb discussion of importance of plants and how to save them from extinction.

Livingston, John. 1981. *The Fallacy of Wildlife Conservation*. London: McClelland & Stewart. Excellent critique of the idea that wildlife resources should be managed to benefit humans.

Myers, Norman. 1983. *A Wealth of Wild Species: Storehouse for Human Welfare*. Boulder, Colo.: Westview Press. Superb presentation of the value of wild species to humans.

National Audubon Society. Annual. *Audubon Wildlife Report*. New York: National Audubon Society. Excellent detailed annual summary published since 1985.

World Resources Institute and International Institute for Environment and Development. Annual. *World Resources*. New York: Basic Books. Excellent annual summary of state of wildlife resources published since 1986.

Chapter 14/Nonrenewable Mineral Resources: Raw Materials from the Earth's Crust

Barnet, Richard J. 1980. *The Lean Years: Politics in an Age of Scarcity*. New York: Simon & Schuster. Superb discussion of the politics and economics of resource use and increasing scarcity.

Borgese, Elisabeth Mann. 1985. *The Mines of Neptune: Minerals and Metals from the Sea*. New York: Abrams. Excellent overview.

Chandler, William U. 1983. *Materials Recycling: The Virtue of Necessity*. Washington, D.C.: Worldwatch Institute. Superb overview.

Dorr, Ann. 1984. *Minerals—Foundations of Society*. Montgomery County, Md.: League of Women Voters of Montgomery County. Outstanding introduction to mineral resources.

Hamrin, Robert D. 1983. *A Renewable Resource Economy*. New York: Praeger Scientific. Excellent overview of resource principles, supplies, and conservation.

Huls, Jon, and Neil Seldman. 1985. *Waste to Wealth*. Washington, D.C.: Institute for Local Self-Reliance. Superb discussion of recycling and reuse.

Office of Technology Assessment. 1985. *Strategic Materials: Technologies to Reduce U.S. Import Vulnerability*. Washington, D.C.: Government Printing Office. Excellent analysis.

Simon, Julian L. 1981. *The Ultimate Resource*. Princeton, N.J.: Princeton University Press. Effective presentation of the cornucopian position.

U.S. Bureau of Mines. 1983. *The Domestic Supply of Critical Minerals*. Washington, D.C.: Government Printing Office. Useful source of data.

Westing, Arthur H. 1986. *Global Resources and International Conflict*. New York: Oxford University Press. Excellent analysis.

Chapter 15/Nonrenewable Energy Resources: Fossil Fuels

See also the readings for Chapter 3.

Brown, Lester R., et al. Annual. *State of the World*. New York: W. W. Norton. Excellent annual overviews of oil use and trends published since 1984.

Congressional Quarterly Editors. 1985. *Energy and Environment: The Unfinished Business*. Washington, D.C.: Congressional Quarterly. Excellent overview.

Edmonds, Jae, and John M. Reilly. 1985. *Global Energy: Assessing the Future*. New York: Oxford University Press. Excellent overview of energy supplies, alternatives, and policy over the next 50 to 75 years.

Flavin, Christopher. 1985. *World Oil: Coping with the Dangers of Success*. Washington, D.C.: Worldwatch Institute. Explains problems of temporary oil glut of the 1980s and suggests ways to avoid serious energy problems in the future.

Gates, David M. 1985. *Energy and Ecology*. Sunderland, Mass.: Sinauer. Excellent detailed analysis of all major energy alternatives.

Hirsch, Robert L. 1987. "Impending United States Energy Crisis." *Science*, vol. 235, 1467–1473. Excellent analysis of what we may face in the 1990s.

Holdren, John. 1982. "Energy Hazards: What to Measure, What to Compare." *Technology Review*, April, 32–38. Excellent guide to comparing risks and benefits of various energy technologies.

Hughes, Barry B., et al. 1985. *Energy in the Global Arena: Actors, Values, Policies, and Futures.* Durham, N.C.: Duke University Press. Outstanding overview and analysis of energy alternatives and policy.

Perry, Harry. 1983. "Coal in the United States: A Status Report." *Science,* vol. 222, no. 4622, 377–394. Excellent overview.

Shahinpoor, Mohsen. 1982. "Making Oil from Sand." *Technology Review,* February-March, pp. 49–54. Excellent summary of oil sands.

Chapter 16/Nonrenewable and Perpetual Energy Resources: Geothermal and Nuclear Energy

See also readings for Chapter 3.

Brown, Lester R., et al. Annual. *State of the World.* New York: W. W. Norton. Excellent annual overviews of economics of nuclear power (1984), decommissioning nuclear power plants (1986), and nuclear power safety (1987).

Cohen, Bernard L. 1983. *Before It's Too Late: A Scientist's Case for Nuclear Power.* New York: Plenum Press. Probably the best available case for nuclear power by an expert.

Flavin, Christopher. 1987. *Reassessing Nuclear Power: The Fallout from Chernobyl.* Washington, D.C.: Worldwatch Institute. Excellent analysis.

Ford, Daniel F. 1986. *Meltdown.* New York: Simon & Schuster. Excellent analysis of nuclear power plant safety by an expert.

Hippenheimer, T. A. 1984. *The Man-Made Sun: The Quest for Fusion Power.* Boston: Little, Brown. Excellent overview.

Hunt, Charles B. 1984. "Disposal of Radioactive Wastes." *Bulletin of the Atomic Scientists,* April, pp. 44–46. Summary by a prominent geologist of problems associated with geologic disposal of radioactive wastes.

Kaku, Michio, and Jennifer Trainer. 1982. *Nuclear Power: Both Sides.* New York: W. W. Norton. Useful collection of pro and con essays.

League of Women Voters Education Fund. 1982. *A Nuclear Power Primer: Issues for Citizens.* Washington, D.C.: League of Women Voters. Outstanding, readable, balanced summary.

League of Women Voters Education Fund. 1985. *The Nuclear Waste Primer.* Washington D.C.: League of Women Voters. Excellent readable, balanced summary.

Lidsky, Lawrence M. 1983. "The Trouble with Fusion." *Technology Review,* October, pp. 32–44. Superb evaluation by one of the world's most prominent nuclear fusion scientists.

Office of Technology Assessment. 1984. *Managing the Nation's Commercial High-Level Radioactive Waste.* Washington, D.C.: Government Printing Office. Excellent sourcebook.

Office of Technology Assessment. 1984. *Nuclear Power in an Age of Uncertainty.* Washington, D.C.: Government Printing Office. Excellent analysis of the future of commercial nuclear power in the United States.

Patterson, Walter C. 1984. *The Plutonium Business and the Spread of the Bomb.* San Francisco: Sierra Club Books. Superb overview.

Union of Concerned Scientists. 1985. *Safety Second: A Critical Evaluation of the NRC's First Decade.* Washington, D.C.: Union of Concerned Scientists. Excellent analysis.

Weinberg, Alvin M. 1985. *Continuing the Nuclear Dialogue.* La Grange Park, Ill.: American Nuclear Society. Thoughtful defense of nuclear power.

Chapter 17/Perpetual and Renewable Energy Resources: Conservation, Sun, Wind, Water, and Biomass

See also the readings for Chapter 3.

Brown, Lester R., et al. Annual. *State of the World.* New York: W. W. Norton. Various chapters in these annual reports published since 1984 give excellent summaries of progress in renewable energy resources and energy conservation.

Deudney, Daniel, and Christopher Flavin. 1983. *Renewable Energy: The Power to Choose.* 1983. New York: W. W. Norton. Superb overview of renewable energy resources.

Heede, H. Richard, et al. 1985. *The Hidden Costs of Energy.* Washington, D.C.: Center for Renewable Resources. Superb summary of federal subsidies provided for development of various energy alternatives.

Kash, Don E., and Robert W. Rycroft. 1984. *U.S. Energy Policy: Crisis and Complacency.* Norman: University of Oklahoma Press. Useful analysis.

Penny, Terry R., and Desikan Bharathan. 1987. "Power from the Sea." *Scientific American,* vol. 286, no. 1, 86–92. Excellent overview.

Pimentel, David, et al. 1984. "Environmental and Social Costs of Biomass Energy." *BioScience,* February, pp. 89–93. Excellent overview.

Pryde, Philip R. 1983. *Nonconventional Energy Resources.* New York: Wiley-Interscience. Excellent overview with emphasis on nonrenewable energy resources.

Purcell, Arthur. 1980. *The Waste Watchers: A Citizen's Handbook for Conserving Energy and Resources.* Garden City, N.Y.: Anchor Press/Doubleday. Superb guide.

Rosenbaum, Walter A.. 1987. *Energy, Politics, and Public Policy.* 2d ed. Washington, D.C.: Congressional Quarterly. Superb analysis.

Sawyer, Stephen W. 1986. *Renewable Energy: Progress, Prospects.* Washington, D.C.: Association of American Geographers. Outstanding evaluation.

Skelton, Luther W. 1984. *The Solar-Hydrogen Economy: Beyond the Age of Fire.* New York: Van Nostrand Reinhold. Excellent overview.

Swan, Christopher C. 1986. *Suncell: Energy, Economy, Photovoltaics.* New York: Random House. Outstanding overview of the potential of this emerging energy alternative.

Chapter 18/Air Pollution

Ember, Lois R., et al. 1986. "Tending the Global Commons." *Chemistry and Engineering News,* Nov. 24, pp. 14–64. Superb analysis of effects of human activities on global climate.

Halpern, Steven, and Louis Savary. 1985. *Sound Health.* New York: Harper & Row. Excellent overview of the effects of noise pollution and ways to minimize exposure to excessive noise.

Lovins, Amory B., et al. 1981. *Least-Cost Energy: Solving the CO_2 Problem.* Andover, Mass.: Brick House. Excellent summary of how the CO_2 problem could be minimized by a combination of energy conservation and a switch to solar, wind, hydro, biomass, and other forms of renewable energy.

McKormick, John. 1985. *Acid Earth: The Global Threat of Acid Pollution.* Washington, D.C.: Earthscan. Excellent analysis.

National Academy of Sciences. 1983. *Changing Climate.* Washington, D.C.: National Academy Press. Excellent overview of the CO_2 problem.

National Academy of Sciences. 1986. *Acid Deposition: Long-Term Trends*. Washington, D.C: National Academy Press. Excellent summary.

Office of Technology Assessment. 1985. *Acid Rain and Transported Air Pollutants: Implications for Public Policy*. New York: Unipub. Excellent analysis.

Postel, Sandra. 1984. *Air Pollution, Acid Rain, and the Future of Forests*. Washington, D.C.: Worldwatch Institute. Superb overview.

Postel, Sandra. 1986. *Altering the Earth's Chemistry: Assessing the Earth's Risks*. Washington, D.C.: Worldwatch Institute. Superb overview of effects of air pollution on ecosystems and human health.

Turiel, Isaac. 1985. *Indoor Air Quality and Human Health*. Stanford, Calif.: Stanford University Press. Superb nontechnical overview.

Chapter 19/Water Pollution

Ashworth, William. 1986. *The Late, Great Lakes: An Environmental History*. New York: Alfred A. Knopf. Excellent overview.

Borgese, Elisabeth Mann. 1986. *The Future of the Oceans*. New York: Harvest House. Excellent overview of ocean pollution and ocean resources.

Conservation Foundation. 1987. *Groundwater Pollution*. Washington, D.C.: Conservation Foundation. Excellent overview.

Goldstein, Jerome. 1977. *Sensible Sludge*. Emmaus, Pa.: Rodale Press. Useful overview of how to recycle sludge from waste treatment plants.

Gordon, Wendy. 1984. *A Citizen's Handbook on Groundwater Protection*. New York: Natural Resources Defense Council. Excellent source of information and ideas.

King, Jonathan. 1985. *Troubled Water*. Emmaus, Pa.: Rodale Press. Excellent overview of pollution of the country's drinking water supplies.

National Academy of Sciences. 1984. *Groundwater Contamination*. Washington, D.C.: National Academy Press. Excellent overview.

National Academy of Sciences. 1985. *Oil in the Sea: Inputs, Fates, and Effects*. Washington, D.C.: National Academy Press. Excellent overview and source of data.

Office of Technology Assessment. 1984. *Protecting the Nation's Groundwater from Contamination*. Washington, D.C.: Government Printing Office. Excellent analysis.

Simon, Anne W. 1985. *Neptune's Revenge: The Ocean of Tomorrow*. New York: Franklin Watts. Superb overview of stresses on the oceans and ways to protect them from excessive abuse.

Chapter 20/Solid Waste and Hazardous Waste

Brown, Lester R., et al. Annual. *State of the World*. New York: W. W. Norton. Excellent annual overviews.

Efron, Edith. 1984. *The Apocalyptics: Cancer and the Big Lie*. New York: Simon & Schuster. Contends that concern over hazardous wastes, pollutants, and other toxic and hazardous materials has been overblown and has little or no scientific foundation.

Environmental Defense Fund. 1985. *To Burn or Not to Burn*. New York: Environmental Defense Fund. Excellent analysis of incineration of garbage.

Epstein, Samuel S., et al. 1982. *Hazardous Waste in America*. San Francisco: Sierra Club Books. Superb analysis of the problems, along with suggested solutions.

Gibbs, Lois. 1982. *The Love Canal: My Story*. Albany: State University of New York Press. Useful description by a former Love Canal area resident who led the fight by homeowners to have the area condemned as unsafe.

Institute for Local Self-Reliance. 1986. *Environmental Review of Waste Incineration*. Washington, D.C.: Institute for Local Self-Reliance. Excellent analysis.

League of Women Voters Education Fund. 1981. *A Hazardous Waste Primer*. Washington, D.C.: League of Women Voters. Excellent overview.

Office of Technology Assessment. 1986. *Serious Reduction of Hazardous Waste*. Washington, D.C.: Government Printing Office. Excellent evaluation.

Pollack, Cynthia. 1987. *Mining Urban Wastes: The Potential for Recycling*. Washington, D.C.: Worldwatch Institute. Superb overview.

Robinson, William D., ed. 1986. *The Solid Waste Handbook*. New York: Wiley. Excellent source of information.

Schroeder, Henry A. 1974. *The Poisons Around Us: Toxic Metals in Food, Air, and Water*. Bloomington: Indiana University Press. Superb summary by an expert toxicologist.

Whelan, Elisabeth M. 1985. *Toxic Terror*. Ottawa, Ill.: Jameson Books. Argues that threats from hazardous wastes have been overblown and that hazardous wastes have been overregulated.

Chapter 21/Pesticides and Pest Control

Carson, Rachel. 1962. *Silent Spring*. Boston: Houghton Mifflin. An environmental classic that provided the first major warning about the dangerous side effects of pesticides.

Dover, Michael J. 1985. *A Better Mousetrap: Improving Pest Management for Agriculture*. Washington, D.C.: World Resources Institute. Excellent analysis.

Dunlap, Thomas R. 1981. *DDT: Scientists, Citizens, and Public Policy*. Princeton, N.J.: Princeton University Press. Excellent discussion of the history of the use of DDT and the problems that led to its banning in the United States.

Gough, Michael. 1986. *Dioxin, Agent Orange: The Facts*. New York: Plenum Press. Excellent source of information.

Graham, Frank, Jr. 1984. *The Dragon Hunters*. New York: E. P. Dutton. Excellent summary of biological control.

Hussey, N. W., and N. Scopes. 1986. *Biological Pest Control*. Ithaca, N.Y.: Cornell University Press. Excellent overview.

National Academy of Sciences. 1986. *Pesticide Resistance: Strategies and Tactics for Management*. Washington, D.C.: National Academy Press. Excellent analysis.

Shrivastava, Paul. 1987. *Bhopal: Anatomy of a Crisis*. New York: Harper & Row. Excellent source of information.

van den Bosch, Robert. 1978. *The Pesticide Conspiracy*. Garden City, N.Y.: Doubleday. Pest management expert exposes political influence of pesticide companies in preventing widespread use of biological controls and integrated pest management.

van den Bosch, Robert, and Mary L. Flint. 1981. *Introduction to Integrated Pest Management*. New York: Plenum Press. Superb presentation.

Glossary

Abiotic nonliving.

Abyssal zone bottom zone of the ocean consisting of dark, deep water. Compare *Bathyal zone, Benthic zone, Euphotic zone.*

Accelerated eutrophication See *Cultural eutrophication.*

Acid deposition combination of wet deposition from the atmosphere of droplets of sulfuric acid and nitric acid dissolved in rain, sleet, and snow (acid precipitation) and dry deposition from the atmosphere of particles of sulfate and nitrate salts. These acids and salts are formed when water vapor in the air reacts with the air pollutants sulfur dioxide (SO_2) and nitrogen dioxide (NO_2).

Acid solution any water solution that contains more hydrogen ions (H^+) than hydroxide ions (OH^-); any water solution with a pH less than 7. Compare *Basic solution, Neutral solution.*

Acid mine drainage dissolving and transporting of sulfuric acid and toxic metal compounds from abandoned underground coal mines to nearby streams and rivers when surface water flows through the mines.

Acid rain See *Acid deposition.*

Acidic See *Acid solution.*

Activated sludge process form of secondary sewage treatment in which wastewater is pumped into a large tank and mixed for several hours with bacteria-rich sludge and air bubbles to increase biodegradation of organic wastes by aerobic bacteria.

Active solar heating system that uses solar collectors to capture energy from the sun as heat and then uses mechanical devices such as pumps and fans to move the captured heat to a storage system (usually an insulated tank of water or bed of rocks) or throughout a dwelling. Compare *Passive solar heating.*

Advanced industrial society highly industrialized society based on extensive use of fossil fuels and efficient machines and techniques for mass production of goods.

Aerobic organism organism that requires oxygen to live. Compare *Anaerobic organism.*

Age structure number or percentage of persons of each sex at each age level or age group (cohort) in a population.

Agriculture-based urban society a society consisting of villages, towns, and cities that produce specialized goods and services which are traded for food and fibers produced by farmers in nearby agricultural land.

Air pollutant chemical whose concentration in the atmosphere builds up to the point of causing harm to human beings, other animals, vegetation, or materials such as metals and stone.

Algae simple one-celled or many-celled plants, usually aquatic, capable of carrying on photosynthesis.

Algal bloom population explosion of algae in surface waters resulting from an increase in plant nutrients such as nitrates and phosphates.

Alkaline solution See *Basic solution.*

Alpha particle positively charged chunks of ionizing radiation consisting of two protons and two neutrons that are spontaneously emitted by the nuclei of some radioisotopes.

Alveoli tiny sacs at the end of the bronchiole tubes in the lungs where oxygen is inhaled, air is transferred to hemoglobin in the blood, and carbon dioxide in the blood is transferred to inhaled air and exhaled.

Ambient air surrounding outdoor air.

Anaerobic organism organism that does not require oxygen to survive. Compare *Aerobic organism.*

Animal feedlot confined area where hundreds or thousands of livestock animals are fattened for sale to slaughterhouses and meat processors.

Animal manure dung (fecal matter) and urine of animals.

Annual population change rate See *Natural change rate.*

Anthracite hard coal with a low to moderate sulfur content, a very low moisture content (2%), and a high heat content. Compare *Bituminous coal, Lignite, Peat, Subbituminous coal.*

Anthropocentric human-centered. Compare *Biocentric.*

Appropriate technology technology that is usually characterized by simple, easy-to-repair, small- to medium-size machines that are used on a decentralized basis, are inexpensive to build and maintain, and that utilize locally available materials and labor.

Aquaculture growing and harvesting of fish and shellfish for human use in freshwater ponds, irrigation ditches, and lakes or fenced-in portions of coastal lagoons and estuaries.

Aquatic pertaining to water. Compare *Terrestrial.*

Aquatic ecosystem any major ecosystem such as a river, pond, lake, and ocean found in the hydrosphere. Compare *Biome.*

Aquifer water-bearing layer of the earth's crust; water in an aquifer is known as groundwater. See *Confined aquifer, Unconfined aquifer.*

Aquifer depletion withdrawal of groundwater from an aquifer faster than it is recharged by precipitation.

Aquifer overdraft See *Aquifer depletion.*

Arable land land that is capable of being cultivated and supporting agricultural production.

Area strip-mining type of surface mining in which minerals such as coal and phosphate are removed by cutting deep trenches in flat or rolling terrain. Compare *Contour strip-mining, Open-pit surface mining.*

Arid dry, parched with heat.

Artesian aquifer   See *Confined aquifer.*

Artesian well   a water well drilled into a pressurized confined aquifer in which the hydraulic pressure is so great that the water flows freely up the bore hole to the level of the water in the earth without any need for pumping.

Atmosphere   layer of air surrounding the earth's surface.

Atoms   extremely small particles that are the basic building blocks of all elements and thus all matter.

Average per capita GNP   the gross national product (GNP) of a country divided by its total population.

Background ionizing radiation   ionizing radiation in the environment from naturally radioactive materials and from cosmic rays entering the atmosphere.

Bacteria   smallest living organisms; with fungi, they comprise the decomposer level of food chains and webs.

Balanced chemical equation   shorthand representation of a chemical change that contains the same number of atoms of each element on each side of the equation.

Barrier beach   gently sloping land along a coastline that normally contains two rows of sand dunes that help protect the land behind them from ravages of the sea. Compare *Rocky shore.*

Barrier islands   thin, sandy islands off the coast, separated from the mainland by bays or lagoons.

Basic   See *Basic solution.*

Basic solution   any water solution containing more hydroxide ions (OH^-) than hydrogen ions (H^+); any water solution with a pH greater than 7. Compare *Acid solution, Neutral solution.*

Bathyal zone   cold, fairly dark zone in an ocean below the euphotic zone in which there is some penetration by sunlight but not enough for photosynthesis. Compare *Abyssal zone, Euphotic zone.*

Benthic zone   bottom of a body of water. Compare *Abyssal zone, Bathyal zone, Euphotic zone, Limnetic zone, Littoral zone.*

Beta particle   swiftly moving electron emitted by the nucleus of a radioisotope.

Biocentric   ecosystem-centered. Compare *Anthropocentric.*

Biodegradable   material that can be broken down into simpler substances (elements and compounds) by bacteria or other decomposers.

Biofuels   gas or liquid fuels (such as ethyl alcohol) made from biomass (plants and trees).

Biogas   mixture of methane (CH_4) and carbon dioxide (CO_2) gases produced when anaerobic bacteria break down plants and organic waste (such as manure).

Biogeochemical cycle   mechanism by which chemicals such as carbon, oxygen, phosphorus, nitrogen, and water are continuously moved through the biosphere to be renewed again and again for use by living organisms.

Biological amplification   increase in concentration of certain fat-soluble chemicals such as DDT in successively higher trophic levels of a food chain or web.

Biological control   use of natural predators, parasites, or disease-causing bacteria and viruses (pathogens) to regulate the population of a pest species.

Biological magnification   See *Biological amplification.*

Biological oxygen demand (BOD)   the amount of dissolved oxygen gas required for bacterial decomposition of organic wastes in water; usually expressed in terms of the parts per million (ppm) of dissolved oxygen consumed over 5 days at 20°C (68°F) and normal atmospheric pressure. See also *Chemical oxygen demand.*

Biomass   total dry weight of all living organisms in a given area. See also *Biomass fuel.*

Biomass fuel   plant and animal matter such as wood that can be burned directly as a source of heat or converted to a more convenient gaseous or liquid biofuel.

Biome   large land ecosystem such as a forest, grassland, or desert. Compare *Aquatic ecosystem.*

Biosphere   total of all the ecosystems on the planet, along with their interactions; parts of the lithosphere, atmosphere, and hydrosphere in which living organisms can be found.

Biotic potential   the maximum rate at which members of a species can reproduce, given unlimited resources and ideal environmental conditions. Compare *Environmental resistance.*

Birth rate   number of live births per 1,000 persons in the population at the midpoint of a given year. Compare *Death rate.*

Bitumen   black, high-sulfur, tarlike heavy oil extracted from tar sands and then upgraded to synthetic fuel oil. See *Tar sands.*

Bituminous coal   soft form of coal with a moderate to high heat content and usu-ally a high sulfur content (2% to 4%). Compare *Anthracite, Lignite, Peat, Subbituminous coal.*

Black lung disease   respiratory disorder that impairs breathing capacity as a result of the accumulation of fine particles of coal dust and other particulate matter in the lungs of coal miners over a prolonged period.

Breeder nuclear fission reactor   nuclear fission reactor that produces more nuclear fuel than it consumes, usually by converting nonfissionable uranium-238 into fissionable plutonium-239.

Bronchitis   See *Chronic bronchitis.*

Bronchiole tubes   tiny ducts entering the lungs that are subdivisions of the bronchus tubes and eventually lead to the alveoli.

Bronchus   either of the two main branches of the trachea, or windpipe, which enter the lungs.

Calorie   amount of energy required to raise the temperature of 1 gram of water 1°C. See also *Kilocalorie.*

Cancer   a group of more than 120 different diseases—one for essentially each major cell type in the human body—all characterized by a tumor in which cells multiply uncontrollably and invade surrounding tissue.

Carbon cycle   cyclic flow of carbon in various chemical forms through various components of the biosphere.

Carcinogen   a chemical or physical agent (such as ionizing radiation) capable of causing cancer.

Carcinogenic   cancer causing.

Carnivore   animal that obtains its food by feeding only on other animals. Compare *Herbivore, Omnivore.*

Carrying capacity   maximum population size of a species that a given ecosystem or area can support indefinitely under a given set of environmental conditions.

Cartel   group of companies or producers of a resource or product who agree to control production so as to keep the price of the resource or product high in order to enhance profits. OPEC is an example of a cartel.

Cell   basic structural unit of all organisms.

Cellular respiration   complex process that occurs in the cells of plants and animals in which food molecules such as glucose ($C_6H_{12}O_6$) combine with oxygen (O_2)

and break down into carbon dioxide (CO_2) and water (H_2O), releasing usable energy. Compare *Photosynthesis*.

Census a count of the population.

Chain reaction series of multiple nuclear fissions taking place within the critical mass of a fissionable isotope and resulting in the emission of an enormous amount of energy.

Chemical any one of the millions of different elements and compounds found in the universe.

Chemical change process in which one or more elements or compounds interact in a way that changes them into one or more different elements or compounds. For example, the element carbon (C) can combine with oxygen (O_2) to form the compound carbon dioxide (CO_2). Compare *Nuclear change, Physical change*.

Chemical cycle See *Biogeochemical cycle*.

Chemical energy potential energy stored in the chemical bonds that hold together the atoms or ions in chemical compounds.

Chemical equation shorthand representation of a chemical change in which symbols are used to represent each element involved. See *Balanced chemical equation*.

Chemical oxygen demand (COD) the measure of the total depletion of dissolved oxygen in polluted water. See also *Biological oxygen demand*.

Chemical reaction See *Chemical change*.

Chemosynthesis process in which certain organisms such as specialized bacteria can convert chemicals obtained from the environment into chemical energy stored in nutrient molecules without the presence of sunlight. Compare *Photosynthesis*.

Chlorinated hydrocarbon insecticides class of synthetic, organic compounds containing chlorine, hydrogen, and carbon that are persistent and fat-soluble and can be biologically amplified in food chains and webs. Examples include DDT, aldrin, and dieldrin.

Chlorination addition of chlorine to drinking water or treated sewage plant effluent to kill germs (disinfection).

Chlorofluorocarbons (CFCs) organic molecules containing varying numbers of chlorine, fluorine, carbon, and hydrogen atoms.

Chronic bronchitis lung disorder characterized by persistent inflammation of the bronchi, excessive mucus buildup, recurrent coughing, and throat irritation.

It appears to be caused and aggravated by smoking and air pollution.

Clay soil low-porosity soil with a high clay content and little, if any, silt and sand.

Clearcutting method of timber harvesting in which all trees in a forested area are removed.

Climate average of day-to-day weather conditions at a given place on earth over a fairly long period, usually 30 years or more. Also includes extremes in weather behavior during the same period.

Climax ecosystem (climax community) a relatively stable, self-sustaining stage of ecological succession; a mature ecosystem with a diverse array of species and ecological niches, capable of using energy and cycling critical chemicals more efficiently than simpler, immature ecosystems.

Coal a solid, combustible material usually containing from 40% to 98% carbon mixed with varying amounts of water (2% to 50%) and small amounts of nitrogen (0.2% to 1.2%) and sulfur (0.6% to 4%) compounds. See *Anthracite, Bituminous coal, Lignite, Peat, Subbituminous coal*.

Coal gasification process in which solid coal is converted to either low heat-content industrial gas or high heat-content synthetic natural gas (SNG).

Coal liquefaction process in which solid coal is converted to liquid hydrocarbon fuel such as methanol or synthetic gasoline.

Coastal wetland land along a coastline that remains flooded with salt water all or part of the year. Compare *Inland wetland*.

Coastal zone See *Neritic zone*.

Cogeneration the production of two useful forms of energy from the same process. In a factory, for example, excess steam produced for industrial processes or space heating is run through turbines to generate electricity, which can be used by the industry or sold to power companies.

Coliform bacteria a normally harmless type of bacteria that resides in the intestinal tract of human beings and other animals and whose presence in water is an indicator that the water may be contaminated with other disease-causing organisms found in untreated human and animal waste.

Coliform bacteria count number of colonies of fecal coliform bacteria present in a 100-milliliter sample of water.

Combustion burning. Any very rapid chemical reaction in which heat and light are produced.

Commensalism interaction between two species in which one species benefits from the association while the other is apparently neither helped nor harmed.

Commercial extinction depletion of the population of a species to a point where it is no longer economical to harvest the species.

Commercial hunting killing of wild animals for profit from sale of their furs or other parts. Compare *Sport hunting, Subsistence hunting*.

Commercial inorganic fertilizer commercially prepared mixtures of plant nutrients such as nitrates, phosphates, and potassium applied to the soil to restore fertility and increase crop yields. Compare *Organic fertilizer*.

Common property resource a resource such as the air, oceans, sunshine, or public land to which a population has free and unmanaged access and thus can be abused. See *Tragedy of the commons*.

Commons See *Common property resource*.

Community (natural) all the populations of plant and animal species living and interacting in a given habitat or area at a given time.

Competition two or more individual organisms of a single species (intraspecific competition) or two or more individuals of different species (interspecific competition) in the same ecosystem attempting to use the same scarce resources.

Competitive exclusion principle no two species in the same ecosystem can occupy exactly the same ecological niche indefinitely.

Compost See *Composting*.

Composting accelerated breakdown of grass clippings, leaves, paper, and other organic solid waste in the presence of oxygen by aerobic (oxygen-needing) bacteria to produce a humuslike product that can be used as a fertilizer or soil conditioner.

Compound substance composed of two or more atoms (molecular compounds) or oppositely charged ions (ionic compounds) of two or more different elements held together in fixed proportions by chemical bonds. Compare *Element*.

Concentration amount of a chemical in a given volume of air, water, or other medium.

Confined aquifer deposit of groundwater sandwiched between two layers of relatively impermeable rock, such as clay or shale. Compare *Unconfined aquifer*.

Coniferous trees cone-bearing trees, mostly evergreens that have needle-shaped or scalelike leaves.

Conservation wise use and careful management of resources, so as to obtain the maximum possible social benefits from them for present and future generations. Methods include preservation, balanced multiple use, reducing unnecessary waste, recycling, reuse, and decreased resource use.

Conservationists people who express their concern for the present and future survival of human beings and other species by not wasting and not irreversibly depleting or degrading the biological, physical, and chemical wealth of the world on which all life depends.

Conservation-tillage farming method of cultivation in which the soil is disturbed very little (minimum-tillage farming) or not at all (no-till farming) in order to reduce soil erosion, decrease labor costs, and save energy.

Consumer organism that relies on other organisms for its food. Generally divided into primary consumers (herbivores), secondary consumers (carnivores), and microconsumers (decomposers).

Consumption overpopulation situation in the world or a given country or region in which a relatively small number of people use resources at such a high rate and without sufficient pollution and land-use controls that the resulting pollution, environmental degradation, and resource depletion can threaten the health and survival of human beings and other species and disrupt the natural processes that cleanse and replenish the air, water, and soil. Compare *People overpopulation*.

Consumption water use water use that results in water being lost by evaporation or transpiration or degraded by pollution so that it is no longer available for reuse in a particular area. Compare *Withdrawal water use*.

Continental shelf submerged sea floor that slopes gradually from the exposed edge or shore of a continent for a variable distance to a point where a much steeper descent to the ocean bottom begins.

Contour farming plowing and planting along rather than up and down the sloped contours of land to reduce soil erosion and conserve water.

Contour strip-mining form of surface mining carried out in hilly or mountainous terrain by cutting out a series of shelves or terraces on the side of a hill or mountain and dumping the overburden from each new terrace onto the one below. Used primarily for coal. Compare *Area strip-mining, Open-pit surface mining*.

Contraceptive any physical, chemical, or biological method used to prevent fertilization of the human ovum by a male sperm.

Control rod neutron-absorbing rods that are raised or lowered in the core of a nuclear reactor to control the rate of nuclear fission.

Coral reef shallow area near the coast of a warm tropical or subtropical ocean consisting of calcium-containing material secreted by photosynthesizing red and green algae and small coral animals.

Core (of the earth) central or innermost portion of the earth. Compare *Crust*.

Cornucopians people, mostly economists, who believe that if present trends continue, economic growth and technological advances based on human ingenuity will produce a less crowded, less polluted world in which most people will be healthier and live longer and will have greater material wealth. Compare *Neo-Malthusians*.

Cosmic rays streams of highly penetrating charged particles composed of protons, alpha particles, and a few heavier nuclei that bombard the earth from outer space.

Cost-benefit analysis technique used to estimate and compare the expected costs or losses associated with a particular project or degree of pollution control with the expected benefits or gains over a given period of time.

Cost-effectiveness analysis technique used to determine how a particular goal, such as pollution control, can be achieved for the least cost.

Critical mass the quantity of fissionable material needed to initiate and maintain a nuclear fission chain reaction.

Crop rotation farming practice that involves planting the same field with a different series of crops from year to year in order to prevent plant nutrient depletion.

Crown fire intensely hot fire that can destroy all or most of the forest vegetation, kill wildlife, and accelerate erosion. Compare *Ground fire*.

Crude birth rate See *Birth rate*.

Crude death rate See *Death rate*.

Crude oil a gooey liquid mixture of hydrocarbon compounds (90% to 95% of its weight) and small quantities of compounds containing oxygen, sulfur, and nitrogen, which can be extracted from underground deposits and then sent to refineries to be converted to useful materials such as heating oil, diesel fuel, gasoline, and tar.

Crust (of the earth) solid, outer layer of the earth. Compare *Core*.

Cultural eutrophication overnourishment of aquatic ecosystems with plant nutrients resulting from human activities such as agriculture, urbanization, and industrial discharge. See *Eutrophication*.

DDT *d*ichloro*d*iphenyl*t*richloroethane, a chlorinated hydrocarbon that has been widely used as a pesticide.

Death rate number of deaths per 1,000 persons in the population at the midpoint of a given year. Compare *Birth rate*.

Decibel (db) unit used to measure sound power or sound pressure.

Deciduous plants plants such as oak and maple that lose all their leaves during part of the year.

Decomposers organisms such as bacteria, mushrooms, and fungi that obtain nutrients by breaking down complex matter in the wastes and dead bodies of other organisms into simpler chemicals, most of which are returned to the soil and water for reuse by producers.

Deforestation removal of trees from an area without adequate replanting.

Degradable See *Biodegradable*. Compare *Nonbiodegradable pollutant*.

Degree of urbanization percentage of a country's population living in areas with a population of 2,500 or more.

Delta built-up deposit of river-borne sediments found near the mouth of a river near the ocean.

Demographic transition the gradual change, supposedly brought about by economic development, from a condition of high birth and death rates to substantially lower birth and death rates for a given country or region.

Demography study of the characteristics and changes of the human population in a particular area.

Dependency load ratio of the number of old and young dependents in a population to the work force.

Depletion time period required to use up a certain fraction—usually 80%—of the known reserves or estimated resources of a mineral at an assumed rate of use.

Desalinization purification of salt or brackish water by removing the dissolved salts.

Desert biome characterized by very low average annual precipitation (less than 25 centimeters, or 10 inches, a year) and sparse, mostly low vegetation.

Desertification conversion of productive grassland, cropland, or forest into desert, usually through a combination of overgrazing, prolonged drought, and climate change.

Desirability quotient number used to determine the desirability of using a particular technology; it is obtained by dividing the estimated short- and long-term benefits of using the technology by its estimated short- and long-term risks. See *Risk-benefit analysis*.

Detritus dead plant material, bodies of animals, and fecal matter.

Detritus consumer organism that feeds on dead plant or animal matter.

Detritus feeder consumers such as vultures, termites, ants, and crayfish that directly feed on dead plant or animal matter. Compare *Decomposers*.

Deuterium (D: hydrogen-2) isotope of the element hydrogen with a nucleus containing one proton and one neutron, thus having a mass number of 2. Compare *Tritium*; see also *Heavy water*.

Developed country See *More developed country*.

Diminishing returns situation in which increased inputs of money, fertilizer, water, energy, or other factors do not lead to increased outputs such as crop productivity.

Dioxins family of at least 75 different highly toxic chlorinated hydrocarbon compounds.

Discount factor measure of how much something may be worth in the future compared to what it is worth now.

Dissolved oxygen (DO) content amount of oxygen gas (O_2) dissolved in a given quantity of water at a given temperature and atmospheric pressure. It is usually expressed in parts per million (ppm).

Diversity physical or biological complexity of a system. Usually a measure of the number of different species in an ecosystem (species diversity).

DNA (*deoxyribonucleic acid*) large molecules found in the cells that carry genetic information.

Doubling time length of time (usually years) it takes for a population to double in size if present annual population growth continues unchanged.

Drainage basin See *Watershed*.

Dredge spoils materials scraped from the bottoms of harbors and rivers to maintain shipping channels.

Dredging surface mining of seabeds and streambeds, primarily for sand and gravel.

Drip irrigation method of irrigation in which small pipes deliver water to plant roots.

Drought prolonged period of dry weather.

Dry farming cultivation of agricultural crops without the use of irrigation.

Early industrial society society in which there is increasing use of inventions such as the coal-burning steam engine, the steam locomotive, the internal combustion engine, and other machines to replace dependence on draft animals and human muscle power for carrying out most tasks. Compare *Advanced industrial society*.

Early-successional species wild animal species found in pioneer communities of plants at the early stage of ecological succession. Compare *Late-successional species, Mid-successional species, Wilderness species*.

Ecological efficiency (food chain efficiency) the percent transfer of useful energy from one trophic level to the next higher trophic level in a food chain.

Ecological equivalents species that occupy the same or similar ecological niches in similar ecosystems located in different parts of the world. For example, cattle in North America and kangaroos in Australia are both grassland grazers; hence, they are ecological equivalents.

Ecological land-use planning method for deciding present and future use for parcels of land in which all major biological and social variables are considered.

Ecological niche description of all the physical, chemical, and biological factors that a species needs to survive, stay healthy, and reproduce in an ecosystem.

Ecological succession process in which communities of plant and animal species are replaced in a particular area over time by a series of different and usually more complex communities. See *Primary succession, Secondary succession*.

Ecology study of the interactions of living organisms with each other and with their environment; study of the structure and function of nature.

Economic resources See *Reserves*.

Ecosphere See *Biosphere*.

Ecosystem self-regulating natural community of plants and animals interacting with one another and with their nonliving environment.

Ectoparasite parasite that lives outside its host organism. Compare *Endoparasite*.

Efficiency See *Ecological efficiency, Energy efficiency*.

Effluent any substance, particularly a liquid, that enters the environment from a point source. Generally refers to wastewater from a sewage treatment or industrial plant.

Electromagnetic radiation radiant energy that can move through a vacuum or through space as waves of oscillating electric and magnetic fields.

Electromagnetic spectrum span of electromagnetic energy ranging from short-wavelength gamma waves to long-wavelength radio waves.

Electron fundamental particle found moving around outside the nucleus of an atom. Each electron has one unit of negative charge and has extremely little mass.

Electrostatic precipitator device for removing particulate matter from smokestack emissions by causing the particles to become electrostatically charged and then attracting them to an oppositely charged plate, where they are removed from the air.

Element chemical such as iron (Fe), sodium (Na), carbon (C), nitrogen (N), or oxygen (O) whose distinctly different atoms serve as the basic building blocks of all matter. Compare *Compound*.

Emergency core cooling system system designed to prevent meltdown if the core of a nuclear reactor overheats by instantaneous flooding of the core with large amounts of water.

Emigration movement of people out of one country to take up permanent residence in another. Compare *Immigration*.

Emigration rate number of people migrating out of a country each year per 1,000 people in its population. Compare *Immigration rate, Net migration rate*.

Emission discharge of one or more gases or liquids into the environment.

Emission standard maximum amount of a pollutant that is permitted by the federal government to be discharged into the air or a body of water from a point source.

Emphysema lung disease in which the alveoli enlarge, fuse together, and lose their elasticity, thus impairing the transfer of oxygen to the blood.

Endangered species a wild species having so few individual survivors that it could soon become extinct in all or most of its natural range. Compare *Threatened species*.

Endoparasite parasite that lives inside its host organism. Compare *Ectoparasite*.

Energy   ability to do work or produce a change by pushing or pulling some form of matter or to cause a heat transfer between two objects at different temperatures.

Energy conservation   reduction or elimination of unnecessary energy use and waste.

Energy crisis   a shortage or a catastrophic price increase for one or more forms of useful energy, or a situation in which energy use is so great that the resulting pollution and environmental degradation threaten human health and welfare.

Energy efficiency   the percentage of the total energy input that does useful work and is not converted into low-quality, essentially useless, low-temperature heat in an energy conversion system or process.

Energy flow pyramid   diagram representing the loss or degradation of useful energy at each step in a food chain. About 80% to 90% of the energy in each transfer is lost as waste heat, and the resulting shape of the energy levels is pyramidal.

Energy quality   ability of a form of energy to do useful work. See *High-quality energy, Low-quality energy*.

Enhanced oil recovery   removal of some of the heavy oil remaining in an oil well after primary and secondary recovery by methods such as pumping in steam or igniting the oil to increase its flow rate so that it can be pumped to the surface. Compare *Primary oil recovery, Secondary oil recovery*.

Entropy   a measure of randomness or disorder. See *Second law of thermodynamics*.

Environment   all of the external conditions that affect an organism or other specified system during its lifetime; everything outside of a specified system.

Environmental degradation   depletion or destruction of some renewable resource by using it at a faster rate than it is naturally replenished. See also *Sustained yield*.

Environmental resistance   all the limiting factors that act together to regulate the maximum allowable size, or carrying capacity, of a population.

Epilimnion   upper layer of warm water with high levels of dissolved oxygen in a stratified lake. Compare *Hypolimnion, Thermocline*.

Erosion   removal of soil by flowing water or wind.

Estuarine zone   area near the coastline that consists of estuaries and coastal saltwater wetlands and that extends out to the edge of the continental shelf.

Estuary   thin zone along a coastline where fresh water from rivers mixes with salty ocean water.

Euphotic zone   surface layer of an ocean, lake, or other body of water through which there is sufficient sunlight for photosynthesis. Compare *Abyssal zone, Bathyal zone*.

Eutrophic lake   lake with a large or excessive supply of plant nutrients (mostly nitrates and phosphates). Compare *Mesotrophic lake, Oligotrophic lake*.

Eutrophication   natural process in which lakes receive inputs of plant nutrients (mostly nitrates and phosphates) as a result of natural erosion and runoff from the surrounding land basin. See also *Cultural eutrophication*.

Evaporation   change of a liquid into vapor.

Evapotranspiration   combination of evaporation and transpiration of liquid water in plant tissue and in the soil to water vapor in the atmosphere.

Evergreen plants   plants such as pines, spruces, and firs that retain some of their leaves or needles throughout the year. Compare *Deciduous plants*.

Evolution   the process by which a population of a species changes its characteristics (genetic makeup) over time in response to changes in environmental conditions. See *Natural selection*.

Exponential growth   growth in which some quantity, such as population size, increases by a constant percentage of the whole during each year or other time period; yields a J-shaped curve.

Extinction   complete disappearance of an entire species.

Fallout   See *Radioactive fallout*.

Family planning   provision of information and contraceptives to help couples choose the number of children and length of intervals of time between children they choose to have.

Famine   situation in which people in a particular area suffer from widespread lack of access to enough food for good health as a result of catastrophic events such as drought, flood, earthquake, or war.

Fertility   the average number of live babies born to women in the population during their normal childbearing years (ages 15–44).

Fertilizer   substance that makes the land or soil capable of producing more vegetation or crops. See *Commercial inorganic fertilizer, Organic fertilizer*.

First law of ecology   when human beings interfere with or modify an ecosystem, there are always numerous short- and long-term effects, many of which are unpredictable.

First law of energy   See *First law of thermodynamics*.

First law of thermodynamics (energy)   in any chemical or physical change, movement of matter from one place to another, or change in temperature, energy is neither created nor destroyed, but merely converted from one form to another. In terms of energy quantity, you can't get something for nothing; you can only break even; there is no free lunch.

Fissionable isotopes   isotopes that are capable of undergoing nuclear fission.

Floodplain   land along a river that is subject to periodic flooding when the river overflows its banks.

Fluidized-bed combustion   process for burning coal more efficiently, cleanly, and cheaply by using a flowing stream of hot air to suspend a mixture of powdered coal and limestone during combustion. About 90% to 98% of the sulfur dioxide produced during combustion is removed by reaction with limestone to produce solid calcium sulfate.

Fly ash   small, solid particles of ash and soot generated when coal, oil, or waste materials are burned.

Food additive   chemical deliberately added to a food, usually to enhance its color, flavor, shelf life, or nutritional characteristics.

Food chain   sequence of transfers of energy in the form of food from organisms in one trophic level to organisms in another trophic level when one organism eats or decomposes another.

Food web   complex network of many interconnected food chains and feeding interactions.

Forest   region with sufficient average annual precipitation of 75 centimeters (30 inches) or more to support various species of trees and smaller forms of vegetation.

Fossil fuel   buried deposits of decayed plants and animals that have been converted to crude oil, coal, natural gas, or heavy oils by exposure to heat and pressure in the earth's crust over hundreds of millions of years.

Freons   See *Chlorofluorocarbons*.

Freshwater fish management   methods used to encourage the growth of populations of desirable commercial and sport freshwater fish species and to reduce or eliminate populations of less desirable species.

Fungicide substance or mixture of substances used to prevent or kill fungi.

Fungus simple or complex organism without chlorophyll. The simpler forms are unicellular; the higher forms have branched filaments and complicated life cycles. Examples are molds, yeasts, and mushrooms.

Game species wildlife animal resources that provide sport for people in the form of hunting and fishing.

Gamma rays high-energy electromagnetic waves emitted by the nuclei of some radioisotopes.

Gasohol vehicle fuel consisting of a mixture of gasoline and ethyl or methyl alcohol that typically contains 10% to 23% by volume alcohol.

Gene pool total genetic information possessed by a given reproducing population.

Genetic adaptation changes in the genetic makeup of organisms of a species that allows the species to reproduce and gain a competitive advantage under changed environmental conditions.

Genetic damage damage by radiation or chemicals to reproductive cells, resulting in mutations that can be passed on to future generations in the form of fetal and infant deaths and physical and mental disabilities.

Geometric growth See *Exponential growth.*

Geothermal energy heat transferred from the earth's intensely hot molten core to underground deposits of dry steam (steam with no water droplets), wet steam (a mixture of steam and water droplets), hot water, or rocks lying relatively close to the earth's surface.

Global net population change difference between the total number of live births and the total number of deaths throughout the world during a given period (usually a year).

GNP See *Gross national product.*

GRAS list list of food additives used in the United States that are *generally recognized as safe.*

Grassland biome found in regions where moderate average precipitation, ranging from 25 to 75 centimeters (10 to 30 inches) a year, is enough to allow grass to prosper but not enough to support large stands of trees.

Greenhouse effect trapping of heat in the atmosphere. Incoming short-wavelength solar radiation penetrates the atmosphere, but the longer-wavelength outgoing radiation is absorbed by water vapor, carbon dioxide, ozone, and several other gases in the atmosphere and is reradiated to earth, causing an increase in atmospheric temperature.

Greenhouse gases gases present in the earth's atmosphere that cause the greenhouse effect.

Green manure fresh or still-growing green vegetation plowed into the soil to increase the organic matter and humus available to support crop growth. Compare *Animal manure.*

Green revolution popular term for the introduction of scientifically bred or selected varieties of a grain (rice, wheat, maize) that with high enough inputs of fertilizer and water can give greatly increased yields per area of land planted.

Gross national product (GNP) total market value of all goods and services produced per year in a country.

Ground fire low-level fire that typically burns only undergrowth. Compare *Crown fire.*

Groundwater water that sinks into the soil, where it may be stored for long times in slowly flowing and slowly renewed underground reservoirs known as aquifers. See *Confined aquifer, Unconfined aquifer.*

Groundwater contamination dissolving of substances at harmful levels in groundwater primarily as a result of human activities.

Growth rate (population) percentage of increase or decrease of a population. It is the number of births minus the number of deaths per 1,000 population, plus net migration, expressed as a percentage.

Gully reclamation using small dams of manure and straw, earth, stone, or concrete to collect silt and gradually fill in channels of eroded soil.

Habitat place or type of place where an organism or community of organisms naturally or normally thrives.

Half-life length of time taken for half the atoms in a given amount of a radioactive substance to emit one or more forms of ionizing radiation and, in the process, change into another nonradioactive or radioactive isotope.

Hazard something that can cause injury, disease, death, economic loss, or environmental deterioration.

Hazardous waste discarded solid, liquid, or gaseous material that may pose a substantial threat to human health or the environment when managed improperly.

Heat form of kinetic energy that flows from one body to another as a result of a temperature difference between the two bodies.

Heavy oil black, high-sulfur, thick oil found in deposits of crude oil, tar sands, and oil shale. See *Enhanced oil recovery.*

Heavy water water (D_2O) in which all the hydrogen atoms have been replaced by deuterium (D).

Herbicide chemical that injures or kills plant life by interfering with normal growth.

Herbivore organism that feeds on plants. Compare *Carnivore, Omnivore.*

High-grade ore an ore that contains a relatively large concentration of a desired metallic element. Compare *Low-grade ore.*

High-quality energy energy that is concentrated and has great ability to perform useful work. Examples include high-temperature heat and the energy in electricity, coal, oil, gasoline, sunlight, and nuclei of uranium-235. Compare *Low-quality energy.*

Host plant or animal fed on by a parasite.

Humus complex mixture of partially decomposed, water-insoluble material found in the topsoil layer; it helps retain water and water-soluble nutrients so they can be taken up by plant roots.

Hunters and gatherers people who obtain their food by gathering edible wild plants and other materials and by hunting wild game and fish from the nearby environment.

Hydrocarbons class of organic compounds containing carbon (C) and hydrogen (H).

Hydroelectric plant electric power plant in which the energy of falling water is used to spin a turbine generator to produce electricity.

Hydrologic cycle biogeochemical cycle that moves and recycles water in various forms through the biosphere.

Hydropower electrical energy produced by falling water.

Hydrosphere region that includes all the earth's moisture as liquid water (oceans, smaller bodies of fresh water, and underground aquifers), frozen water (polar ice caps, floating ice, and frozen upper layer of soil known as permafrost), and small amounts of water vapor in the earth's atmosphere.

Hypolimnion bottom layer of cold, more dense water in a lake. Compare *Epilimnion, Thermocline.*

Identified resources specific bodies of a particular mineral-bearing material whose location, quantity, and quality are known or have been inferred from geologic evidence and measurements. Compare *Reserves, Resources,* and *Undiscovered resources.*

Immigration process of entering one country from another to take up permanent residence. Compare *Emigration.*

Immigration rate number of people migrating into a country each year per 1,000 people in its population. Compare *Emigration rate, Net migration rate.*

Incineration controlled process by which combustible solid or liquid wastes are burned and changed into gases.

Industrialized agriculture supplementation of solar energy with large amounts of energy derived from fossil fuels (especially oil and natural gas) to produce large quantities of crops and livestock for sale within the country where it is grown and to other countries. Compare *Subsistence agriculture.*

Industrial smog air pollution, primarily from sulfur dioxide and suspended particulate matter, produced by the burning of coal and oil by industries and in power plants. Compare *Photochemical smog.*

Inertia stability ability of a living system to resist being disturbed or altered. See also *Resilience stability.*

Infant mortality rate number of deaths of infants under 1 year of age in a given year per 1,000 live births in the same year.

Inland wetland land such as a swamp, marsh, or bog found inland that remains flooded all or part of the year with fresh water. Compare *Coastal wetland.*

Inorganic compounds substances that consist of chemical combinations of two or more elements other than those used to form organic compounds. Compare *Organic compounds.*

Inorganic fertilizer See *Commercial inorganic fertilizer.*

Input pollution control any method that prevents potential pollutants from entering the environment or sharply reduces the amount entering the environment. Compare *Output pollution control.*

Insecticide substance or mixture of substances intended to prevent, destroy, or repel insects.

Integrated pest management (IPM) use of a combination of biological, chemical, and cultivation methods in proper sequence and timing in order to keep pest population sizes just below the level of economic loss.

Intensive forest management clearing an area of all vegetation, planting it with a single tree species, then fertilizing and spraying the resulting even-aged stand of trees with pesticides.

Interspecific competition two or more species in the same ecosystem attempting to use the same scarce resources.

Intraspecific competition two or more individual organisms of a single species in an ecosystem attempting to use the same scarce resources.

Intrauterine device (IUD) small plastic or metal device inserted into the uterus to prevent contraception.

Inversion See *Thermal inversion.*

Ionizing radiation fast-moving alpha or beta particles or high-energy electromagnetic radiation emitted by radioisotopes that have enough energy to dislodge one or more electrons from atoms it hits to form charged ions, which can react with and damage living tissue.

Ions atoms or groups of atoms with one or more net positive (+) or negative (−) electrical charges.

Isotopes two or more forms of a chemical element that have the same number of protons but different mass numbers or numbers of neutrons in their nuclei.

IUD See *Intrauterine device.*

J-shaped curve curve with the shape of the letter J that depicts exponential or geometric growth.

Kerogen solid, waxy mixture of hydrocarbons that is intimately mixed with a fine-grained sedimentary rock. When the rock is heated to high temperatures, the kerogen is vaporized and much of the vapor can be condensed to yield shale oil, which can be refined to give petroleum-like products. See also *Oil shale, Shale oil.*

Kilocalorie (kcal) unit of energy equal to 1,000 calories. See *Calorie.*

Kilowatt (kw) unit of electrical power equal to 1,000 watts. See *Watt.*

Kinetic energy energy that matter has because of its motion and mass. Compare *Potential energy.*

Kwashiorkor nutritional deficiency (malnutrition) disease that occurs in infants and very young children when they are weaned from mother's milk to a starchy diet that is relatively high in calories but low in protein. See also *Marasmus.*

Lake large natural body of standing fresh water formed when water from precipitation, land runoff, or groundwater flow fills depressions in the earth created by glaciation, earthquakes, volcanic activity, and crashes of giant meteorites.

Landfarming spreading and mixing of hazardous or other solid or liquid wastes with surface soil to allow biodegradation to less hazardous or nonhazardous materials.

Landfill land waste disposal site that is located without regard to possible pollution of groundwater and surface water resulting from runoff and leaching; waste is covered intermittently with a layer of earth to reduce scavenger, aesthetic, disease, and air pollution problems. Compare *Open dump, Sanitary landfill, Secured landfill.*

Land-use planning process for deciding the best use of each parcel of land in an area. See also *Ecological land-use planning.*

Laterite soil found in some tropical areas in which an insoluble concentration of such metals as iron and aluminum is present; soil fertility is generally poor.

Late-successional species wild animal species found in moderate-size mature forest habitats. Compare *Early-successional species, Mid-successional species, Wilderness species.*

Law of conservation of energy See *First law of thermodynamics.*

Law of conservation of matter in any ordinary physical or chemical change, matter is neither created nor destroyed but merely changed from one form to another.

Law of tolerance the existence, abundance, and distribution of a species are determined by whether the levels of one or more physical or chemical factors fall above or below the levels tolerated by the species.

LDC See *Less developed country.*

Leaching process in which various soil components found in upper layers are dissolved and carried to lower layers and in some cases to groundwater.

Less developed country (LDC) country that typically has low to moderate industrialization, a very low to moderate average GNP per person, a high rate of population growth, a large fraction of its labor force employed in agriculture, and a high level of adult illiteracy. Compare *More developed country*.

Life-cycle cost initial cost plus lifetime operating costs.

Life expectancy average number of years a newborn can be expected to live.

Lifetime cost See *Life-cycle cost*.

Light-water reactor (LWR) a nuclear reactor in which ordinary water, called light water, is used as a moderator inside its core to slow down the neutrons emitted by the fission process and thus sustain the chain reaction.

Lignite form of coal with a low heat content and usually a low sulfur content. Compare *Anthracite, Bituminous coal, Peat, Subbituminous coal*.

Limiting factor factor such as temperature, light, water, or a chemical that limits the existence, growth, abundance, or distribution of an organism.

Limiting factor principle the single physical or chemical factor that is most deficient in an ecosystem determines the presence or absence and population size of a particular species.

Limnetic zone open-water surface layer of a lake through which there is sufficient sunlight for photosynthesis. Compare *Benthic zone, Littoral zone, Profundal zone*.

Liquefied natural gas (LNG) natural gas which is converted to liquid form by cooling to a very low temperature and then transported by sea in specially designed, refrigerated tanker ships.

Liquefied petroleum gas (LPG or LP-gas) the mixture of liquefied propane and butane gas removed from a deposit of natural gas.

Lithosphere region of soil and rock consisting of the earth's upper surface or crust and the upper portion of the mantle of partially molten rock beneath this crust.

Littoral zone shallow waters near the shore of a body of water. Compare *Benthic zone, Limnetic zone, Profundal zone*.

Loams medium-porosity soils consisting of almost equal amounts of sand and silt and somewhat less clay; considered to be the best soil for growing crops.

Longwall method method of subsurface mining for a mineral such as coal in which a narrow tunnel is cut and then supported by movable metal pillars. After the coal or ore is removed, the roof supports are moved forward, allowing the earth behind them to collapse. Compare *Room-and-pillar method*.

Low-grade ore an ore that contains a relatively low concentration of a desired metallic element. Compare *High-grade ore*.

Low-quality energy form of energy such as low-temperature heat that is dispersed or diluted and has little ability to do useful work. Compare *High-quality energy*.

LP-gas See *Liquefied petroleum gas*.

Lung cancer abnormal, accelerated growth of cells in the mucous membranes of the bronchial passages.

Magma molten rock material within the earth's core.

Malnutrition condition in which quality of diet is inadequate and an individual's minimum daily requirements for proteins, fats, vitamins, minerals, and other specific nutrients necessary for good health are not met. Compare *Overnutrition, Undernutrition*.

Malthusian theory of population the theory of economist Thomas Malthus that human population growing exponentially eventually outgrows food supply. The conclusion is that human beings are destined to misery and poverty unless population growth is controlled.

Manure See *Animal manure, Green manure*.

Marasmus nutritional deficiency disease that results from a diet low in both calories and protein. See also *Kwashiorkor*.

Mass number sum of the number of neutrons and the number of protons in the nucleus of an atom. It is a measure of the approximate mass of that atom.

Mass transit transportation systems (such as buses, trains, and trolleys) that use vehicles that carry large numbers of people.

Matter anything that has mass and occupies space.

Matter-recycling society society based on significant recycling of nonrenewable, nonfuel matter resources. Compare *Sustainable-earth society, Throwaway society*.

Maximum sustained yield See *Sustained yield*.

MDC See *More developed country*.

Megawatt (Mw) unit of electrical power equal to 1,000 kilowatts, or 1 million watts. See *Watt*.

Meltdown complete melting of the fuel rods and core of a nuclear reactor.

Mesotrophic lake lake with a moderate supply of plant nutrients. Compare *Eutrophic lake, Oligotrophic lake*.

Metabolic reserve lower half of rangeland grass plants that can grow back as long as it is not consumed by herbivores.

Metallic mineral inorganic substance found in the earth's crust that contains a useful metallic element such as aluminum, iron, or uranium.

Metastasis release of malignant (cancerous) cells from a tumor into other parts of the body.

Microconsumer See *Decomposers*.

Microorganism generally, any living thing of microscopic size; examples include bacteria, yeasts, simple fungi, some algae, slime molds, and protozoans.

Mid-successional species wild species found around abandoned croplands and partially open areas characterized by vegetation at the middle stages of ecological succession. Compare *Early-successional species, Late-successional species, Wilderness species*.

Migration rate difference between the numbers of people leaving and entering a given country or area per 1,000 people in its population at midyear.

Mineral an inorganic substance (element or compound) occurring naturally in the earth's crust. See *Metallic mineral, Nonmetallic mineral*.

Mineral deposit any naturally occurring concentration in the lithosphere of a free element or compound in solid form.

Mineral resource nonrenewable chemical element or compound usually in solid form that is used by people. Mineral resources are classified as metallic (such as iron and tin) or nonmetallic (such as fossil fuels, sand, and salt).

Minimum-tillage farming planting crops by disturbing the soil as little as possible and keeping crop residues and litter on the ground instead of turning them under by plowing.

Molecule chemical combination of two or more atoms of the same chemical element (such as O_2) or different chemical elements (such as H_2O). Compare *Atoms*.

Monoculture cultivation of a single crop (such as maize or cotton) to the exclusion of other crops on a piece of land.

More developed country (MDC) a country with significant industrialization, a high average GNP per person, a low rate of population growth, a small fraction of its labor force employed in agriculture, a low level of adult illiteracy, and a strong economy. Compare *Less developed country.*

Mortality   the death rate.

Multiple use   principle for managing a forest so that it is used for a variety of purposes, including timbering, mining, recreation, grazing, wildlife preservation, and soil and water conservation.

Municipal waste   combined residential and commercial waste materials generated in a given municipal area.

Mutagen   any substance capable of increasing the rate of genetic mutation of living organisms.

Mutation   inheritable changes in the DNA molecules found in genes as a result of exposure to various environmental factors such as radiation and certain chemicals or during cell division (asexual reproduction) and when a sperm and egg cell fuse (sexual reproduction).

Mutualism   interaction between species in which all species involved are benefited.

National ambient air quality standard (NAAQS)   federal standard that specifies the maximum allowable level, averaged over a specific time period, for a certain pollutant in outdoor (ambient) air. Compare *Emission standard.*

Natural change rate   how fast a population is growing or decreasing per year; usually expressed with a percentage and obtained by subtracting the death rate from the birth rate and dividing the result by 10.

Natural community   See *Community.*

Natural eutrophication   See *Eutrophication.*

Natural gas   underground deposits of gases consisting of 50% to 90% methane (CH_4) and small amounts of heavier gaseous hydrocarbon compounds such as propane (C_3H_8) and butane (C_4H_{10}).

Natural increase (or decrease)   difference between the birth rate and the death rate in a given population during a given period.

Natural ionizing radiation   See *Background ionizing radiation.*

Natural radioactivity   a nuclear change in which unstable nuclei of atoms spontaneously shoot out "chunks" of mass, energy, or both at a fixed rate.

Natural resource   anything obtained from the physical environment to meet human needs.

Natural selection   mechanism for evolutionary change in which individual organisms in a single population die off over time because they cannot tolerate a new stress and are replaced by individuals whose genetic traits allow them to cope with the stress and reproduce successfully to pass these adaptive traits on to their offspring. See also *Evolution.*

Neo-Malthusians   people who believe that if present trends continue, the world will become more crowded and more polluted, leading to greater political and economic instability and increasing the threat of nuclear war as the rich get richer and the poor get poorer. Compare *Cornucopians.*

Neritic zone   relatively warm, nutrient-rich, shallow portion of the ocean that extends from the high-tide mark on land to the edge of the continental shelf. Compare *Open sea.*

Net energy   See *Net useful energy.*

Net migration rate   for a given place on earth, the difference between the numbers of people immigrating and emigrating during a given period (usually a year) per 1,000 people in the population at midyear.

Net population change   difference between the total number of live births and the total number of deaths throughout the world or a given part of the world during a specified period (usually a year).

Net primary productivity   rate at which all the plants in an ecosystem produce net useful chemical energy. It is equal to the difference between the rate at which the plants in an ecosystem produce useful chemical energy and the rate at which they use some of this energy through cellular respiration.

Net useful energy   total useful energy available from an energy resource or energy system minus the useful energy used, lost, and wasted in finding, processing, concentrating, and transporting it to a user.

Neutral solution   water solution containing an equal number of hydrogen ions (H^+) and hydroxide ions (OH^-); has a pH of 7. Compare *Acid solution, Basic solution.*

Neutron (n)   elementary particle present in the nuclei of all atoms (except hydrogen-1). It has a relative mass of 1 and no electric charge.

Niche   See *Ecological niche.*

Nitrogen cycle   biogeochemical cycle in which nitrogen is converted into various forms and transported through the biosphere.

Nitrogen fixation   process in which bacteria and other soil microorganisms convert atmospheric nitrogen into nitrates, which become available to growing plants.

Nonbiodegradable pollutant   material, such as toxic mercury and lead compounds, that is not broken down to an acceptable level or form in the environment by natural processes. Compare *Rapidly biodegradable pollutant, Slowly biodegradable pollutant.*

Nonmetallic mineral   inorganic substance found in the earth's crust that contains useful nonmetallic compounds such as those in sand, stone, and nitrate and phosphate salts used as commercial fertilizers. Compare *Metallic mineral.*

Nonpoint source   source of pollution in which wastes are not released at one specific, identifiable point but from a number of points that are spread out and difficult to identify and control. Compare *Point source.*

Nonrenewable resource   resource that is available in a fixed amount (stock) in various places in the earth's crust and either is not replenished by natural processes or is replenished more slowly than it is used, so that it can ultimately be totally depleted or depleted to the point at which extracting and processing it for human use is too expensive. Compare *Perpetual resource, Renewable resource.*

No-till cultivation   See *Conservation-tillage farming.*

Nuclear autumn effect   moderate drop in atmospheric temperature and degree of light penetration to the earth's surface as a result of large amounts of smoke, soot, dust, and other debris lifted into the atmosphere by a limited nuclear war. Compare *Nuclear winter effect.*

Nuclear change   process in which the isotope of an element changes into one or more different isotopes by altering the number of protons, neutrons, or both in its nucleus. Compare *Chemical change.*

Nuclear energy   energy released when atomic nuclei undergo fission or fusion.

Nuclear fission   nuclear change in which the nuclei of certain heavy isotopes with large mass numbers such as uranium-235 are split apart into two lighter nuclei when struck by slow- or fast-moving neutrons; this process also releases more neutrons and a substantial amount of energy. Compare *Nuclear fusion.*

Nuclear fusion   nuclear change in which two nuclei of light elements such as hydrogen are forced together at high temperatures of 100 million to 1 billion °C until they fuse to form a heavier nucleus with the release of a substantial amount of energy. Compare *Nuclear fission.*

Nuclear winter effect significant drop in atmospheric temperature and degree of light penetration to the earth's surface as a result of massive amounts of smoke, soot, dust, and other debris lifted into the atmosphere by a limited nuclear war. Compare *Nuclear autumn effect.*

Nucleus the extremely tiny center of an atom, which contains one or more positively charged protons and in most cases one or more neutrons with no electrical charge. The nucleus contains most of an atom's mass.

Nutrient element or compound needed for the survival, growth, and reproduction of a plant or animal.

Ocean thermal energy conversion (OTEC) use of a floating power plant located in a suitable tropical ocean area to produce electricity by taking advantage of the temperature difference between warm surface waters and cold bottom waters in an ocean.

Oil See *Crude oil.*

Oil shale underground formation of a fine-grained rock that contains varying amounts of a solid, waxy mixture of hydrocarbon compounds known as kerogen. When the rock is heated to high temperatures, the kerogen is converted to a vapor that can be condensed to form a slow-flowing heavy oil called shale oil. See *Kerogen, Shale oil.*

Oligotrophic lake a lake with a low supply of plant nutrients. Compare *Eutrophic lake, Mesotrophic lake.*

Omnivore organism such as a pig, rat, cockroach, or human being that can use both plants and other animals as food sources. Compare *Carnivore, Herbivore.*

Open dump land disposal site where wastes are deposited and left uncovered with little or no regard for control of scavenger, disease, air pollution, or water pollution problems or for aesthetics. Compare *Sanitary landfill, Secured landfill.*

Open-pit surface mining surface mining of materials (primarily stone, sand, gravel, iron, and copper) that creates a large pit.

Open sea the part of an ocean beyond the continental shelf.

Ore mineral deposit containing a high enough concentration of at least one metallic element to permit the metal to be extracted and sold at a profit. See *High-grade ore, Low-grade ore.*

Organic compounds molecules that typically contain atoms of the elements carbon and hydrogen; carbon, hydrogen, and oxygen; or carbon, hydrogen, oxygen, and nitrogen. Compare *Inorganic compounds.*

Organic farming method of producing crops and livestock naturally by using organic fertilizer (manure, legumes, compost, crop residues), crop rotation, and natural pest control (bugs that eat harmful bugs, plants that repel harmful bugs, and environmental controls such as crop rotation) instead of using commercial fertilizer and synthetic pesticides and herbicides.

Organic fertilizer organic material such as animal manure, green manure, and compost applied to cropland as a source of plant nutrients. Compare *Commercial inorganic fertilizer.*

Organism any form of life.

Organophosphates diverse group of nonpersistent synthetic chemical insecticides that act chiefly by breaking down nerve and muscle responses; examples are parathion and malathion.

Output pollution control method for reducing the level of pollution once a pollutant has entered the environment. Compare *Input pollution control.*

Overburden layer of soil and rock overlying a mineral deposit that is removed during surface mining.

Overfishing harvesting so many fish, especially immature individuals, of a species that not enough breeding stock is left for adequate annual renewal.

Overgrazing excessive grazing of rangeland by livestock to the point at which it cannot be renewed or is renewed slowly because of damage to the root system.

Overnutrition diet so high in calories, saturated (animal) fats, salt, sugar, and processed foods, and so low in vegetables and fruits that the consumer runs high risks of diabetes, hypertension, heart disease, and other health hazards. Compare *Malnutrition, Undernutrition.*

Overpopulation impairment of the life-support systems in a country, a region, or the world when its people use nonrenewable and renewable resources to such an extent that the resource base is degraded or depleted and air, water, and soil are severly polluted. See *Consumption overpopulation, People overpopulation.*

Oxygen cycle biogeochemical cycle in which oxygen is converted into various forms and transported through the biosphere.

Oxygen-demanding wastes organic water pollutants that are usually degraded by aerobic (oxygen-consuming) bacteria if there is sufficient dissolved oxygen (DO) in the water. See also *Biological oxygen demand.*

Ozone layer layer of gaseous ozone (O_3) in the upper atmosphere that protects life on earth by filtering out harmful ultraviolet radiation from the sun.

Package sewage treatment plant small plant sometimes used for treatment of small quantities of wastewater for shopping centers, apartment complexes, villages, and small housing subdivisions.

Parasite primary, secondary, or higher consumer that feeds on a plant or animal, known as a host, over an extended period of time. See *Ectoparasite, Endoparasite.*

Paratransit transit system such as carpools, vanpools, jitneys, and dial-a-ride systems that carry a relatively small number of passengers per vehicular unit.

Particulate matter solid particles or liquid droplets suspended or carried in the air.

Parts per billion (ppb) number of parts of a chemical found in one billion parts of a particular gas, liquid, or solid mixture.

Parts per million (ppm) number of parts of a chemical found in one million parts of a particular gas, liquid, or solid mixture.

Parts per trillion (ppt) number of parts of a chemical found in one trillion parts of a particular gas, liquid, or solid mixture.

Passive solar heating system that captures sunlight directly, usually through large windows or an attached greenhouse, and distributes and stores some of it without the use of fans, pumps, or other mechanical devices. Compare *Active solar heating.*

Pathogen organism that produces disease.

PCBs (polychlorinated biphenyls) mixture of at least 50 widely used organic compounds containing chlorine that can be biologically magnified in food chains and food webs with unknown effects.

Peat a fuel with a low heat content and a high moisture content (70% to 95%) that is the first step in the formation of various types of coal. Compare *Anthracite, Bituminous coal, Lignite, Subbituminous coal.*

People overpopulation condition in which there are more people in a particular country, region, or the world than the available supplies of food, water, and other vital resources can support, or where the rate of population growth so exceeds the rate of economic growth and the equitable distribution of wealth that a number of people are too poor to grow or buy sufficient food. Compare *Consumption overpopulation*.

Permafrost water permanently frozen year-round in thick underground layers of soil found in tundra.

Peroxyacyl nitrates group of chemicals (photochemical oxidants) also known as PANs, found in photochemical smog.

Perpetual resource a resource such as solar energy that comes from an essentially inexhaustible source and thus will always be available on a human time scale regardless of whether or how it is used. Compare *Nonrenewable resource, Renewable resource*.

Pest unwanted organism that directly or indirectly interferes with human activities.

Pesticide any chemical designed to kill weeds, insects, fungi, rodents, and other organisms that humans consider to be undesirable.

Pesticide treadmill situation in which the costs of using pesticides increases while their effectiveness decreases, primarily as a result of genetic resistance to the chemicals by target organisms.

Petrochemicals chemicals obtained by refining (distilling) crude oil that are used as raw materials in the manufacture of most industrial chemicals, fertilizers, pesticides, plastics, synthetic fibers, paints, medicines, and numerous other products.

Petroleum See *Crude oil*.

pH numeric value that indicates the relative acidity or alkalinity of a substance on a scale of 0 to 14, with the neutral point at 7.0. Acid solutions have pH values lower than 7.0 and basic solutions have pH values greater than 7.

Phosphorus cycle biogeochemical cycle in which phosphorus is converted into various chemical forms and transported through the biosphere.

Photochemical smog complex mixture of air pollutants produced in the atmosphere by the reaction of hydrocarbons and nitrogen oxides under the influence of sunlight. Especially harmful components include ozone, peroxyacyl nitrates (PANs), and various aldehydes. Compare *Industrial smog*.

Photosynthesis complex process that occurs in the cells of green plants whereby radiant energy from the sun is used to combine carbon dioxide (CO_2) and water (H_2O) to produce oxygen (O_2) and simple sugar or food molecules, such as glucose ($C_6H_{12}O_6$). Compare *Cellular respiration, Chemosynthesis*.

Photovoltaic cell (solar cell) device in which radiant (solar) energy is converted directly into electrical energy.

Physical change process that alters one or more physical properties of an element or compound without altering its chemical composition. Examples include changing the size and shape of a sample of matter (crushing ice and cutting aluminum foil) and changing a sample of matter from one physical state to another (boiling and freezing water). Compare *Chemical change*.

Phytoplankton free-floating, mostly microscopic aquatic plants.

Pioneer community first successfully integrated set of plants, animals, and decomposers that is found in an area undergoing primary ecological succession.

Plankton microscopic floating plant and animal organisms of lakes, rivers, and oceans. See also *Phytoplankton*.

Point source source of pollution that involves discharge of pollutants from an identifiable point, such as a smokestack or sewage treatment plant. Compare *Nonpoint source*.

Pollution a change in the physical, chemical, or biological characteristics of the air, water, or soil that can affect the health, survival, or activities of human beings or other living organisms in a harmful way.

Polychlorinated biphenyls See *PCBs*.

Population group of individual organisms of the same species that occupy particular areas at given times.

Population crash extensive deaths over a relatively short time resulting when a population exceeds the ability of the environment to support it.

Population density number of organisms in a particular population per square kilometer or other unit of area.

Population distribution variation of population density over a given country, region, or other area.

Porosity See *Soil porosity*.

Potential energy energy stored in an object as a result of its position or the position of its parts. Compare *Kinetic energy*.

Power tower See *Solar furnace*.

ppb See *Parts per billion*.

ppm See *Parts per million*.

ppt See *Parts per trillion*.

Precipitation water in the form of rain, sleet, hail, and snow that falls from the atmosphere onto the land and bodies of water.

Predation situation in which an organism of one species (the predator) captures and feeds on an organism of another species (the prey).

Predator organism that captures and feeds on parts or all of an organism of another species (the prey).

Preservationists people who believe that large areas of public land should be protected and preserved from mining, lumbering, and other forms of development by establishing parks, wilderness areas, and wildlife refuges that can be enjoyed by present generations and passed on unspoiled to future generations. Compare *Scientific conservationists*.

Prey organism that is captured and serves as a source of food for an organism of another species (the predator).

Primary air pollutant chemical that has been added directly to the air and occurs in a harmful concentration. Compare *Secondary air pollutant*.

Primary consumer See *Herbivore*.

Primary oil recovery pumping out all of the crude oil that will flow by gravity into the bottom of an oil well. Compare *Enhanced oil recovery, Secondary oil recovery*.

Primary succession sequential development of communities in a bare or soil-less area that has never been occupied by a community of organisms. Compare *Secondary succession*.

Primary sewage treatment mechanical treatment in which large solids, like old shoes and sticks of wood, are screened out, and suspended solids settle out as sludge. Compare *Secondary sewage treatment, Tertiary sewage treatment*.

Prime reproductive age years between ages 20 and 29 during which most women have most of their children. Compare *Reproductive age*.

Producer organism that uses solar energy (green plant) or chemical energy (some bacteria) to manufacture its own organic substances (food) from inorganic nutrients. Compare *Consumer, Decomposers*.

Profundal zone deep-water region of a lake, which is not penetrated by sunlight. Compare *Benthic zone, Limnetic zone, Littoral zone*.

Proton (p) positively charged particle found in the nuclei of all atoms. Each proton has a relative mass of 1 and a single positive charge.

Pyramid of biomass diagram representing the biomass, or total dry weight of all living organisms, that can be supported at each trophic level in a food chain.

Pyramid of numbers diagram representing the number of organisms of a particular type that can be supported at each trophic level from a given input of solar energy in food chains and food webs.

Pyrolysis high-temperature decomposition of material in the absence of oxygen.

Radiation propagation of energy through matter and space in the form of fast-moving particles (particulate radiation) or waves (electromagnetic radiation).

Radioactive fallout radioactive dirt and debris that falls back to earth after being released into the atmosphere as a result of the detonation of a nuclear weapon or an accident at a nuclear power plant or other facility handling radioactive materials.

Radioactive isotope See *Radioisotope.*

Radioactive waste end products of nuclear power plants, research, medicine, weapons production, or other processes involving radioisotopes.

Radioactivity See *Natural radioactivity.*

Radioisotope isotope of an atom whose unstable nuclei spontaneously emit fast-moving particles (such as alpha or beta particles), high-energy electromagnetic radiation in the form of gamma rays, or both to form nonradioactive or radioactive isotopes of a different kind.

Rangeland land on which the vegetation is predominantly grasses, grasslike plants, or shrubs such as sagebrush and that is capable of providing forage for grazing or browsing animals.

Range of tolerance range or span of chemical and physical conditions that must be maintained for populations of a particular species to stay alive and grow, develop, and function normally.

Rapidly biodegradable pollutant substance such as human sewage that can be rapidly broken down in the environment into an acceptable level or form by natural processes. Compare *Nonbiodegradable pollutant, Slowly biodegradable pollutant.*

Rate of natural change measure of population change obtained by finding the difference between the birth rate and the death rate.

Rate of population change difference between the birth rate and the death rate plus net migration rate for a particular country or area.

Recharge area area in which an aquifer is replenished with water by the downward percolation of precipitation through soil and rock.

Recharging replenishment of water in an aquifer.

Recycling collecting and remelting or reprocessing a resource so it can be used again, as when used glass bottles are collected, melted down, and made into new glass bottles. Compare *Reuse.*

Renewable resource resource that can be depleted in the short run if used or contaminated too rapidly but normally will be replaced through natural processes in the long run. Compare *Nonrenewable resource, Perpetual resource.*

Replacement-level fertility number of children a couple must have to replace themselves; the average for a country or the world is usually slightly higher than 2 (2.1 in the United States) because some children die before reaching their reproductive years.

Reproductive age ages 15 to 44, when most women have all their children. Compare *Prime reproductive age.*

Reproductive potential See *Biotic potential.*

Reserves identified deposits of a particular resource in known locations that can be extracted profitably at present prices and with current mining technology. Compare *Identified resources, Resources, Undiscovered resources.*

Reservoir large and deep, human-created body of standing fresh water often filled behind a dam. Compare *Lake.*

Resilience stability ability of a disturbed living system to restore itself to its condition before the disturbance. See also *Inertia stability.*

Resource See *Natural resource.*

Resource conservation developing and protecting natural resources for the greatest good of the greatest number of people for the longest length of time by reducing unnecessary resource use and waste.

Resource recovery extraction of useful materials or energy from waste materials. This may involve recycling or conversion into different and sometimes unrelated products or uses. Compare *Recycling, Reuse.*

Resource recovery plant centralized facility in which mixed urban solid waste is shredded and automatically separated to recover glass, iron, aluminum, and other valuable materials. The remaining paper, plastics, and other materials are incinerated to produce steam, hot water, or electricity.

Resources identified and unidentified deposits of a particular mineral that cannot be recovered profitably with present prices and mining technology but may be converted to reserves when prices rise or mining technology improves. Compare *Identified resources, Reserves, Undiscovered resources.*

Respiration See *Cellular respiration.*

Reuse to use a product again and again in the same form, as when returnable glass bottles are washed and refilled. Compare *Recycling.*

Risk probability that something undesirable will happen from deliberate or accidental exposure to a hazard.

Risk assessment process of determining the short- and long-term adverse consequences to individuals or groups from the use of a particular technology in a particular area.

Risk-benefit analysis estimating the short- and long-term societal benefits and risks involved in using a particular technology; the benefits can be divided by the risks to find a desirability quotient that can be used to help determine whether the technology should be used. See also *Cost-benefit analysis.*

Risk management the administrative, political, and economic actions taken to decide how, and if, a particular societal risk is to be reduced to a certain level and at what cost.

River fairly wide and deep-flowing body of water that usually empties into an ocean.

Rocky shore steep, rock-laden coastline. Compare *Barrier beach.*

Rodenticide substance that can kill rodents.

Room-and-pillar method type of subsurface mining in which a mineral deposit such as coal is removed in a manner that creates a series of rooms supported by pillars of unremoved mineral. Compare *Longwall method.*

Ruminant animals animals such as cattle, sheep, goats, and buffalo with four-chambered stomachs that digest cellulose.

Runoff surface water entering rivers, freshwater lakes, or reservoirs from land surfaces.

Rural area area in the United States with a population of fewer than 2,500 people.

S-shaped curve leveling off of an exponential, or J-shaped, curve.

Salinity amount of dissolved salts (especially sodium chloride) in a given volume of water.

Salinization accumulation of salts in soils that can eventually render the soil incapable of supporting plant growth.

Saltwater intrusion movement of salt water into freshwater aquifers in coastal areas as groundwater is withdrawn faster than it is recharged by precipitation.

Sandy soil highly porous soil containing a large amount of sand and little, if any, silt and clay. Compare *Clay soil, Loams.*

Sanitary landfill land waste disposal site located to minimize water pollution from runoff and leaching; waste is spread in thin layers, compacted, and covered with a fresh layer of soil each day. Compare *Open dump, Secured landfill.*

Scientific conservationists people guided by the belief that public land resources should be used to enhance economic growth and national strength and protected from depletion and degradation by being managed according to the principles of sustained yield and multiple use. Compare *Preservationists.*

Scrubber common antipollution device that uses a liquid spray to remove pollutants from a stream of air.

Secondary air pollutant harmful chemical formed in the atmosphere through a chemical reaction among air components. Compare *Primary air pollutant.*

Secondary consumer See *Carnivore.*

Secondary oil recovery injection of water to force some of the remaining crude oil from a well after primary recovery. Usually primary and secondary recovery remove about one-third of the crude oil in a well. Compare *Enhanced oil recovery, Primary oil recovery.*

Secondary sewage treatment second step in most waste treatment systems, in which aerobic bacteria break down biodegradable organic wastes in wastewater; usually accomplished by bringing the sewage and bacteria together in trickling filters or in the activated sludge process. Compare *Primary sewage treatment, Tertiary sewage treatment.*

Secondary succession sequential development of communities in an area in which natural vegetation has been removed or destroyed, but the soil or bottom sediment is not destroyed. Compare *Primary succession.*

Second law of energy See *Second law of thermodynamics.*

Second law of thermodynamics **(1)** in any conversion of heat energy to useful work, some of the initial energy input is always degraded to a lower-quality, more dispersed, less useful form of energy, usu-

ally low-temperature heat that flows into the environment; you can't break even in terms of energy quality. **(2)** Any system and its surroundings (environment) as a whole spontaneously tend toward increasing randomness, disorder, or entropy; if you think things are mixed up now, just wait.

Secured landfill a land site for the storage of hazardous solid and liquid wastes that are normally placed in containers and buried in a restricted-access area that is continually monitored. Such landfills are located above geologic strata that are supposed to prevent the leaching of wastes into groundwater. Compare *Open dump, Sanitary landfill.*

Sediment soil particles, sand, and minerals washed from the land into aquatic systems as a result of natural and human activities.

Seed-tree cutting removal of nearly all trees on a site in one cut, with a few of the better commercially valuable trees left uniformly distributed as a source of seed to regenerate the forest. Compare *Clearcutting, Selective cutting, Shelterwood cutting, Whole-tree harvesting.*

Selective cutting cutting of intermediate-aged or mature or diseased trees in an uneven-aged forest stand either singly or in small groups to encourage younger trees to grow and to produce an uneven-aged stand with trees of different species, ages, and sizes. Compare *Clearcutting, Seed-tree cutting, Shelterwood cutting, Whole-tree harvesting.*

Septic tank underground receptacle for wastewater from a home in rural and suburban areas. The bacteria in the sewage decompose the organic wastes, and the sludge settles to the bottom of the tank. The effluent flows out of the tank into the ground through a field of drain pipes.

Sewage sludge See *Sludge.*

Shale oil a slow-flowing, dark brown, heavy oil obtained when kerogen in shale oil rock is vaporized at high temperatures and then condensed. Shale oil can be refined to yield petroleum products. See *Kerogen, Oil shale.*

Shelterbelt See *Windbreaks.*

Shelterwood cutting removal of all mature trees in an area in a series of cuts over one or more decades. Compare *Clearcutting, Seed-tree cutting, Selective cutting, Whole-tree harvesting.*

Shifting cultivation clearing and planting a plot of ground in a forest for two to five years, until no further cultivation is worthwhile because of a reduction of soil fertility or invasion by a dense growth of

vegetation, and then clearing a new plot to grow crops by slash-and-burn cultivation. See *Slash-and-burn cultivation.*

Silviculture cultivation and management of forests to produce a renewable supply of timber.

Slash-and-burn cultivation in many tropical areas, the practice of clearing a patch of forest, leaving the cut vegetation on the ground to dry, burning the dried residue to add nutrients to the soil, and planting crops. Ideally the patch is abandoned after two or five years of cultivation to prevent depletion of soil fertility. See *Shifting cultivation.*

Slowly biodegradable pollutant a pollutant such as DDT that is broken down in the environment by natural processes to an acceptable level or form at a relatively slow rate. Compare *Nonbiodegradable pollutant, Rapidly biodegradable pollutant.*

Sludge gooey solid mixture of bacteria- and virus-laden organic matter, toxic metals, synthetic organic chemicals, and solid chemicals removed from wastewater at a sewage treatment plant.

Smog originally a combination of *smoke* and *fog*; now applied also to the photochemical haze produced by the action of sun and atmosphere on automobile and industrial exhausts. Compare *Industrial smog, Photochemical smog.*

Soil complex mixture of inorganic minerals (mostly clay, silt, and sand), decaying organic matter, water, air, and living organisms.

Soil conservation methods used to reduce soil erosion and prevent depletion of soil nutrients.

Soil erosion movement of soil components, especially topsoil, from one place to another, usually by exposure to wind, flowing water, or both.

Soil horizons horizontal layers that make up a particular type of soil.

Soil porosity number of pores and the average distances between pores in a given sample of soil.

Soil profile cross-sectional view of the horizons or horizontal layers in a soil.

Solar cell device that converts radiant energy from the sun directly into electrical energy.

Solar collector device for collecting radiant energy from the sun and converting it into heat.

Solar energy direct radiant energy from the sun plus indirect forms of energy—such as wind, falling or flowing water (hydropower), ocean thermal gradients, and biomass—that are produced when solar energy interacts with the earth.

Solar furnace　device for collecting and concentrating radiant energy from the sun to reach temperatures high enough to melt metals and carry out other high-temperature operations.

Solar pond　relatively small body of fresh water or salt water in which stored solar energy can be extracted as a result of the temperature difference between the surface layer and bottom layer.

Solid waste　any unwanted or discarded material that is not a liquid or a gas.

Speciation　splitting of a single species over thousands to millions of years into two or more different species in response to new environmental conditions.

Species　all organisms of the same kind; a group of plants or animals potentially capable of breeding with other members of its group but normally not with organisms outside its group.

Species diversity　number of different species and their relative abundances in a given area.

Sport hunting　killing of animals for recreation. Compare *Commercial hunting, Subsistence hunting.*

Stability　ability of a living system to withstand or recover from externally imposed changes or stresses. See *Inertia stability, Resilience stability.*

Strategic materials　fuel and nonfuel minerals vital to the industry and defense of a country. Ideally supplies are stockpiled to cushion against supply interruptions and sharp price increases.

Strip cropping　planting regular crops and close-growing plants such as hay or nitrogen-fixing legumes in alternating rows or bands.

Strip-mining　See *Surface mining.*

Subatomic particles　extremely small particles such as electrons, protons, and neutrons that make up the internal structure of atoms.

Subbituminous coal　form of coal with a low heat content and usually a low sulfur content. Compare *Anthracite, Bituminous coal, Lignite, Peat.*

Subsidence　sinking down of part of the earth's crust resulting from underground excavation—such as a coal mine—or removal of groundwater.

Subsistence agriculture　supplementation of solar energy with energy from human labor and draft animals to produce enough food to feed one's self and family

members; occasionally some may be left over to sell or put aside for hard times. Compare *Industrialized agriculture.*

Subsistence hunting　killing of animals to provide enough food and other materials for survival. Compare *Commercial hunting, Sport hunting.*

Subsurface mining　underground extraction of a metal ore or fuel resource such as coal. Compare *Surface mining.*

Succession　See *Ecological succession.*

Succulent plants　plants such as cacti that store water and produce the food they need in the thick, fleshy tissue of their green stems and branches.

Superinsulated house　house that contains massive amounts of insulation, is extremely airtight, typically uses active or passive solar collectors to heat water, and has an air-to-air heat exchanger to prevent buildup of excessive moisture and indoor air pollutants.

Surface mining　the process of removing the overburden of topsoil, subsoil, and other strata to permit the extraction of underlying mineral deposits. See *Area strip-mining, Contour strip-mining, Open-pit surface mining;* compare *Subsurface mining.*

Surface water　precipitation that does not infiltrate into the ground or return to the atmosphere and becomes runoff that flows into nearby streams, rivers, lakes, wetlands, and reservoirs. Compare *Groundwater.*

Surroundings (environment)　everything outside a specified system or collection of matter.

Sustainable-earth agriculture　method of growing crops and raising livestock that places heavy reliance on organic fertilizers, soil conservation, water conservation, biological control of pests, and minimal use of nonrenewable fossil fuel energy.

Sustainable-earth society　society based on working with nature by recycling and reusing discarded matter, conserving matter and energy resources by reducing unnecessary waste and use, and building items that are easy to recycle, reuse, and repair. Compare *Matter-recycling society, Throwaway society.*

Sustainable-earth world view　the belief that the earth is a place with finite room and resources so that continuing population growth, production, and consumption inevitably put severe stress on natural processes that renew and maintain the resource base of air, water, and soil that support all life. To prevent environmental overload and resource depletion, people should work with—not against—nature by controlling population growth and

reducing unnecessary use and waste of matter and energy resources. Compare *Throwaway worldview.*

Sustained yield　the highest rate at which a renewable resource can be used without impairing or damaging its ability to be fully renewed. See also *Environmental degradation.*

Synergism　interaction in which the total effect is greater than the sum of two effects taken independently.

Synergistic effect　result of the interaction of two or more substances or factors that cause a net effect greater than that expected from adding together their independent effects.

Synfuels　synthetic gaseous and liquid fuels produced from coal or sources other than natural gas or crude oil.

Synthetic natural gas (SNG)　gaseous fuel containing mostly methane that is produced from solid coal.

System　any collection of matter under study. Compare *Surroundings.*

Tailings　rock and other waste materials removed as impurities when minerals are mined and mineral deposits are processed. These materials are usually piled on the ground or dumped into ponds.

Tar sands　swamplike deposits of a mixture of fine clay, sand, water, and variable amounts of a tarlike heavy oil known as bitumen. The bitumen or heavy oil can be extracted from the tar sand by heating and purified and upgraded to synthetic crude oil. See *Bitumen.*

Temperature inversion　See *Thermal inversion.*

Teratogen　substance that, if ingested by a pregnant female, causes malformation of the developing fetus.

Terracing　planting crops on a long, steep slope that has been converted into a series of broad, nearly level terraces with short vertical drops from one to another, and following the slope of the land in order to retain water and reduce soil erosion.

Terrestrial　pertaining to the land. Compare *Aquatic.*

Terrestrial ecosystem　See *Biome.*

Tertiary sewage treatment　series of specialized chemical and physical processes that reduce the quantity of specific pollutants still left in wastewater after primary and secondary sewage treatment. Compare *Primary sewage treatment, Secondary sewage treatment.*

Thermal enrichment beneficial effects in an aquatic ecosystem as a result of a rise in water temperature. Compare *Thermal pollution.*

Thermal inversion layer of cool air trapped under a layer of less dense warm air, thus reversing the normal situation. In a prolonged inversion, air pollution near the earth's surface may rise to harmful levels.

Thermal pollution increase in water temperature that has harmful ecological effects on an aquatic ecosystem. Compare *Thermal enrichment.*

Thermal shock harmful ecological effects in an aquatic ecosystem as a result of a sharp rise or drop in water temperature.

Thermocline fairly thin transition zone in a lake that separates an upper warmer zone (*epilimnion*) from a lower colder zone (*hypolimnion*).

Threatened species a wild species that is still abundant in its natural range but is considered likely to become endangered within the foreseeable future because of a decline in numbers. Compare *Endangered species.*

Threshold effect a harmful or fatal effect that does not occur until the level of a particular physical or chemical factor exceeds the limit of tolerance of an organism.

Throwaway society society such as that found in most advanced industrialized countries in which ever-increasing economic growth is sustained by maximizing the rate at which matter and energy resources are used with little emphasis on resource conservation. Compare *Matter-recycling society, Sustainable-earth society.*

Throwaway worldview belief held by cornucopians that the earth is a place of unlimited resources. Any type of resource conservation that hampers short-term economic growth is unnecessary because if we pollute or deplete the resource in one area, we will find substitutes, control the pollution through technology, and if necessary obtain additional resources from the moon and asteroids in the "new frontier" of space. Compare *Sustainable-earth worldview.*

Tolerance limit point at and beyond which a chemical or physical condition (such as heat) becomes harmful to a living organism.

Total fertility rate (TFR) estimate of the number of children the average woman will bear during her reproductive years, assuming she lives to age 44.

Total resources total amount of a particular mineral that exists on earth.

Tragedy of the commons depletion or degradation of a resource such as clean air or clean water to which a population has free and unmanaged access.

Transpiration transfer of water from exposed parts of plants through leaf pores to the atmosphere.

Trickling filters form of secondary sewage treatment in which aerobic bacteria biodegrade organic wastes in wastewater as it seeps through a large vat filled with crushed stones covered with bacterial growths.

Tritium (T: hydrogen-3) isotope of hydrogen with a nucleus containing one proton and two neutrons, thus having a mass number of 3. Compare *Deuterium.*

Trophic level all organisms that consume the same general types of food in a food chain or food web. For example, all producers belong to the first trophic level and all primary consumers belong to the second trophic level in a food chain or a food web.

Troposphere innermost layer of the atmosphere, which contains about 95% of the earth's air and extends about 8 to 12 kilometers (5 to 7 miles) above the earth's surface.

Unconfined aquifer water-bearing layer of the earth's crust when groundwater collects above a layer of relatively impermeable rock or compacted clay. Compare *Confined aquifer.*

Undernutrition condition characterized by an insufficient quantity or caloric intake of food to meet an individual's minimum daily energy requirement. Compare *Malnutrition, Overnutrition.*

Undiscovered resources potential supplies of a particular mineral resource believed to exist on the basis of broad geologic knowledge and theory, although location, quality, and amounts are unknown. Compare *Identified resources, Reserves, Resources.*

Upwelling area along a steep coastal area where winds blow surface water away from the shore and allow cold, nutrient-rich bottom water to rise to the surface.

Urban area place with a population of 2,500 or more.

Urban growth rate of growth of urban population.

Urban heat island buildup of heat in the atmosphere above an urban area as a result of the dense concentration of cars, buildings, factories, and other heat-producing activities.

Urbanization the percentage of the total population of the world or a country concentrated in urban areas.

Water cycle See *Hydrologic cycle.*

Waterlogging saturation of soil with irrigation water so that the water table rises close to the surface.

Water pollution any physical or chemical change in surface water or groundwater that can adversely affect living organisms.

Watershed land area that delivers runoff water, sediment, and dissolved substances to a major river and its tributaries.

Water table top of the water-saturated portion of an unconfined aquifer.

Water table aquifer See *Unconfined aquifer.*

Watt unit of power, or rate at which electrical work is done.

Wavelength distance between the crest (or trough) of one wave of electromagnetic radiation and that of the next.

Weather moment-to-moment and day-to-day variation in atmospheric conditions.

Weathering process in which bedrock is gradually broken down into small bits and pieces that make up most of a soil's inorganic material as a result of exposure to physical and chemical processes.

Wetland land that remains flooded all or part of the year with fresh or salt water. See *Coastal wetland, Inland wetland.*

Whole-tree harvesting use of machines to pull entire trees from the ground and reduce them to small chips.

Wilderness area where the earth and its community of life have not been seriously disturbed by human beings and where human beings are only temporary visitors.

Wilderness species wild animal species that flourish only in relatively undisturbed climax vegetational communities such as large areas of mature forest, tundra, grassland, and desert. Compare *Early-successional species, Late-successional species, Mid-successional species.*

Wildlife all free, undomesticated species of plants and animals on earth.

Wildlife conservation the worldwide social movement to bring about the protection, preservation, management, and study of wildlife and wildlife resources.

Wildlife management manipulation of populations of wild species and their habitats for human benefit, the welfare of other species, and the preservation of threatened and endangered wildlife species.

Wildlife resources species of wildlife that are actually or potentially useful to humans. See also *Game species.*

Windbreaks rows of trees or hedges planted in a north-to-south direction to partially block wind flow and reduce soil erosion on cultivated land that is exposed to high winds.

Wind farm clusters of a number of small to medium-size wind turbines located in windy areas to capture wind energy and convert it to electrical energy.

Withdrawal water use use of water when it is taken from a surface or ground source and conveyed to its place of use. Compare *Consumption water use.*

Work what happens when a force is used to push or pull a sample of matter over some distance. Energy is defined as the capacity to do such work.

Zero population growth (ZPG) state in which the birth rate (plus immigration) equals the death rate (plus emigration) so that population is no longer increasing.

Index

Note: Page numbers appearing in **boldface** indicate where definitions of key terms can be found in the text; these terms also appear in the glossary. Page numbers in *italics* indicate illustrations, tables, and figures.

Roger K. Burnard

Tropical savana biome

Kenneth W. Fink Ardea. London

Shortgrass grassland biome

Arctic tundra biome

Chaparral biome

Taiga biome